Contemporary studies on fish feeding

Developments in environmental biology of fishes 7

Series Editor
EUGENE K. BALON

Contemporary studies on fish feeding: the proceedings of GUTSHOP '84

Papers from the fourth workshop on fish food habits held at the Asilomar Conference Center, Pacific Grove, California, U.S.A., December 2–6, 1984

Edited by
CHARLES A. SIMENSTAD & GREGOR M. CAILLIET

Reprinted from *Environmental biology of fishes* 16 (1–3), 1986 with addition of eight more papers and a summary of the workshop

1986 **DR W. JUNK PUBLISHERS**
a member of the KLUWER ACADEMIC PUBLISHERS GROUP
DORDRECHT / BOSTON / LANCASTER

Distributors

for the United States and Canada: Kluwer Academic Publishers, 190 Old Derby Street, Hingham, MA 02043, USA
for the UK and Ireland: Kluwer Academic Publishers, MTP Press Limited, Falcon House, Queen Square, Lancaster LA1 1RN, UK
for all other countries: Kluwer Academic Publishers Group, Distribution Center, P.O. Box 322, 3300 AH Dordrecht, The Netherlands

Library of Congress Cataloging in Publication Data

Contemporary studies on fish feeding.

(Developments in environmental biology of fishes ; 7)
"Reprinted from Environmental biology of fishes 16 (1-3), 1986 with addition of eight more papers and a summary of the workshop."
Includes indexes.
1. Fishes--Food--Congresses. 2. Fishes--Physiology--Congresses. I. Simenstad, Charles A. II. Cailliet, Gregor M., 1943- III. Series.
QL639.1.C64 1986 597'.01'3 86-10398

ISBN 90-6193-615-2 (this volume)
Cover design: Max Velthuijs

PRINTED IN THE NETHERLANDS

Contents

Papers on pages 7 to 223 have been reprinted from Environmental Biology of Fishes vol. 16, no. 1–3.

Preface

GUTSHOP '84 was the fourth in a series of workshops on various aspects of fish feeding (Table 1). Initially, the organizers merely invited regional (Pacific Northwest) fisheries scientists to share, and possibly develop mutual solutions to, the many technical problems associated with trying to obtain meaningful, quantitative information from fish stomach contents, and the subsequent statistical treatment and interpretation of the multivariate data. Since then, although not explicitly based upon any internal cycle, these scientists and increasingly more and more dispersed colleagues continued to congregate for workshop deliberations every two or three years. From the 49 attendees at the first workshop, the number of participants had grown to 65 at GUTSHOP '78, and 107 at GUTSHOP '81. By the third workshop, we were drawing scientists from across the U.S. and Canada, and from as far away as Norway.

The topical content of the workshops has also evolved from the predominantly technical aspects of fish collection and stomach contents processing techniques, statistical analysis, and data manipulation and presentation to considerations of theoretical ecology, bioenergetics, and behavior.

Based upon discussions during GUTSHOP '81 and on results of a questionnaire distributed at the end of that workshop, the consensus of the participants was that: (1) a fourth workshop should be scheduled for 1983–1984; (2) it again be located in central California; (3) the format permit more time for discussions between sessions; (4) the session leaders play a greater role in selection of papers, perhaps including an introductory plenary paper; (5) preference be given to papers written specifically for the workshops objectives and topics; and, (6) session topics include:

(a) methodology, sampling design, and statistics;
(b) the bioenergetics and physiology of feeding and digestion;
(c) behavioral aspects of fish foraging and prey evasion;
(d) competition, resource partitioning, and community structure, including prey switching;
(e) ontogenetic changes in food habits, including feeding in fish larvae; and,
(f) the effects of fish predation in structuring prey assemblages.

Assuming the coordinating role more by default than by fiat, in winter 1983–1984 we began discussions about organising the next workshop, and with some of our more ardent GUTSHOP protagonists about chairing one of the recommended sessions. Five session topics were chosen: I. Methodology and Statistics, chaired by Robert Feller (University of South Carolina, Columbia); II. Physiology of Feeding and Digestion, chaired by George Boehlert (NOAA-NMFS, Honolulu); III. Foraging Behavior, chaired by Brian Marcotte (McGill University, Montreal); IV. Competition and Resource Partitioning, chaired by Al Ebeling (University of California, Santa Barbara); and, V. Predation Effects on Prey Behavior and Ecology, chaired by Mark Hixon (Oregon State University, Corvallis). An additional 'open microphone' session, chaired by Greg Cailliet, was scheduled to accommodate overflow papers and inspirations of the moment. These chairmen were given the responsibility of selecting their participants from among the responses to a call for papers, and for plenary papers to open the five main sessions. Unique to GUTSHOP '84 was the potential for publication of the workshops proceedings by a peer-reviewed primary journal, which had been arranged with the editor and publisher of the *Environmental Biology of Fishes* (EBF). The orientation and reputation of this journal has promoted submission of original, creative papers for consideration, a formidable task for most workshop organizers in this era of severe time constraints upon research and manuscript preparation.

The consequence of these preparations, GUTSHOP '84, commenced with the ritual social mixer on Sunday, 2 December, continued through three and a half days of sessions and discussions, and concluded with a general discussion session on 6 December. As in the past workshops, the format consisted of half-day sessions including formal presentation of papers and concluding discussion. The related manuscripts were turned over to us at the end of the workshop or soon thereafter and sent out for review according to established EBF procedures. In addition to ourselves, for each paper we typically included one established EBF reviewer and one of the workshop attendees, with considerable reliance upon the session chairmen. During the review process, we also prepared transcriptions of the taped discussions and synopsized these with our summary of the workshop.

Based upon the reviews, we recommended to the editor of EBF for some accepted and revised papers to be compiled into a dedicated issue (1986; Vol. 16, No. 1–3), covering the most significant contributions to the workshop, and Volume 7 of the *Developments in Environmental Biology of Fishes* series (DEBF) (Simenstad & Cailliet [ed.] 1986), which includes additional papers of consequence, transcriptions of the discussions, and our final summary. Appropriately, our editorial mentor and counsellor, EBF editor Eugene Balon, had final say in the selection and organization of these papers.

Support of the workshops has continued to occur through the benevolence of the sponsoring institutions, the participants themselves, and the Washington Sea Grant Program at the University of Washington, which funded the preparation, printing, and mailing of all pre-workshop materials and the publication of the workshop proceedings. In the case of GUTSHOP '84, we are deeply indebted (literally) to Robert Burgner, Director of Fisheries Research Institute, Louie Echols, Director of the Washington Sea Grant Program, and John Martin, Director of the Moss Landing Marine Laboratories, for their contributions.

Table 1. Chronology of GUTSHOP series of workshops of fish food habits studies.

Workshop	Dates	Location	Sponsors*	Proceedings
GUTSHOP '76	13–15.10.1976	Astoria, Oregon	NWAFC/FRI	Simenstad & Lipovsky (1977). Washington Sea Grant Pub. WSG-WO 77-2. 193 pp.
GUTSHOP '78	10–13.10.1978	Lake Wilderness Conference Center, Maple Valley, Washington	FRI/WSGP	Lipovsky & Simenstad (1979). Washington Sea Grant Pub. WSG-WO 79-1. 22 pp.
GUTSHOP '81	6–9.12.1981	Asilomar Conference Center, Pacific Grove, California	MLML/FRI	Cailliet & Simenstad (1982). Washington Sea Grant Pub. WSG-WO 82-2. 312 pp.
GUTSHOP '84	2–6.12.1984	Asilomar Conference Center, Pacific Grove, California	FRI/WSGP/MLML	This volume

* NWAFC = NOAA-NMFS, Northwest and Alaska Fisheries Center, Seattle, FRI = Fisheries Research Institute, University of Washington, Seattle. WSGP = Washington Sea Grant Program, University of Washington, Seattle. MLML = Moss Landing Marine Laboratories, Moss Landing.

In addition to this institutional support, there were a number of individuals who contributed considerable time and energy through their assistance in the planning, preparation, and coordination of GUTSHOP '84 and the compilation of these proceedings. Sally Lawrence, Laura Mason, Vicki Miles, Andrea Jarvela, and Patricia Peyton, at Washington Sea Grant Communications, provided more than the expected support in preparing the pre-workshop mailings and the program. The ichthyology students of Moss Landing Marine Laboratories again responded with unfailing stamina in conducting registration, arranging transportation, assisting with the audiovisual equipment, and helping us keep our cool in the most confusing times; we particularly acknowledge Debbie Molnar, Allan Fukuyama, Kevin Hill, Kevin Lohman, Brian Sak, Jim Brennan, Mark Silberstein, Doris Small, Bruce Welden, Mike Haberland, and Dave Ebert. We especially thank Sandy Lipovsky and Bob Emmett for their effervescent assistance and encouragement. Last, and certainly not least, we extend our appreciation to the session chairmen, participants, and reviewers (see acknowledgements at the end of the EBF volume) of the manuscripts for their persistent dedication to the often mystifying question of why and how fish feed the way they do.

1 July 1985 *Charles A. Simenstad and Gregor M. Cailliet*

On blank pages, after articles ending on odd numbered pages, are reproductions of printed lithographic plates from the U.S. National Museum. Most of these plates are poor and inaccurate copies by unknown lithographers of fine original pencill-and-wash illustrations by J.H. Richard for the U.S. Pacific Railroad Reports and Mexican Boundary Survey.

Types of feeding in the ontogeny of fishes and the life-history model

Eugene K. Balon
Department of Zoology, College of Biological Science, University of Guelph, Guelph, Ontario N1G 2W1, Canada

Keywords: Nutrient acquisition, Life history, Evolution, Saltatory ontogeny, Endogenous, Absorptive, Exogenous, Mixed feeding

Synopsis

Different types of ontogenies in fishes – indirect and direct – are correlated with different nutrient availability and feeding during early life history. A comprehensive life-history model, developed earlier, facilitates the understanding of decisive events in the life of an organism. Embryos with insufficient endogenous food supply (yolk) to build a definitive phenotype directly need the transient form of a nutrient-gathering larva. They represent an indirect development. In contrast, a large endogenous supply of nutrients enables the definitive adult phenotype to develop directly, avoiding an intervening larva and the cost of metamorphosis. The larger and more advanced an individual at the onset of exogenous feeding, the better are its chances to survive. This can be achieved by heterochronies related to feeding. Different types of feeding during the early ontogeny of fishes – endogenous, exogenous, absorptive, and a combination of all (mixed) – are demonstrated and integrated into the life-history model.

I felt it to be useful to oblige the call for papers on 'ontogenetic changes in food habits' by Cailliet & Simenstad (1982, p. 305) in the previous 'Gutshop '81'. Organisms can hardly be understood considering their adult forms only, and to study their early ontogeny may be more revealing and therefore important, a plea many (e.g. Orton 1953) made before.

There are four types of food acquisition in fishes – endogenous, absorptive, mixed and exogenous (Fig. 1.) Before we attempt to define each of these types and proceed in presenting examples, we must ask 'what is the real significance of feeding?' This should become obvious after we return to an earlier answer on the significance of ontogeny: 'The *specialized* multicellular organism is reduced by each act of reproduction into a large number of non-specialized single cells, available for natural selection to act upon. Epigenesis creates new phenotypes under the 'instructions' given by the genome. The latter results from selection by the past environments, but the phenotype is formed by an interaction with the present environment. What a

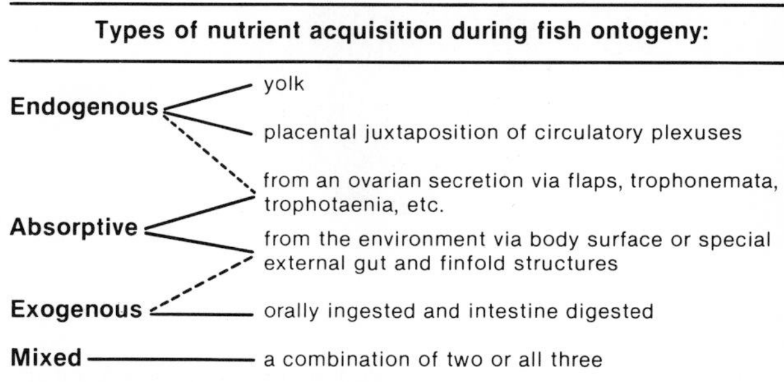

Fig. 1. The different ways of feeding, some or all of which can occur during a life history of a fish.

marvelous mechanism for extension of survival by maintenance of flexibility and an ultimate ability for change (Balon 1983). Each time ontogeny must begin anew, as if returned to the drawing board (Dawkins 1982)' (Balon 1984a, p. 182).

To create each time a new phenotype from a single unspecialized cell requires not only the 'instructions' passed with the germ material, but 'building material' in a form of nutrients (Calow 1977). In other words, the fate of the cell at the beginning of ontogeny depends as much on germ line instructions or physicochemical variables as on the vagaries in nutrient provision. For me, the pluralistic perception of the new evolutionary paradigms (e.g. Ho & Saunders 1984, Pollard 1984) starts somewhere here.

The endogenous feeding

The egg cell in various species is supplied with various amounts of yolk, containing mainly nutrients.[1] Moreover, various clutches or individuals of the same species can be supplied with different amounts of yolk. Mesolecithal eggs have moderate amounts of yolk, telolecithal an enormous amount in relation to the amount of cytoplasm, and alecithal eggs remain without yolk (see e.g. Cohen & Massey 1982). Eggs with little or no yolk occur only in livebearing fishes and are supplied with nutrients via maternal-embryonic exchange of molecules, i.e. by juxtaposition of circulatory plexuses of the two organisms. Nutrients provided through yolk via vitellogenesis (i.e. lecithotrophy, Wourms 1981) or directly (Fig. 2) via placenta (i.e. placental matrotrophy, Soin 1968, Korsgaard 1983, Hartvig & Weber 1984) represent *endogenous* feeding. [Let me, however, note that low yolk supply in livebearers is also substituted by absorptive uptake of an ovarian secretion (Veith 1980) which I call absorptive matrotrophy and return to later.]

The endogenous food supply is the most important, for it determines and changes the entire remaining life history. In other words, the ultimate form of a phenotype is determined during the earliest intervals of ontogeny when endogenous feeding is the main source of nutrients.

As already said by Orton (1953, p. 63) 'inadequate provision for embryonic nutrition is the primary factor governing the occurrence of a larval stage'. Low amount of yolk is insufficient to build the definitive phenotype and a temporary larva needs to become part of the life history as an external nutrient-acquiring device (highly efficient feeding machine, Wassersug 1975, 1984), until there is enough building material to proceed with the for-

[1] Yolk may also contain respiratory carotenoids assisting in endogenous oxidative metabolism (see Balon 1977, 1981a, Karnaukhov 1979), oil globules for buoyancy, and other structures of no nutritive value (e.g. Grodziński 1949, 1968).

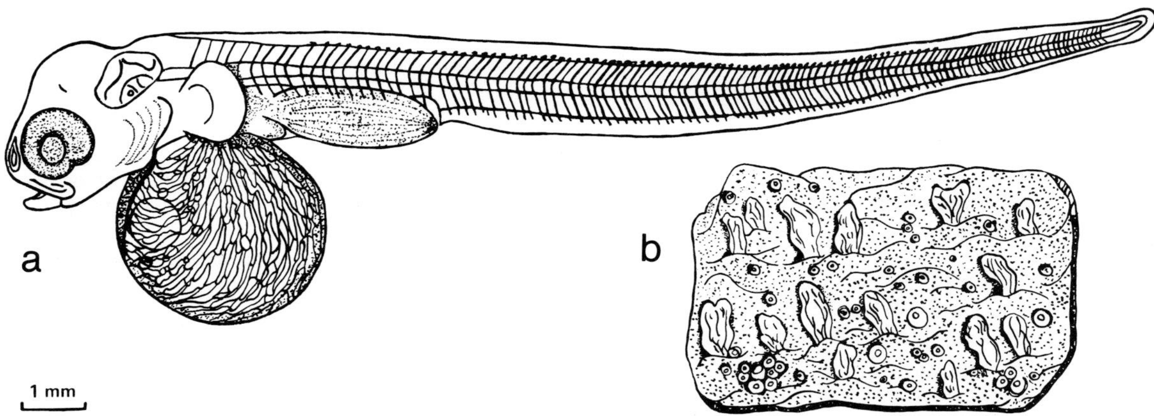

Fig. 2. In some livebearers the endogenous nutrient supply of yolk is partly replaced by direct nutrient transfer via placental circulatory plexuses: *a* – the livebearing eelpout, *Zoarces viviparus* embryo taken out of the ovary, and *b* – detail of the ovarian wall with heavily vascularized folds and flaps that align closely with heavily vascularized yolksac and hypertrophied (e.g. Kimura 1973) intestine walls (after Soin 1968).

mation of the definitive phenotype. The initial size of such a temporary nutrient-gathering larva depends directly on the amount of yolk available. A large amount of yolk enables elimination of the temporary larva (Garstang 1929) and direct development of a definitive phenotype, fully differentiated into a 'small adult' (= juvenile) at the time of first external feeding (Fig. 3).

It is the amount of yolk, not necessarily the egg size, that determines the further development. Eggs of the same size may have different amounts of yolk, one egg having a larger perivitelline space and consequently less yolk. Even the volume of yolk can be misleading, since there are also differences in density (Balon 1977). Eggs of the same size with approximately the same volume, but low or high density of yolk, can lead to either indirect or direct development, respectively. The significance of egg size alone is, therefore, of limited value, but rather an arbitrary variable which is used because of convenience and which only in some cases corresponds to the amount of yolk (e.g. Marshall 1953, Kendall 1981, Bone & Marshall 1982).

Indirect and direct ontogenies

A natural consequence of the theory of saltatory ontogeny (Balon 1981b) is the hierarchical life-history model of embryo, larva, juvenile, adult and

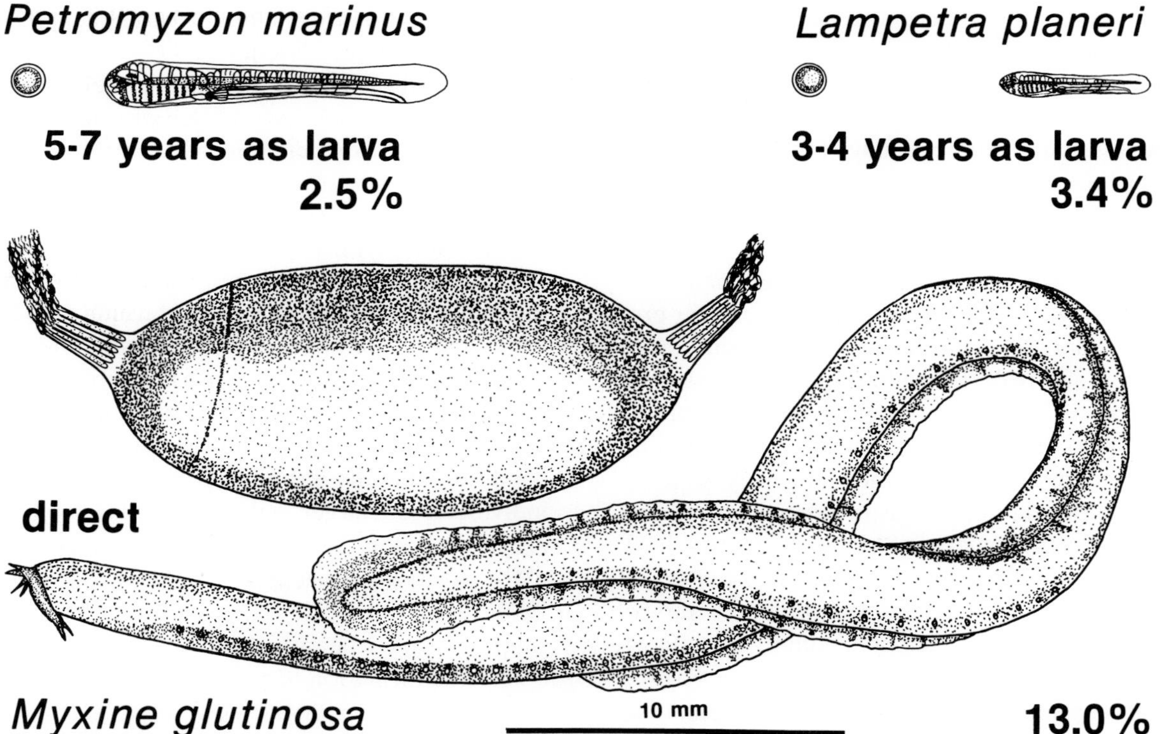

Fig. 3. The volume of yolk in extreme cases determines the egg size and so fishes with small eggs – although yolk supply is the real decisive feature – seem to undergo indirect development with nutrient gathering larva, and fishes with large eggs direct development without a need of a larva: The sea lamprey, *Petromyzon marinus,* develops from an egg of 1 mm in diameter to hatch at 4 mm and to first feed orally at 10 mm, i.e. 2.5% of an average adult female size, living as a larva (ammocoete) for 5–7 years; the nonparasitic brook lamprey, *Lampetra planeri,* with an egg diameter of 1.1 mm only slightly improves on its indirect ontogeny by growing into an adult, definitive phenotype much smaller (131 versus 400 mm in sea lamprey), hatching when 2.3 and first feeding at 4.4 mm (3.4% of an average female size), living as a larva 3–4 years and dying after reproduction (after Schultze 1856, Balon 1957, Piavis 1971). The hagfish, *Myxine* sp., on the other hand, produces a few large eggs 17 mm long, hatching at 65 mm as a fully formed definitive phenotype 13% of an average female's size, feeding immediately like and adult; the large yolk volume made the elimination of a larva and a direct development from embryo to juvenile possible (after Walvig 1963 and Brodal & Fänge 1963). All drawings are to the same scale.

senescence periods, each separated by a 'decisive' threshold (which equals often the 'critical intervals' of some), and each consisting of a sequence of saltatory intervals – homeorhetic states called steps – separated by less steady thresholds. In comparative studies such a model provides the possibility to recognize and interpret correctly shifts in thresholds, which often result in new life-history patterns. For example, it elucidates the ecological significance of *not* having a larva (Balon 1977, 1984b) as well as the importance of having a larva in spite of the cost of metamorphosis (Balon 1984a, 1985). The model should also help to identify 'prejuveniles' (e.g. Lewis et al. 1972, Brothers & McFarland 1981) for what they really are – larvae undergoing metamorphosis.

The first, embryo period of ontogeny is characterized by mainly endogenous feeding. The transition to exogenous feeding, i.e. to the orally ingested and intestinally digested acquisition of nutrients, marks the beginning of the next period of life history, be it larva in case of indirect, or juvenile in case of direct ontogeny.

Larvae are very vulnerable and restricted to small food particles by virtue of their small sizes. Eggs of low yolk amount, however, can be produced in larger quantities and so compensate for the low survival. In addition, most larvae are adapted to inhabit different niches (e.g. planktonic) than their definitive phenotypes, so they do not compete for food. The need for early dispersal may be an additional reason for having indirect ontogeny with temporary larvae (Garstang 1929, Barlow 1981, Doherty et al. 1985), and the dispersal distance and time, from spawning to nursery grounds (= first exogenous feeding), may be assisted by yolk supply (Kiørboe & Munk 1985).

There is another price besides high mortality to be paid for having a larva. The numerous caenogenetic (temporary) structures of larvae specialized into separate habitats and niches (Whittaker et al. 1973) need be remodelled into the permanent organs and shapes at substantial energy cost. This process of remodelling, the metamorphosis, terminates the larva period (e.g Fostner et al. 1983). In some cases (e.g. elopomorphs, stomatioids) much of size gained during the larva period must be sacrificed for the remodelling (Fig. 4), thus decreasing the survival value of larger size. En passant it provides clear circumstantial evidence that the main purpose of a larva is the acquisition of external nutrients when the endogenous supply is insufficient.

In contrast, when sufficient endogenous food supply is provided at the cost of a lower number of eggs, elimination of the vulnerable larva and the costly metamorphosis facilitates the development into a juvenile that is comparatively advanced at the time of first feeding – a clear survival advantage superior to larger size. Moreover, low egg number, larger yolk size and density (negative buoyancy), prolonged development in egg envelopes and sessile stages of free embryos enable further protection by parental care (Balon 1975a, 1981a, c, 1984b, Baylis 1981, Blumer 1982).

It is possible that the increase in vitellogenesis, responsible for the larger amount of yolk, is mediated by the environment (Gerbilsky 1956) via endocrine mechanisms (e.g. Campbell & Idler 1976, Ng & Idler 1978, van Bohemen et al. 1982, van Tienhoven 1983). I rather doubt, however, that such factors as temperature, salinity, and water-to-land invasion can cause directly, as Matsuda (1982, p. 735) claims, 'the increase in yolk content per egg (. . .) and there seems to [be] an endocrine mechanism that responds to environmental stimuli in producing such eggs . . .'. More likely, competitive conditions and the resulting specialization of some individuals on larger, more nutritious food (see APHS in Balon 1985), enhance vitellogenesis and more precocial progeny (e.g. Balon 1980, Goto 1980, 1982). Of course, temperature and similarly simple physiochemical factors may initiate epigenetic formation of more specialized individuals (Balon 1983, p. 2049).

Direct ontogeny, therefore, seems to be a more specialized type of development, and the evolutionary trends should proceed in this direction in an attempt to shorten the most vulnerable period (e.g. Wassersug 1975) and, even more importantly, to improve competitiveness when the community structure eventually becomes more diverse and complex through speciation and invasion (Balon 1983, 1985). Consequently, the 'adult pressure' and

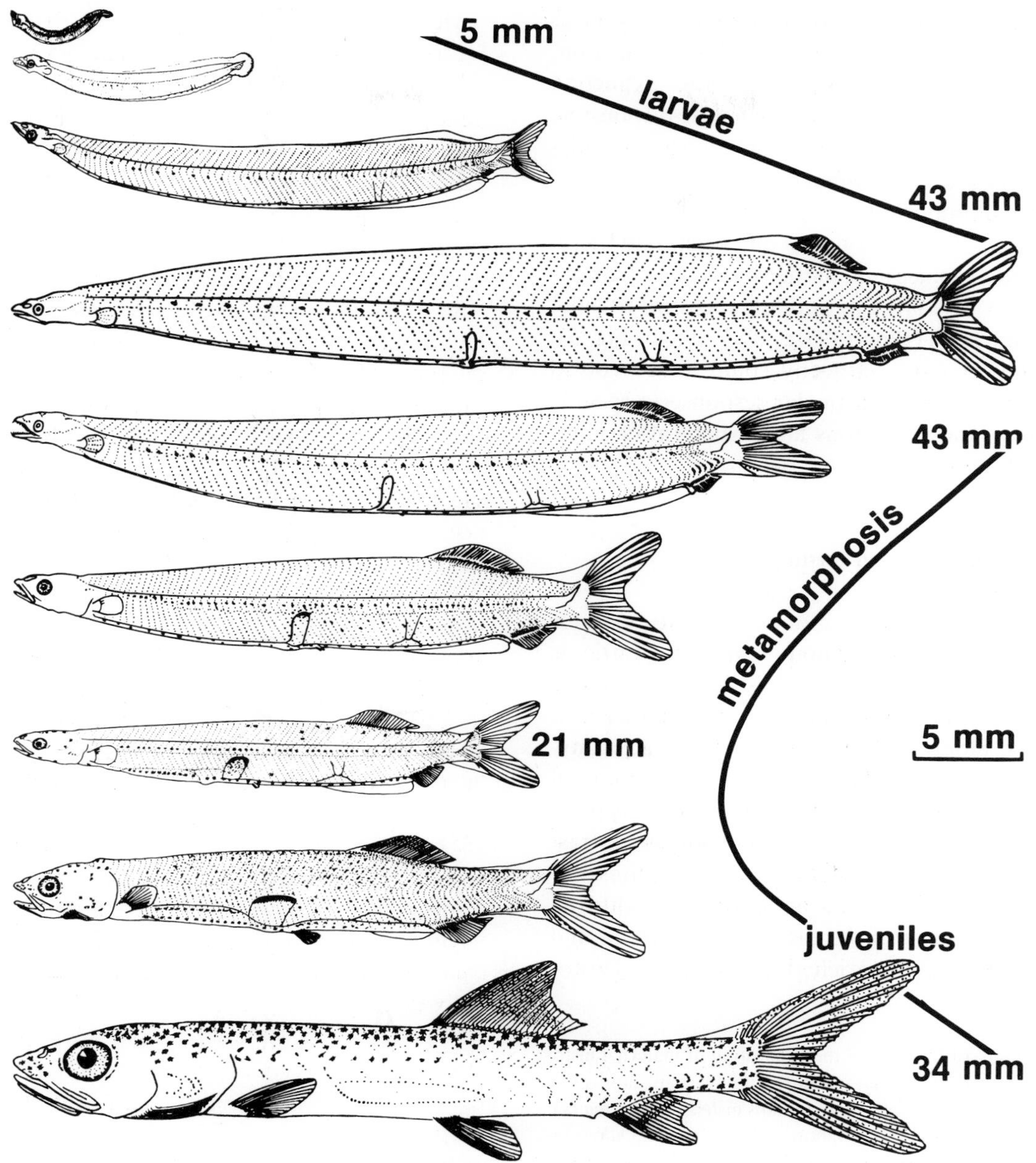

Fig. 4. The cost of metamorphosis, i.e. a complete remodelling of a number of temporary larval structures into structures of the definitive phenotype, can be demonstrated on the developmental sequence for the ladyfish *Elops saurus* (from Gehringer 1959): From the time of first oral feeding at about 5 mm length the food gathering planktonic larva with specialized temporary structures grows to 43 mm before enough building material is available to proceed with the remodelling into a small definitive phenotype. This remodelling, akin to metamorphosis, costs any advantage gained in size, for the larva shrinks to 21 mm before it is able to grow again.

'adultation' of Jägersten (1972) are nothing more than different formulations of the same phenomenon, and so is the 'effect hypothesis' (Vrba 1980) only a pattern caused by the same trend. The reduction of chromosome numbers (Gold et al. 1980), for example, between indirect ontogeny in lamprey (168 chromosomes) and direct in hagfish (42) (Fig. 3) indicates loss of genetic variation, a pattern seen earlier in other vertebrates and explained elsewhere (Balon 1985). Since overspecialization leads to extinction, in some circumstances paedomorphosis, a process of despecialization, counteracts this tendency, delaying extinction (Balon 1983). The same process may be responsible for a formation of indirect ontogeny from an ametamorphic direct one when dispersal, for example, becomes of importance (e.g. Norris 1983, Doherty et al. 1985). This, of course, contradicts and therefore partly replaces my earlier hypothesis that ontogenies with a larva period are ancestral to ametamorphic ones (Balon 1981c, 1983, 1984b, Fig. 5), for either type of ontogeny could have evolved independently as a particular response to specific environmental conditions and structural or physiological constraints.

The absorptive type of feeding

Often, absorptive acquisition of nutrients operates along with the endogenous nutrition. Little is known about this type of feeding in oviparous fishes, however, and the proportions of nutrients it provides, along with endogenous feeding, is anybody's guess. Notwithstanding, the circumstantial evidence is strong enough to expect absorptive feeding to play some role (Pfeiler 1986). In contrast to endogenous, absorptive feeding acquires nutrients from the external environment. Both the aquatic habitat, with its dissolved nutrients, and the ovarian fluid, enriched by secretion (histo- →

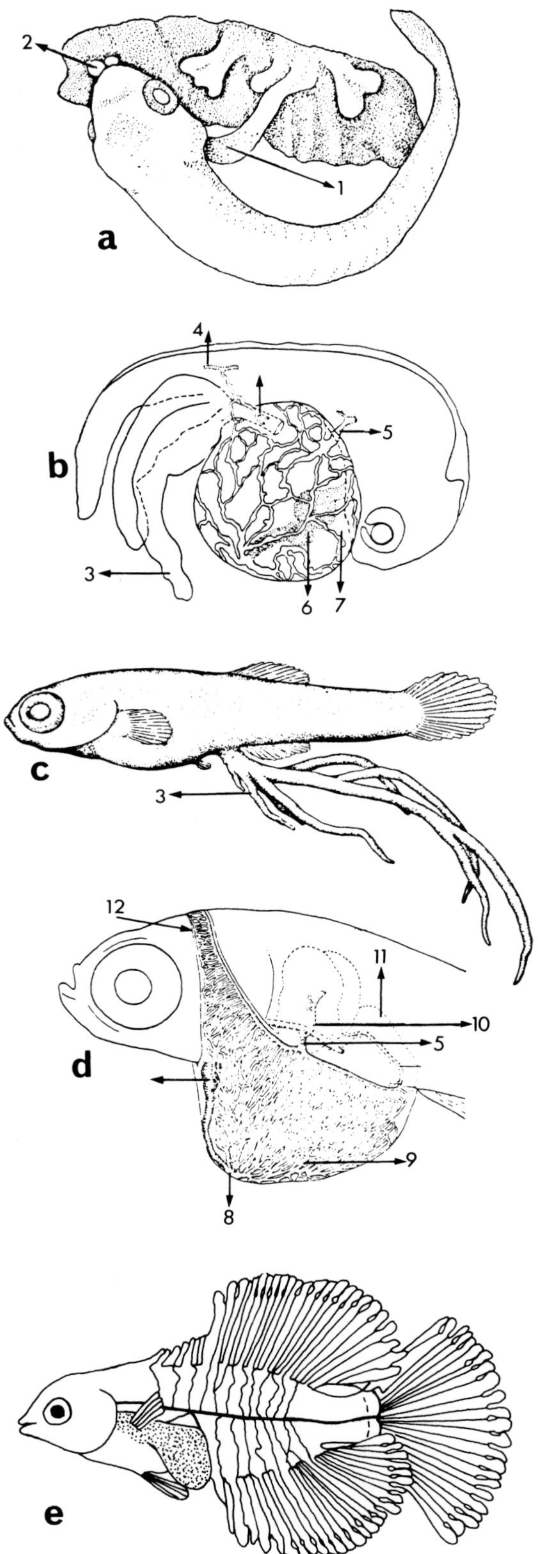

Fig. 5. In teleost livebearers the absorption of maternal histotrophe (ovarian milk) is facilitated by various absorptive 'organs' like, for example, ovarian flaps in *Jenynsia lineata* (*a*), pericardial plexuses and trophotaeniae in *Girardichthys innominatus* (*b*), the prolongation of an upturned hindgut so well developed in *Zoogonecticus cuitzooensis* and other goodeids (*c*), similar in origin and function to that of planktonic larvae in Figure 6, pericardial serosa and neck strap in the embryo of *Poecilia reticulata* (*d*), and the extensive plexuses in fins and their spatulate extensions in addition to the hypertrophied gut in the embiotocid *Embiotoca lateralis* at middle gestation (*e*). 1 = trophonema, flap of ovarian tissue, 2 = tip of the ovarian flap inserted through gill slit, 3 = trophotaenia, 4 = caudal vein, 5 = duct of Cuvier, 6 = heart, 7 = pericardial sac, 8 = sinus venosus or collecting vitelline vein, 9 = portal system or yolksac plexus, 10 = hepatic vein, 11 = intestine, 12 = neck strap. From Balon (1975a) and Webb & Brett (1972).

trophe or uterine milk, Veith 1980, Wourms 1981, Boehlert & Yoklavich 1984), should be considered environments for absorptive feeding.

Therefore, absorptive feeding could be as common in oviparous as it is in viviparous fishes (Fig. 5). Its peak function in oviparous species can be expected in the embryo and larva period for at that time the surface area to volume ratio is optimal and nutrient acquisition of uppermost importance. As a rule, embryos of fishes with indirect ontogeny are small and so are the resulting larvae; their elongated and thin bodies are well suited for epidermal uptake of food molecules dissolved in the 'planktonic soup' they most often inhabit. The ability of the rectal epitelium in fish larvae to ingest soluble substances into the cells by micropinocytosis has been demonstrated, for example, by Watanabe (1984). Increasing the larval body surfaces by hypertrophy of fins and spines may not only improve defense and buoyancy, as is usually suggested (e.g. Moser 1981), but also increase nutrient uptake. Hypertrophied and trailing guts – extreme rectal extensions – suggest even more enhanced absorptive feeding (Fig. 6). I would not be surprised if some of these trailing guts were turned inside out with the mucosal epithelium facing the environment, as do trophotaeniae in livebearers (Wourms & Lombardi 1979).

A similar analogy can be demonstrated between the hypertrophied finfold of the planktonic myctophid larva *Loweina rara* (Fig. 6) and the enlarged and vascularized fins of the viviparous embiotocid *Embiotoca lateralis* (Fig. 5), for the latter amply demonstrated by Blanco (1938), Webb & Brett (1972) and Ingermann & Terwilliger (1984), for the former implied by its shape and the niche inhabited. Again, earlier interpretations that such hypertrophied finfolds in planktonic larvae act mainly as hydrostatic organs and for the storage of surplus water from yolk digestion in the absence of excretory organs (Kryzhanovsky et al. 1953) may be incorrect or incomplete (e.g. Mendoza 1958, 1972). As said earlier, there are only scarce direct data on epidermal uptake of nutrients by oviparous organisms (e.g. Koller 1930, Terner 1968, Moore & Potter 1976, Richards 1980, Stephens 1982) and obviously more has to be learned to prove the case (Hulet 1978). The analogy to the epidermal uptake assisted by villi and ridges in livebearers (e.g. Veith 1980) is, however, too obvious to pass unnoticed.

The exogenous feeding

The main type of food acquisition in the definitive phenotype is of course exogenous feeding, but it should not be treated as the only feeding mode in fishes (e.g. Liem 1980, Moyle & Cech 1982, Bone & Marshall 1982, Lauder 1985). True, it provides most of the building material for tissue production at sizes of commercial interest, for locomotion and reproduction, in short, for the maintenance and raison d'être of the definitive phenotype which cannot be substantially modified anymore.

The final form of a phenotype and its life history are determined during early ontogeny, however, at a time when other types of feeding operate. New phenotypes 'evolve by introducing drastic sudden changes into the developmental program of early ontogenetic stages, which are more plastic than later ontogenetic stages' (Reif 1983, p. 181). The definitive form of an organism, twisted jaws or not (Liem & Stewart 1976), can merely select within the range of accessible foods, i.e. partition the resource or play the game of predator and prey (e.g. Noakes et al. 1983, Crowder 1986). Not always has it been realized that these concepts exist only at a given community structure and are not a property of an individual organism, especially throughout its entire life history. Even the extremely specialized 'left eye pickers' or 'scale eaters' (Fryer 1959, Liem & Osse 1975, Sazima 1983) turn into 'anything available gobblers' when displaced from the given community and density, both in nature and in captivity (Ribbink et al. 1983, p. 295).

The ever changing community structures and densities, environment therewithal however, require new specializations which can be provided only by new organisms developing from single cells deposited by organisms of the previous generation. Only these unspecialized cells (Løvtrup 1974) have the chance to specialize into the current environment. Their genome represents a 'memory' of the

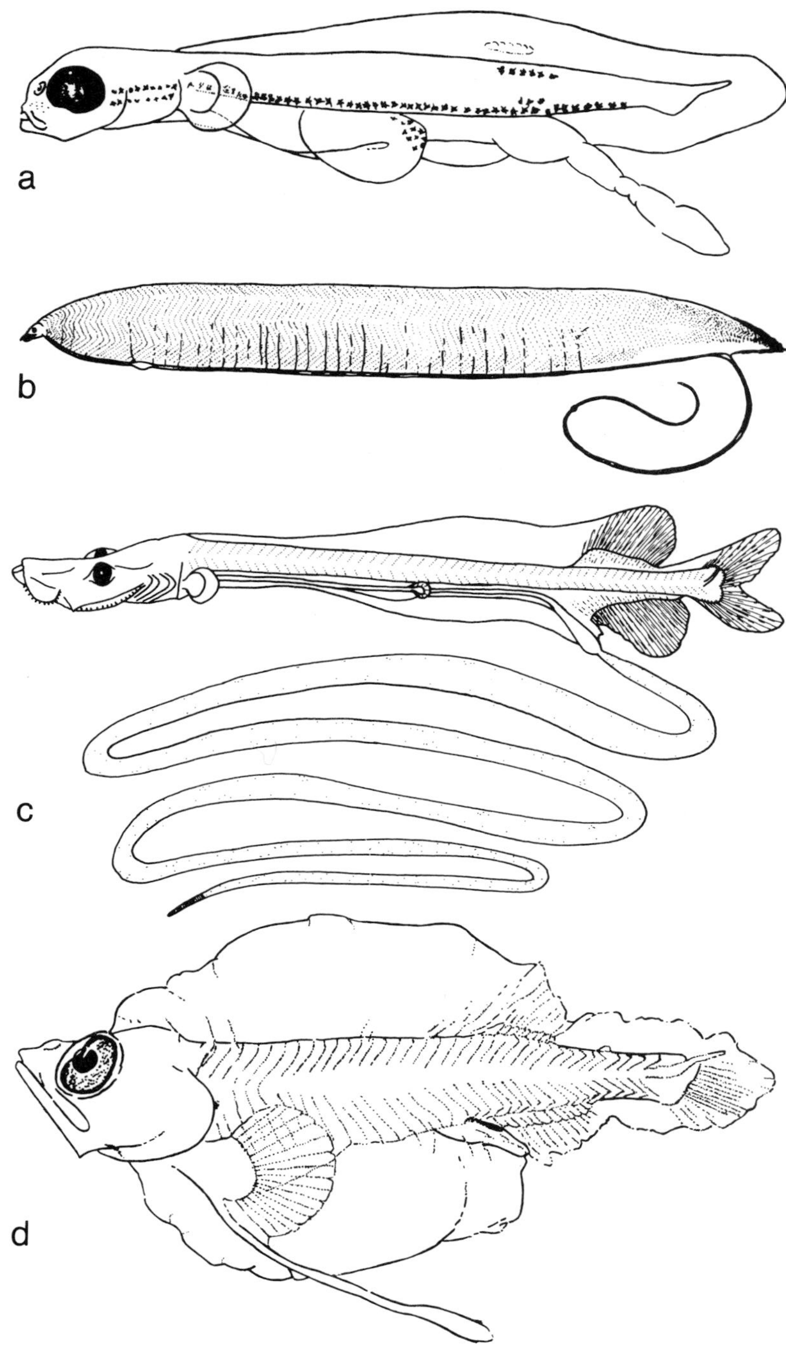

Fig. 6. Larvae of oviparous fish species with caenogenetic structures analogous to the ones in livebearers of Figure 5. The endogenous and/or exogenous nutrient supply can be supplemented by absorption from the environment: *a* – argentinid *Microstoma microstoma* with a hypertrophied and trailing gut; *b* – congrid, Bathymirinae, with a trailing gut that may possibly be turned inside-out; *c* – melanostomiatid with an even longer trailing gut; and *d* – myctophid *Loweina rara* with hypertrophied finfolds. After (*a, b*) Fahay (1983), (*c*) Moser (1981), and (*d*) Moser & Ahlstrom (1970).

past environments and the processes of epigenesis a chance to respond to the latest environment. Therefore, epigenesis is the real creative force which fulfills the purpose of the constant seesawing from one cell to a specialized multicellular mortal (Løvtrup 1974, Balon 1983). For me, these processes of *change* are as worthy of attention as the abilities the definitive phenotype has for selection and manipulation of food for maintenance, growth and reproduction.

Mixed feeding

A single type of feeding, e.g. endogenous, may operate for a long time alone or may always be supplemented by absorptive nutrient uptake. What really matters is the main type of feeding at any given time, and that can be endogenous or exogenous. In special cases yet to be documented, only absorptive feeding may occur (e.g. leptocephali, Hulet 1978, Pfeiler 1986).

The ability to delay or accelerate the start or cessation of each feeding type during epigenesis is of enormous significance. Overlaps, varying times of mixed feeding, may then facilitate the required changes. If dispersal is of value or cues for metamorphosis and settlement are not present (e.g. Marliave 1981, Flüchter 1982, Burke 1984), prolongation of one type of feeding may delay differentiation or size dependent metamorphosis. An earlier start with a new type of feeding, in contrast, will create a longer duration of mixed feeding, causing increased nutrient intake and accelerated differentiation, i.e. specialization.

While the commencement of oral ingestion and intestinal digestion is a sign of the beginning of the larva period in case of indirect, or juvenile in the case of direct ontogeny, some overlap of endogenous and exogenous feeding is nearly always present especially in the first case of lesser developed larva. Most probably, this interval of mixed feeding is retained as a back-up system while exogenous acquisition of food is learned by the newly developed, not yet fully efficient hunter (e.g. Ivanova & Lopatko 1984), or the development of suitable food organisms is delayed, or a suitable patch of prey is missed (e.g. Hunter 1981).

Mixed feeding, therefore, is an important feature of a developing organism. The flexibility provided by varying durations of mixed feeding can have direct consequences for survival or can facilitate changes in the entire life history.

The succession of feeding types within consecutive intervals of life history

An organism should always be considered over its entire ontogeny, from the single cell at activation to death. Focusing on the later part of ontogeny represented by the definitive phenotype, be it a juvenile, an adult or the senescent, restricts us to studies of only form and function. The processes which create this bewildering diversity of forms and functions can be revealed mainly through studies of early ontogeny.

From its initial formulation (Balon 1960, 1971, 1975b) the life-history model has undergone a series of modifications and simplifications (Balon 1984a). The decisive events and processes during the ontogeny of fishes (Fig. 7) seem to be activation, onset of oral feeding, metamorphosis in taxa with indirect ontogeny, maturation of gametes leading to reproduction, and death. Some significant accompanying processes, often erroneously considered to be instantaneous events, are fertilization, hatching, parturition, and on occasion, secondary metamorphosis (see Balon 1985).

Løvtrup (1984) distinguishes three phases of epigenesis: (1) form creation, (2) differential growth, and (3) allometric growth. They correlate well with our model. Form creation is the prevailing property of the embryo period (1), differential growth the property of the larva period in case of indirect ontogeny or the embryo and free embryo phases in case of direct ontogeny (2), and allometric growth is the property of the definitive phenotype (3) starting with a juvenile period. Of course, as elements of differential growth may occur during the secondary metamorphosis, so form creation and differential growth may overlap, start earlier or be delayed in much the same manner as the different feeding types. The variety of life-

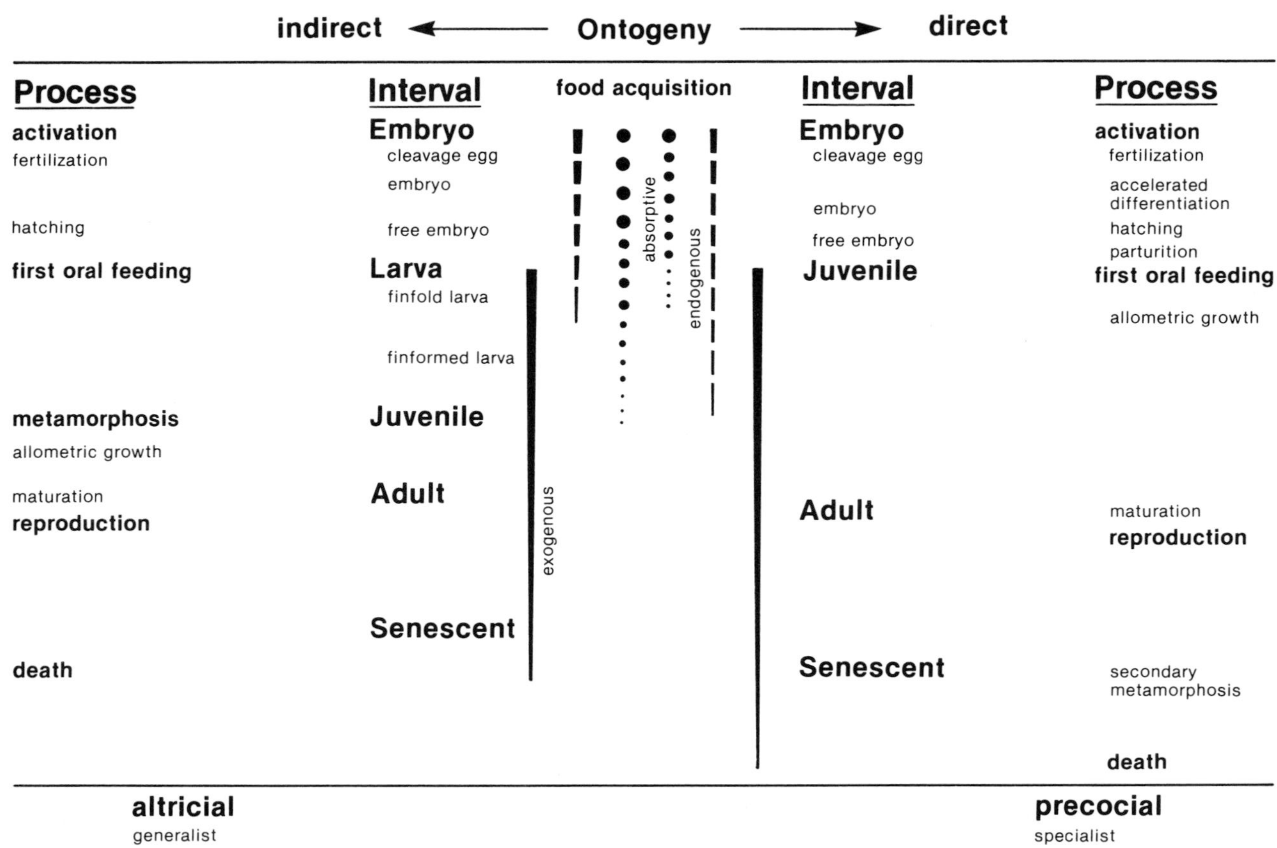

Fig. 7. Comparison of the types of food acquisition within the indirect (left) and direct development of the life-history model (intermediate and extreme ontogenies are ignored). The decisive (bold) and some accompanying events (processes) in either type of ontogeny are listed in the columns on the extreme left and right, respectively. Solid vertical line = exogenous, dashed line = endogenous, and dotted line = absorptive nutrient uptake.

history styles is such that knowledge of only some (e.g., fishes with planktonic larvae, amphibians with tadpoles) will render well meant interpretations demonstrably false. It should always be realized that 'creative processes of epigenesis', 'nutrient acquisition' and 'environments' interact in a multitude of ways, and that individual variation and homeorhetic canalization (Balon 1985) will further complicate our resolutions.

In the case of indirect ontogeny, the embryo period, which depends entirely on an endogenous and/or absorptive nutrient supply, is comparatively short because of few nutrients available. This short embryo period appears as if extended by the larva period, which is characterized by exogenous feeding prior to the formation of a definitive phenotype (Fig. 7). In more 'specialized' taxa (Balon 1981a, 1983) with direct ontogeny, the embryo period can be prolonged due to a large endogenous nutrient supply, that enables the embryo to differentiate and develop directly into the definitive phenotype. In some cases, prolongation of endogenous feeding and acceleration of exogenous feeding causes an extremely long interval of mixed feeding. As a result, a 'giant' young is released from the parental care – one might call it a 'yolksac juvenile'. This is best illustrated by the gluttonous embryo of lamnoid sharks (Compagno 1977) or also livebearing *Latimeria chalumnae* (Balon 1977, 1984b), and of mouthbrooding *Cyphotilapia frontosa* (Balon 1985).

The prolongation of the embryo period in ametamorphic species, accompanied by a trade-off between nutrient supply and number of eggs, is the prime candidate for parental care, a sensible security system in case of such an expensive 'real estate', according to the metaphores used elsewhere (Balon 1981a). With parental care increasing – from precisely selecting the site for egg deposition to hiding the zygotes, from tending the clutch on a substrate to guarding a nest, and from external to internal bearing (Balon 1975b, 1981b) – the features typical of direct ontogeny become more pronounced and as a result the alevins or juveniles at first feeding are ever larger and, more importantly, better developed. This trend is also apparent in many invertebrates (e.g. Jägersten 1972) and other vertebrates (e.g. Wassersug 1975, 1984, Townsend & Stewart 1985). The vagaries of resource partitioning are limited mostly to only the last part of life history. Nevertheless they affect gametogenesis. Better fed parents will have enhanced vitellogenesis and consequently produce larger and better developed offspring (see p. 14). And this may well be one of the mechanisms by which the constant tendency to specialize is driven.

In short, within a more unpredictably perturbed environment the slowly differentiating generalist – with its larvae ready to scatter through large areas and feed on any peaks in density of food organisms – has better chances for survival than a specialist, finely tuned into an unperturbed (or predictably perturbed) and therefore highly competitive environment. No wonder that the 'altricial ⇌ precocial homeorhetic states' (Balon 1985) may be the most significant property of living organisms, as if two answers were constantly given by epigenesis to any new question an environment might ask (Balon 1981b, 1983). Of course, more than two answers are possible, but only two needed. The time of ontogeny when the above must be decided is the embryo period, for 'If according to Samuel Butler's aphorism, '*a hen is only an egg's way of making another egg'*, then a woman or a man is only an embryo's way of making another embryo' (Weightman 1984, p. 55).

Acknowledgements

The ideas presented in this essay came together gradually mainly when I tried to convince my students that textbook information is rarely complete, sometimes wrong and always out of date. Hardest to convince was my closest collaborator Christine Flegler-Balon and she deserves my greatest thanks for stimulating explantations. Numerous suggestions for improvement were made by Mike Bruton, Karel Liem and Charles Simenstad. John Govoni directed very timely my attention to Grace Orton's papers enhancing my awareness that little we think of is entirely new, and later commented on the manuscript. It is therefore somewhat modified from the talk given in Asilomar.

References cited

Balon, E.K. 1957. Growth and biometry of larvae of the brook lamprey from the Lučina River in Silesia. Věst. Čs. spol. zool. 21: 193–202 (In Slovak).

Balon, E.K. 1960. Über die Entwicklungsstufen des Lebens der Fische und ihre Terminologie. Zeitschr. wiss. Zool. 164: 294–314.

Balon, E.K. 1971. The intervals of early fish development and their terminology (review and proposals). Věst. Čs. spol. zool. 35: 1–8.

Balon, E.K. 1975a. Reproductive guilds of fishes: a proposal and definition. J. Fish. Res. Board Can. 32: 821–864.

Balon, E.K. 1975b. Terminology of intervals in fish development. J. Fish. Res. Board Can. 32: 1663–1670.

Balon, E.K. 1977. Early ontogeny of *Labeotropheus* Ahl, 1927 (Mbuna, Cichlidae, Lake Malawi), with a discussion on advanced protective styles in fish reproduction and development. Env. Biol. Fish. 2: 147–176.

Balon, E.K. (ed.). 1980. Charrs: salmonid fishes of the genus *Salvelinus*. Perspectives in Vertebrate Science 1, Dr W. Junk Publishers, The Hague. 928 pp.

Balon, E.K. 1981a. Additions and amendments to the classification of reproductive styles in fishes. Env. Biol. Fish. 6: 377–389.

Balon, E.K. 1981b. Saltatory processes and altricial to precocial forms in the ontogeny of fishes. Amer. Zool. 21: 567–590.

Balon, E.K. 1981c. About processes which cause the evolution of guilds and species. Env. Biol. Fish. 6: 129–138.

Balon, E.K. 1983. Epigenetic mechanisms: reflections on evolutionary processes. Can. J. Fish. Aquat. Sci. 40: 2045–2058.

Balon, E.K. 1984a. Reflections on some decisive events in the early life of fishes. Trans. Amer. Fish. Soc. 113: 178–185.

Balon, E.K. 1984b. Patterns in the evolution of reproductive styles in fishes. pp. 35–53. *In*: C.W. Potts & R.J. Wootton (ed.) Fish Reproduction: Strategies and Tactics, Academic Press, London.

Balon, E.K. 1985. Early life histories of fishes: new developmental, ecological and evolutionary perspectives. Developments in Env. Biol. Fish. 5, Dr W. Junk Publishers, Dordrecht. 280 pp.

Barlow, G.W. 1981. Patterns of parental investment, dispersal and size among coral-reef fishes. Env. Biol. Fish. 6: 65–85.

Baylis, J.R. 1981. The evolution of parental care in fishes, with reference to Darwin's rule of male sexual selection. Env. Biol. Fish. 6: 223–251.

Blanco, G.J. 1938. Early life history of the viviparous perch *Taeniotoca lateralis* Agassiz. Philippine J. Sci. 67: 379–391.

Blumer, L.S. 1982. A bibliography and categorization of bony fishes exhibiting parental care. Zool. J. Linn. Soc. 76: 1–22.

Boehlert, G.W. & M.M. Yoklavich. 1984. Reproduction, embryonic energetics, and the maternal-fetal relationship in the viviparous genus *Sebastes* (Pisces: Scorpaenidae). Biol. Bull. 167: 354–370.

Bone, Q. & N.B. Marshall. 1982. Biology of fishes. Blackie, Glasgow. 253 pp.

Brodal, A. & R. Fänge (ed.). 1963. The biology of myxine. Universitetsforlaget, Oslo. 588 pp.

Brothers, E.B. & W.N. McFarland. 1981. Correlations between otolith microstructure, growth and life history transitions in newly recruited French grunts [*Haemulon glavolineatum* (Desmarest), Hamulidae]. Rapp. P-v. Reun. Cons. int. Explor. Mer. 178: 369–374.

Burke, R.D. 1984. Pheromonal control of metamorphosis in the Pacific sand dollar, *Dendraster excentricus*. Science 225: 442–443.

Cailliet, G.M. & C.A. Simenstad (ed.). 1982. Gutshop '81. Fish food habits studies. Proceedings of the Third Pacific Workshop, Washington Sea Grant Publication, Seattle. 312 pp.

Calow, P. 1977. Conversion efficiencies in heterotrophic organisms. Biol. Rev. 52: 385–409.

Campbell, C.M. & D.R. Idler. 1976. Hormonal control of vitellogenesis in hypophysectomized winter flounder (*Pseudopleuronectes americanus* Walbaum). Gen. Comp. Endocrinol. 28: 143–150.

Cohen, J. & B. Massey. 1982. Living embryos. Third Edition, Pergamon Press, Oxford. 173 pp.

Compagno, L.J.V. 1977. Phyletic relationships of living sharks and rays. Amer. Zool. 17: 303–322.

Crowder, L.B. 1986. Ecological and morphological shifts in Lake Michigan fishes: glimpses of the ghost of competition past. Env. Biol. Fish. 16: 147–157 (this volume).

Doherty, P.J., D. McB. Williams & P.F. Sale. 1985. The adaptive significance of larval dispersal in coral reef fishes. Env. Biol. Fish. 12: 81–90.

Fahay, M.P. 1983. Guide to the early stages of marine fishes occurring in the western north Atlantic Ocean, Cape Hatteras to the southern Scotian Shelf. J. Northwest Atlantic Fishery Sci. 4, Dartmouth. 423 pp.

Flüchter, J. 1982. Substance essential for metamorphosis of fish larvae extracted from *Artemia*. Aquaculture 27: 83–85.

Fostner, H., S. Hinterleitner, K. Mähr & W. Wieser. 1983. Towards a better definition of 'metamorphosis' in *Coregonus* sp.: biochemical, histological, and physiological data. Can. J. Fish. Aquat. Sci. 40: 1224–1232.

Fryer, G. 1959. The trophic interrelationships and ecology of some littoral communities of Lake Nyasa with especial reference to the fishes, and a discussion of the evolution of a group of rock-frequenting Cichlidae. Proc. zool. Soc. Lond. 132: 153–281.

Garstang, W. 1929. The origin and evolution of larval forms. Rep 96th Meeting British Ass. Adv. Sci. London: 77–98.

Gehringer, J.W. 1959. Early development and metamorphosis of the tenpounder *Elops saurus* Linnaeus. U.S. Fish. Bull. 155: 617–647.

Gerbilsky, N.L. 1956. The role of nervous system in determining the process of change in fish spawning conditions. Trudy Karelskogo fil. AN SSSR 5: 6–12 (In Russian).

Gilbert, L.I. & E. Frieden (ed.). 1981. Metamorphosis. A problem in developmental biology. Plenum Press, New York. 578 pp.

Gold, J.R., W.J. Karel & M.R. Strand. 1980. Chromosome formulae of North American fishes. Prog. Fish – Cult. 42: 10–23.

Goto, A. 1980. Geographic distribution and variations of two types of *Cottus nozawae* in Hokkaido and morphological characteristics of *C. amblystomopsis* from Sakhalin. Jap. J. Ichthyol. 27: 97–105.

Goto, A. 1982. Reproductive behavior of a river sculpin, *Cottus nozawae*. Jap. J. Ichthyol. 28: 453–457.

Grodziński, Z. 1949. Fat drops in the yolk of the sea-trout *Salmo trutta* L. Bull. Acad. Sc. Cracovie Ser. B. 2: 59–78.

Grondziński, Z. 1968. The yolk of holostean fishes. Acta Biol. Cracov. Ser. Zool. 11: 315–323.

Hartvig, M. & R.E. Weber. 1984. Blood adaptations for maternal-fetal oxygen transfer in the viviparous teleost, *Zoarces viviparus* L. pp. 17–30. *In*: R.S. Seymour (ed.) Respiration and Metabolism of Embryonic Vertebrates, Perspectives in Vertebrate Science 3, Dr W. Junk Publishers, Dordrecht.

Ho, M.-W. & P.T. Saunders. 1984. Beyond neo-Darwinism. An introduction to the new evolutionary paradigm. Academic Press, London. 376 pp.

Hulet, W.H. 1978. Structure and functional development of the eel leptocephalus *Ariosoma balearicum* (De LaRoche, 1809). Phil. Trans. R. Soc. London, Ser. B 282: 107–138.

Hunter, J.R. 1981. Feeding ecology and predation of marine fish larvae. pp. 33–77. *In*: R. Lasker (ed.) Marine Fish Larvae, Washington Sea Grant Program, University of Washington Press, Seattle.

Ingermann, R.L. & R.C. Terwilliger. 1984. Facilitation of maternal-fetal oxygen transfer in fishes: anatomical and molecular specialization. pp. 1–15. *In*: R.S. Seymour (ed.) Respiration and Metabolism of Embryonic Vertebrates, Perspectives in Vertebrate Science 3, Dr W. Junk Publishers, Dordrecht.

Ivanova, M.N. & A.M. Lopatko. 1984. Feeding behavior of pike, *Esox lucius* (Esocidae), larvae from the progeny of a single pair of spawners. J. Ichthyol. 23: 171–173.

Jägersten, G. 1972. Evolution of the metazoan life cycle. Academic Press, London. 276 pp.

Karnaukhov, V.N. 1979. The role of filtrator molluscs rich in carotenoid in the self-cleaning of fresh waters. Symp. Biol. Hung. 19: 151–167.

Kendall, A.W. Jr. 1981. Early life history of eastern north Pacific fishes in relation to fisheries investigations. Washington Sea Grant Program Technical Report, Seattle. 7 pp.

Kimura, S. 1973. On the fetus of the eel pout *Enchelyopus elongatus*. Jap. J. Ichthyol. 20: 123–124.

Kiørboe, T., P. Munk & J.G. Støttrup. 1985. First feeding by larval herring *Clupea harengus* L. Dana 5: 95–107.

Koller, G. 1930. Versuche an marinen Wirbellosen über die Aufnahme gelöster Nahrstoffe. Zeitschr. vergl. Physiol. 11: 437–447.

Korsgaard, B. 1983. The chemical composition of follicular and ovarian fluid of the pregnant blenny (*Zoarces viviparus* (L.)). Can. J. Zool. 61: 1101–1108.

Kryzhanovsky, S.G., N.N. Disler & E.N. Smirnova. 1953. Ecomorphological principles of development in percids. Tr. Inst. Morph. Zhiv. Severtsova 10: 3–138 (In Russian).

Lauder, G.V. 1985. Aquatic feeding in lower vertebrates. pp. 210–229, 397–399. *In*: M. Hildebrand, D.M. Bramble, K.F. Liem & D.B. Wake (ed.) Functional Vertebrate Morphology, The Belknap Press of Harvard University Press, Cambridge.

Lewis, R.M., E.P.H. Wilkens & H.R. Gordy. 1972. A description of young Atlantic menhaden, *Brevoortia tyrannus*, in the White Oak River estuary, North Carolina. U.S. Fish. Bull. 70: 115–118.

Liem, K.F. 1980. Acquisition of energy by teleosts: adaptive mechanisms and evolutionary patterns. pp. 299–334. *In*: M.A. Ali (ed.) Environmental Physiology of Fishes, Plenum Publishing Corp., New York.

Liem, K.F. & J.W.M. Osse. 1975. Biological versatility, evolution, and food resource exploitation in African cichlid fishes. Amer. Zool. 15: 427–454.

Liem, K.F. & D.J. Stewart. 1976. Evolution of the scale-eating cichlid fishes of Lake Tanganyika: a generic revision with a description of a new species. Bull. Mus. Comp. Zool. Harvard 147: 319–350.

Løvtrup, S. 1974. Epigenetics – a treatise on theoretical biology. John Wiley and Sons, London. 548 pp.

Løvtrup, S. 1984. Ontogeny and phylogeny. pp. 159–190. *In*: M.-W. Ho & P.T. Saunders (ed.) Beyond Neo-Darwinism, An Introduction to the New Evolutionary Paradigm, Academic Press, London.

Marliave, J.B. 1981. Vertical migrations and larval settlement in *Gilbertidia sigalutes*, F. Cottidae. Rapp. P.-V. Reun. Cons. int. Explor. Mer. 178: 349–351.

Marshall, N.B. 1953. Egg size in Arctic, Antarctic and deep-sea fishes. Evolution 7: 328–341.

Matsuda, R. 1982. The evolutionary process in talitrid amphipods and salamanders in changing environments, with a discussion of 'genetic assimilation' and some other evolutionary concepts. Can. J. Zool. 60: 733–749.

Mendoza, G. 1958. The fin fold of *Goodea luitpoldi*, a viviparous cyprinodont teleost. J. Morphol. 103: 539–560.

Mendoza, G. 1972. The fine structure of an absorptive epithelium in a viviparous teleost. J. Morphol. 136: 109–115.

Moore, J.W. & I.C. Potter. 1976. A laboratory study on the feeding of larvae of the brook lamprey *Lampretra planeri* (Bloch). J. Anim. Ecol. 45: 81–90.

Moser, H.G. 1981. Morphological and functional aspects of marine fish larvae. pp. 89–131. *In*: R. Lasker (ed.) Marine Fish Larvae, Washington Sea Grant Program, University of Washington Press, Seattle.

Moser, H.G. & E.H. Ahlstrom. 1970. Development of lanternfishes (family Myctophidae) in the California Current. Part. I. Species with narrow-eyed larvae. Bull. Los Angeles County Mus. Nat. Hist. Sci. 7, Los Angeles. 145 pp.

Moyle, P.B. & J.J. Cech. 1982. Fishes: an introduction to ichthyology. Prentice-Hall, Englewood Cliffs. 593 pp.

Ng, T.B. & D.R. Idler. 1978. 'Big' and 'little' forms of plaice vitellogenic and maturational hormones. Gen. Comp. Endocrinol. 34: 408–420.

Noakes, D.L.G., D.G. Lindquist, G.S Helfman & J.A. Ward (ed.). 1983. Predators and prey in fishes. Developments in Env. Biol. Fish. 2, Dr W. Junk Publishers, The Hague. 228 pp.

Norris, D.O. 1983. Evolution of endocrine regulation of metamorphosis in lower vertebrates. Amer. Zool. 23: 709–718.

Orton, G.L. 1953. The systematics of vertebrate larvae. Syst. Zool. 2: 63–75.

Pfeiler, E. 1986. Towards an explanation of the developmental strategy in leptocephalous larvae of marine teleost fishes. Env. Biol. Fish. 15: 3–13.

Piavis, G.W. 1971. Embryology. pp. 361–400. *In*: M.W. Hardisty & I.C. Potter (ed.) The Biology of Lampreys, Vol. 1, Academic Press, London.

Pollard, J.W. (ed.) 1984. Evolutionary theory: paths into the future. John Wiley & Sons, Chichester. 271 pp.

Reif, W.-E. 1983. Evolutionary theory in German paleontology. pp. 173– 203. *In*: M. Grene (ed.) Dimensions of Darwinism, Cambridge University Press, Cambridge.

Ribbink, A.J., B.A. Marsh, A.C. Marsh, A.C. Ribbink & B.J. Sharp. 1983. A preliminary survey of the cichlid fishes of rocky habitats in Lake Malawi. S. Afr. J. Zool. 18: 149–310.

Richards, J.E. 1980. The freshwater biology of the anadromous Pacific lamprey (*Lampetra tridentata*). M.Sc. Thesis, University of Guelph, Guelph. 99 pp.

Sazima, I. 1983. Scale-eating in characoids and other fishes. Env. Biol. Fish. 9: 87–101.

Schultze, M.S. 1856. Die Entwickelungs-Geschichte von *Petromyzon planeri*. Natuurkundige Verh. Holl. Maatsch. Wetenschappen Haarlem 2: 1–51.

Soin, S.G. 1968. Some peculiarities of development of the eelpout [*Zoarces viviparus* (L.)] related to livebearing. Voprosy Ichtiologii 8: 283–293 (In Russian).

Stephens, G.C. 1982. Recent progress in the study of 'Die Ernährung der Wassertiere und der Stoffhaushalt der

Gewässer'. Amer. Zool. 22: 611–619 (and other papers of the Symposium *The Role of Uptake of Organic Solutes in Nutrition of Marine Organisms*).

Terner, C. 1968. Studies of metabolism in embryonic development. I. The oxidative metabolism of unfertilized and embryonated eggs of the rainbow trout. Comp. Biochem. Physiol. 24: 933–940.

Townsend, D.S. & M.M. Steward. 1985. Direct development in *Eleutherodactylus coqui* (Anura: Leptodactylidae): a staging table. Copeia 1985: 423–436.

Van Bohemen, J.G.D., H.J. Lambert, T. Goos & P.G.W. van Oordt. 1982. Estrone and estradiol participation during exogenous vitellogenesis in the female rainbow trout, *Salmo gairdneri*. Gen. Comp. Endocrinol. 46: 81–92.

Van Tienhoven, A. 1983. Reproductive physiology of vertebrates. Cornell University Press, Ithaca. 491 pp.

Veith, W.J. 1980. Viviparity and embryonic adaptations in the teleost *Clinus superciliosus*. Can. J. Zool. 58: 1–12.

Vrba, E.S. 1980. Evolution, species and fossils: how does life evolve? S. Afr. J. Sci. 76: 61–84.

Walvig, F. 1963. The gonads and the formation of the sexual cells. pp. 530–580. *In*: Broadal & R. Fänge (ed.) The Biology of Myxine, Universitetsforlaget, Oslo.

Wassersug, R.J. 1975. The adaptive significance of the tadpole stage with comments on the maintenance of complex life cycles in anurans. Amer. Zool. 15: 405–417.

Wassersug, R.J. 1984. Why tadpoles love fast food. Natural History 93(4): 60–69.

Watanabe, Y. 1984. Digestion and absorption in fish larvae. Aquabiology 6: 191–197 (In Japanese).

Webb, P.W. & J.R. Brett. 1972. Respiration adaptations of prenatal young in the ovary of two species of viviparous seaperch, *Rhacochilus vacca* and *Embiotoca lateralis*. J. Fish. Res. Board Can. 29: 1525–1542.

Weightman, J. 1984. Voltaire to Medawar. Encounter 63: 53–56.

Whittaker, R.H., S.A. Levin & R.B. Root. 1973. Niche, habitat, and ecotope. Amer. Nat. 107: 321–338.

Wourms, J.P. 1981. Viviparity: the maternal-fetal relationship in fishes. Amer. Zool. 21: 473–515.

Wourms, J.P. & J. Lombardi. 1979. Cell ultrastructure and protein absorption in the trophotaenial epithelium, a placental analogue of viviparous fish embryos. J. Cell Biol. 83: 399.

Received 8.3. 1985 *Accepted 2.9.1985*

Foraging behaviour in fishes: perspectives on variance

Brian M. Marcotte & Howard I. Browman[1]
McGill University, Institute of Oceanography, 3620 University Street, Montreal, Quebec H3A 2B2, Canada
[1] *Present address: Department of Systematics and Ecology, The University of Kansas, Lawrence, KS 66045, U.S.A.*

Keywords: Optimal foraging, Prey selection, Learning, Memory, Perception, Cognition, Neuroethology, *Salmo salar*

Synopsis

The positive relationship between size of prey and frequency of ingestion by predators has been a focal point of investigations in foraging ecology. Field studies compare the frequency distribution of prey sizes in the predator's gut with that in the environment. Laboratory and field (enclosure) studies are based upon comparison of the frequency distributions of prey sizes in controlled environments, before and after the introduction of a predator. 'Optimal' caloric return for foraging effort (i.e. the theory of optimal foraging) has been widely used as a guiding principle in attempts to explain what a fish consumes. There is a body of information, however, which seems to indicate that the perceptual potentialities and cognitive abilities of a predator can account for both the direction of the prey size versus ingestion frequency relationship and the variance surrounding it. Part of this variance may be evidence of 'systematic ambiguity', a property of cognitive skills causing predators to respond to the same stimulus in different ways and to different stimuli in the same way. More extensive examination of cognitive skills (minimally defined as learning, remembering and forgetting) in fish may permit causal interpretations (immediate and ultimate) of variance in predatory skills. In such a paradigm of foraging behaviour, environmental stimulus is not taken as the predator's object of response (percept); a cognitive representation connects mind to stimulus and this is the criterion for the act of perception. Cognition, here considered as a formal system which acts upon representations, connects mind to response and thus to adaptation. Studies of the relationships among rates of learning, long and short-term memory, rates of forgetting, prey behavior, size and population turnover rates, lateralization of brain functions, diel fluctuations in predator activity levels and sleep, experience, and 'critical periods' in the development of the predator's nervous system should be examined in relation to foraging behaviour.

Introduction

Predators frequently select the largest manageable prey item available to them. Mean prey size increases with increasing predator size and with increasing predator-prey pursuit distance (Schoener 1979). Such patterns in predator-prey relations have been explained with theories of 'optimal foraging', which argue that predators should optimize their energy (caloric) return per unit effort (Pyke et al. 1977, Krebs 1979, Schoener 1979). Optimal foraging is a subset of theories concerned with adaptive responses (strategies) in evolutionary biology. The principles of optimal foraging have been reviewed (Krebs 1979, Schoener 1979), tested (e.g. Diamond 1984) and critically evaluated (Cody 1974, Levins 1975a, b). Because every adaptive response becomes a factor in all subsequent adaptations, a multivariate approach to organism-environment relations (i.e. fitness) is necessary. Within such systems of interconnected adaptative traits, some traits may be at their selective optima while others are at their pessima. There is no need to assume that all adaptive responses are ultimately or immediately optimal (Levins 1975a, b). Accordingly, the nature and manner in which adaptive traits – e.g. perceptual abilities, cognitive skills and the ethological consequences of neurological development – are interconnected are proper, and perhaps essential, objects of study in trophic biology.

To focus attention on our subject we will consider the food size choice made by Atlantic salmon, *Salmo salar* alevins (Fig. 1). A clear positive trend, and some variance, is observed. If optimal foraging is invoked to explain these data, how are offlyers from this relationship to be explained; stochasticity alone? What is the biological basis of this foraging strategy? How does the predator determine the prey's caloric content? Is this ability hard-wired (genetically imprinted opon the animal's nervous system), learned, or some combination of the two? If learned, how rapidly? How is the learned response retained and for how long? How do these responses deteriorate when the prey population is

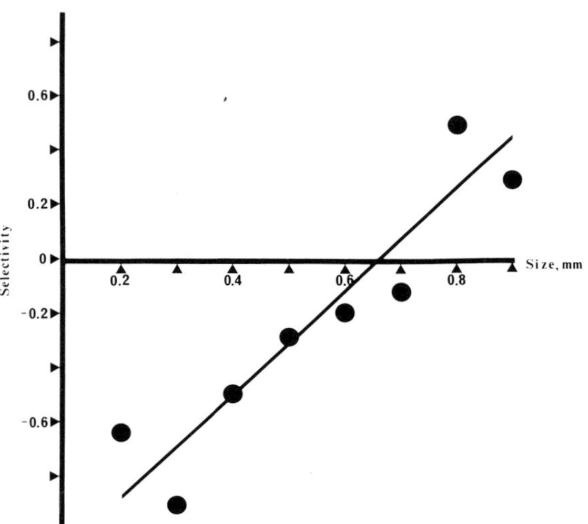

Fig. 1. Relationship between feeding electivity (Standardized Forage Ratio) and prey size for *Salmo salar* alevins. Line fitted by eye.

no longer abundant or when the predator grows and must pursue different prey? If a skill is hard-wired, how does it differ from perception per se (e.g. Brawn 1969, Maiorana 1981)? How does a hard-wired response remain selectively advantageous in a variable environment (e.g. short and long term climatic fluctuations)?

We introduce below the notions of perception, cognition and developmental neuroethology as substrates of foraging behaviour in juvenile fish. Our purpose is general, our literature review necessarily selective. We concentrate attention on the biology of Atlantic salmon, with which we are most familiar, but introduce salient insights from studies on other vertebrates in order to shed light on the meaning of patterns observed in the foraging behaviours of fish.

Feeding in Atlantic salmon alevins

Diurnal rhythms have been demonstrated in the feeding activity of juvenile fish (e.g. Oliphan 1957, Pinskii 1967, Godin 1981). Browman & Marcotte (1986) have demonstrated diurnal rhythms in the intensity and skill with which Atlantic salmon alevins feed. Mean values of the fishes' total behavioural activity, feeding success and feeding error peaked at 0600–0900, 1100–1400 and 1900–2000 h. Variance (standard deviations around the mean) in total activity and feeding error was positively related to the mean. Mean frequency of ingestion peaked at 0600 h and declined asymptotically after 0800 h, although the total number of prey items in the fishes' gut remained unchanged throughout the day. The variance of ingestion frequency, however, followed the three peak trend of the other behavioural variables. The intensity of prey size selection generally increased during the day, with lows associated with periods of higher activity except in the evening. Selection of the largest prey (>0.8 mm) was associated with low light levels and high activity. These fluctuations in foraging behaviour and dietary selectivity appeared homologous to, and in almost perfect temporal agreement with, diurnal rhythms in wakefulness and sleep (which may be interpreted as levels of cognitive/perceptual awareness) observed for other fish species (Fig. 2).

Feeding rates, capture success, ingestion rate and total activity in juvenile fish have been shown to change with the abundance of prey (Hunter 1980, Werner & Blaxter 1980, Sharp 1981, Morgan & Ritz 1984). Prey abundance affects the foraging behaviour of juvenile Atlantic salmon as well (Browman 1985). Above prey densities of 400 items per litre, 'sensory overload' led to optic tetanus (nystagmus, or uncontrolled repetition of ocular fixations) and caused success to decline (Fig. 3). The total number of prey items in the fishes' guts also followed this bell-shaped response pattern. There was some suggestion that selectivity of prey sizes was negatively affected by increased prey abundance, perhaps as a result of perceptual confusion (see below).

Background contrast affected the colour of prey selected by Atlantic salmon alevins (Browman 1985). Red prey were preferred under blue background conditions, blue prey against aqua-green backgrounds and unstained prey against red backgrounds. These studies clearly indicate the importance of controlling for time of day, prey density, and prey and background contrast in fish feeding experiments.

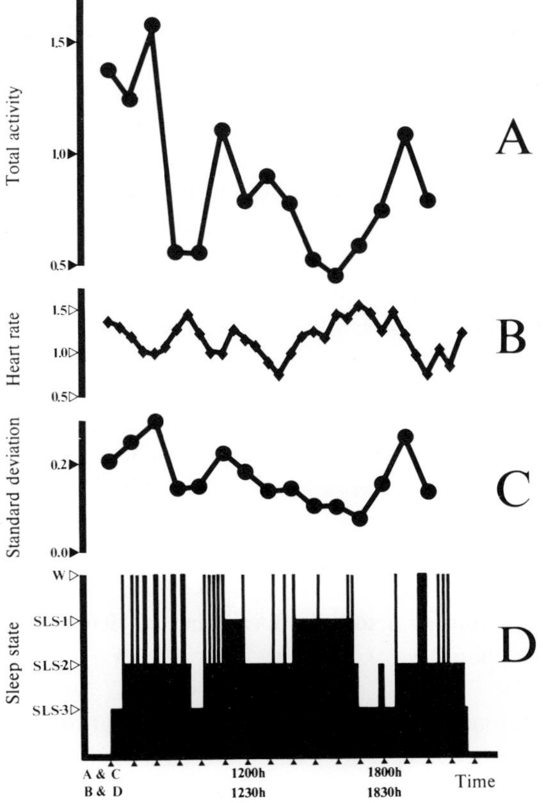

Fig. 2. A – *Salmo salar* alevins. The relationship between mean (for 10 fish) Total activity (#Eye fixations + #Moves + #Bites + #Misses + #Spits + #Social + #Ambiguous) per 3 min observation period (normalized for tank zooplankton abundance) and time of day (from Browman & Marcotte 1986). *B* – Diurnal periodicity in the heart rate of the catfish. Ordinate = time interval (seconds) between two subsequent heartbeats (redrawn from Karmanova et al. 1981). *C* – Variance in Total activity (Fig. 9A) as represented by its standard deviation. *D* – Diurnal periodicity in the wakefulness-primary sleep cycle of the catfish. SLS = sleep-like-state; SLS-1 = characterized by immobility with plastic muscle tone; SLS-2 = characterized by immobility with rigid muscle tone; SLS-3 = characterized by immobility with muscle relaxation; W = wakefulness (redrawn from Karmanova et al. 1981).

Perception as a source of variance in fish foraging behaviour

The perceptual abilities/limitations of fish necessarily affect their foraging behaviour (see, for ex-

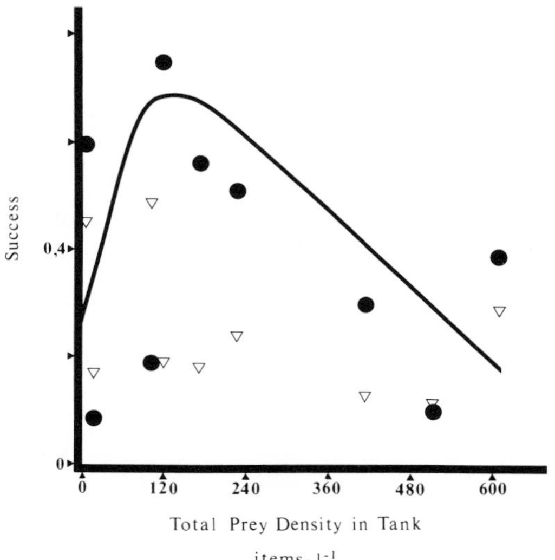

Fig. 3. Salmo salar alevins. The relationship between mean (for 10 fish) feeding success (#Ingestions/#Bites) and tank zooplankton density. Triangles are standard deviations around the means. Curve fitted by eye.

ample Dill 1983, Ringler 1983). It is to this aspect of fish biology that we now address ourselves.

Vision in fish is sensitive to differential prey movements against a kinematic background (e.g. Bateson 1889, Ingle 1968). Object orientation and direction of movement, form, texture and colour are also important in visual prey detection (e.g. Cronley-Dillon 1964, Jacobson & Gaze 1964, Sutherland 1968, Trevarthen 1968). In addition, the perceptual field of a fish's retina can be limited by environmental (or experiential) restrictions (or deprivations) during early development (Blaxter 1970, Ahlbert 1976). It is important, therefore, to control for these variables when considering trends in food selectivity by fish. It is possible that fish perceive the choreographies and second derivatives (i.e. accelerations and decelerations) of prey motion and not their size or taxonomic identity per se (Marcotte 1983).

Abundant prey may confuse fish predators (e.g. Williams 1964, Neill & Cullen 1974). Because large prey occur infrequently in most ecosystems, fish may minimize this confusion at intermediate prey densities by concentrating their attention on rare items. Thus, it is possible that fish 'select' large prey at intermediate densities for perceptual reasons, and not necessarily to optimize energetic returns. At very high densities (rare occurrences in natural environments, but the norm in hatchery settings), optic tetanus may render the animal incapable of making foraging decisions and dietary selection may tend toward randomness.

When humans observe two identical objects placed in the 'foreground' and 'background' of a flat polar projection (or other background which indicates a distant horizon) the object in the 'background' appears larger. This optical illusion applies until the image of the object subtends less than one degree of arc on the retina, at which point the reality of identity is perceived (Kaufman & Rock 1962, Ross et al. 1980). Juvenile fish have eyes on which, for any given object, a greater arc of retinal surface is subtended than for older, larger fish (Hairston et al. 1982). If this optical illusion occurs in fish, juveniles may be victimized more often than their older counterparts. As a consequence, a juvenile's prey size selection will be more variable and less intense. Further, tests of the 'apparent prey size hypothesis' (e.g. Luecke & O'Brien 1981) which have used background polar projections to allow observers to estimate prey distances, may have inadvertantly created this optical illusion. Given the similarity of vertebrate visual systems (see Ali 1975, Northmore et al. 1978), there is no strong reason to neglect such an idea as impossible or even unlikely for fish.

Images falling on the retina must flicker to be perceived (Cohen 1969). Flicker may be generated by the perceiver's motion (e.g. eye tremor) or may be environmentally induced, e.g. prey movement or wave generated flicker (McFarland & Loew 1983). Turbidity in water decreases the distance a predator can see, increases luminosity, decreases contrast, defocuses and dissipates wave generated flicker and changes the wave lengths of light transmitted through water (Marcotte, unpublished). All of these changes will affect a predator's ability to see distant prey, judge pursuit distances, and determine prey size, shape and colour. Previous studies have also indicated the need to control for effects of turbidity and the amount and kind of environmental lighting in experiments of foraging behaviour (e.g. Vinyard & O'Brien 1976, O'Brien 1979).

Cognition as a source of variance in fish foraging behaviour

Once a fish's perceptual abilities/limitations have been accounted for, cognitive skills can be considered to enable further explanation of trends and variance in dietary selections. For our purposes here, cognition can be minimally defined as learning, remembering and forgetting.

Cognition is to perception what syntax is to semantics (see Marcotte 1983); it is that mental function through which the organism creates mental representations of real stimuli, such as prey items. These representations possess attributes to which the organism responds – animals do not necessarily respond to external stimuli per se. For this reason, an animal can respond to the same stimuli in different ways, and to different stimuli in the same way. The result could be a systematic ambiguity in observed behaviour which may be mistaken for stochastic variance in experimental examinations of food selection. Perception connects mind to stimulus; cognition connects mind to action. Perception is on the cutting edge of selection pressure; cognition is on the cutting edge of adaptation. Organisms do not adapt to an environment, they define it through mental representations and cannot adapt to an unperceived or unrepresented cause.

Two cognitive variables can be considered in the context of foraging behaviour and dietary selectivity, learning and memory: (1) Learning ability in fish has been adequately demonstrated (e.g. Ware 1971) and so we must ask: how fast does a fish learn appropriate responses to evasive prey and how does this change ontogenetically? (2) How does a fish remember learned responses? What are the durations of long and short-term memory (see Peeke et al. 1972), with and without reinforcements at various temporal frequencies. How do these durations change with development? Phrased differently; what is the turnover rate of 'search images' and how is this turnover related to development – especially given the continuous growth of the nervous system in fish (e.g. Johns & Easter 1975, Easter et al. 1977).

Learning is critical for juvenile fish. Hatchery-reared salmon do not feed as effectively upon release as do their wild counterparts and it may take up to two months for fish reared on artificial hatchery food to learn how to feed successfully on living prey (MacCrimmon 1954, Fenderson et al. 1968, Blaxter 1970, 1975, Sosiak et al. 1979, Shustov et al. 1980, Dickson & MacCrimmon 1982). Congruence, or lack of it, between developmental changes in the speed with which fish learn and forget (the dynamics of their memories) and the life histories and ecologies of available prey probably affect diet. The internal ecology of the predator must keep pace with that of its surroundings. For example, small prey are characterized by high reproductive potentials, short generation times and therefore high turnover frequencies and often short durations of population maxima (Fig. 4). Visual predators which feed on small prey must either feed non-selectively (e.g. filter feed) or must have the capacity to learn and forget 'search images' quickly (e.g. daily) to keep pace with the prey's turnover rate.

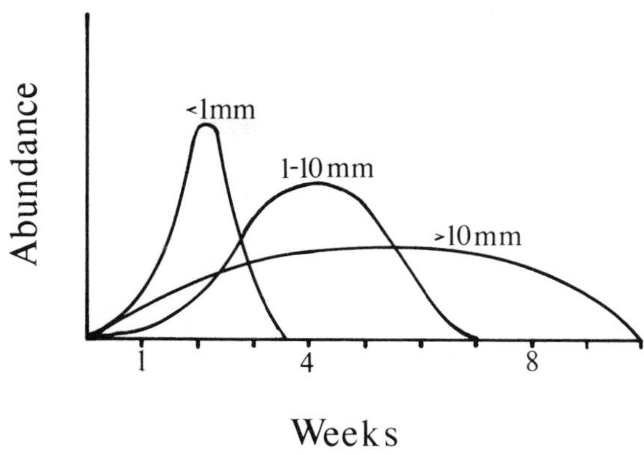

Fig. 4. Schematic representation of generational durations for hypothetical prey of three different sizes.

Developmental neuroethology and fish foraging behaviour

How learned responses can be made to resemble hard-wired stimulus-response arcs may hold a key to the developmental neuroethology of fish. Larval fish 'practice' feeding (e.g. Fortier 1983), implying that the 'critical period' for the development of learned skills and associated parts of the nervous system may be during the late embryonic and/or early larval periods. During these developmental intervals visual stimuli enter the fish's brain and comprehension (leading to learning) and response determination may be handled in a locus separate from that of the primary visual-motor center (Fig. 5). An effective response pathway may be selected in the motor nerve routes (e.g. trigeminal nerve), perhaps by neuronal competition (sensu Edelman & Finkel 1985). The dendritic arborizations of nerves in these selected pathways may become reduced and the axons may subsequently become myelinated. Later, the centres for comprehension, having executed their neuronal selection function, may divest themselves of memory functions as the visual-motor pathway is short-circuited (Fig. 5) and a true stimulus-response arc (Fig. 5B) is established (see also Luria 1978, Marcotte 1983). A behavioural suite, controlled by some established (selected)

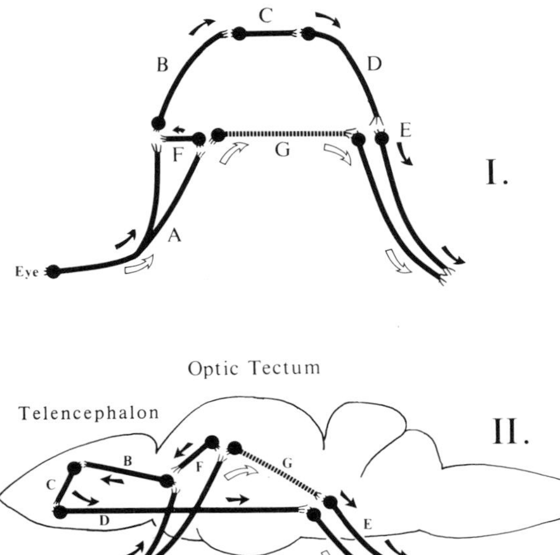

Fig. 5. I Schematic representation of hypothetical neuronal network involving sensory input from the eye via (A) to the sensorimotor center of the brain (F), to the center of comprehension (B, C, D), then to the area of neuronal selection (E) – follow solid arrows. A possible short-circuit of this pathway (via G) is represented by the open arrows. *II* Neuronal pathway of *I* superimposed upon schematic representation (lateral view) of a generalized fish brain.

behavioural pathway, may be less plastic: development (e.g. sprouting) and selection of new pathways may be necessary to handle new experiences (a relatively long process). The pace of this development and selection may control the duration of the observed delay in trophic recovery of newly released fish, cited above. Establishing the congruence between the rate of neuronal development (e.g. neuron 'birthdays') and the speed of prey switching behaviour may form a useful test of this hypothesis. Comparing foraging behaviour and brain development in viviparous and oviparous fish may be a useful guide to research of this type.

The brains of fish have specialized areas for long and short-time memory, comprehension and sensory-motor function. It may be possible to study the neuroethological consequences of brain lesions in fish, the results of which may mimic human diseases such as Korsakov's Syndrome and aphasias following stroke. Thus, fish may provide a useful prototype (straight-brain) model for the study of human brain disorders (Campbell & Hodos 1970, Bullock 1983a, 1983b, 1984a, 1984b).

The diurnal rhythms in feeding behaviour presented above imply that subtle aspects of neuronal activity may also affect foraging success and diet choice. Klein & Armitage (1979) have reported 90–100 min oscillations in the abilities of humans to solve verbal and spatial tasks which require mental abilities residing, to some extent, on opposite sides of the brain. These oscillations imply rhythms in brain hemispheric dominance of behaviour. Preliminary studies carried out in our laboratory indicate that handedness (evaluated through assymetric directions of foraging sorties) occurs in fish, implying hemispheric dominance of behaviour. Diurnal rhythms in feeding error, i.e. error in a behavioural task requiring spatial skill (which may be a unilaterally distributed brain function, as it is in humans), implies that oscillations in hemispheric dominance also occur in fish. These neurological properties may be a source of variance in trophic studies in which timing is uncontrolled.

Diurnal behavioural rhythms may also be manifestations of oscillations in brain function – analogous to those which occur nocturnally during sleep (Kleitman 1963, 1969, Broughton 1975, Klein & Armitage 1979). Wake-sleep rhythms may be an underlying cause of variance in fish behaviour. Sleep in higher vertebrates seems to be associated with the brain's attempt to consolidate memory and eliminate extraneous connections in neuronal circuitry (e.g. Crick & Mitchison 1983). Sleep is of paramount importance in the neuroethology of higher vertebrates and sleep deprivation can have profound consequences. The same may be true of fish; again, fish may present a straight-brain model system for testing hypotheses relating sleep, neuronal structure and ethology. If so, it may be crucial to the proper development of fish that hatchery environments mimic the diel light rhythms of nature. Attempts to enhance growth by keeping fish larvae in perpetual light may frustrate sleep rhythms (but probably not override them entirely) and may promote dysfunctional neuroethological development.

Prolepsis

Fish are not automata, black boxes or stochastic generators of time series data (i.e. behaviours). They, like all animals, are a union of cognitive skills and perceptual ability/limitation (Marcotte 1983). The plasticity of brain development with regard to experience is fundamental to this union as well as establishing an important feedback loop within it.

Three objections are often raised to the inclusion of ideas from the cognitive sciences in ecological theories: (1) it violates simplicity, (2) it violates intuition or (3) it is an anthropocentric imposition. Cognition stands between stimulus and behavioural response. It does not complicate behavioural science; it simplifies it by providing meaning to both the trends and variances (heretofore often considered stochastic, meaningless noise) in behavioural data sets.

Finally, cognition is a plesiomorphic or 'primitive' property of nervous systems – from that of an earthworm, to fish and man (Thorpe 1969). Invocation of cognitive abilities to explain patterns in fish behaviour does not constitute an anthropocentric imposition. It is simply the recognition of the primitive as the primitive: that humans share certain basic processes with other creatures. It is the recognition of the existence of an internal ecology which is responsive to environmental experiences and that these responses alter reactions to future experiences (Hubel & Wiesel 1970, Blakemore 1976, Keating 1976, Blakemore 1977, Cooke & Horder 1977, Keating 1977, Edelman & Finkel 1985). It is the recognition of the reflexive/feed-back relationship between an animal's internal and external ecologies which lies at the heart of adaptive strategies.

References cited

Ahlbert, I.B. 1976. Organization of the cone cells in the retinae of salmon, *Salmo salar* and trout *Salmo trutta* in relation to their feeding habits. Acta Zool. 57: 13–35.

Ali, M.A. (ed.). 1975. Vision in fishes: new approaches in research. Plenum Press, New York. 836 pp.

Bateson, W. 1889. Notes on the senses and habits of some crustacea. J. Mar. Biol. Assoc. UK 1: 211–224.

Blakemore, C. 1976. Modification of visual function by early visual experience. Bull. Schweiz. Akad. Med. Wiss. 32: 13–28.

Blakemore, C. 1977. Genetic instructions and developmental plasticity in the kitten's visual cortex. Phil. Trans. R. Soc. Lond. B 278: 425–434.

Blaxter, J.H.S. 1970. Sensory deprivation and sensory input in rearing experiments. Helgol. Wiss. Meeresunters. 20: 642–654.

Blaxter, J.H.S. 1975. Reared and wild fish. How do they compare? Proc. 10th Europ. Symp. Mar. Biol. 1: 11–26.

Brawn, V.M. 1969. Feeding behaviour of cod. J. Fish. Res. Board Can. 26: 583–596.

Broughton, R.J. 1975. Biorhythm variations in consciousness and psychological functions. Can. Psychol. Rev. 16: 217–239.

Browman, H.I. 1985. Feeding behaviour in fry of Atlantic salmon, *Salmo salar* L. M.Sc. Thesis. McGill University, Montreal. 141 pp.

Browman, H.I. & B.M. Marcotte. 1986. Diurnal feeding and prey size selection in Atlantic salmon (*Salmo salar* L.) alevins. Dev. Env. Biol. Fish. 7: 269–284.

Bullock, T.H. 1983a. Why study fish brains? Some aims of comparative neurology today. pp. 361–368. *In*: R.E. Davis & R.G. Northcutt (ed.) Fish Neurobiology, Vol. 2, Higher Brain Areas and Functions, University of Michigan Press, Ann Arbor.

Bullock, T.H. 1983b. Neurobiological roots and neuroethological sprouts. pp. 403–412. *In*: F. Huber & H. Markl (ed.) Neuroethology and Behavioral Physiology, Springer-Verlag, New York.

Bullock, T.H. 1984a. Comparative neuroscience holds promise for quiet revolutions. Science 225: 473–478.

Bullock, T.H. 1984b. The future of comparative neurology. Amer. Zool. 24: 693–700.

Campbell, C.B.G. & W.W. Hodos. 1970. The concept of homology and the evolution of the nervous system. Brain Behav. Evol. 3: 353–367.

Cody, M.L. 1974. Optimization in ecology. Science 183: 1156–1164.

Cohen, J. 1969. Sensation and perception I. Vision. Rand McNally & Co., Chicago. 76 pp.

Cook, J.E. & T.J. Horder. 1977. The multiple factors determining retinotopic order in the growth of optic fibers into the optic tectum. Phil. Trans. R. Soc. Lond. B 278: 261–276.

Crick, F. & G. Mitchison. 1983. The function of dream sleep. Nature 304: 111–114.

Cronley-Dillon, J.R. 1964. Units sensitive to direction of movement in goldfish optic tectum. Nature 203: 214–215.

Diamond, J.M. 1984. Optimal foraging theory tested. Nature 311: 603–604.

Dickson, T.A. & H.R. MacCrimmon. 1982. Influence of hatchery experience on growth and behavior of juvenile Atlantic salmon (*Salmo salar*) within allopatric and sympatric stream populations. Can. J. Fish. Aquat. Sci. 39: 1453–1458.

Dill, L.M. 1983. Adaptive flexibility in the foraging behavior of fishes. Can. J. Fisher. Aquat. Sci. 40: 398–408.

Easter, S.S., P.R. Johns & L.R. Baumann. 1977. Growth of the adult goldfish eye I: optics. Vision Res. 17: 469–477.

Edelman, G.M. & L.H. Finkel. 1985. Neuronal group selection in the cerebral cortex. pp. 00–00. *In*: G.M. Edelman, W.M. Cowan & W.E. Gall (ed.) Dynamic Aspects of Neocortical Function, Wiley, New York (in press).

Fenderson, O.C., W.H. Everhart & K.M. Muth. 1968. Comparative agonistic and feeding behavior of hatchery-reared and wild salmon in aquaria. J. Fish. Res. Board Can. 25: 1–14.

Fortier, L. 1983. Environmental and behavioral control of large-scale distribution and local abundance of ichthyoplankton in the St. Lawrence estuary. Ph.D. Thesis, McGill University, Montreal. 176 pp.

Glasser, J.W. 1984. Is conventional foraging theory optimal? Amer. Nat. 124: 900–905.

Godin, J.-G.J. 1981. Daily patterns of feeding behaviour, daily rations, and diets of juvenile Pink salmon (*Oncorhynchus gorbuscha*) in two marine bays of British Columbia. Can. J. Fish. Aquatic Sci. 38: 10–15.

Hairston, N., T. Li Kao & S.S. Easter. 1982. Fish vision and the detection of planktonic prey. Science 218: 1240–1242.

Hubel, D.H. & T.N. Wiesel. 1970. The period of susceptibility to the physiological effects of unilateral eye closure in kittens. J. Physiol., London 206: 419–436.

Hunter, J.R. 1980. The feeding behavior and ecology of marine fish larvae. pp. 287–330. *In*: J.E. Bardach, J.J. Magnuson, R.C. May & J.M. Reinhart (ed.) Fish Behavior and Its Use In the Capture and Culture of Fishes, ICLARM, Manila.

Ingle, D. 1968. Spatial dimensions of vision in fish. pp. 51–59. *In*: D. Ingle (ed.) The Central Nervous System and Fish Behaviour, University of Chicago Press, Chicago. 272 pp.

Jacobson, M. & R.M. Gaze. 1964. Types of visual response from single units in the optic nerve of the goldfish. Quart. J. Expt. Physiol. 49: 199–209.

Johns, P.R. & S.S. Easter. 1977. Growth of the adult goldfish eye II: increase in retinal cell number. J. Comp. Neurol. 176: 331–342.

Karmanova, I.G., A.I. Belich & S.G. Lazarev. 1981. An electrophysiological study of wakefulness and sleep-like states in fish and amphibians. pp. 181–200. *In*: P.R. Laming (ed.) Brain Mechanisms of Behavior in Lower Vertebrates, Cambridge University Press, Cambridge.

Kaufman, L. & I. Rock. 1962. The moon illusion. Scient. Amer. 207: 120–130.

Keating, M.J. 1976. The formation of visual neuronal connections: an appraisal of the present status of the theory of neuronal plasticity. pp. 59–110. *In*: G. Gottlieb (ed.) Neuronal and Behavioral Specificity, Academic Press, New York.

Keating, M.J. 1977. Evidence for plasticity of intertectal neuronal connections in adult *Xenopus*. Phil Trans. R. Soc. Lond. B 278: 277–294.

Klein, R. & R. Armitage. 1979. Rhythms in human performance: $1^1/_2$ hour oscillations in cognitive style. Science 204: 1326–1328.

Kleitman, N. 1963. Sleep and wakefulness. Univ. Chicago Press, Chicago. 552 pp.

Kleitman, N. 1969. Basic rest-activity cycle in relation to sleep and wakefulness. pp. 33–38. *In*: A. Kales (ed.) Sleep Physiology and Pathology, Lippincott, Philadelphia.

Krebs, J.R. 1979. Foraging strategies and their social significance. pp. 225–270. *In*: P. Marler & J.G. Vandenbergh (ed.) Handbook of Behavioral Neurobiology, Vol. 3, Social Behavior and Communication, Plenum Press, New York.

Lasker, R. 1981. Marine fish larvae. Morphology, ecology, and relation to fisheries. University of Washington Press, Seattle. 131 pp.

Levins, R. 1975a. Evolution of communities near equilibrium. pp. 16–50. *In*: M.L. Cody & J.M. Diamond (ed.) Ecology and Evolution of Communities, Harvard University Press, Cambridge.

Levins, R. 1975b. The limits of optimization. pp. 49–60. *In*: Proceedings Can. Mathematical Congress, Mathematics in the Life Sciences.

Luecke, C. & W.J. O'Brien. 1981. Prey location volume of a planktivorous fish: a new measure of prey vulnerability. Can. J. Fish. Aquat. Sci. 38: 1264–1270.

Luria, A.R. 1973. The working brain. Penguin, London. 398 pp.

MacCrimmon, H.R. 1954. Stream studies in planted Atlantic salmon. J. Fish. Res. Board Can. 11: 362–403.

Maiorana, V.C. 1981. Prey selection by sight: random or economical. Amer. Nat. 118: 450–451.

Marcotte, B.M. 1983. Imperatives of copepod diversity: perception, cognition, competition and predation. pp. 47–72. *In*: F. Schram (ed.) Crustacean Phylogeny, Balkema, Rotterdam.

McFarland, W.N. & E.R. Loew. 1983. Wave produced changes in underwater light and their relations to vision. Env. Biol. Fish. 8: 173–184.

Morgan, W.L. & D.A. Ritz. 1984. Effect of prey density and hunger state on capture of krill, *Nyctiphanes australis* S., by Australian salmon, *Arripis trutta*. J. Fish. Biol. 24: 51–58.

Neill, S.R. & J.M. Cullen. 1974. Experiments on whether schooling by their prey affects the hunting behavior of cephalopods and fish predators. J. Zool. Lond. 172: 549–569.

Northmore, D., F.C. Volkmann & D. Yager. 1978. Vision in fishes: colour and pattern. pp. 79–136. *In*: D.I. Mostofsky (ed.) The Behaviour of Fish and Other Aquatic Animals, Academic Press, New York.

O'Brien, W.J. 1979. The predator-prey interaction of planktivorous fish and zooplankton. Amer. Scient. 67: 572–581.

Oliphan, V.I. 1957. On the diel rhythm of feeding among Baikal grayling fry, and on diel rhythms among young fish in general. Doklady Akademii Nauk (USSR) 114: 669–672.

Peeke, H.V., S.C. Peeke & J.S. Williston. 1972. Long term memory deficits for habituation of predatory behavior in the forebrain ablated goldfish (*Carassius auratus*). Exp. Neurol. 36: 288–294.

Pinskii, F. Ya. 1967. Daily feeding rhythm and diets of young of the salmon (*Salmo salar* L.) when raised in ponds. Fish. Res. Board. Can. Translation Ser. No. 114.

Pyke, G.H., H.R. Pulliam & E.L. Charnov. 1977. Optimal foraging: a selective review of theory and tests. Quart. Rev. Biol. 52: 137–154.

Ringler, N.H. 1983. Variation in foraging tactics of fishes. pp. 159–171. *In*: D.L.G. Noakes et al. (ed.) Predators and Prey in Fishes, Dev. in Env. Biol. Fish. 2, Dr W. Junk Publishers, the Hague.

Ross, J., B. Jenkins & J.R. Johnstone. 1980. Size constancy fails below half a degree. Nature 283: 473–474.

Schoener, T.W. 1979. Generality of the size-distance relation in models of optimal foraging. Amer. Nat. 114: 902–914.

Shustov, Yu.A., I.L. Shchurov & Yu.A. Smirnov. 1980. Adaptation times of hatchery salmon, *Salmo salar,* to river conditions. J. Ichthyol. 20: 156–159.

Sosiak, A.J., R.G. Randall & J.A. McKenzie. 1979. Feeding by hatchery-reared and wild Atlantic salmon (*Salmo salar*) parr in streams. J. Fis. Res. Board Can. 36: 1408–1412.

Sutherland, N.S. 1968. Shape discrimination in the goldfish. pp. 33–50. *In*: D. Ingle (ed.) The Central Nervous System and Fish Behaviour, University of Chicago Press, Chicago.

Thorpe, W.H. 1969. Learning and instinct in animals. Harvard University Press, Cambridge. 558 pp.

Trevarthen, C. 1968. Vision in fish: the origins of the visual frame of action in vertebrates. pp. 61–94. *In*: D. Ingle (ed.) The Central Nervous System and Fish Behaviour, University of Chicago Press, Chicago.

Vinyard, G.I. & W.J. O'Brien. 1976. Effects of light and turbidity on the reactive distance of Bluegill sunfish (*Lepomis machrochirus*). J. Fish. Res. Board Can. 33: 2845–2849.

Ware, D.M. 1971. Predation by rainbow trout (*Salmo gairdneri*): the effect of experience. J. Fish. Res. Board Can. 28: 1842–1852.

Werner, R.G. & J.H.S. Blaxter. 1980. Growth and survival of larval herring (*Clupea harengus*) in relation to prey density. Can. J. Fish. Aquat. Sci. 37: 1063–1069.

Williams, G.C. 1964. Measurement of co-association among fishes and comments on the evolution of schooling. Publ. Mus. Michigan State Univ. 2: 351–383.

Received 28.1.1985 *Accepted 9.9.1985*

SALMO QUINNAT, Richards.

Mythical[1] models of gastric emptying and implications for food consumption studies

Malcolm Jobling
Fisheries Institute, University of Tromsø, P.O. Box 3083, Guleng, N-9001 Tromsø, Norway

Keywords: Gastrointestinal motility, Regulatory feedback mechanisms, Mathematical models, Daily food intake

Synopsis

Two of the most important factors governing gastric emptying are meal volume and food composition. These factors have also been demonstrated to influence the secretion of gastric acid, digestive enzymes and gut hormones in both fish and mammals. In mammals, feedback loops involving gastrointestinal hormones have been implicated in the control of gastric motility and enzyme secretion. These findings are briefly reviewed and it is demonstrated how the functioning of the feedback loops could lead to changes in gastric emptying patterns. A simulation exercise was carried out incorporating these physiological observations into the emptying model. The results predicted that an exponential function would best describe the emptying of small, easily digested low energy food particles from the stomach, but that a linear expression would give a better fit to the emptying data when food consisted of high energy large sized particles. These predictions were supported by results obtained in a number of experimental studies. These results are discussed in terms of selecting methods for the estimation of daily food consumption of fish species.

Introduction

In recent years, many ecologists have attempted to estimate food consumption by fish using direct methods based on stomach contents analysis and rates of gastric emptying. The two methods most commonly used are those proposed by Bajkov (1935) (used either in its original or a modified form) and Elliott & Persson (1978). An essential requirement of both these methods is that reliable gastric emptying rate data are available.

Rates of gastric emptying are generally calculated by fitting data to a mathematical representation of the gastric emptying curve, but problems arise in choosing the most appropriate mathematical expression. Mammalian physiologists have attempted to describe gastric emptying using a wide range of mathematical expressions (Stubbs 1977, Smith et al. 1984) but in work with fishes, attention has focused upon three types of equations – linear, square-root and exponential. None of these equations consistently gives the best fit to empirical data and the mathematical expression having the best predictive value may vary with species, food type and experimental conditions (Jobling 1981, McDonald et al. 1982, Brodeur 1984, From &

[1] Myth (mith) n., a traditional story offering an explanation of some fact or phenomenon; a story with a veiled meaning; a figment. Adj., mythical: relating to myths; fabulous; untrue. (Gr. mythos)

Chambers 20th Century English Dictionary

Rasmussen 1984, Talbot et al. 1984). This suggests that all of the methods of expression contain inherent weaknesses. It must also be remembered that a statistically acceptable mathematical relationship does not, per se, purvey information about biological/physiological processes. In the present study (1) the literature relating to the physiological control of gastric motility and emptying is reviewed in an attempt to provide a physiological explanation for the lack of consistency in predictive value of the different mathematical expressions; (2) an attempt is made to predict which mathematical expression will give the best fit to empirical data under different conditions, and (3) findings are examined and the consequences for food consumption studies discussed.

Thus, the main purpose of the study was to discuss the physiological mechanisms involved in the emptying of food from the stomach and to investigate how these mechanisms could influence gastric emptying patterns under different conditions. No attempts are made to present numerical values for rates of gastric emptying, nor are the influences of environmental factors, such as temperature, on rates of gastric emptying and intestinal transit times discussed.

The physiological background

Non-nutrient bulk and saline solutions are emptied rapidly from the stomach with gastric emptying being the result of the propulsive activity in response to the volume of the contents. Under these circumstances, the emptying pattern is volume dependent (McHugh & Moran 1978, Smith et al. 1984). Results obtained in experiments on plaice, *Pleuronectes platessa,* suggest that activity of the gastric musculature may be influenced by the volume of the stomach contents. Recordings of muscular contractions were made using intragastric balloons coupled to a pressure transducer and chart recorder. The results show that contraction frequency was relatively constant, irrespective of volume, but the amplitude of the contractions was highly dependent upon the volume of the stomach contents (Table 1). It appears, therefore, that the stomach can be envisaged as acting as a constant frequency pump with variable stroke volume. The stroke volume of the pump is assumed to be governed by the volume of the stomach contents (Hunt & Knox 1968). Thus, the emptying pattern of liquid or well-homogenised, non-nutrient bulk is curvilinear (Fig. 1).

Replacement of inert bulk by a variety of nutrient solutions results in slowing of gastric emptying in mammals (Hunt 1980). Increases in the energy density and changes in biochemical composition of the stomach content have been reported to influence gastric emptying in both mammals and fish (Windell 1966, Elliott 1972, Grove et al. 1978, Flowerdew & Grove 1979, Jobling 1980). In some cases it has been assumed that the slowing of gastric emptying was brought about by a modification of the volume dependent emptying pattern (Hunt 1980, Jobling 1980, Table 2), but other workers have suggested that the volume dependent mechanism was completely overridden such that emptying was constant in terms of energy emptied per minute (McHugh & Moran 1978). These observa-

Table 1. The effect of intragastric volume upon frequency and amplitude of stomach contractions in plaice, *Pleuronectes platessa.* Stimulus is defined as (intragastric volume (ml) × 100) (body weight (g))$^{-1}$. Values given are means ± s.d.

Fish weight (g)	Intragastric volume (ml)	Stimulus	Intracontraction time (min)	Contraction frequency n h^{-1}	Contraction amplitude arb. units
200	2	1.0	1.61 ± 0.22	37	8.9 ± 2.2
450	5	1.1	1.58 ± 0.13	38	32.0 ± 10.7
350	5	1.4	1.46 ± 0.24	41	37.0 ± 10.7
350	10	2.8	1.59 ± 0.16	38	50.0 ± 13.8

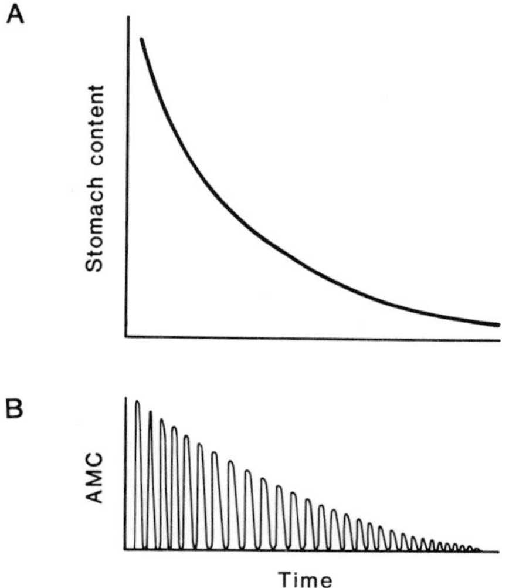

Fig. 1. Schematic diagrams showing intrinsic pattern of gastric emptying (A) and the effects of volume of stomach content on muscular contractions (B); AMC = Amplitude of muscular contractions.

tions on the influence of energy on emptying suggest a feedback control which is present during the emptying of nutrients but absent during bulk/saline emptying.

Results from an increasing number of investigations, including those with 'crossover' parabiotic rats (Lepkovsky et al. 1975) and those in which duodenal infusions have been employed (Meeroff et al. 1975, Stephens et al. 1975, 1976, Hunt 1980 and references therein) suggest that the receptors giving rise to this feedback control are located in the upper intestine rather than in the stomach. In mammals, duodenal infusions of low pH solutions, amino acids, carbohydrates and fatty acids have all been shown to cause a delay in gastric emptying and it has been hypothesised that there are duodenal receptors tuned to respond to variations in pH, osmotic pressure, fatty acid anions and certain amino acids (Cooke 1974, Bell & Mostaghni 1975, Ochia 1976, Barker et al. 1978, Bell et al. 1981). It is not known whether cells capable of responding to these stimuli are found in the fish intestine, but if receptors are present it is difficult to envisage how all types could be functionally operative in all species. The stomachs of a number of fish species secrete acid and, although there is some uncertainty as to the cell types and secretory mechanisms involved (McKenzie & Gibson 1970, Fange & Grove 1979, Holstein & Haux 1982, Vigna 1983 and references therein), it is clear that functional pH receptors could be involved in the regulation of gastric emptying in these species. On the other hand, pH receptors would be of little value to the fish species which are not capable of acid secretion. Similarly, it is difficult to envisage how receptors responding to osmotic properties could provide an effective regulation of gastric emptying in marine species where the osmotic properties of the gastric chyme would be expected to be influenced by the drinking of sea water. Thus, the importance of upper intestinal receptors in the regulation of gastric emptying in fish remains open to speculation, but the fact that gastric emptying in fish is influenced by dietary energy content (Grove et al. 1978, Flowerdew & Grove 1979, Jobling 1980) must mean that some species of fish are capable of assessing the amount of energy contained in the chyme. Many questions relating to this problem could be answered by the conducting of experiments in which intestinal infusion of nutrients and the monitoring of gastric contractile activity were performed simultaneously.

In mammals, feedback control of gastric empty-

Table 2. Influence of dietary dilution on rates of gastric evacuation in rainbow trout, *Salmo gairdneri* (calculated from Grove et al. 1978).

Food type	Dietary energy kcal g^{-1}	Gastric evacuation time (h)	Data fitted to	
			$\sqrt{\text{weight}} = c_w - m_w$ time	$\sqrt{\text{energy}} = c_e - m_e$ time
Salmon Pellet	4.78	15.5	Y = 28.28 – 1.825 X	Y = 1.956 – 0.1262 X
Food 75% Kaolin 25%	4.20	14.5	Y = 28.28 – 1.950 X	Y = 1.833 – 0.1264 X
Food 67% Kaolin 33%	3.23	11.8	Y = 28.28 – 2.397 X	Y = 1.608 – 0.1362 X
Food 50% Kaolin 50%	2.18	10.0	Y = 28.28 – 2.828 X	Y = 1.321 – 0.1321 X

ing is at least partially accomplished by effects on the contractions of the smooth musculature of the stomach wall, which results in alterations in gastric motility patterns and reductions in the force exerted by the antral pump (Bell & Grivel 1975, Roze et al. 1977). Gastric emptying may also be slowed by contractions of the pyloric and duodenal musculature, which increases the resistance to flow of the gastric chyme (Weisbrodt et al. 1969). The precise mechanisms by which these effects are exerted are not well understood, but both nervous and hormonal/humoral links may be involved. In the case of the interactions between the duodenal receptors and gastric muscular effectors, the pathway(s) must be either via extrinsic nerves or be hormonal (Bell & Grivel 1975).

In mammals, a wide range (to date in excess of twenty) of peptides produced by cells scattered throughout the gastrointestinal tract have been reported as displaying hormonal action but knowledge about the stimuli required for release and about the physiological functions of some of these peptides is incomplete. The situation for fish is even less clear and there is considerable uncertainty about the array of gastrointestinal hormones found in fish (Langer et al. 1979, Holmgren et al. 1982, Crim & Vigna 1983, Vigna 1983, Rombout & Reinecke 1984 and references therein). Much of this confusion arises from doubts about the specificity of assay techniques and questions about whether or not the application of purified mammalian peptides can be expected to produce physiological responses in fish species.

Following the ingestion of a meal, a number of different gastrointestinal peptide hormones may be released into the circulation and both the amount and time course of release may be affected by meal composition and rate of gastric emptying. In turn, gastric emptying may be influenced by individual hormones or, more probably, by interactions between series of hormones. The gastrointestinal hormones whose effects on gastric emptying have been most extensively studied are the gastrins, cholecystokinins (CCK), secretin, motilin, glucagon and vasoactive intestinal peptide (VIP). From the point of view of negative regulatory feedback of gastric emptying peptides of the gastrin-CCK family appear to be the most interesting.

CCK is best known for its effects upon gall bladder (Vigna & Gorbman 1977) and pancreas but recent studies have also shown the peptide to be instrumental in causing a slowing of gastric emptying (Debas et al. 1975, Yamagishi & Debas 1978, Strunz 1979, Moran & McHugh 1982, Mangel & Koegel 1984). Regulation of gastric emptying is effected by a number of different mechanisms including (i) lowering of intragastric pressure, (ii) contraction of the pylorus, and (iii) changes in motility of the duodenum and upper intestine, leading to increased resitance to flow of gastric chyme into the duodenum.

Thus, the effects observed when CCK is administered are very similar to the effects reported to occur when nutrients are infused in to the duodenum. There is increasing evidence to suggest that many of the substances known to slow gastric emptying (solutions of low pH, amino acids, fatty acids) also cause CCK to be released from the I cells of the small intestinal mucosa. Amongst the most potent stimuli of CCK release are the two aromatic amino acids tryptophan and phenylalanine (Strunz 1979, Dockray 1982).

In the light of this, it is interesting to note that the gastric enzyme pepsin has a specificity for the peptide links formed between aromatic (tryptophan, phenylalanine, tyrosine) and dicarboxylic (glutamic, aspartic) amino acids. Thus, this endopeptidase hydrolyses only a few of the bonds in a large protein molecule, but the bonds hydrolysed lead to the exposure of the amino acids known to be potent stimuli of CCK release. Peptic activity has been found to be present in the stomach of several fish species (Fange & Grove 1979).

When viewed together, all this evidence suggests the existence of a relatively simple feedback loop for the control of gastric muscular activity. The challange is to incorporate this into a model which provides an integrated approach to gastric emptying.

During the course of a meal a small amount of food will be emptied into the upper intestine and the nutrients contained within the chyme will stimulate the duodenal receptors. This, in turn, leads to release of an array of peptides, amongst them

CCK. The release of CCK results in the stimulation of gall bladder contraction, the secretion of pancreatic enzymes and influences the muscular contractions of the stomach, duodenum and pylorus. The consequence is a cessation of, or a reduction in, flow of chyme into the duodenum.

The release of bile salts from the gall bladder into the upper intestine will neutralise the acid chyme and thereby remove the stimulus for receptors responsive to low pH. In addition, the pancreatic secretions contain trypsin(ogens) which provide a negative feedback signal for CCK secretion (Schneeman & Lyman 1975). Thus, the release of bile salts and pancreatic juices leads to inhibition of CCK secretion and thereby allows the entry of a new 'pulse' of gastric chyme into the upper intestine. This, in turn, begins a new and sequential cycle of secretion and inhibition.

The emptying of food from the stomach is envisaged not as a continuous smooth process, but as one occurring in small pulses. The timing of the emptying of these pulses of chyme into the upper intestine is determined by a series of feedback loops ultimately under the influence of the nutrient concentration of the gastric chyme (Fig. 2). Consequently, gastric emptying is slow when the stomach contains nutrient-rich food and is more rapid when the stomach content is less energy-rich.

Most of the knowledge about gastrointestinal regulatory mechanisms has been gained from studies of the emptying of homogenous nutrient solutions from the mammalian stomach. Most food is neither homogenous nor liquid and any hypothesis regarding gastric emptying must take into account the fact that when solids and liquids are ingested simultaneously, the emptying pattern of the various components may be different (Windell 1966, Pandian 1967, Kionka & Windell 1972, Hinder & Kelly 1977, Kelly 1980, Meyer 1980). Liquids are usually emptied rapidly but digestible solids must be reduced to a particle size of 0.5 mm or smaller before they are passed through the pylorus. Thus, the emptying of solids is usually slower than that of liquids (Hinder & Kelly 1977, Meyer 1980). When dogs were fed liver as a mixture of homogenate, paté and large chunks, the small food particles emptied relatively rapidly from the stomach but the larger chunks were retained in the stomach until they had been reduced to suitable size (Hinder & Kelly 1977, Kelly 1980). Undoubtedly, both chemical and mechanical processes play roles in bringing about this reduction in size but it has been suggested that the gastric muscular contractions are particularly important in sorting and grinding the food particles (Kelly 1980, Meyer 1980). Additional results from studies on dogs showed that the nutrient (energy) content of the liquid component affected the rapidity with which solids were emptied (Hinder & Kelly 1977). This finding is not surprising in light of the facts that feedback signals from the upper intestinal receptors influence both gastric and pyloric muscular contractions and that this feedback is dependent upon the composition of the gastric chyme. Both the speed at which food is fragmented into particles capable of being passed through the pylorus and the speed of emptying may be regulated by feedback signals emanating from the upper intestinal receptors. Consequently, the slowing of emptying caused by large food particle size and the slowing brought about by feedback loops between the intestine and stomach should not be viewed as being separate processes. They must be considered to interact with each other to produce the patterns of emptying observed in experimental studies.

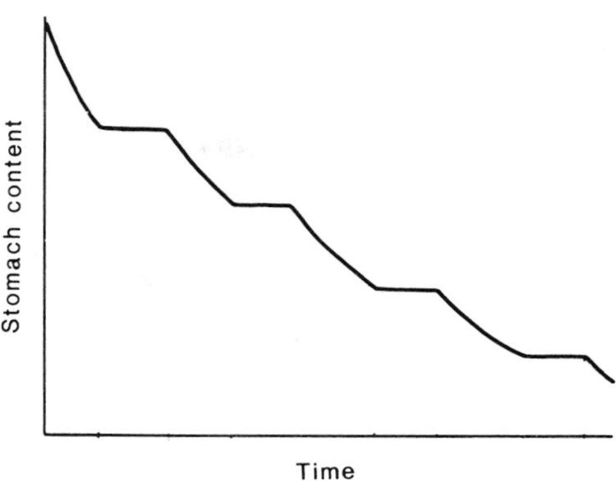

Fig. 2. Schematic representation of discontinuous emptying of food from the stomach when muscular contractions are influenced by feedback signals from the upper intestine.

The observation that food particles must be broken down to a small size before they are passed from the stomach, through the pylorus and into the intestine has important consequences for predictions concerning the pattern of emptying to be expected when large food items are consumed. Many fish species eat food items, such as other fish, which are relatively large in comparison with their own body size. When such food items are consumed, the time required to break down the majority of the food into fragments of suitable size for passage through the pylorus may be relatively long. Consequently, there may be a 'time-lag' or initial emptying delay before there is any substantial diminution in the quantity of food remaining in the stomach. 'Time-lags' in the initial emptying of the stomach are apparent in the results of a number of experiments in which large prey items were used as food (Markus 1932, Jones 1974, McDonald et al. 1982).

Mathematical modelling of the emptying pattern

The emptying pattern of non-nutrient bulk is curvilinear and may be described mathematically by an equation of the form $dV/dt = -r\ V^b$ where V is volume, t is time, r is the rate of gastric emptying and b is a constant. The stomach empties more slowly when non-nutrient bulk is replaced by food. This slowing could result either from a modification of the volume dependent emptying curve or by a change in the pattern of emptying.

In two of the most frequently used mathematical models (square-root and exponential) the slowing of emptying is implied to occur via modification of the volume dependent emptying curve such that the value of r in the expression $dV/dt = -r\ V^b$ changes with changes in food composition (Elliott 1972, Hunt 1980, Jobling 1980). The value of b is 1.0 in the exponential and 0.5 in the square root model. The linear emptying model implies that the volume dependent mechanism is overridden in the presence of food such that $dV/dt = -r$, but r may change with changing food composition. Implicit assumptions in all of these mathematical expressions are that gastric emptying is a continuous smooth process and that the slowing of emptying in the presence of food is brought about by feedback mechanisms acting as a brake. The examination of the physiological factors known to affect gastric motility suggested that the slowing of emptying may not be the result of a continuous braking action but could be explained on the basis of the gastric chyme being delivered into the upper intestine in pulses.

An additional weakness with all the mathematical expressions is that they do not incorporate a 'time-lag' in order to account for the emptying delay likely to be encountered when large prey are consumed.

The question arises – How good are the mathematical models as representations of a pulsed emptying pattern in which there may also be an initial emptying delay? This was examined by simulating emptying curves based upon the physiological model and then fitting the various mathematical expressions. This allowed predictions to be made about the expression which would best describe empirical data collected under different experimental conditions.

Simulation of gastric emptying and predictions arising therefrom

In simulating gastric emptying it was assumed that the intrinsic emptying pattern (i.e. that seen in the absence of feedback control) was exponential and could be described by the expression $dV/dt = -r\ V^{1.0}$ (Smith et al. 1984). The value of r was arbitrarily set at 0.2 and the initial volume was assumed to be 100 units. The physiological emptying model is based upon an assumption that when nutrient-rich chyme is emptied from the stomach it stimulates receptors located in the upper intestine. These receptors, in turn, relay signals to the gastric and pyloric musculature and thereby cause a cessation of emptying. For the purpose of the simulation, it was assumed that there was a time-lag in this feedback loop such that the stomach pumped for a given time period (taken as 1 time unit) before feedback signals took effect. It was further assumed that the duration of the pause in emptying resulting from intestinal feedback was directly proportional to the amount of energy emptied from the stomach during the previous pumping period.

This feedback pause was assumed to be 0.1 time units per energy unit.

The characteristics of a diet affect the way in which it is emptied from the stomach. Consequently, the effects of changes in dietary energy content (0.5, 1.0 and 2.0 energy units per unit volume) were simulated.

For some time following ingestion, food particles may still be too large to pass through the pylorus into the upper intestine. The consequences of these effects for the overall emptying pattern were examined by the introduction of an initial emptying delay (0, 2 or 5 time units) into the simulation.

In order that the simulation should represent practical experiments, the data used for curve-fitting of the three mathematical expressions (linear, square-root and exponential) were collected using three sampling schemes, with the data points being selected at different times into the simulation run: (1) after 2, 4, 8, 12, 14, 18, 22 and 24 time units; (2) after 3, 6, 9, 12, 15, 18, 21 and 24 time units; (3) after 2, 3, 8, 10, 14, 16, 20 and 24 times units. This gave a total of 27 different combinations of energy content, initial delay and sampling scheme.

When using mathematical expressions, it is normal practice to transform the data before using least squares regression analysis to obtain an estimate of goodness of fit. This proceedure was initially carried out on the simulated data. The first thing noted was that least squares correlation coefficients were high for all three mathematical expressions ($n = 8$; R^2 ranges for linear 0.878–0.998, square-root 0.972–0.998 and exponential 0.958–0.994). According to accepted practice, it would be concluded that all three mathematical expressions described the data very well. A problem arose, however, since the purpose of the simulation was to predict which of the expressions would give the best fit to the data under a given set of conditions. The fact that the data had been transformed in order to perform linear regression analysis precluded the comparison of correlation coefficients; an alternative method, based upon untransformed data, had to be found. In the present study, the following deviation index was used:

$$\sqrt{\frac{1}{n} \sum (Y_{obs} - Y_{pred})^2},$$

where n is the number of observations, Y_{obs} is the observed value of Y for a given X and Y_{pred} is the value of Y predicted at the same X by the mathematical expression. In addition, results were examined in terms of how accurately the different mathematical expressions predicted the value of Y when $X = 0$.

The major conclusions drawn from the simulation exercise were: (1) increasing the initial emptying delay and/or feedback pause (by increasing dietary energy content) produced an increase in the deviation index for the exponential, showing that this function gave a progressively poorer fit to the data (Table 3a) column 5); (2) a perturbation of the intrinsic emptying pattern, resulting in a slowing of gastric emptying, produced an overestimation of the initial stomach content when the exponential function was fitted (Table 3b column 6) and (3) the introduction of delay functions led to a progressively better fit to the data with a linear expression (Table 3a column 3).

Thus, in cases where the emptying pattern is not too far removed from the intrinsic (i.e. no initial emptying delay, low energy diets) the exponential expression will give the best fit to the data. Conversely, under conditions where gastric emptying has been interupted by prolonged delays the overall pattern of emptying will be better described by a linear function. At intermediate stages in this transition other forms of expression (in this case the square-root) are to be preferred (Table 3). Based upon the above, the following predictions can be made about the patterns of emptying expected to be found in experiments: (1) when food consists of small, relatively easily digested particles of low energy content, the exponential function will give a better fit to the data than either the linear or square-root expression;
(2) if food consists of a single or a small number of relatively large particles, which in addition are difficult to digest and have a high energy content, the best fit to the gastric emptying data will be provided by a linear expression, and
(3) between these two extremes, other forms of mathematical expression will give the best fit to the empirical data.

Table 3. The effects of different conditions upon the accuracy of various mathematical functions in describing simulations of a physiological model of gastric emptying (see text for details of simulation and statistical methods).

Conditions		Stomach content known from simulation	Mathematical expression		
Diet	Delay		Linear	Square root	Exponential
(a) *Deviation index*					
0.5	0		9.67	4.68	4.57
1.0	2		6.84	5.30	5.37
2.0	5		2.69	2.51	5.48
(b) *Predicted stomach content at time zero*					
0.5	0	100	72.9	82.7	129.3
1.0	2	100	96.8	107.6	140.1
2.0	5	100	108.8	114.2	151.9

Experimental data and the accuracy of the predictions

Plaice, *Pleuronectes platessa,* of mean weight approximately 30 g were held at 10° C and fed 0.5 ml of moist paste diets which varied in energy content from 5.20 to 10.97 kJ ml^{-1}. At given times after force-feeding, fish were killed and their stomach contents removed and dried to constant weight at 60° C (Jobling & Davies 1979). Data were fitted to linear, square root and exponential functions.

Results from the plaice experiment show that the emptying pattern of the low energy diet was best described by an exponential function (Table 4a, Fig. 3 A3). As the energy content of the diet increased, the other mathematical expressions gave a better fit to the empirical data. When high energy diets were fed, there was little difference in goodness of fit between the square-root and linear expressions (Table 4a, Fig. 3 C1–C2), but the square-root function gave a slightly more accurate prediction of the stomach content at time 0 (Table 2b). It

Table 4. Effect of dietary composition upon the predictive accuracy of various mathematical functions in describing gastric emptying in plaice, *Pleuronectes platessa* L. Best fit indicated by underlining.

Dietary energy KJ ml^{-1}		Stomach content (mg dry weight) known from amount fed	Mathematical expression		
			Linear	Square root	Exponential
(a) *Deviation index*					
5.20	Low		25.19	17.16	<u>12.14</u>
7.81			46.20	<u>39.03</u>	69.80
9.46	Medium		26.86	<u>16.16</u>	31.21
9.80			31.61	<u>16.17</u>	49.29
10.46	High		40.62	<u>39.86</u>	48.48
10.97	High		17.73	<u>16.12</u>	29.38
(b) *Predicted stomach content at time zero*					
5.20	Low	157	131.5	135.7	<u>150.8</u>
7.81		331.5	290.1	<u>326.9</u>	507.3
9.46	Medium	295.8	250.1	<u>275.2</u>	377.7
9.80		296.4	238.7	<u>268.3</u>	437.5
10.46	High	296.9	256.8	<u>280.2</u>	352.1
10.97	High	291.1	267.9	<u>289.7</u>	347.2

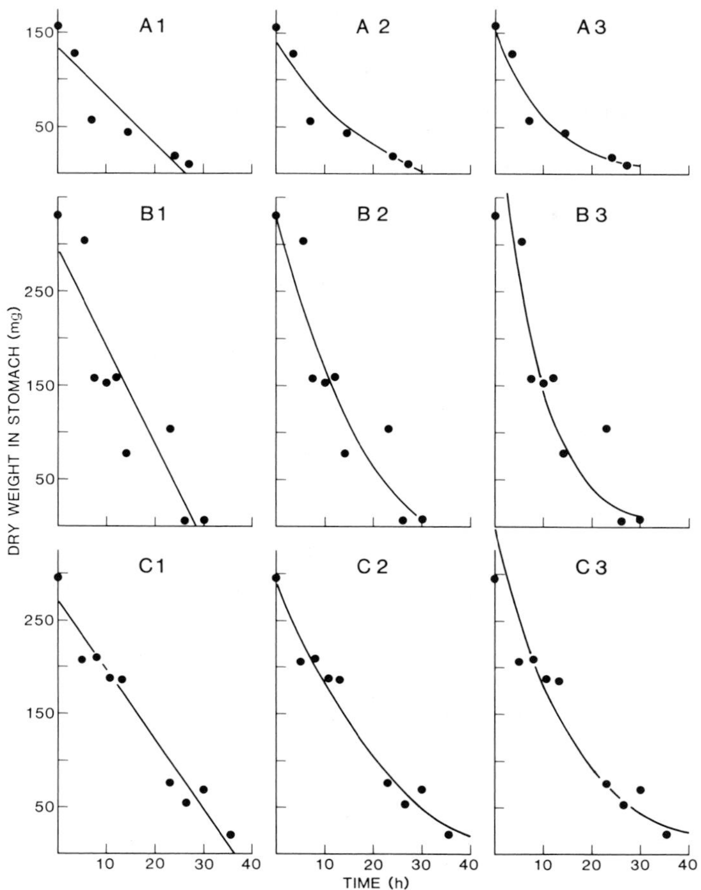

Fig. 3. Effects of dietary energy level on the emptying of food from the stomach of plaice, *Pleuronectes platessa.* Data fitted to three different mathematical functions: (1) linear, (2) square root, (3) exponential. Dietary energy level (A) 5.20 kJ ml^{-1}, (B) 7.81 kJ ml^{-1}, (C) 10.97 kJ ml^{-1}.

is also noteworthy that, as predicted, the fitting of an exponential function usually led to an overestimate of stomach content at time 0. The diets used in the plaice study were finely homogenised pastes and the differences in emptying patterns were, therefore, most probably caused by differences in energy contents between diets, rather than by energy and particle size interactions. A re-analysis of over 80 published data sets obtained in experiments conducted on 17 different fish species (Jobling unpublished; examples in Tables 5, 6, 7) also upheld the predictions made on the basis of simulations of the physiological model.

In a series of papers, Elliott and Persson (Elliott 1972, Elliott & Persson 1978, Persson 1979, 1981) reported that an exponential function described the pattern of gastric emptying in brown trout, *Salmo trutta* and perch, *Perca fluviatilis.* The adequacy of this form of expression was tested and confirmed by conducting multiple meal experiments. The predictive power of the exponential fit was found to be excellent. It should be noted that the results were expressed in terms of the emptying of digestible dry organic matter i.e. total dry matter – (ash weigh + chitin) and, therefore, the results refer only to the emptying of the digestible components of a meal, rather then to the meal as a whole. Thus, the contribution of the indigestible components to the overall emptying pattern were ignored, a fact which is unfortunate, since there is evidence that the indigestible components are the last to be emptied from the stomach (Pandian 1967, Kionka

Table 5. Effect of dietary type upon the predictive accuracy of various mathematical functions in describing gastric emptying in rainbow trout, *Salmo gairdneri* (re-calculated from Windell & Norris 1969, Windell et al. 1969, 1972, From & Rasmussen 1984). Best fit indicated by underlining.

Food/prey type	Mathematical expression		
	Linear	Square root	Expo-nential
(a) *Deviation index*			
Oligochaetes	14.67	8.21	5.77
Moist pellet	16.55	7.59	6.59
Moist pellet	14.90	4.93	5.85
Dry pellet	1.81	0.48	3.32
Dry pellet	1.34	1.96	3.75
(b) *Predicted stomach content at time zero* (expected = 100)			
Oligochaetes	85	130	155
Moist pellet	75	100	105
Moist pellet	80	100	105
Dry pellet	90	100	115
Dry pellet	90	100	110

& Windell 1972, Gannon 1976). In all of Elliott and Persson's studies, the experimental fish were fed small, natural food organisms (zooplankton, small invertebrates and fish larvae) which presumably also had a relatively low energy content. As illustrated in this paper, under these feeding conditions, the exponential would be expected to give the best fit to the empirical data.

From & Rasmussen (1984) suggested that the exponential function should be used for describing the pattern of emptying of moist feeds from the stomach of rainbow trout, *Salmo gairdneri,* but Talbot et al. (1984) stated that the square-root function gave the best representation of the gastric emptying of moist pellets in Atlantic salmon, *Salmo salar*. Similarly, the square-root function was more accurate than either the exponential or linear in predicting the results of multiple meal experiments with bluegill sunfish, *Lepomis macrochirus,* fed oligochaetes (data of El-Shamy 1976 analysed by Jobling 1981). Use of the square-root function also gave accurate predictions of food consumption by cod, *Gadus morhua,* fed on a range of food organisms (chopped capelin, Iceland scallop, benthic invertebrates) (Jobling 1982). Other workers (Pandian 1967, Jones 1974) have found that when fish were fed large food items the mathematical expression which best described the emptying pattern appeared to be the rectilinear.

Thus, the mathematical expression to be preferred for describing gastric emptying varies according to prevailing conditions. In fact, between the two extremes of exponential and direct linearity, many different mathematical equations may give an adequate fit to the data (Magnuson 1969, Stubbs 1977, McDonald et al. 1982, Fletcher et al. 1984).

The advantages of volume and energy based regulation of gastric emptying

One of the consequences of a volume dependent regulation of gastric emptying is that the amount of food delivered to the upper intestine changes markedly with time. That is, at the start of emptying large quantities of food are passed from the stomach to the intestine, but as emptying proceeds the amount delivered per unit time declines. Thus, if a volume dependent mechanism were the only

Table 6. Effect of dietary type upon the predictive accuracy of various mathematical functions in describing gastric emptying in perch, *Perca fluviatilis* and *Perca flavescens* (re-calculated from Griffiths 1976, Persson 1979, 1981, Garber 1983). Best fit indicated by underlining.

Food/prey type	Mathematical expression		
	Linear	Square root	Expo-nential
(a) *Deviation index*			
Gammarids	9.78	4.89	1.41
Gammarids	6.65	3.56	1.75
Fish larvae	13.00	4.40	1.00
Fish	7.43	6.71	6.76
Dry pellet	4.82	2.64	4.91
Dry pellet	4.98	8.50	10.95
(b) *Predicted stomach content at time zero* (expected = 100)			
Gammarids	85	93	100
Gammarids	90	100	100
Fish larvae	85	100	100
Fish	80	93	110
Dry pellet	95	107	125
Dry pellet	115	130	145

Table 7. Effect of small and large food items on the predictive accuracy of various mathematical functions in describing gastric emptying in fish (re-calculated from Markus 1932, Windell 1966, Elliott 1972, Jones 1974). Best fit indicated by underlining.

Fish species	Food type	Mathematical expression		
		Linear	Square root	Exponential
(a) Deviation index				
Salmo trutta	Gammarids	16.97	6.15	<u>0.84</u>
Lepomis macrochirus	Chironomids/Oligochaetes	17.40	6.91	<u>4.58</u>
Micropterus salmoides	Fish	4.74	<u>4.53</u>	6.61
Melanogrammus aeglefinus	Fish	<u>8.13</u>	11.46	14.70
Merlangius merlangus	Fish	<u>7.74</u>	8.07	8.94
(b) Predicted stomach content at time zero (expected = 100)				
Salmo trutta	Gammarids	75	93	<u>100</u>
Lepomis macrochirus	Chironomids/Oligochaetes	75	<u>100</u>	<u>100</u>
Micropterus salmoides	Fish	<u>125</u>	138	160
Melanogrammus aeglefinus	Fish	<u>115</u>	122	120
Merlangius merlangus	Fish	<u>105</u>	107	115

one regulating emptying, the intestinal digestive and absorptive processes would be subjected to wide variations in loading during the course of processing a single meal. Even greater variations in loading would be expected to occur over a longer time period since fish may consume food items of varying composition and energy content. The introduction of a regulatory mechanism based upon feedback signals generated by the transfer of energy from the stomach to the upper intestine would serve to reduce the range in variation in loading of the digestive and absorptive processes. The question now arises – under what circumstances would the possession of an energy based regulatory mechanism be advantageous?

For fish consuming low energy (nutrient) – high bulk diets large amounts of food must be processed per unit energy gain. In these species, high rates of food turnover are of the essence (Pandian & Vivekanandan 1985). Fishes which consume diets of this type (detritus, aquatic plants, calcareous corals) are either microphagous or possess mechanisms, such as the pharyngeal mill, for grinding the food into small particles before it reaches the stomach (Fange & Grove 1979, Pandian & Vivekanandan 1985). For these fishes a gastric emptying strategy based upon volume dependent emptying, coupled with the maintenance of a high degree of stomach fullness, would ensure maximum rates of turnover. This would lead to the delivery of large quantities of small particles to the intestine. However, since the food contains only small amounts of energy (nutrient) per unit bulk it is unlikely that the digestive and absorptive capacities of the intestine would be overloaded.

On the other hand, the adoption of this type of emptying strategy by a fish consuming high energy-low bulk diets would be extremely wasteful. The delivery of large quantities of nutrient-rich food may overload the digestive and absorptive capacities of the intestine resulting in a low absorption efficiency. For fish consuming high energy diets it would be advantageous if gastric emptying were precisely regulated such that the amounts of energy reaching the small intestine remained constant over time. This would enable digestive and absorptive processes to occur at a constant and predetermined rate. It is this type of regulation that appears to be one of the most important functions of the upper intestinal receptors found in mammals (McHugh & Moran 1979, Nomura et al. 1985). In fish, however, the regulation of gastric emptying does not appear to be precise enough to ensure a constant rate of delivery of energy (nutrient) to the intestine. Some regulation does occur in response to changes in energy content of the food, but even when relatively high energy diets are fed the mathematical expression giving the best fit to the pattern of emptying is a curvilinear one (Tables 4, 5, 6).

When fish consume large food organisms, the mathematical expression giving the best fit to the pattern of emptying appears to be the linear one (Pandian 1967, Jones 1974, Jobling unpublished, Table 7, but cf. Table 6). If the emptying of the large food organisms were governed solely by the rate at which gastric enzymes and acid attacked the exposed surface of the food, the emptying pattern would be curvilinear and data would fit a mathematical expression of the form $dV/dt = -r\ V^{0.67}$. The linearity of the emptying suggests, therefore, that feedback mechanisms may have a role to play during the emptying of large food organisms and that this role is more subtle than merely acting as an energy regulator. The review of the physiological literature presented earlier suggested that the feedback mechanisms effected their regulation of gastric emptying by influencing both the gastric musculature and the pylorus. When food is delivered from the stomach to the intestine the feedback signals ensure that the pyloric opening is small. The gastric muscular contractions serve both to retain large food particles in the stomach and also to grind them into smaller fragments (Hinder & Kelly 1977, Meyer 1980). Thus, the operation of the feedback signals, particularly via the effects on the pylorus, is envisaged as preventing the 'dumping' of large food particles into the upper intestine.

In a series of elegant experiments conducted on dogs, Hinder & Kelly (1977) demonstrated the importance of the feedback signals in controlling the emptying of mixed meals consisting of liquids, digestible solids and indigestible solids. Liquids were emptied rapidly, digestible solids were emptied following reduction to particles of small size, but for as long as digestible foodstuffs remained in the stomach few or none of the large indigestible particles were emptied. Once all traces of liquids and digestible solids had been emptied, and feedback signals from the upper intestine were no longer operating to restrict the size of the pyloric opening, the indigestible solids were emptied rapidly into the intestine. This demonstrates that feedback signals are necessary for prevention of the passage of large particles through the pylorus.

Several workers have observed that indigestible skeletal fragments were retained in the stomachs of fish until after the digestible material had disappeared (Darnell & Meierotto 1962, Windell 1966, Pandian 1967, Kionka & Windell 1972, Gannon 1976) and it is tempting to suggest that this retention of indigestible matter was brought about by mechanisms similar to those known to be present in mammals (Hinder & Kelly 1977). Macrophagous fish species generally lack the pharyngeal mill but may sometimes consume low energy food items. For example, the marine flatfishes dab, *Limanda limanda* and plaice, *Pleuronectes platessa,* occassionally feed on whole bivalves or *Arenicola marina* 'tails', both of which have a high ash content. These food items may occur in the intestine in a relatively undigested condition and bivalves may pass through the entire alimentary tract and remain unscathed. (Kuipers 1975, personal observation). Thus, when indigestible material is consumed alone large fragments may be emptied fairly rapidly from the stomach, but when indigestible and digestible material are consumed together the large fragments of indigestible material are retained in the stomach until most or all of the digestible matter has been emptied. Taken together, these observations suggest that some form of feedback, based upon energy and/or nutrient reception, occurs in some fishes and the operation of this feedback may be prerequisite to prevent transfer of large food items from the stomach to the intestine.

It is hypothesised, therefore, that feedback mechanisms may have evolved in fish as a means of preventing the transfer of large nutrient-rich food particles from the stomach to the intestine. Further, it is suggested that the receptors responsible for this feedback are located in the upper intestine and that they are stimulated by the breakdown products of peptic enzyme activity. Peptic activity results in the breakdown of proteins to (poly) peptides which have aromatic and dicarboxylic amino acids at the terminal ends. The aromatic amino acids, particularly tryptophan, are potent stimuli for the release of CCK and the infusion of tryptophan solutions in to the upper intestine has been shown to cause a marked slowing of gastric emptying in mammals (Stephens et al. 1975, 1976). In mammals, the proposed original function of the feedback mechanisms has been retained but an increased sensitivity of, and/or evolution of, additional types of receptors has given rise to a system which is also an extremely efficient energy regulator.

The physiological emptying model, the mathematical expression of emptying data and consequences for food consumption studies

Bajkov (1935) described a method for the estimation of daily rates of food consumption in which it was assumed that the rate of gastric emptying was constant ($dV/dt = -r$) but the method was later modified to incorporate an exponential function ($dV/dt = -r V^{1.0}$) (Doble & Eggers 1978). Another method, which has been widely adopted for use in fish ecological research, was introduced by Elliott & Persson (1978). This method is also based upon the assumption that gastric emptying occurs exponentially. The accuracy of the latter method has been tested under laboratory conditions and has been shown to give excellent results for fish (brown trout, *Salmo trutta* and roach, *Rutilus rutilus*) fed on small natural food organisms. The method is, therefore, to be recommended for use in the estimation of food consumption of fish species which feed on plankton or relatively small benthic organisms. When using the Elliott & Persson (1978) method the stomach contents data should be expressed in terms of digestible organic matter rather than total dry weights. This is advisable because the accuracy of the method was tested using emptying rates calculated in terms of digestible organic matter and the possible influences of the indigestible components of the diet on emptying pattern have been inadequately investigated.

It has been shown that the use of an exponential function is less appropriate than a linear one when describing the pattern of emptying of large and/or high energy food particles. This leads to the suggestion that neither the Elliott & Persson (1978) method nor the Doble & Eggers (1978) modification of the Bajkov (1935) method should be employed under these circumstances. On the contrary, it is to be expected that, in its original form, the Bajkov method is the one to be used when estimating the food consumption of either predatory fish which consume small numbers of large food items or of fish held under culture conditions and fed high energy, dry pellet foods.

Extreme caution should, therefore, be exercised when evaluating methods to be used in food consumption studies. Care must be taken to select the method most suitable for the type of fish and diets under study and it would be advisable to conduct preliminary trials under laboratory conditions before adopting a particular method for use in the field.

Concluding comments

It will be obvious from the account given in this paper that the present knowledge of factors controlling gastrointestinal motility and digestive processes in fish is incomplete. The physiological mechanisms involved in the control of gastrointestinal function in mammals have been more extensively investigated and the hypotheses presented in the current paper are largely based upon knowledge of mammalian systems. The model of gastric emptying derived from the hypotheses describes the way in which a 'single meal' is emptied from the stomach but provides the basis for speculation about the gastric emptying strategies likely to be most advantageous for fish from different ecological groups. Further, when the aim is the estimation of daily ration, the physiological model can be used to predict the type of simple mathematical expression to be prefered for the calculation of gastric emptying rate for different species fed different food types. The application of the physiological model gives clear indications as to the mathematical expressions best suited for the calculation of gastric emptying rates of microphagous/planktivorous fish which feed more or less continuously and of 'sit-and-wait' predators which tend to consume large meals. However, the majority of fish species do not fit into either of these distinct ecological categories. Many fish feed opportunistically and, at any given time, the stomach may contain food items in various stages of digestive breakdown. Under these conditions, changes in the degree of stomach fullness and patterns of gastric emptying are likely to be complex, which suggests that estimates of gastric emptying rates derived from 'single meal' experiments carried out in the laboratory may have limited application to field studies of food consumption. It is important, therefore, that more laboratory studies be directed towards the investigation of feeding periodicity and patterns of gastric emptying under different levels of food availability. Thus, many problems remain and the onus is on physiologists and ecologists to undertake better designed studies if the complexities of the system are to be unraveled. One of the major stumbling blocks to an advancement in knowledge seems to be the lack of communication and co-operation between the physiological and ecological disciplines, despite the fact that it should be obvious that each has an important role to play if the goal of a fuller understanding of the way in which food is processed is to be achieved.

Acknowledgements

Thanks to George Boehlert who, in sending the invitation to participate in Gutshop '84, got me re-thinking about alimentary alchemy. Thanks also to Arild Saasen for advice about and help with handling the data and to Per-Arne Amundsen whose comments led me to make some minor alterations to the text, in an attempt to reach the non-physiologists amongst our fishy fraternity.

References cited

Bajkov, A.D. 1935. How to estimate the daily food consumption of fish under natural conditions. Trans. Amer. Fish. Soc. 65: 288–289.

Barker, G.R., G.McL. Cochrane, G.A. Corbett, J.F. Dufton, J.N. Hunt & S.K. Roberts. 1978. Glucose, glycine and diglycine in test meals as stimuli to a duodenal osmoreceptor slowing gastric emptying. J. Physiol. 283: 341–346.

Bell, F.R. & M-L. Grivel. 1975. The effect of duodenal infusion on the electromyogram of gastric muscle during activation and inhibition of gastric emptying. J. Physiol. 248: 377–391.

Bell, F.R. & K. Mostaghni. 1975. Duodenal control of gastric emptying in the milk-fed calf. J. Physiol. 245: 387–407.

Bell, F.R., M. Nouri & D.E. Webber. 1981. The interplay between hydrogen ions, bicarbonate ions and osmolarity in the anterior duodenum modulating gastric function in the concious calf. J. Physiol. 314: 331–341.

Brodeur, R.D. 1984. Gastric evacuation rates for two foods in the black rockfish, *Sebastes melanops* Girard. J. Fish Biol. 24: 287–298.

Cooke, A.R. 1974. Duodenal acidification: role of first part of duodenum in gastric emptying and secretion in dogs. Gastroenterology 67: 85–92.

Crim, J.W. & S.R. Vigna. 1983. Brain, gut and skin peptide hormones in lower vertebrates. Amer. Zool. 23: 621–638.

Darnell, R.M. & R.M. Meierotto. 1962. Determination of feeding chronology in fishes. Trans. Amer. Fish. Soc. 91: 313–320.

Debas, H.T., O. Farooq & M.I. Grossman. 1975. Inhibition of gastric emptying is a physiological action of cholecystokinin. Gastroenterology 68: 1211–1217.

Doble, B.D. & D.M. Eggers. 1978. Daily feeding chronology, rate of gastric evacuation, daily ration and prey selection in Lake Washington juvenile sockeye salmon (*Oncorhynchus nerka*). Trans. Amer. Fish. Soc. 107: 36–45.

Dockray, G.J. 1982. The physiology of cholecystokinin in brain and gut. Br. Med. Bull. 38: 253–258.

Elliott, J.M. 1972. Rates of gastric evacuation in brown trout, *Salmo trutta* L. Freshwater Biology 2: 1–18.

Elliott, J.M. & L. Persson. 1978. The estimation of daily rates of food consumption for fish. J. Anim. Ecol. 47: 977–991.

El-Shamy, F.M. 1976. Analyses of gastric emptying in bluegill (*Lepomis macrochirus*). J. Fish. Res. Board Can. 33: 1630–1633.

Fange, R. & D. Grove. 1979. Digestion. pp. 161–260. *In:* W.S. Hoar, D.J. Randall & J.R. Brett (ed.) Fish Physiology 8, Academic Press, New York.

Fletcher, D.J., D.J. Grove, R.A. Basimi & A. Ghaddaf. 1984. Emptying rates of single and double meals of different quality from the stomach of the dab, *Limanda limanda* (L.) J. Fish Biol. 25: 435–444.

Flowerdew, M.W. & D.J. Grove. 1979. Some observations of the effects of body weight, temperature, meal size and quality on gastric emptying time in the turbot, *Scophthalmus maximus* (L.) using radiography. J. Fish Biol. 14: 229–238.

From, J. & G. Rasmussen. 1984. A growth model, gastric evacuation and body composition in rainbow trout, *Salmo gairdneri* Richardson, 1836. Dana 3: 61–139.

Gannon, J.E. 1976. The effects of differential digestion rates of zooplankton by alewife, *Alosa pseudoharengus,* on determinations of selective feeding. Trans. Amer. Fish. Soc. 105: 89–95.

Garber, K.J. 1983. Effect of fish size, meal size and dietary moisture on gastric evacuation of pelleted diets by yellow perch, *Perca flavescens.* Aquaculture 34: 41–49.

Griffiths, W.E. 1976. Feeding and gastric evacuation in perch (*Perca fluviatilis* L.). Mauri Ora 4: 19–34.

Grove, D.J., L.G. Loizides & J. Nott. 1978. Satiation amount, frequency of feeding and gastric emptying rate in *Salmo gairdneri*. J. Fish Biol. 12: 507–516.

Hinder, R.A. & K.A. Kelly. 1977. Canine gastric emptying of solids and liquids. Amer. J. Physiol. 223: E335–E340.

Holmgren, S., C. Vaillant & R. Dimaline. 1982. VIP-, substance P-, gastrin/CCK-, bombesin-, somatostatin- and glucagon-like immunoreactivities in the gut of the rainbow trout, *Salmo gairdneri*. Cell Tissue Res. 223: 141–153.

Holstein, B. & C. Haux. 1982. Inhibition of gastric acid secretion by intestinal and parenteral administration of a mixture of L-amino acids in the Atlantic cod, *Gadus morhua*. Acta Physiol. Scand. 116: 141–145.

Hunt, J.N. 1980. A possible relation between the regulation of gastric emptying and food intake. Amer. J. Physiol. 239: G1–G4.

Hunt, J.N. & M.T. Knox. 1968. Regulation of gastric emptying. pp. 1917–1935. *In:* C.F. Code (ed.) Handbook of Physiology, Alimentary Canal IV, American Physiological Society, Washington.

Jobling, M. 1980. Gastric evacuation in plaice, *Pleuronectes platessa* L.: effects of dietary energy level and food composition. J. Fish Biol. 17: 187–196.

Jobling, M. 1981. Mathematical models of gastric emptying and the estimation of daily rates of food consumption for fish. J. Fish Biol. 19: 245–257.

Jobling, M. 1982. Food and growth relationships of the cod, *Gadus morhua* L., with special reference to Balsfjorden, north Norway. J. Fish Biol. 21: 357–371.

Jobling, M. & P.S. Davies. 1979. Gastric evacuation in plaice, *Pleuronectes platessa* L.: effects of temperature and meal size. J. Fish Biol. 14: 539–546.

Jones, R. 1974. The rate of elimination of food from the stomachs of haddock, *Melanogrammus aeglefinus,* cod, *Gadus morhua,* and whiting, *Merlanguis merlangus*. J. du Conseil 35: 225–243.

Kelly, K.A. 1980. Gastric emptying of liquids and solids: roles of proximal and distal stomach. Amer. J. Physiol. 239: G71–G76.

Kionka, B.C. & J.T. Windell. 1972. Differential movement of digestible and indigestible food fractions in rainbow trout, *Salmo gairdneri*. Trans. Amer. Fish. Soc. 101: 112–115.

Kuipers, B. 1975. Experiments and field observations on the daily food intake of juvenile plaice, *Pleuronectes platessa* L. pp. 1–12. *In:* H. Barnes (ed.) Proc. 9th Europ. Mar. Biol. Symp., Aberdeen University Press, Aberdeen.

Langer, M., S. van Noorden, J.M. Polak & A.G.E. Pearse. 1979. Peptide hormone-like immunoreactivity in the gastrointestinal tract and endocrine pancreas of eleven teleost species. Cell Tissue Res. 199: 493–508.

Lepkovsky, S., M.K. Dimick, F. Furuta, S.E. Feldman & R. Park. 1975. Stomach and upper intestine of the rat in the regulation of food intake. J. Nutr. 105: 1491–1499.

MacDonald, J.S., K.G. Waiwood & R.H. Green. 1982. Rates of digestion of different prey in Atlantic cod (*Gadus morhua*), ocean pout (*Macrozoarces americanus*), winter flounder (*Pseudopleuronectes americanus*) and American plaice (*Hippoglossoides platessoides*). Can. J. Fish. Aquat. Sci. 39: 651–659.

MacKenzie, K. & D.I. Gibson. 1970. Ecological studies of some parasites of plaice, *Pleuronectes platessa* L. and flounder, *Platichthys flesus* (L.). pp. 1–42. *In:* A.E.R. Taylor & R. Muller (ed.) Aspects of Fish Parasitology, Blackwell Scientific, Oxford.

Magnuson, J.J. 1969. Digestion and food consumption by skipjack tune (*Katsuwonus pelamis*). Trans. Amer. Fish. Soc. 98: 379–392.

Mangel, A.W. & A. Koegel. 1984. Effects of peptides on gastric emptying. Amer. J. Physiol. 246: G342–G345.

Markus, H.C. 1932. The extent to which temperature changes influence food consumption in largemouth bass. Trans. Amer. Fish. Soc. 62: 202–210.

McHugh, P.R. & T.H. Moran. 1978. Accuracy of the regulation of caloric ingestion in the rhesus monkey. Amer. J. Physiol. 235: R29–R34.

McHugh, P.R. & T.H. Moran. 1979. Calories and gastric emptying: a regulatory capacity with implications for feeding. Amer. J. Physiol. 236: R254–R260.

Meeroff, J.C., V.L.W. Go & S.F. Phillips. 1975. Control of gastric emptying by osmolality of duodenal contents in man. Gastroenterology 68: 1144–1151.

Meyer, J.H. 1980. Gastric emptying of ordinary food: effect of antrum on particle size. Amer. J. Physiol. 239: G133–G135.

Moran, T.H. & P.R. McHugh. 1982. Cholecystokinin suppresses food intake by inhibiting gastric emptying. Amer. J. Physiol. 242: R491–R497.

Nomura, M., G.R. Greenberg, A. Bahoric, B. Zinman & A.M. Albisser. 1985. How laboratory dogs accomodate meals of different size but similar composition. Amer. J. Physiol. 248: E101–E107.

Ochia, B.A. 1976. Influence of amino acids on gastric emptying in young pigs. Acta Physiol. Hung. 48: 41–50.

Pandian, T.J. 1967. Transformation of food in the fish *Megalops cyprinoides* I. Influence of quality of food. Marine Biology 1: 60–64.

Pandian, T.J. & E. Vivekanandan. 1985. Energetics of feeding and digestion. pp. 99–124. *In:* P. Tytler & P. Calow (ed.) Fish Energetics – New Perspectives, Croom Helm, London.

Persson, L. 1979. The effects of temperature and different food organisms on the rate of gastric evacuation in perch (*Perca fluviatilis*). Freshwater Biology 9: 99–104.

Persson, L. 1981. The effects of temperature and meal size on the rate of gastric evacuation in perch (*Perca fluviatilis*) fed on fish larvae. Freshwater Biology 11: 131–138.

Persson, L. 1982. Rate of food evacuation in roach (*Rutilus rutilus*) in relation to temperature and the application of evacuation rate estimates for studies on the rate of food consumption. Freshwater Biology 12: 203–210.

Rombout, J.H.W.M. & M. Reinecke. 1984. Immunohistochemical localization of (neuro) peptide hormones in endocrine cells and nerves of the gut of a stomachless teleost fish, *Barbus conchonius* (Cyprinidae). Cell Tissue Res, 237: 57–65.

Roze, C., D. Couturier, J. Chariot & C. Debray. 1977. Inhibition of gastric electrical and mechanical activity by intraduodenal agents in pigs and the effects of vagotomy. Digestion 15: 526–539.

Schneeman, B.O. & R.L. Lyman. 1975. Factors involved in the intestinal feedback regulation of pancreatic enzyme secretion in the rat. Proc. Soc. Exp. Biol. Med. 148: 897–903.

Smith, J.L., C.L. Jiang & J.N. Hunt. 1984. Intrinsic emptying pattern of the human stomach. Amer. J. Physiol. 246: R959–R962.

Stephens, J.R., R.F. Woolson & A.R. Cooke. 1975. Effects of essential and nonessential amino acids on gastric emptying in the dog. Gastroenterology 69: 920–927.

Stephens, J.R., R.F. Woolson & A.R. Cooke. 1976. Osmolyte and tryptophan receptor controlling gastric emptying in the dog. Amer. J. Physiol. 231: 848–853.

Stubbs, D.F. 1977. Models of gastric emptying. Gut 18: 202–207.

Strunz, U. 1979. Hormonal control of gastric emptying. Acta Hepato-Gastroenterol. 26: 334–341.

Talbot, C., P.J. Higgins & A.M. Shanks. 1984. Effects of pre- and post-prandial starvation on meal size and evacuation rate of juvenile Atlantic salmon, *Salmo salar* L. J. Fish Biol. 25: 551–560.

Vigna, S.R. 1983. Evolution of endocrine regulation of gastrointestinal function in lower vertebrates. Amer. Zool. 23: 729–738.

Vigna, S.R. & A. Gorbman. 1977. Effects of cholecystokinin, gastrin and related peptides on coho gallbladder contraction *in vitro*. Amer. J. Physiol. 232: E485–E491.

Weisbrodt, N.W., J.N. Wiley, B.F. Overholt & P. Bass. 1969. A relation between gastroduodenal muscle contractions and gastric emptying. Gut 10: 543–548.

Windell, J.T. 1966. Rate of digestion in the bluegill sunfish. Invest. Indiana Lakes Streams 7: 185–214.

Windell, J.T. & D.O. Norris. 1969. Gastric digestion and evacuation in rainbow trout. Prog. Fish-Cult. 31: 20–26.

Windell, J.T., J.D. Hubbard & D.L. Horak. 1972. Rate of gastric evacuation in rainbow trout fed three pelleted diets. Prog. Fish-Cult. 34: 156–159.

Windell, J.T., D.O. Norris, J.F. Kitchell & J.S. Norris. 1969. Digestive response of rainbow trout, *Salmo gairdneri*, to pellet diets. J. Fish. Res. Board Can. 26: 1801–1812.

Yamagishi, T. & H.T. Debas. 1978. Cholecystokinin inhibits gastric emptying by acting on both proximal stomach and pylorus. Amer. J. Physiol. 234: E375–E378.

Received 6.12.1984 *Accepted 11.7.1985*

Patterns of food evacuation in fishes: a critical review

Lennart Persson
Fish Ecology Research Group, Institute of Limnology, University of Lund, P.O. Box 65, S-221 00 Lund, Sweden

Keywords: Rate models, Nutrition, Lag phase, Experimental data, Exponential, Square root

Synopsis

Discussion concerning the fitting of different models to experimental data on food evacuation in fishes has largely revolved around two models, the exponential and the square root. In this review it is argued that, contrary to a previous review, the exponential model will in general give a better description of the rate of food evacuation than the square root model. It is also suggested that the rationale behind the square root model is questionable. In several laboratory studies (often when large meals were given), a lag phase in digestion was observed, a phenomenon which is not predicted by either model. In these cases other models could be applicable.

Introduction

The number of studies on food evacuation in fishes has increased rapidly during the last twenty years. Although earlier workers did not fit mathematical models to their data, many of these studies showed that the relation between food remaining in the gut and time was curvilinear (e.g. Windell 1966, 1967, Pandian 1967). In some studies the decrease in meal content with time was, however, almost linear (Windell 1966, Swenson & Smith 1973) suggesting that the pattern of evacuation could take different forms. Early studies also demonstrated profound effects of starvation (Windell 1966, Western 1971), force feeding (Windell 1966, 1967, Western 1971) and fat content of the food (Windell 1966, Windell & Kitchell 1972) on the rate of food evacuation.

The studies of Magnuson (1969), Tyler (1970) and Brett & Higgs (1970) were the first in which mathematical models were fitted to the experimental data. Magnuson (1969) used a second degree polynomial, while Tyler (1970) and Brett & Higgs (1970) employed an exponential model. The adequacy of using the exponential model to describe the pattern of food evacuation in fishes has subsequently been supported by a number of studies (Elliott 1972, Persson 1979, 1981, 1982, Andersson 1984, From & Rasmussen 1984). Though other models have been developed (see Olson & Mullen 1986), the most commonly used model for the estimation of food consumption rates of fish in the field (Elliott & Persson 1978) also assumes an exponential rate of food evacuation.

In a study of gastric evacuation in plaice, *Pleuronectes platessa,* Jobling & Davies (1979) fitted an alternative weight (volume) dependent model, the square root model originally proposed by Hopkins (1966), to the experimental data. Later Jobling (1981) reviewed and reanalysed older data and concluded that most studies showed that gastric evacuation was better described by the square root model than by the exponential model.

In this review I will argue that, contrary to Jobling (1981), the exponential model generally gives a better approximation of the pattern of food evacuation for most fishes studied than the square root model. I suggest that Jobling's conclusion was based partially on the use of data from studies of limited value for the purpose of comparing different evacuation models. I also argue that the rationale for using the square root model is based on a questionable assumption. Jobling's (1981) quantitative comparison was limited to the above mentioned weight (volume)-dependent models, although a surface-dependent model was also discussed. In some cases where the square root model provides a better description of the pattern of food evacuation, a surface-dependent model appears to perform equally well.

In several of the studies where the square root model gave the better fit (usually when large meal sizes were given to the fish), a lag phase in digestion was observed. This situation is not predicted by any of the above mentioned models, hence other, more complex models appear to be more appropriate.

Mathematical models and the rate of food evacuation

Three mathematical models that have been discussed in relation to food evacuation in fishes, the exponential, square root and surface area models can all be written in the same general form:

$$\frac{dW}{dt} = -rW^{b}, \tag{1}$$

where *W* is the weight (or volume) of the food, *r* is the rate constant and *b* is a constant that is dependent on the mathematical model in question (Jobling 1981, From & Rasmussen 1984). If the rate of food evacuation is linearly related to the amount of food in the stomach as evacuation progresses, the result will be an exponential decay:

$$\frac{dW}{dt} = -rW. \tag{2}$$

Hopkins (1966) suggested that the pattern of food evacuation was better described by a square root model than an exponential model. The rationale for the model was based on the Law of Laplace. Radial gastric distension will stimulate peristaltic contractions in the stomach. The circumferential tension so developed is proportional to the radius of the stomach (the stomach is compared to a cylinder of constant length) and thus, to the square root of the weight (volume) of the stomach content:

$$\frac{dW}{dt} = -rW^{0.5}. \tag{3}$$

Fänge & Grove (1979) (see also Tyler 1970) suggested another model based on the assumption that the rate of food evacuation is proportional to the surface area of the meal, based on the argument that the enzymes and acids secreted will mainly affect the surface of the meal. The result will be a model which assumes that rate of food evacuation is proportional to two thirds of the weight (or volume) of the stomach contents:

$$\frac{dW}{dt} = -rW^{0.67}. \tag{4}$$

The predicted rate of food evacuation of a given meal based on these three models is shown in Figure 1. The amount of food in the stomach was arbitrarily set equal at time *t* for all models. It can be seen that the exponential model will predict the highest rate of food evacuation during the first part of digestion, while the square root model predicts the highest rate at the end of digestion. The prediction of the surface area dependent model is intermediate between the predictions made by the exponential and the square root models.

A reexamination of previous data

Jobling (1981) reanalyzed several studies on food evacuation of fishes to determine whether the exponential or the square root model gave the best fit to the data. Jobling's conclusion was that the majority of studies showed that the application of the square root model in general provided the better fit to the data.

In this review I have included 7 recent studies not

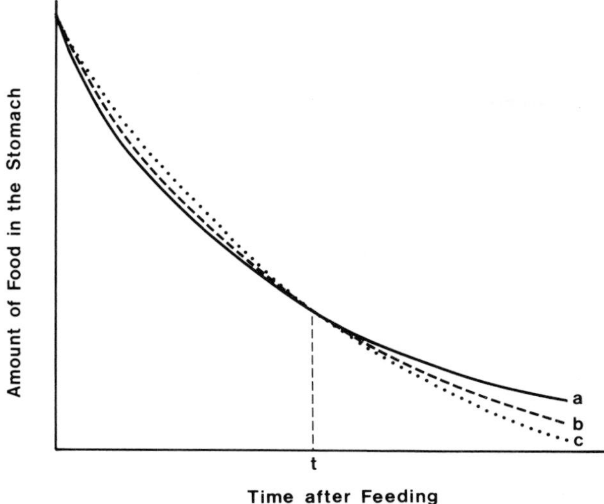

Fig. 1. The predicted pattern of food evacuation of a given meal size assuming the exponental (a), the surface area (b), and the square root (c) models respectively. The amount of food evcacuated of a specific meal at time *t* is assumed to be the same for all models.

included by Jobling and have also carried out an evaluation of them based on 3 criteria. Firstly, and most importantly, I have considered whether the experimental fish were fed voluntarily or force fed. In a number of studies (Windell 1966, Western 1971, Swenson & Smith 1973, Griffiths 1976) force feeding has been shown to negatively affect the rate of food evacuation, especially during the early stages of digestion. This means that force feeding fish could introduce a bias resulting in a better fit for the square root model compared to the exponential (see Fig. 1). Secondly, I have considered the number of observations upon which the final statistical analysis was based as this will affect the accuracy and adequacy with which a certain model can be fitted to the data. Thirdly, I have also considered if a lag phase in digestion was observed. I have defined a lag phase in digestion as the situation when the amount of food evacuated per unit time early in digestion is lower than later on. A lag phase defined in this way is not predicted by any of the above models (see Fig. 1), suggesting all three models to be inappropriate when a lag phase is observed.

The coefficient of determination (r^2) was chosen as a criterion to evaluate which of the two models gave the best fit. Brodeur (1984) and From & Rasmussen (1984) presented r^2 values for both models. In the other studies r^2's were recalculated using either raw data (Griffiths 1976, Person 1979, 1981, 1982) or mean values if raw data were not available. In the situations where different units had been used to estimate food evacuation in the same experiment, weight was preferred over volume and dry weight over wet weight (see Persson 1981).

The exponential model gave a better fit in 10 of the 22 studies considered (Tables 1, 2). The fish were fed voluntarily in all the studies where the exponential model gave a better fit (Table 1). The numbers of observations were higher in these studies than in those where the square root model provided a better fit (Table 1, 2). In 4 of the latter

Table 1. Studies on food evacuation where the exponential model provided a better fit than the square root model.

Author	Species	Feeding technique	Sample size
Brett & Higgs (1970)	*Oncorhynchus nerka*	Voluntary	110
Elliott (1972)	*Salmo trutta*	Voluntary	⩾60
Windell et al. (1976)*	*Salmo gairdneri*	Voluntary	⩾60
Persson (1979)	*Perca fluviatilis*	Voluntary	⩾40
Grove & Crawford (1980)	*Blennius flesus*	Voluntary	⩾16
Persson (1981)*	*Perca fluviatilis*	Voluntary	⩾40
Persson (1982)*	*Rutilus rutilus*	Voluntary	⩾50
Andersen (1984)*	*Pomatoschistus microps*	Voluntary	⩾21
Brodeur (1984)*	*Sebastes groenlops*	Voluntary	30
From & Rasmussen (1984)*	*Salmo gairdneri*	Voluntary	⩾42

* Not included in Jobling's (1981) review

Table 2. Studies on food evacuation where the square root model provided a better fit than the exponential model.

Author	Species	Feeding technique	Sample size	Comment
Hunt (1960)	*Micropterus salmoides* *Chaenobryttus gulosus* *Lepisosteus platyrhincus*	Force fed	⩾ 18	
Herting & Witt (1968)	*Amia calva*	Force fed	22	
Jobling & Davies (1979)	*Pleuronectes platessa*	Force fed	⩽ 9	
Jobling (1980)*	*Pleuronectes platessa*	Force fed	⩽ 9	
Windell (1966)	*Lepomis macrochirus*	Voluntary	⩾ 16	
Kitchell & Windell (1968)	*Lepomis gibbosus*	Voluntary	⩾ 27	Lag phase observed
Windell et al. (1969)	*Salmo gairdneri*	Voluntary	⩾ 18	Lag phase observed
Gerald (1973)	*Ophiocephalus punctatus*	Voluntary	⩾ 15	Lag phase observed
Swenson & Smith (1973)	*Stizostedion vitreum vitreum*	Voluntary	⩾ 9	Lag phase observed
Bagge (1977)	*Myoxocephalus scorpius*	Voluntary	⩾ 16	
Tyler (1970)**	*Gadhus morhua*	Voluntary	⩾ 10	
Griffiths (1976)**	*Perca fluviatilis*	Voluntary	27	

* Not included in Jobling's (1981) review.
** Exponential model fitted in original paper.

studies the fish were force fed (Table 2). As mentioned above, this will bias the results towards the predictions given by the square root model, hence these studies should not be used when comparing evacuation models. A distinct lag phase was observed in at least 4 of the remaining studies in which the square root model gave a better fit (Kitchell & Windell 1968, Windell et al. 1969, Gerald 1973, Swenson & Smith 1973). In two cases (Windell et al. 1969, Swenson & Smith 1973) this was attributed to large meals given to the experimental fish.

To summarize, in the studies where the exponential model gave the better fit to the data, the experimental fish were all fed voluntarily and sample sizes were generally high. This was not the case for many of the studies were the square root model gave the best fit. The conclusion drawn by Jobling (1981) that the square root model usually gives a better approximation of the rate of food evacuation is thus not supported. Rather, this reanalysis suggests that the exponential model is more appropriate for the majority of these studies. Finally, in several of the studies where the square root model gave a better prediction, both models seem to be inadequate due to an initial lag phase in digestion.

Evaluating the assumptions of the square root model

The appropriateness of evacuation models may be evaluated also by considering their underlying rationale. As mentioned above, a physical explantaion based on the Law of Laplace has been put forward in support of the square root model. It is assumed that the stomach of the fish can be approximated to a cylinder of constant length (l), while the radius (r) of the stomach is dependent on the amount of food it contains (Fig. 2a). However, it has never been shown that the length of the stomach is independent of the amount of food it contains. I therefore measured length and diameter of the stomachs of 30 perch, *Perca fluviatilis* (mean total length 114.5 mm, range 112–119 mm) that had been fed different amounts of food. The diameter (b) was measured at its broadest point, and the length was measured from the attachment of the intestine (Fig. 2b). The justification for measuring stomach length in this way is that it is difficult to measure total length reliably because the anterior end of the stomach grades indistinctly into the oesophagus.

The diameter of the stomach increased with meal size, as predicted by the square root model, al-

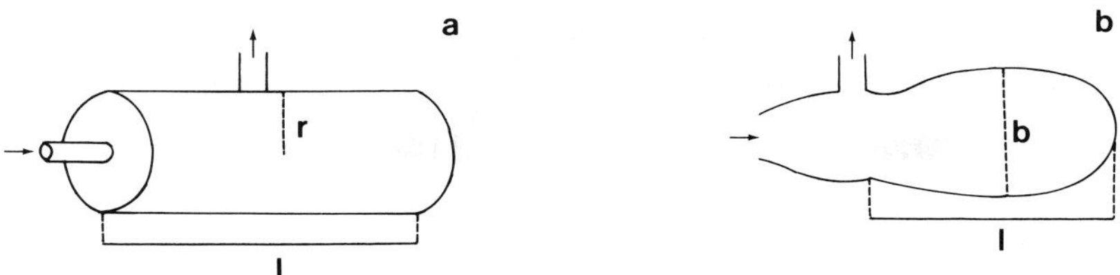

Fig. 2. a – A schematic diagram of the stomach of a fish as assumed by the square model. The radius (r) is assumed to vary with stomach contents while the length (l) is assumed to be independent of stomach contents. *b* – Diagram of the measurements carried out on perch stomachs to study the effect of stomach contents on stomach length (l) and diameter (b) respectively.

though the increase was not linear (Fig. 3) (power function $r^2 = 0.77$, $p<0.001$, $F = 95$). The length of the stomach was, however, not independent of meal size ($P<0.001$) (Fig. 3, regression of stomach length on meal size $r^2 = 0.93$, $F = 361$). Thus, at least in the case of 110–120 mm perch, the assumption of constant stomach length for the square root model is incorrect.

Evacuation of large meals and lag phases in digestion

The exponential model is expected to give a good approximation of the rate of food evacuation assuming muscle contractions, acid secretion and enzyme secretion are directly related to the amount of food in the stomach. Digestion, however, will also be dependent on the degree that acid and enzymes penetrate the food, factors which are themselves influenced by the surface area and content of the meal. Therefore, an effect of meal size on both the rate and pattern of evacuation should be expected. Despite this, no effect of meal size on the rate of food evacuation was found by Elliott (1972) and Persson (1981), with the exception of one experiment. This could, however, have been a result of relatively small meal sizes and/or that the meal consisted of several small food items, into which acid and enzymes penetrated more or less instantaneously. Persson (1981) also indicated that the exponential model failed to predict the pattern of food evacuation when the ratio between meal size and fish size was high. This was due to the fact that the evacuation rate at the beginning of digestion was lower than that predicted by the exponential model (Persson 1981, Fig. 4a), which was also the case in the study of Griffiths (1976). The low

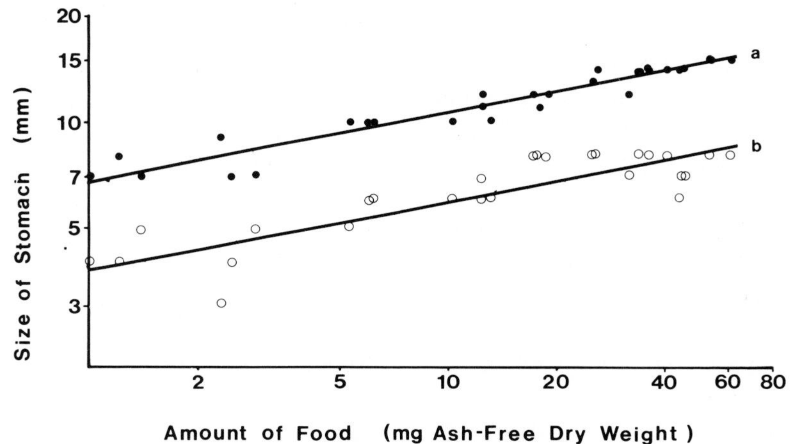

Fig. 3. Relation between (a) length and (b) diameter of the stomach respectively and amount of food (mg ash-free dry weight) in the stomach. A power function was fitted to the data in both cases.

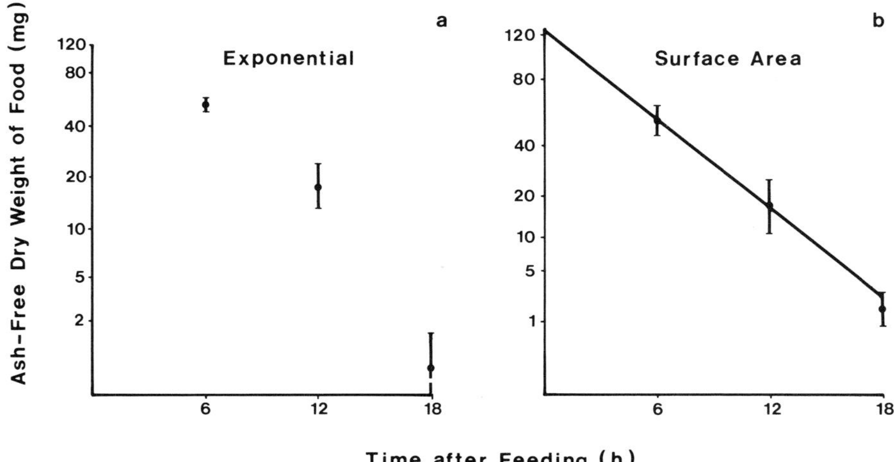

Fig. 4. Relation between amount of food recovered in the stomach of perch (total length 89–104 mm) and time (h). *a* – Data presented on a semilogarithmic scale showing the stomach contents after 6, 12 and 18 h respectively; *b* – fitted to the surface area model. Means with 95% CL are given. Data are from Persson (1981).

rate of food evacuation in the early stages of digestion suggests that digestion may be limited by the penetrating ability of acid and enzymes. Thus, the surface-dependent model could provide a better description of the rate of food evacuation.

To examine the appropriateness of the surface-dependent model in these circumstances, equation 4 was fitted to original data. The model gave a good description of the pattern of food evacuation (Griffiths 1976, $r^2 = 0.87$, $F = 158$; Persson 1981 $r^2 = 0.96$, $F = 863$, Fig. 4b). As the data of the study (Griffiths 1976) were fitted to the square root model by Jobling (1981), a comparison between the square root model and the surface area model could be made. The coefficient of determination in the Persson (1981) study was similar when fitting either model (0.96). The initial meal size was, however, better predicted by the surface area model (122.7 mg) than by the square root model (115.8 mg). Fitting the square root model to Griffiths' (1976) data gave a higher r^2 value (0.90) than fitting the surface area model ($r^2 = 0.87$). The initial meal size was again, more accurately predicted by the surface area model (95.1% compared to 89.7%).

The higher r^2 value obtained when fitting the square root model to the data of Griffiths (1976) was due to the surface area model predicting an excessively low rate of food evacuation during the later stages of digestion (see amount of food recovered at $t = 18$ in Fig. 4b). This underestimation could be explained by the fact that the reduction in meal size during digestion also involves a disintegration of the meal. The surface area-weight (volume) relation is therefore contrary to the assumption of equation 4 not constant. This will substantially undermine the applicability of the surface-dependent model (see also Jobling 1981).

None of equations 2–4 will predict that the amount of food evacuated per unit time should be lower during the beginning of digestion than later on (Fig. 1). Several studies (Kitchell & Windell 1968, Windell et al. 1969, Gerald 1973, Swenson & Smith 1973), however, found such results. Hence the justification for fitting any of above models to those data is questionable.

Two models have been suggested to handle this pattern of food evacuation, an exponential model with a time lag (Fänge & Grove 1979) and the logistic model (Persson 1981, MacDonald et al. 1982) but laboratory data where these models have been tested is lacking. Available data do, however, suggest that the presence of lag phases is dependent both on fish size and meal size (Jobling 1981, Persson 1981).

Patterns of food evacuation and the estimation of food consumption rates in the field

For ecologists, the main reason for carrying out studies of food evacuation is to apply the results in the estimation of food consumption rates in the field. The major conclusion to be drawn from the present review is that of the three models, the exponential, the square root and the surface area model, the exponential model is generally the most appropriate. It is also preferred due to its simple mathematical properties. In these cases, the food consumption rate model developed by Elliott & Persson (1978) is suitable. The pattern of food evacuation may, however, take other forms. The food evacuation for fish fed nutritious food items has also been described by a linear function (Windell 1966, Swenson & Smith 1973, Olson & Mullen 1986). In these cases, the assumption of an exponential evacuation in the Elliott & Persson model is obviously violated and other models must be used. Olson & Mullen (1986) proposed a general model that makes no assumption about the pattern of food evacuation and thus should be applicable in many situations. One precarious assumption of this model is, however, that every food item is assumed to be evacuated independently of other food items. This assumption is in conflict with the experimental results of Persson (1984). In view of Persson's (1984) results, care must be taken when using this general model for estimating consumption of food items ingested at different times.

Acknowledgements

I thank Ebbe Lundgren for valuable discussions concerning the assumptions of the square root model. Ian Winfield gave useful comments and also improved the English. While writing this review the author was supported by a grant from the Swedish Natural Science Research Council. The Swedish Natural Science Research Council also provided a travelling grant for the author to the Fourth Workshop on Fish Food Habits in California, where this review was presented.

References cited

Andersen, N.G. 1984. Depletion rates of gastrointestinal content in common goby (*Pomatoschistus microps* (Kr.)). Effects of temperature and fish size. Dana 3: 31–42.

Bagge, O. 1977. Meal size and digestion in cod (*Gadus morhau* L.) and sea scorpion (*Myoxocephalus scorpius* L.). Meddr. Danm. Fisk.- og Havsunders. 7: 437–446.

Brett, J.R. & D.A. Higgs 1970. Effect of temperature on the rate of gastric digestion in fingerling sockeye salmon. J. Fish. Res. Board Can. 27: 1767–1779.

Brodeur, R.D. 1984. Gastric evacuation rates for two foods in the black rockfish, *Sebastes melanops*, Girard. J. Fish Biol. 24: 287–298.

Elliott, J.M. & L. Persson 1978. The estimation of daily rates of food consumption for fish. J. Anim. Ecol. 47: 977–991.

From, J. & G. Rasmussen. 1984. A growth model, gastric evacuation and body composition in rainbow trout, *Salmo gairdneri* Richardson, 1836. Dana 3: 61–139.

Fänge, R. & D.J. Grove. 1979. Digestion. pp. 161–260. *In:* W.S. Hoar, D.J. Randall & J.R. Brett (ed.), Fish Physiology, Vol. 7, Academic Press, New York.

Gerald, V.M. 1973. Rate of digestion in *Ophiocephalus punctatus* Bloch. Comp. Bochem. Physiol. 46A: 195–205.

Griffiths, W.E. 1976. Feeding and gastric evacuation in perch (*Perca fluviatilis* L.). Mauri Ora 4: 19–34.

Grove, D.J. & C. Crawford. 1980. Correlation between digestion rate, frequency of feeding in the stomachless teleost, *Blennius pholis* L. J. Fish Biol. 16: 235–247.

Herting, G.E. & A.W. Witt. 1968. Rate of digestion in the bowfin. Prog. Fish. Cult. 30: 26–28.

Hopkins, A. 1966. The patttern of gastric emptying: a new view of old results. J. Physiol. 182: 144–149.

Hunt, B.P. 1960. Digestion rate and food consumption of Florida gar, warmouth and largemouth bass. Trans. Amer. Fish. Soc 89: 206–211.

Jobling, M. 1980. Gastric evacuation in plaice, *Pleuronectes platessa* L.: effects of dietary energy level and food composition. J. Fish Biol. 17: 187–196.

Jobling, M. 1981. Mathematical models of gastric emptying and the estimation of daily rates of food consumption for fish. J. Fish Biol. 19: 245–257.

Jobling, M. & S.P. Davies. 1979.Gastric evacuation in plaice, *Pleuronectes platessa* L.: effects of temperature and meal size. J. Fish Biol. 14: 539–546.

Kitchell, J.F. & J.T. Windell. 1958. Rate of gastric digestion in pumpkinseed sunfish (*Lepomis gibbosus*). Trans. Amer. Fish. Soc. 97: 489–492.

MacDonald, J.S., K.G. Waiwood & R.H. Green. 1982. Rates of digestion of different prey in Atlantic cod (*Gadus morhua*), ocean pout (*Macrozoarces americanus*), winter flounder (*Pseudopleuronectes americanus*) and American plaice (*Hippoglossoides platessoides*). Can. J. Fish. Aquat. Sci. 39: 651–659.

Magnuson, J.J. 1969. Digestion and food consumption by skipjack tuna (*Katzuwonus pelamis*). Trans. Amer. Fish. Soc. 98:

379–392.

Olson, R.J. & A.J. Mullen. 1986. Recent developments for making gastric evacuation and daily ration determinations. Env. Biol. Fish. 16: 183–191 (this volume).

Pandian, T.J. 1967. Intake, digestion, absorption and conversion of food in the fishes, *Megalops cyprinoides* and *Ophiocephalus striatus*. Mar. Biol. 1: 16–32.

Persson, L. 1979. The effects of temperature and different food organisms on the rate of gastric evacuation in perch (*Perca fluviatilis*). Freshwat. Biol. 9:99–104.

Persson, L. 1981. The effects of temperature and meal size on the rate of gastric evacuation in perch (*Perca fluviatilis*) fed on fish larvae. Freshwat. Biol. 11: 131–138.

Persson, L. 1982. Rate of food evacuation in roach (*Rutilus rutilus*) in relation to temperature, and the application of evacuation rate estimates for studies on the rate of food consumption. Freshwat. Biol. 12:203–210.

Persson, L. 1984. Food evacuation and models for multiple meals in fishes. Env. Biol. Fish. 10: 305–309.

Swenson, A.W. & L.L. Smith Jr. 1973. Gastric digestion, food consumption, feeding periodicity and food conversiona efficiency in walleye (*Stizostedion vitreum vitreum*). J. Fish. Res. Board Can. 30: 1327–1336.

Tyler, A.V. 1970. Rates of gastric emptying in young cod. J. Fish. Res. Board Can. 27: 1177–1189.

Western, J.R.H. 1971. Feeding and digestion in two cottid fishes, the freshwater *Cottus gobio*, L. and the marine *Enophrus bubalis* (Euphrasen). J. Fish Biol. 3: 225–246.

Windell, J.T. 1966. Rate of digestion in bluegill sunfish. Invest. Indiana Lakes & Streams 7: 185–214.

Windell, J.T. 1967. Rates of digestion in fishes. pp. 151–174. *In:* S.D. Gerking (ed.) The Biological Basis of Freshwater Fish Production, Blackwell's Scientific Publications, Oxford.

Windell, J.T. & J.F. Kitchell. 1972. Rate of gastric digestion in pumpkinseed sunfish (*Lepomis gibbosus*). Trans. Amer. Fish. Soc. 97: 489–492.

Windell, J.T., D.O. Norris, J.F. Kitchell & J.S. Norris. 1969. Digestive response of rainbow trout, *Salmo gairdneri*, to pelleted diets. J. Fish. Res. Board Can. 26: 1801–1812.

Windell, J.T., J.F. Kitchell, D.O. Norris, J.S. Norris & J.W. Foltz. 1976. Temperature and rate of gastric evacuation by rainbow trout, *Salmo gairdneri*. Trans. Amer. Fish. Soc. 105: 713–717.

Received 15.12.1984 *Accepted 10.9.1985*

The physiology of digestion in fish larvae

John J. Govoni[1], George W. Boehlert[2] & Yoshirou Watanabe[3]

[1] *National Marine Fisheries Service, NOAA, Southeast Fisheries Center, Beaufort Laboratory, Beaufort, NC 28516, U.S.A.*

[2] *National Marine Fisheries Service, NOAA, Southwest Fisheries Center, Honolulu Laboratory, Honolulu, HI 96812, U.S.A.*

[3] *Tohoku Regional Fisheries Research Laboratory, Shiogama, Miyagi 985, Japan*

Keywords: Alimentary canal, Morphology, Histology, Histochemistry, Enzymology, Assimilation, Development

Synopsis

The acquisition, digestion, and assimilation of food is critical for the growth and survival of fish larvae; a fish larva either grows or it perishes. Fish larvae are characterized by digestive systems and diets that differ from adults. Larvae undergo a pattern of trophic ontogeny, changing diet with increasing size, and these changes result in differences in digestive requirements. At first feeding, the larval alimentary canal is functional, but is structurally and functionally less complex than that of adults. The larval alimentary canal remains unchanged histologically during the larval period before transformation. During transformation, major changes that result in the development of the adult alimentary canal occur. The ontogeny of the alimentary canal differs in different taxa, and experimental evidence suggests that functional differences exist as well. Assimilation efficiency may be lower in larvae than it is in adult fishes, due to a lack of a morphological and functional stomach in larvae, but the question of improving assimilation efficiencies during larval development before transformation remains unresolved.

Introduction

The alimentary canal of fish larvae is morphologically, histologically, and physiologically less elaborate than the alimentary canal of adult fishes. Unlike the gradual development of some other organ systems, e.g., the integumentary, visual, musculature, and acoustic-lateralis systems (see review in O'Connell 1981), the development of the alimentary canal from the simple, undifferentiated, straight incipient gut of the yolk-sac larva to the complex, segmented alimentary canal of the adult proceeds by periodically rapid changes rather than continuous gradation. The incipient gut remains unchanged during yolk and oil-globule absorption (several days or weeks), then changes rapidly just before first feeding (1 to 3 d). In the main, the larval alimentary canal remains unchanged during the long larval period (several months to a year), then changes rapidly into the adult alimentary canal during transformation of the larval to the juvenile fish (weeks or months). [We use the terminology of Kendall et al. (1984) in describing early life history stages].

Concomitant with growth and the changing complexity of the alimentary canal are marked differences in diets. Most fish larvae are visual, raptorial planktivores (Hunter 1981) regardless, of whether their adult counterparts are indiscriminant filter-feeders, pelagic carnivores, or benthic pickers. Larvae begin feeding on large phyto- and small zooplankters and follow by feeding on increasingly larger zooplankters (Hunter 1981). For some species,changes in diets are required. For example, northern anchovy, *Engraulis mordax*, larvae survive on the dinoflagellate *Gymnodinium splendens*, but do not grow well beyond 6 mm in length unless larger food organisms are included in their diets (Hunter 1977). The diets of fish larvae change again during transformation.

The alimentary canal of fish larvae also shows considerable diversity among taxa and several general dichotomies are valid. The alimentary canal of precocial larvae that generally hatch from large, negatively buoyant eggs, is more developmentally advanced than the canal of altricial larvae that hatch from small, positively buoyant eggs (terminology from Balon 1979, 1981). The larvae of most 'lower' fishes (e.g., clupeoids, salmonids, and cyprinoids) have a straight alimentary canal, whereas the larvae of 'higher' fishes (e.g., paracanthopterygians and percomorphs have a looped canal). Notable exceptions to the latter generality are some ostariophysans and salmonids that have looped guts (Iwai 1969) and some stomiatoids that have unusual, trailing, exterilial guts (Moser 1981). Other peculiarities include the occluded guts of elopomorph leptocephali (Hulet 1978, Smith 1984).

The varied structural adaptations of the fish larvae alimentary canal and the changes of these adaptations with development are characteristic of differing functional adaptations to diets and prey concentrations. There is some evidence that ingestion and digestion rates as well as assimilation efficiencies are adapted to maximize larval growth and that these adaptations differ among taxa (Houde & Schekter 1980, 1983). Within taxa, digestion rates and assimilation efficiencies may change with prey availability and ration size (Werner & Blaxter 1980, Boehlert & Yoklavich 1984a) as well as with development (Laurence 1977, Buckley & Dillmann 1982). Here we consider the morphological and histological development of the alimentary canal, review histochemical and enzymological assays of larval alimentary canal function, and describe physiological studies of larval digestion and assimilation. While this paper is a review, we also include original observations and offer recommendations for future research.

Methods

Morphology and histology

Standard histological techniques, light microscopy, transmission electron microscopy (TEM), and scanning electron microscopy (SEM), have contributed to the understanding of the structure and function of the alimentary canal of fish larvae. Standard paraffin procedures have been used with some success for light microscopy (O'Connell 1981), but techniques that use glycol methacrylate as an embedding medium (Govoni 1984) are preferred. Procedures used for TEM of larval tissues are standard (e.g., see Iwai 1969). Boehlert (1984) offers techniques for the preparation of soft larval tissues for examination with SEM.

Histochemistry

The use of horseradish peroxidase (HRP) as a histochemical tracer has resolved the morphological sites and cytological mechanisms of protein absorption in fish larvae (Stroband et al. 1979, Stroband & Kroon 1981, Watanabe 1981, 1982a, 1984a). An advantage of the HRP technique is that in addition to allowing the observation of protein absorption, the technique affords an assessment of enzymatic hydrolysis. For the present observations, 0.5% HRP solution in physiological saline was injected into the larval alimentary canal through a glass capillary inserted into either the foregut or anus. The injection volume was adjusted to fill the entire lumen. Following incubation in vivo, tissues were fixed for 4 h in cold 5% glutaraldehyde in 0.1 molar phosphate buffer (pH 7.6) and then were washed at least overnight in the same cold buffer containing 5% sucrose. For light microscope observations, the tissues were frozen and cut into 10 μm thick sections. The sections were mounted on glass slides and incubated for 5–15 min at room temperature in 0.05% 3,3-diaminobenzidine tetrahydrochloride (DAB) in 0.05 M tris-HCl buffer (pH 7.6) containing 0.01% H_2O_2 as a substrate of HRP. They then were washed in 70% alcohol and observed. Peroxidase activity was detected with light microscopy as dark brown deposits. For TEM, the tissues were cut into small pieces, incubated in DAB, washed in distilled water, and postfixed for an hour in cold 1% OsO_4 solution in phosphate buffer (pH 7.3) containing 0.54% sucrose. This was followed by standard TEM preparations, but sections were observed without staining; peroxidase activity can be detected as electron-opaque deposits. Controls for both light and TEM were designed to show that there was no reaction product when either HRP or its substrate was absent. Groups of larvae or tissues were administered only physiological saline and observed, while another group was administered HRP solution, but without H_2O_2 substrate.

Ikeda (1959) and Prakash (1961) examined the histochemistry of alkaline phosphatase in the alimentary canal of fish larvae. Both used standard paraffin histochemical methods, although improved techniques with glycol methacrylate sections are now developed (Higuchi et al. 1979).

Enzymology

The digestive enzymes of fish larvae have been studied by testing enzyme activities of tissue homogenates against dissolved substrates (Tanaka et al. 1972, Kawai & Ikeda 1973a, 1973b, Dąbrowski 1982), whereby reaction products or the disappearance of substrates are measured photometrically; by testing the enzyme activities of tissue sections against substrate films, whereby the disappearance of substrate is assessed with light microscopy after histochemical staining (Szlaminska 1980, Vu 1983); and, by radio-immunoassay (Hjelmeland et al. 1983). The activities of amylase, maltase, pepsin, trypsin, chymotrypsin, and aminopeptidase have been assayed with the tissue homogenate method; the activities of amylase, pepsin, trypsin and chymotrypsin, and lipase with the substrate-film method. Trypsin and trypsinogen have been radio-immunoassayed.

Peristalsis and digestion rates

Peristalsis and evacuation rates of the alimentary canal have been studied by gross observation of live larvae (Blaxter & Hempel 1961, Chitty 1981, Pedersen 1984) and by collecting feces after larvae were fed. Enhanced resolution of evacuation rates can be obtained by feeding larvae alternately with stained and unstained food organisms (Laurence 1971) or with radioactively labeled food and unlabeled food. The latter technique also affords continuous as well as pulsed feeding experiments.

Assimilation

The use of the physiological energetic approach (Brett & Groves 1979) and of radiotracer techniques (Sorokin 1966) have provided assessments of assimilation efficiency of the larval alimentary canal. Assimilation is normally a difficult physiological parameter to measure (Johannes & Satomi 1967, Conover 1978). This problem is particularly difficult for fish larvae because of their small size, planktonic habitat, and lack of discrete feces production. The alimentary canal of fish larvae discharges both liquid and solid feces, and it is often impossible to separate excreted products of catabolism from defecated dissolved feces. As a result, assimilation efficiency is often estimated by difference, i.e., by the subtraction of other measurable energetic parameters from the matter or energy of ingested food. Assimilation efficiency as calculated by the differences in energy budget parameters is subject to the cumulative biases of these measurable parameters. Radiotracer methods offer a direct, short-term measure of assimilation, but are also subject to error. Based on the retention of metabolically active tracers (e.g., ^{14}C, ^{3}H, or ^{32}P), estimates of assimilation can be biased if the specific activity of the tracer in the food changes during the experiment, or if there are metabolic pools of unknown boundaries and rapid turnover rates that result in short-term loss of tracer from the animal (Conover & Francis 1973). Applications of radiotracer methods to fish larvae are discussed in Govoni et al. (1982) and Boehlert & Yoklavich (1984a).

Results and discussion

The development of the alimentary canal

The development of the alimentary canal of fishes encompasses morphological, histological, and functional changes that are aligned with major changes in gross morphology (Fig. 1). Here we describe the development of the alimentary canal of a generalized altricial larva with exceptions noted where marked variations occur. Embryonically, the alimentary canal develops from the involution of columnar endodermal cells that lie above the yolk (Devillers 1961). At hatching, the alimentary canal is a straight tube lying dorsal to the yolksac, is closed at the mouth and anus in some species, and is histologically undifferentiated along its length (Engen 1968, Tanaka 1969a, Fukusho 1972, Umeda & Ochiai 1973, Vu 1976, Govoni 1980, O'Connell 1981). The incipient gut remains unchanged until the completion of yolk- and oil-globule absorption when the undifferentiated tube becomes segmented by muscular valves into a buccopharynx, fore-, mid-, and hindgut (Kostomarova 1962, Engen 1968, Tanaka 1969a, 1969b, Fukusho 1972, Umeda & Ochiai 1973, Govoni 1980, O'Connell 1981). With the exception of some precocial young that develop from negatively buoyant eggs, most notably the salmonids, gasterosteids, and mouth brooding cichlids, fish larvae lack both a morphological and functional (secretory) stomach; the posterior region of the foregut (Watanabe & Sawada 1985) and the midgut (Govoni 1980) can expand and function to store food in some larvae. The larval alimentary canal remains largely unchanged until the onset of transformation. The development of a stomach and pyloric caeca from the posterior foregut accompanies transformation and constitutes the last major morphological change of the alimentary canal. The liver and pancreas (along with their ducts) are formed at hatching and are functional by the end of yolk absorption (Tanaka 1969a, 1969b, Vu 1976, Govoni 1980, O'Connell 1981, Watanabe & Sawada 1985).

An interesting exception to the above pattern of early gut development occurs in rockfish, *Sebastes*, embryos. This group is characterized by small eggs and high fecundity and has typically been referred to as being a primitive ovoviviparous genus (Wourms & Cohen 1975). Boehlert & Yoklavich (1984b), however, demonstrated that embryos receive additional nutrition during gestation. The uptake of nutritional substances by *S. melanops* embryos does not occur across epidermal surfaces, as in many other viviparous species (Veith 1980). Epidermal microridges on *Sebastes* embryos, characteristic of fish larvae epidermis (Yamada 1968, Roberts et al. 1973), are present, but microvilli, characteristic of epidermal absorptive surfaces on embryos of viviparous fishes (Veith 1980), are absent (Fig. 2a). Instead, uptake occurs across the alimentary canal mucosa, which develops while the embryo is still within the egg envelope and a significant yolk mass remains (Boehlert & Yoklavich 1984b). Histological observations on the alimentary canal of *S. melanops* embryos demonstrate that in early embryos the foregut is not open,

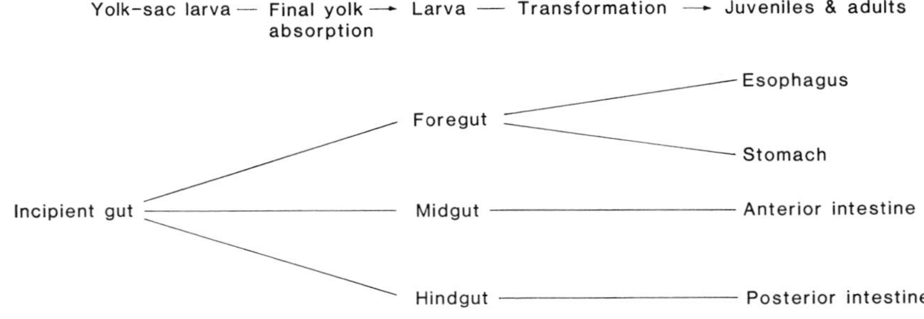

Fig. 1. The derivation, sequence, and timing of alimentary canal organs in typical larval fishes.

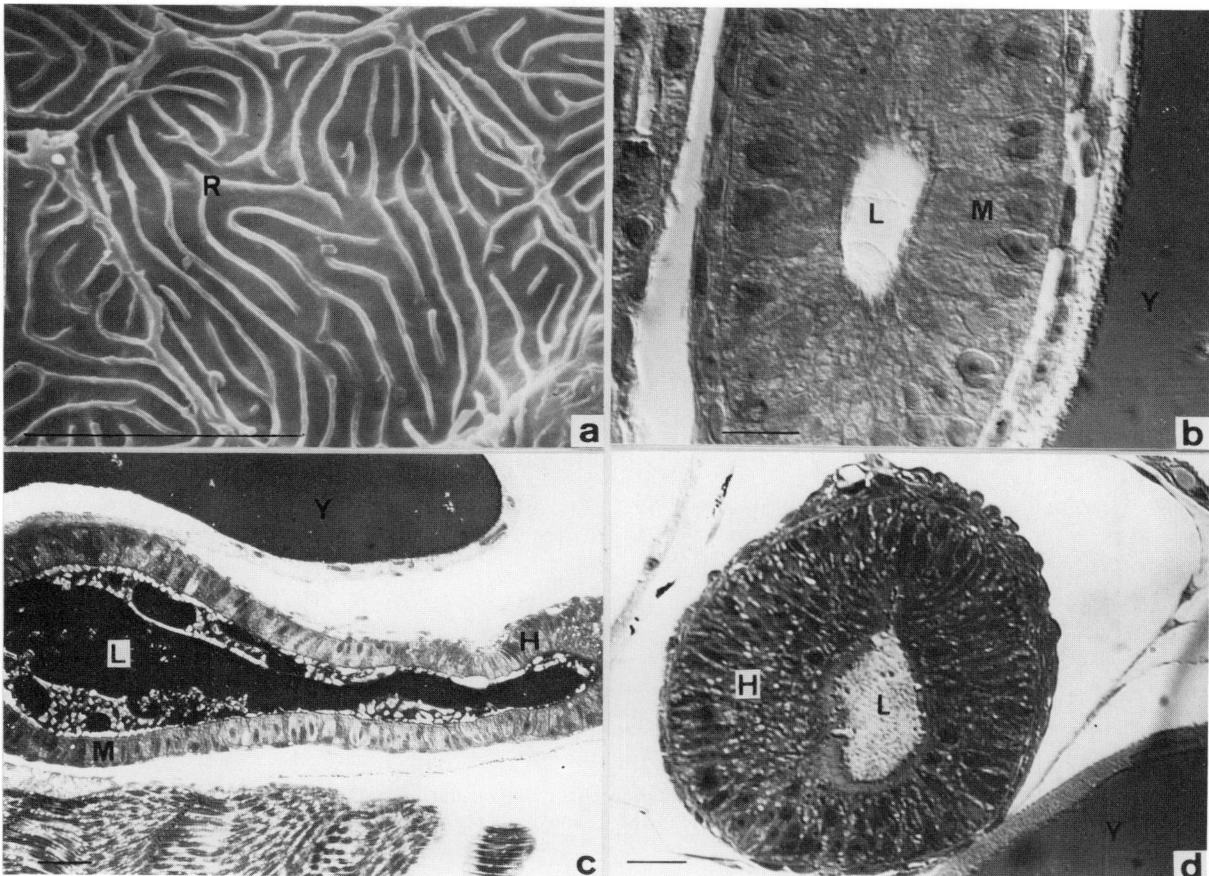

Fig. 2. Micrographs of the epidermis and gut tissue of developing embryonic states from the viviparous species *Sebastes melanops* showing the mechanism of nutrition of pre-parturition embryos. Scanning electron micrograph (*a*) of a dorsal epidermal cell just posterior of the cranium at Oppenheimer (1937) stage 31 (29 d past-fertilization, approximately 7 d before parturition). Note the well developed microridges (R) which characterize the embryonic epidermis throughout development and the absence of absorptive microvilli (scale bar is 10 μm). Photomicrograph of cross section (*b*) taken just dorsal to the yolksac (Y) of a stage 21 embryo (13 d post-fertilization). Although the midgut epithelium (M) is well developed, the lumen (L) is devoid of material (scale bar is 20 μm). Photomicrograph of a longitudinal section (*c*) of the mid- (M) and hindgut (H) of a stage 28 (11 d later than *b*). At this stage, the gut is complete and the mouth open. Note the densely stained, acidophilic substance through lumen of the gut (scale bar is 50μm). Photomicrograph of a cross-section of the hindgut of a stage 31 embryo (*d*). Note that the lumen contains an amorphous, granular substance, and that the epithelial cells have supranuclear granules (scale bar is 20 μm).

and the gut epithelium is relatively narrow (Fig. 2b). With development past stage 27 (of Oppenheimer 1937), however, observation of whole embryos and serial sections show that the gut is open and functional. The lumen now contains acidophilic, amorphous material (Fig. 2c), and the hindgut epithelium becomes markedly deeper. Additionally, there are now large, supranuclear cellular inclusions and some vacuolation (Fig. 2d). These inclusions are discussed below and are generally indicative of a functional alimentary canal. Additional physiological energetic evidence (Boehlert & Yoklavich 1984b) indicates that the embryos ingest and assimilate ovarian fluid.

The larval fore-, mid-, and hindgut are histologically and functionally distinct (Richards & Dove 1971, Tanaka 1971, Theilacker 1978, Umeda & Ochiai 1973, O'Connell 1976, 1981, Vu 1976,

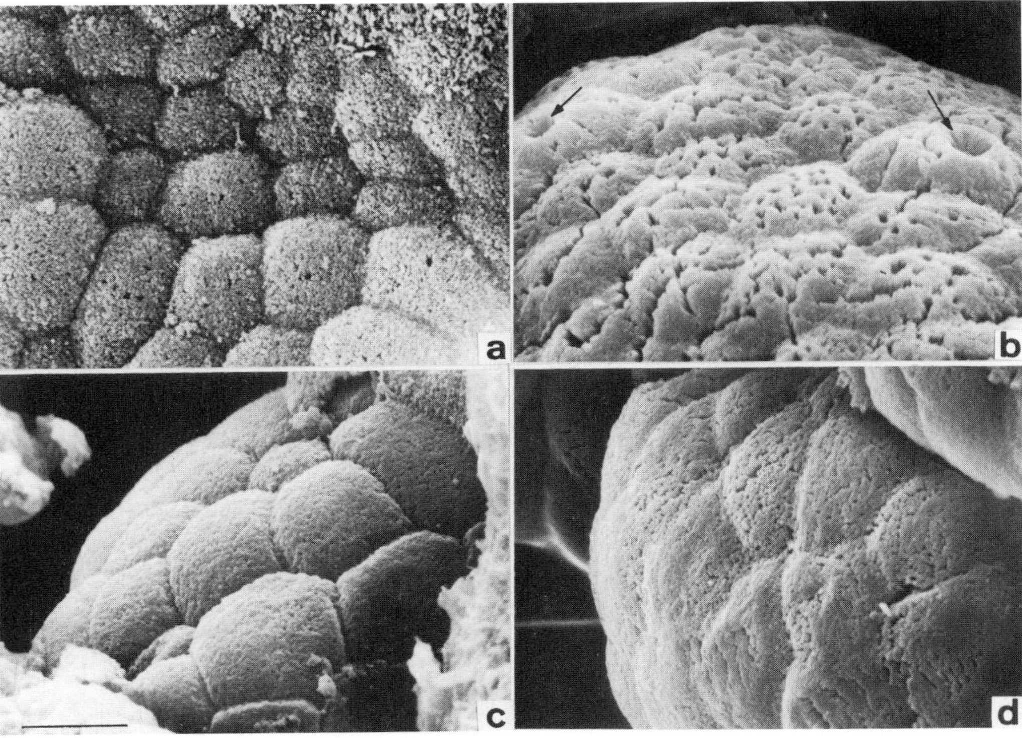

Fig. 3. Scanning electron micrographs of the apical epithelial surface of the mid- (*a*) and hindgut (*b*) of *Brevoortia patronus* larva and the mid- (*c*) and hindgut (*d*) of *Leiostomus xanthurus* larva. Note the striated border of microvilli and discharging mucous cells (arrows). (Scale bar is 5 μm; electromicrographs courtesy of J.T. Turner).

Govoni 1980, Watanabe 1981). A single layer of cuboidal epithelial cells interspersed with mucous cells lines the larval foregut. Mucous cells are more densely distributed in the anterior than in the posterior foregut of some species (Watanabe & Sawada 1985). A single layer of columnar epithelial cells with a well defined striated border of microvilli (Fig. 3) subtended by a terminal web (Fig. 4) lines the larval midgut (Iwai 1968a, 1968b, Iwai & Tanaka 1968a, 1968b, Tanaka 1971, Umeda & Ochiai 1973, Vu 1976, Stroband & Dąbrowski 1979, Stroband & Kroon 1981, Govoni 1980, O'Connell 1981, Watanabe 1982b). A single layer of columnar epithelia with microvilli (Fig. 3) and no subtending terminal web (Fig. 4) lines the larval hindgut. Mucous cells are seen in the hindgut of some species, for example gulf menhaden, *Brevoortia patronus,* while they are absent in others, for example spot, *Leiostomus xanthurus* (Fig. 3). Similarly the striated border of the hindgut in gulf menhaden is deeply furrowed, whereas it is without furrows in spot (Fig. 3).

Although not directly involved with the digestion and absorption of food, other structures, mainly the buccopharynx and the tunicae that envelop the alimentary canal, effect the processing and transport of food. At yolk absorption, the buccopharynx is lined with squamous epithelium along with scattered mucous cells and taste buds (Tanaka 1971, Govoni 1980). In relation to changes in the diets of fish larvae, taste buds become more numerous and functional as larvae grow (Twongo & MacCrimmon 1977, Appealbaum et al. 1983). Teeth develop in the areolar connective tissue underlying the buccopharyngeal epithelium, subsequently erupting during the larva period (Twongo & MacCrimmon 1977, Govoni 1980).

The dentition of fish larvae is often different

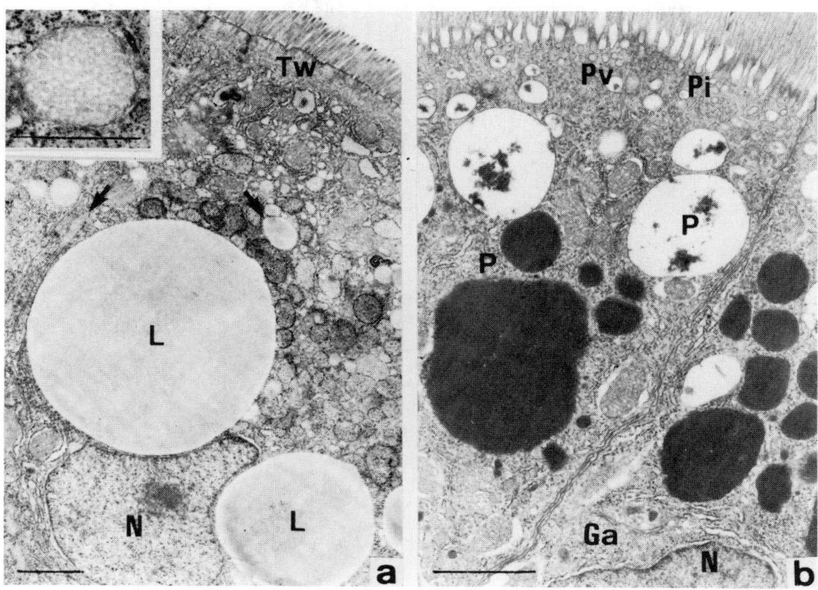

Fig. 4. Transmission electron micrographs of the mid- (*a*) and hindgut (*b*) epithelium of *Chaenogobius annularis* larva showing protein inclusion bodies (P), microvilli (MV), and terminal web (TW). Note the well defined terminal web, small, electron-lucent lipid particles (arrows), and large electron-lucent, supranuclear lipid droplets (L) in the midgut; inset is a magnification of a lipid particle. The lack of a terminal web, the presence of pinocytotic invaginations (PI) and vesicles (PV), and electron-opaque, supranuclear protein inclusion bodies are apparent in the hindgut. The lumenal, mucosal surface is in the upper right corner of each micrograph. Other abbreviations: GA, Golgi apparatus; N, nucleus. (Scale bars are 2 μm for *a* and *b*; 1 μm for inset; after Watanabe & Sawada 1985).

from that of adults, reflecting differences in feeding. For example, spot larva have retrorse conical teeth on the upper (premaxillary) and lower (dentary) jaws, but premaxillary teeth are lost in adults. Larval teeth are used for grasping rather than masticating food as prey are typically ingested whole.

Only three of the four enveloping tunicae typical of the adult fish alimentary canal (Kapoor et al. 1975, Reifel & Travill 1977, 1978, 1979) are present in larvae. A mucosa of absorptive epithelium, a muscularis consisting of a single layer of circular smooth muscle, and a serosa of fibrous connective tissue is present, but a muscularis mucosa or submucosa is absent until transformation. Longitudinally oriented smooth muscle in the muscularis is absent in larvae and does not develop until transformation. Whereas the longitudinal folds of the larval foregut compare well with the convolutions of the esophagus of adult fishes, the mid- and hindgut mucosa lack the distinct villi and crypt-like indentations of adults (Reifel & Travill 1979). Instead, the mucosa forms shallow rugae.

Digestive mechanisms

The function of the larval mid- and hindgut has received considerable attention inasmuch as their functions are in some ways analogous to the function of the anterior and posterior intestines of adult stomachless fishes. Moreover, there has been some controversy over the mechanisms of digestion and absorption in the mid- and hindgut (O'Connell 1976, Govoni 1980, O'Connell 1981). Cytological evidence suggests that the large supranuclear, vacuolar, electron-lucent structures of the midgut mucosal epithelial cells are the result of lipid absorption after lumenal hydrolysis to fatty acids and monoglycerides and intracellular resynthesis to lipids (Iwai 1968a, 1968b, 1969, Iwai & Tanaka 1968a, 1968b, Tanaka 1972a, 1972b, Umeda & Ochiai 1975, Stroband & Kroon 1981, Watanabe & Sawada 1985). In contrast, the acidophilic, granular, electron-opaque, supranuclear inclusion bodies of the hindgut epithelial cells are the result of pinocytotic absorption of macromolecules from the gut lumen. In addition, the terminal web underlying the striated border of the midgut shows no signs of interruption during digestion, whereas the apical plasma membrane of the hindgut shows numerous invaginations especially in recently fed larvae (Iwai 1969). Tanaka (1972b) showed that inclusion bodies become indistinct with the development of a stomach with gastric glands during transformation, but remain present in adult stomachless fishes. Govoni (1980), however, found similar supranuclear acidophilic inclusion bodies in juvenile spot that possess a functional stomach as well as in larvae that did not. Furthermore, O'Connell (1981) did not report lipid deposits in the midgut epithelial cells and found that the granular, acidophilic inclusion bodies of the hindgut epithelium stained heavily with OsO_4, a reagent that stains lipids black (O'Connell 1976).

Recently, Watanabe (1981, 1982a, 1984a), using HRP as a histochemical tracer of protein, observed pinocytotic absorption and intracellular digestion of macromolecular protein in the hindgut mucosal epithelium and revised Iwai's (1969) proposed mechanism of protein absorption. When larvae were injected with the HRP solution and their hindgut tissues incubated with H_2O_2, the epithelial cells showed HRP reaction products within inclusion bodies (Fig. 5). Control larvae showed no reaction products. It was clear that HRP was absorbed into the cells without losing enzymatic activity, i.e., without hydrolysis into peptides or amino acids.

Absorbed HRP molecules were digested in the epithelial cells through 5 successive stages: pinocytosis, transport, accumulation, digestion, and extinction (Fig. 6). Pinocytosis of HRP molecules occurred along the intermicrovillous plasma membrane. Within the cell, membrane-bounded, pinocytotic vesicles moved toward the nucleus. HRP was then accumulated in supranuclear inclusion bodies by the coalescence of vesicles. Lysosomes, presumably derived from Golgi, became associated with these inclusion bodies soon after their formation. HRP molecules in the supranuclear inclusion bodies finally lost their enzymatic activity as indicated by a waning of reaction products, probably as a result of lysosomal hydrolysis, and

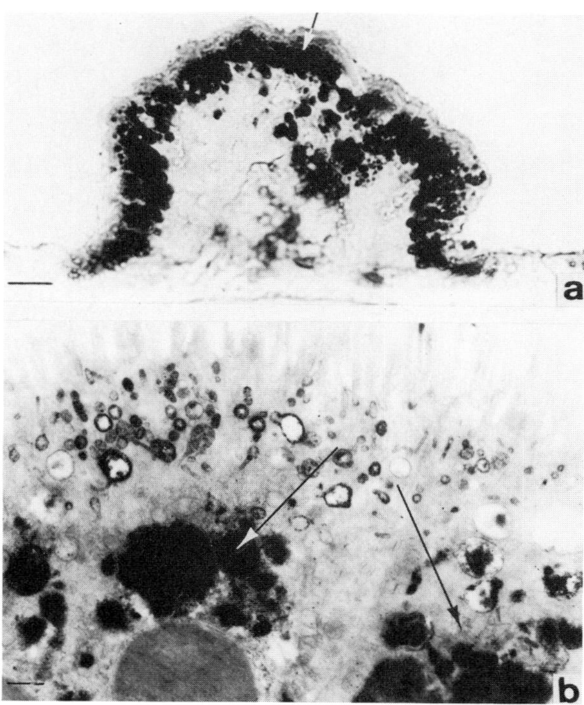

Fig. 5. Photomicrograph (*a*) and transmission electron micrograph (*b*) of the horseradish peroxidase (HRP) histochemistry of the hindgut epithelium of larval *Cottus nozawae*. Note granular and electron-opaque reaction products of HRP in pinocytotic vesicles and supranuclear inclusion bodies (arrows). (Scale bars are 30 μm for a; 1 μm for b).

eventually became extinct.

The time required for complete digestion of HRP differed with species from 10 to 24 h in pond smelt, *Hypomesus transpacificus nipponensis,* to 1 to 2 weeks in cherry salmon, *Oncorhynchus masou.* The intracellular digestion time also depends on the developmental interval of fish, e.g., it was shorter in larvae than in juveniles.

Most fish larvae lack a stomach with functional gastric glands until the completion of transformation, while the larvae of some taxa develop a functional stomach before yolk absorption and first oral feeding. Secretions of gastric glands, pepsin and HCl, effect preliminary protein digestion, thereby facilitating the complete hydrolysis of proteins to peptides and amino acids through the action of trypsin, chymotrypsin, and aminopeptidase in the mid- and hindgut. One might expect greater pi-

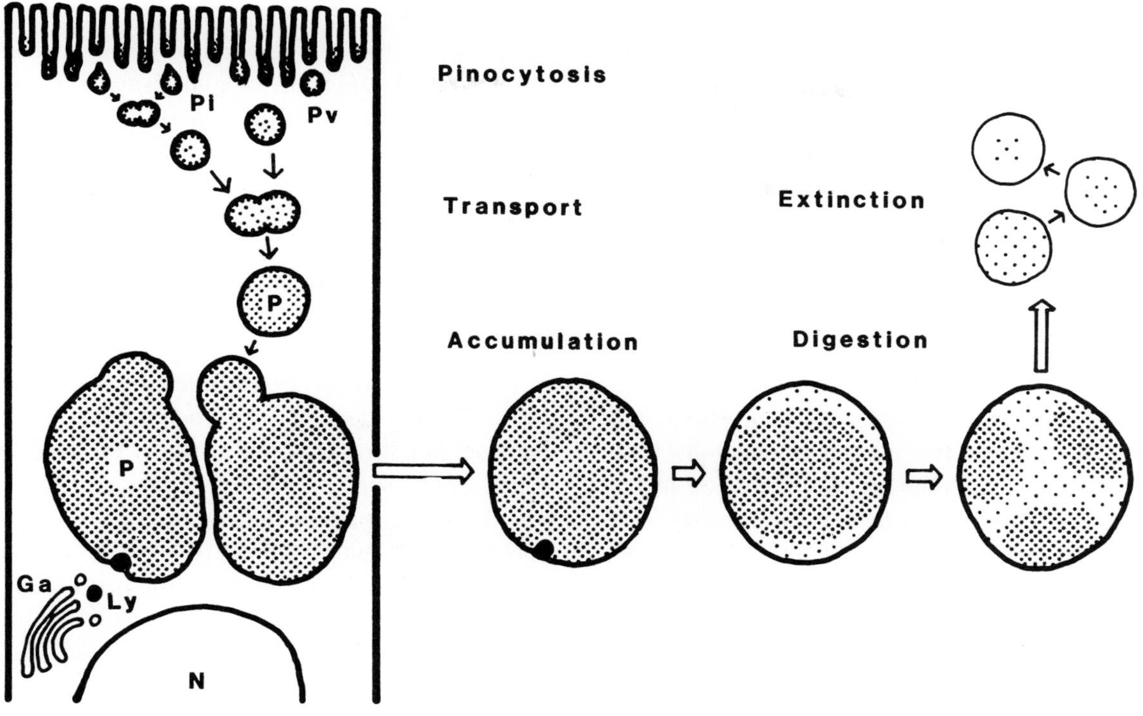

Fig. 6. Successive, 5 stages of protein absorption and intracellular digestion by the hindgut epithelial cells of fish larvae. Organelles include pinocytotic invaginations (Pi), pinocytotic vesicles (Pv), protein inclusion bodies (P), Golgi apparatus (Ga), lysosomes (Ly), and the nucleus (N). (After Watanabe 1984a).

nocytotic absorption of partially digested macromolecular proteins by the hindgut epithelial cells in fish larvae and adults that lack a stomach and preliminary protein digestion. Cherry salmon may serve as an example of a fish larvae with a functional stomach (Watanabe 1984c); pond smelt may serve as an example of a more typical fish larva without a stomach (Watanabe 1984b). Based on the absorption of HRP, the capacity to pinocytotically absorb proteins appeared before yolk absorption and first feeding in both species and persisted long after the development of gastric glands (Fig. 7). The amount of protein absorbed from food by pinocytosis, as indicated by the number of electron-opaque granules in epithelial inclusion bodies, was much greater in pond smelt larvae before gastric development than in cherry salmon larvae (Fig. 7). Pinocytotic absorption diminished in pond smelt after transformation. Pond smelt illustrated the typical pattern of fish larvae, the mechanism of protein digestion and absorption changes from pinocytosis and intracellular digestion to extracellular digestion and membrane transport with the development of gastric glands, whereas in cherry salmon, the digestive mechanism of larvae is similar to that of adults. Pinocytotic

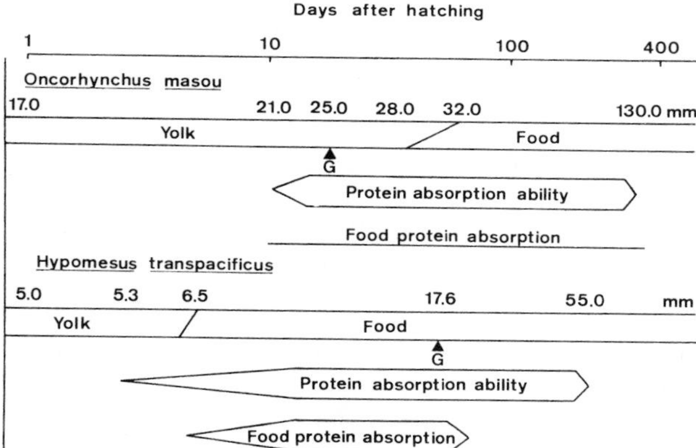

Fig. 7. Post-hatching changes in digestive mechanisms of *Oncorhynchus masou* and *Hypomesus transpacificus nipponensis*. (After Watanabe 1984d).

absorption and intracellular digestion of proteins lends a possible adaptive advantage. This mechanism of digestion may compensate for incomplete digestion by allowing for the assimilation of macromolecular proteins and is apparent in the undeveloped alimentary canals of some other chordates, including some fetal and infant mammals (see reviews in Yamamoto 1982, Gauthier & Landis 1972) as well as in larvae and adult stomachless fishes (Yamamoto 1966, Kapoor et al. 1975, Noaillac-Depeyre & Gas 1976, Weinberg 1976, Stroband 1977, Stroband & Van Der Veen 1981).

Lipids are apparently digested to fatty acids and monoglycerides in the midgut lumen, absorbed into the midgut epithelium, resynthesized in the agranular endoplasmic reticulum, and deposited in large lipid droplets (sensu Watanabe & Sawada 1985) of the mucosal epithelial cells (Iwai 1969), but the exact mechanism of lipid absorption is not known. Similarly, lipids are digested and absorbed in the anterior intestine of adult fishes (Barrington 1957, Kapoor et al. 1975, Fange & Grove 1979). Lipids observed in the midgut of fish larvae and the anterior intestine of adult stomachless fishes (Noaillac-Depeyre & Gas 1974, 1976, Stroband & Debets 1978) are thought to be temporary storage (Tanaka 1972a, Stroband & Dąbrowski 1979, Watanabe & Sawada 1985).

The morphological position and mechanism of carbohydrate absorption is unexplored in fish larvae. The high densities of mitochondria in the midgut epithelial cells, some of which are associated with lipid droplets (Iwai & Tanaka 1968a, Iwai 1968b, 1969), suggest that these cells are energetically active and capable of active transport. Whether effected by non-saturable influx or by

Table 1. Digestive enzyme activities reported for fish larvae. Positive signs (+) indicate presence of enzyme activities; negative signs (−) absence; NA indicates that the enzyme was not assayed.

Species	Age range	Pepsin	Trypsin	Chymo-trypsin	Amino pepti-dase	Lipase	Amylase	Maltase	Authors
Salmo gairdneri	first feeding to 110 d	NA	+	+	+	NA	NA	NA	Lauff & Hofer (1984)
S. gairdneri	first feeding to 66 d	+	+	NA	NA	NA	NA	NA	Dąbrowski (1982)
S. gairdneri	hatching to 60 d	+	+	NA	NA	NA	+	+	Kawai & Ikeda (1973a)
Coregonus hybrids	first feeding to 130 d	NA	+	+	+	NA	NA	NA	Lauff & Hofer (1984)
Coregonus pollan	first feeding to 42 d	+	+	NA	NA	NA	NA	NA	Dąbrowski (1982)
Plecoglossus altivelis	yolksac to transformation	+	+	NA	NA	NA	+	NA	Tanaka, Kawai & Yamamoto (1972)
Esox lucius	18 d	+	+	NA	NA	−	−	NA	Szlaminska (1980)
Cyprinus carpio	hatching to 125 d	+	+	NA	NA	NA	+	+	Kawai & Ikeda (1973b)
Gadus morhua	hatching to 30 d	NA	+	NA	NA	NA	NA	NA	Hjelmeland et al. (1983)
Acanthopagrus schlegellii	hatching to 35 d	+	+	NA	NA	NA	+	NA	Kawai & Ikeda (1973b)
Dicentrarchus labrax	hatching to 30 d	+	+	+	NA	NA	NA	NA	Alliot, Pastoureaud & Treller (1977)
D. labrax	2 to 60 d	+	+	+	NA	NA	NA	NA	Vu (1983)

sodium-dependent active transport (Crane 1975, Ferraris & Ahearn 1984), the mechanism of carbohydrate absorption is unknown.

Enzymology

Enzyme assays of the alimentary canal in fish larvae indicate that pepsin, trypsin, chymotrypsin, and amylase activities are apparent in several freshwater and marine fish larvae (Table 1). Maltase and aminopeptidase activities are present in the two species so far examined. Surprisingly, lipase activities have not been reported in the larval alimentary canal, although lipase has been assayed in only one species. The liver and the mucosal epithelium are likely the sources of lipase secretion (Kapoor et al. 1975), but zymogen granules, the histological evidence of secretory enzyme precursors, have not been reported in hepatocytes or mucosal epithelial cells of fish larvae. The pancreas is well developed at hatching in northern anchovy and spot, and its acinar cells are invested with conspicuous, acidophilic, zymogen granules, presumably the precursors of trypsin and chymotrypsin (O'Connell 1976, Govoni 1980). Aminopeptidase, another secretion of the mucosal epithelium has not been observed as zymogen. The origin of maltase is unknown in fishes (Kapoor et al. 1975). Pepsin, a secretion of the gastric mucosa, is not apparent (in those fish larvae that lack a stomach) until gastric glands in the developing stomach become functional during transformation (Tanaka et al. 1972, Kawai & Ikeda 1973b, Alliot et al. 1977, Vu 1983).

There is some indication that the activities of digestive enzymes are low at first feeding and increase during the larva period before transformation. Kawai & Ikeda (1973b) observed that activities of amylase, maltase, and trypsin increased in bulk assays of larvae of increasing age after yolk absorption. In contrast, Hjelmeland et al. (1983), using the more sensitive and specific radio-immunoassay technique, observed that after an initial increase betweeen hatching and yolk absorption, trypsin and trypsinogen activities decreased to low levels for 14 d, then increased.

There are two possible explanations for the possible increase in carbohydrolytic and proteolytic enzyme activities in fish larvae after yolk absorption and first feeding. First, the enzymes that are inherent in the food of fish larvae may increase as a result of increasing ration size. Second, enzyme production by the larval liver, pancreas, and mucosal epithelium may be stimulated in response to initial food consumption and increasing ration size. Lauff & Hofer (1984) found that exogenous trypsin activated by the high pH (9) of the hindgut of whitefish larvae, *Coregonus,* accounted for a high percentage of the total tryptic activity. Enzyme stimulation has not been observed, but the deterioration of pancreatic zymogen granules (O'Connell 1976, Theilacker 1978) and of proteolytic enzyme activities in the alimentary canal (Dąbrowski 1982, Hjelmeland et al. 1983) of starved fish larvae implies that enzyme production is variable.

Histochemistry

Beyond the histochemical application of HRP to define the mechanism of protein digestion and absorption, few studies have examined the histochemistry of the developing alimentary canal. Prakash (1961) found increasing activities of alkaline phosphatase in the striated border of the hindgut of rainbow trout, *Salmo gairdneri,* larvae with the greatest changes in the intensity of activity during transformation. Ikeda (1959) related the activities of alkaline phosphatase to the development of a functional stomach in Japanese killifish, *Oryzias latipes.* Inasmuch as alkaline phosphatase activity is associated with sodium-mediated active transport in absorptive tissues (Ugolev 1965), changes in the intensity of alkaline phosphatase reactions may indicate increased absorption through active transport.

Peristalsis and digestion rates

Constriction of the single layer of smooth circular muscle in the muscularis that progresses posteriad in a wave-like manner effects the transport of food toward the anus of fish larvae. Peristalsis is apparent in young Atlantic menhaden (*B. tyrannus*), spot, and rockfish embryos. Blaxter (1969) and

Rosenthal & Hempel (1970) reported that *Artemia* nauplii move rapidly and undigested through the midgut to the hindgut of Atlantic herring, *Clupea harengus harengus* larvae. Pedersen (1984), in contrast, observed that ingested copepods passed through the foregut of herring within seconds, passed through the anterior midgut within minutes, and stopped in the posterior midgut at the ileocaecal valve. Copepods remained in the posterior midgut and hindgut for several hours. Constrictions of the ileocaecal valve also check the transport of food from the mid- to the hindgut in Atlantic menhaden and spot larvae.

Fish larvae lack the layer of longitudinally aligned, smooth muscle in the muscularis that is characteristic of adults, but some have ancillary means of transporting food. Iwai (1964, 1967a, 1967b), Iwai & Rosenthal (1981), and Watanabe (1984b) have observed ciliated cells and cilia movement that aid in the transport of food in the alimentary canals of some plecoglossid, osmerid, clupeid, and salangid larvae. Ciliated cells have also been observed in some adult fishes with poorly developed musculari (Ferraris & Ahearn 1984).

The passage of food through the larval alimentary canal (Table 2) occurs at rates that are somewhat faster than rates observed in adult fishes (Kapoor et al. 1975, Fange & Grove 1979). Extremely

Table 2. Evacuation rates and times of fish larvae as determined in the laboratory.

Species	Age or length	Feeding protocol	Food	Temperature (° C)	Evacuation rate	Evacuation time	
Clupea harengus	12 mm		*Artemia salina*			19	Kurata (1959)
C. harengus	10 to 14 mm	Continuous feeding	*A. salina*			4 to 10	Rosenthal & Hempel (1970)
C. harengus	8 to 12 wk		*A. salina*		−0.09 to −0.84	3 to 7+	Werner & Blaxter (1980)
C. harengus	8–22 d	Single feeding	Copepod nauplii and polychaete larvae	6		12.5 to 22.5 h	Fossum (1983)
C. harengus	22–52 d	Continuous	zooplankton	9.5		40 min to 3 h	Pedersen (1984)
Coregonus clupeaformis	19–20 mm	Single feeding	Copepod nauplii	14		16	Hoagman (1974)
Cyprinus carpio						1 to 8	Chiba (1961)
Abramis brama	12–18 mm	Single feeding	*Bosmina longirostris*			5	Sorokin & Panov (1966)
Micropterus salmoides	2–8 d	Continuous feeding	zooplankton	17 to 23		2.0 to 2.8	Laurence (1971)
M. salmoides	2–8 d	Single feeding	zooplankton	17 to 23		3.8 to 5.2	Laurence (1971)
Lagodon rhomboides	15 to 18 mm	Single feeding	copepods	12	0.8		Kjelson & Johnson (1976)
Leiostomus xanthurus	16 to 20 mm	Single feeding	copepods	17	0.24		Kjelson & Johnson (1976)
L. xanthurus	7 to 47 d 1.8 to 9.4 mm	Single feeding	*Brachionus plicatilis*	20		5 h	Govoni et al. (1982)
Scomber japonicus	3–5 d	Single feeding	*B. plicatilis*	19	0.5 to 2		Hunter & Kimbrell (1980)
Pseudopleuronectes americanus		Continuous feeding	zooplankton	8		5.1 to 8.4	Laurence (1977)

rapid rates have been observed with some clupeoid larvae. In one report, bay anchovy, *Anchoa mitchilli,* defecated within minutes after eating (Chitty 1981). In Atlantic menhaden and spot larvae, however, defecation begins 1 to 2 h after eating. Larval age does not markedly affect the evacuation rate of herring larvae (Pedersen 1984). Evacuation rates are faster for larvae fed continuously than for larvae fed a single ration (Blaxter 1965, Laurence 1971) and are positively related to ration size (Werner & Blaxter 1980). Evacuation rates are also positively related to temperature (Laurence 1971). In any case, digestion and assimilation in fish larvae are rapid. By visual inspection, Fossum (1983) observed the digestion of copepod nauplii and polychaete larvae within 1.5 h following their ingestion by Atlantic herring larvae. Govoni et al. (1982) observed that larvae respired $^{14}CO_2$ within 3 h following the ingestion of ^{14}C-labeled food, which indicates that food is digested, assimil-

Table 3. Assimilation efficiencies and the coefficient of utilization of fish larvae.

Species	Age, length or weight	Food	Food concentration	Method of determination	Assimilation efficiency (%)	Coefficient of utilization (%)	Author
Anchoa mitchilli	10 to 100 μg dry weight	Copepod nauplii	1000 l^{-1}	gravimetric		24 to 41	Houde & Schekter (1983)
Clupea harengus	6 wk	*Artemia salina*	10^{-6} to 10^{-3} l^{-1}	radiotracer (^{14}C)		38 to 68	Boehlert & Yoklavich (1984a)
Cyprinus carpio	1–5 mg dry weight	zooplankton	10 to 100 mg l^{-1}	calorimetric	74 to 87		Filatov (1972)
Abramis brama	12 to 18 mm	*Bosmina longirostris*	20 to 2000 l^{-1}	radiotracer (^{14}C)	74 to 83		Sorokin & Panov (1966)
Micropterus salmoides		zooplankton	200 to 1200 l^{-1}	calorimetric		80	Laurence (1971)
Archosargus rhomboidalis	10 to 100 μg dry weight	copepod nauplii	500 l^{-1}	gravimetric	44 to 75		Houde & Schekter (1983)
Leiostomus xanthurus	7 to 47 d 1.8 to 9.4 mm 0.022 to 2300 μg dry weight	*Brachionus plicatilis*	10^5 to 10^6 l^{-1}	radiotracer (^{14}C)	67 to 99	9 to 65	Govoni et al. (1982)
Paralichthys dentatus	1 to 56 d	zooplankton	unlimited	nitrogen assay		27 to 68	Buckley & Dillman (1982)
Limanda limanda	0 to 32 d	oligochaetes unlimited calorimetric	unlimited		99		Calculated from Edwards et al. (1969)
Pleuronectes platessa	0 to 45 d	ologochaetes unlimited calorimetric	unlimited		96 to 99		Calculated from Edwards et al. (1969)
Pseudo-pleuronectes americanus	16 to 668 μg dry weight	copepod nauplii	2000 to 3000 l^{-1}	calorimetric		69 to 74	Laurence (1977)
P. americanus	16 to 23 d	zooplankton	>2000 l^{-1}	nitrogen assay		20	Cetta & Capuzzo (1982)
Achirus faciatus	10 to 100 μg dry weight	copepod	1000 l^{-1}	gravimetric		34 to 54	Houde & Scheckter (1983)

ated, and metabolized within this short period.

Assimilation

Mass and energy budgets (physiological energetics) as well as radiotracer experiments indicate that the assimilation efficiency of fish larvae (Table 3) is near or somewhat below the assimilation efficiency of adult fishes (Kapoor et al. 1975, Conover 1978, Brett & Groves 1979). Assimilation efficiency (the percentage of a food ration that is assimilated after loss to feces (Brett & Groves 1979)) ranges from 67 to 99% for fish larvae, whereas assimilation efficiency ranges from 80 to 90% for adults. Coefficients of utilization (the fraction of matter or energy retained after losses to defecation of feces as well as metabolic excretion of urine (Winberg 1956)) for fish larvae are also generally lower than they are for adults. The coefficients of utilization ranges from 9 to 80% for larvae, whereas coefficients average about 70% for adults (Ware 1975).

The effects of food type and ration size on assimilation efficiencies are poorly known for fish larvae. In stomachless pipefish larvae, *Syngnathus fuscus*, Ryer & Boehlert (1983) noted slower evacuation with smaller ration and suggested that assimilation efficiency would increase at low ration when the residence time of individual food particles increased. In Atlantic herring larvae, Werner & Blaxter (1980) observed that high rations resulted in food that was apparently less digested when defecated; indeed, at the highest ration, *Artemia* nauplii passed rapidly through the alimentary canal and were defecated alive. This observation coupled with the extremely rapid evacuation times in bay anchovy larvae (Chitty 1981) implies that large rations may result in low assimilation efficiencies, especially for clupeoid larvae with straight alimentary canals. This has been experimentally confirmed for Pacific herring larvae, *C. harengus pallasi*; the uptake and retention of ^{11}C-labeled food decreased significantly with increasing food concentration and, therefore, ration (Boehlert & Yoklavich 1984a). Herring larvae have straight guts. It would be of interest to repeat these experiments to determine if percoid larvae with coiled guts show the same change in assimilation efficiency, inasmuch as the feeding strategies of species with differing gut morphologies may differ, as demonstrated by the work of Houde & Schekter (1980, 1983). The clupeoids are adapted to utilize contagiously distributed prey in high concentrations; although assimilation efficiency and gross growth efficiency may decrease, the total energy intake increases (Fig. 8), resulting in a net energetic gain to the larva. Similarly, comparing a wide variety of fish larvae, Checkley (1984) noted that gross growth efficiency showed a peak at moderate food densities and decreased thereafter, a pattern similar to adult fishes (Brett & Groves 1979).

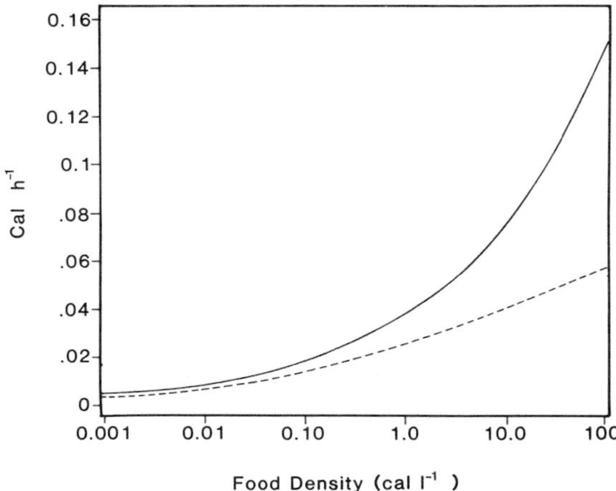

Fig. 8. The relationship of calories consumed (solid line) and calories assimilated (dashed line) as a function of food density for 6 wk-old *Clupea harengus pallasi* larva feeding on *Artemia* nauplii. Ingestion rate increases with food density for this species (Werner & Blaxter 1981). Although assimilation efficiency decreases with increasing ingestion rate, the magnitude of decrease with increasing ingestion rate is more than compensated for by increased ingestion such that the total energy assimilation continues to increase. (From Boehlert & Yoklavich 1984a).

Some work has indicated that the low assimilation efficiencies of some fish larvae improve during the larva period, while other research has failed to demonstrate improved efficiencies until transformation with its attendant elaboration of alimentary canal morphology. By subtracting the amount of energy in growth and metabolism from the energy in food, Laurence (1977) reported increasing assimilation efficiencies during the larva period before transformation. In contrast, Sorokin & Panov (1966) and Govoni et al. (1982), by using ^{14}C as a tracer, and Houde & Schekter (1983), by subtracting the energy of growth and metabolism from the energy of food, reported no change in assimilation with development before transformation in several other species. Buckley & Dillmann (1982) measured the nitrogen excreted as ammonia in urine and defecated as primary amines in feces by summer flounder larvae, *Paralichthys dentatus*. They reported that the rate of defecation of primary amine, representing the nonassimilated fraction of ingested nitrogen, decreased with age, while the rate of ammonia, representing the assimilated and metabolized fraction, decreased until 35 h after hatching, then increased. More importantly, the coefficient of nitrogen utilization increased with age during the larva period. Cetta & Capuzzo (1982) found a similar pattern of nitrogen loss in the larvae of another flatfish, *Pseudopleuronectes americanus*. Pinocytosis and intracellular digestion of protein macromolecules is possibly a less efficient mechanism of protein digestion and assimilation than hydrolysis to amino acids and active transport across the lumenal plasma membrane (Gardner 1984). While this might account for the overall lower assimilation efficiencies of the fish larvae, what accounts for changes in assimilation during the larva period before transformation? Some of proteolytic enzymes indicate that enzyme activities increase during the larva period before transformation, and Buckley & Dillmann (1982) invoked these increases to explain increasing assimilation. This explanation is plausible, but there is no corresponding cytological evidence of an increase in the capacity of the pancreas to produce these enzymes; the pancreas has conspicuous zymogen granules at yolk absortion and these remain obvious throughout development. This lack of apparent morphological or cytological change that could cause a shift in the efficiency of digestion and assimiltion is puzzling.

Improvement of assimilation efficiency is likely with the development of a functional stomach with gastric glands and pyloric caeca at transformation due the digestive contribution of pepsin and hydrochloric acid and the increased absorptive surface area of the alimentary canal. Mironova (1974) reported increasing assimilation efficiency of *Oreochromis mosambicus* that was concomitant with transformation.

Recommendations for future research

We know that the alimentary canal of fish larvae is less complex than that of adults and that digestion and assimilation in larvae may be less efficient as a result. The activities of proteolytic enzymes are apparently lower in larvae, but may increase as larvae grow before transformation. In some flatfish larvae, nitrogen assimilation improves during the larva period, but measures of calorie and carbon assimilation with larvae other than flatfish have indicated no improved assimilation before transformation. For most fish larvae there is no apparent elaboration of alimentary canal tissues after yolk absorption and before transformation; alimentary canal development is not well studied among flatfish.

Research on the physiology of digestion in fish larvae has faced several technical difficulties. The small size of larvae, their mostly planktonic habitat, and the unavailability of natural prey sources to experimenters, among other problems, have all contributed variability to the range of results reported herein. Consequently, several questions remain to be answered. Inasmuch as it has bearing on models of larval growth and survival, the question of changing digestive and assmilative abilities with larval development before transformation warrants the most immediate attention. Several avenues of research to address this issue are possible.

The length as well as the absorptive surface area of the alimentary canal should be measured and related to larval length. The morphometric methods of Hughes (1984) could be used for this purpose. Such comparisons would indicate disproportionate increases in the absorptive surface area with development and might offer an explanation for improving assimilation.

A histochemical examination of developmental stages of various species should also contribute to the resolution of changing assimilation. Such an examination should focus on lipase and esterase because histochemical evidence of these enzymes would indicate the capability of larvae to digest and absorb lipids. The histochemistry of alkaline phosphatase, because of its association with sodium-mediated active transport, and acid phosphatase, because of its association with supranuclear inclusion bodies and pinocytotic absorption (Ugolev 1965, Watanabe 1984a), should also be considered. We recommend the histochemical methods of Higuchi et al. (1979) with the intensity of histochemical reactions quantified with image-analysis densitometry.

The question of enzyme stimulation should be resolved by conducting carefully controlled experiments wherein developmental stages of different species are starved and then fed for various intervals, and the activities of digestive enzymes measured. Because of its sensitivity in distinguishing specific antibody and antigen reactions and its capacity to detect inhibited as well as active trypsin, we recommend radio-immunoassay (Hjelmeland 1983, Hjelmeland et al. 1983) for measurement of trypsin activities. Perhaps radio-immunoassays could be developed for other digestive enzymes.

A comparative approach to these recommendations would be clearly advantageous, for morphological and physiological differences among taxa are great. Studies on digestive physiology elucidate the feeding strategies and growth potential of fish larvae and thereby contribute to our understanding of fish larvaė survival.

Acknowledgments

We acknowledge with gratitude L.J. Buckley (National Marine Fisheries Service, Narragansett, Rhode Island), P.A. Tester (National Marine Fisheries Service, Beaufort, North Carolina), and Eugene K. Balon for their critiques and exchange of information and ideas.

References cited

Alliot, E., A. Pastoureaud & J. Trellur. 1977. Evolution des activitiés enzymatiques dans le tube digestif au cours de la vie larvaire du bar (*Dicentrarchus labrax*). Variations des protéinogrammes et des zymogrammes. Colloq. Int. Cent. Nat. Rech. Sci 4: 85–91.

Appealbaum, S., J.W. Adron, S.G. George, A.M. Mackie & B.J.S. Pirie. 1983. On the development of the olfactory and the gustatory organs of the dover sole, *Solea solea*. J. Mar. Biol. Assoc. U.K. 63: 97–108.

Balon, E.K. 1979. The juvenilization process in phylogeny and the altricial to precocial forms in the ontogeny of fishes. Env. Biol. Fish. 4: 193–198.

Balon, E.K. 1981. Saltatory processes and altricial to precocial forms in the ontogeny of fishes. Amer. Zool. 21: 573–596.

Barrington, E.J.W. 1957. The alimentary canal and digestion. pp. 109–161. *In:* M.E. Brown (ed.) The Physiology of Fishes, Vol. 1, Academic Press, New York.

Blaxter, J.H.S. 1965. The feeding of herring larvae and their ecology in relation to feeding. Calif. Coop. Oceanic. Fish. Invest. Rep. 10: 79–88.

Blaxter, J.H.S. 1969. Development of eggs and larvae. pp. 178–252. *In:* W.S. Hoar & D.J. Randall (ed.) Fish Physiology, Vol. 3, Academic Press, New York.

Blaxter, J.H.S. & G. Hempel. 1961. Biologische Beobachtungen bei der Aufzucht von Herringsbrut. Helgol. Wiss. Meeresunters. 7: 260–283.

Boehlert, G.W. 1984. Scanning electron microscopy. pp. 43–48. *In:* H.G. Moser, W.J. Richards, D.M. Cohen, M.P. Fahay, A.W. Kendall & S.L. Richardson (ed.) Ontogeny and Systematics of Fishes, Amer. Soc. Ichthyol. Herpetol., Spec. Publ. 1, Lawrence.

Boehlert, G.W. & M.M. Yoklavich. 1984a. Carbon assimilation as a function of ingestion rate in larval Pacific herring, *Clupea harengus pallasi* Valenciennes. J. Exp. Mar. Biol. Ecol. 79:251–262.

Boehlert, G.W. & M.M. Yoklavich. 1984b. Reproduction, embryonic energetics, and the maternal-fetal relationship in the viviparous genus *Sebastes*. Biol. Bull. (Woods Hole) 167: 354–370.

Brett, J.R. & T.D.D. Groves. 1979. Physiological energetics. pp. 279–352. *In:* W.S. Hoar, D.J. Randall & J.R. Brett (ed.) Fish Physiology, Vol. 8, Academic Press, New York.

Buckley, L.J. & D.W. Dillmann. 1982. Nitrogen utilization of larval summer flounder, *Paralichthys dentatus* (Linnaeus). J. Exp. Mar. Biol. Ecol. 59: 243–256.

Cetta, C.M. & J.M. Capuzzo.1982. Physiological and biochemical aspects of embryonic and larval development of the winter flounder, *Pseudopleuronectes americanus*. Mar. Biol. 71: 327–337.

Checkley, D.M. 1984. Relation of growth to ingestion for larvae of Atlantic herring *Clupea harengus* and other fish. Mar. Ecol. Prog. Ser. 18: 215–224.

Chiba, K. 1961. The basic study on the production of fish seedling under possible control. I. The effect of food in quality and quantity on the survival and growth of the common carp fry. Bull. Freshwater Fish. Res. Lab. (Tokyo) 11: 105–132.

Chitty, N. 1981. Behavioral observations of feeding larvae of bay anchovy, *Anchoa mitchilli*, and bigeye anchovy, *Anchoa lamprotaenia*. Rapp. P.-V. Reun. Cons. Int. Explor. Mer 178: 320–321.

Conover, R.J. 1978. Transformation of organic matter. pp. 221–

500. *In:* O. Kinne (ed.) Marine Ecology, Vol. 4, Wiley, New York.

Conover, R.J. & V. Francis. 1973. The use of radioactive isotopes to measure the transfer of materials in aquatic food chains. Mar. Biol. 10: 272–283.

Crane, R.K. 1975. A digestive-absorptive surface as illustrated by the intestinal cell brush border. Trans. Amer. Microsc. Soc. 94: 529–544.

Dąbrowski, K. 1982. Proteolytic enzyme activity decline in starving fish alevins and larvae. Env. Biol. Fish. 7: 73–76.

Devillers, C. 1961. Structural and dynamic aspects of the development of the teleostean egg. Adv. Morphol. 1: 379–428.

Edwards, R.R.C., D.M. Finlayson & J.H. Steele. 1969. The ecology of O-group plaice and common dabs in Lock Ewe. II. Experimental studies of metabolism. J. Exp. Mar. Biol. Ecol. 3: 1–17.

Engen, P.C. 1968. Organogenesis in the walleye surfperch, *Hyperprosopon argenteum* (Gibbons). Calif. Fish Game 54: 156–169.

Fange, R. & D. Grove. 1979. Digestion. pp. 161–260. *In:* W.S. Hoar, D.J. Randall & J.R. Brett (ed.) Fish Physiology, Vol. 8, Academic Press, New York.

Ferraris, R.P. & G.A. Ahearn. 1984. Sugar and amino acid transport in fish intestine. Comp. Biochem. Physiol. A. Comp. Physiol. 77: 397–413.

Filatov, V.I. 1972. Effectiveness of the utilization of natural foods by carp [(*Cyprinus carpio* (L)] larvae. J. Ichthyol. 12: 812–818.

Fossum, P. 1983. Digestion rate of food particles in the gut of larval herring. Fiskeridir Skr. Ser. Havunders. 17: 347–357.

Fukusho, K. 1972. Organogenesis of digestive system in the mullet, *Liza haematocheila,* with special reference to gizzard. Jpn. J. Ichthyol. 19: 283–294 (In Japanese).

Gardner, M.L.G. 1984. Intestinal assimilation of intact peptides and proteins from the diet – a neglected field? Biol. Rev. 59: 289–331.

Gauthier, G.F. & S.C. Landis. 1972. The relationship of ultrastructural and cytochemical features to absorptive activity in the goldfish intestine. Anat. Rec. 172: 675–702.

Govoni, J.J. 1980. Morphological, histological, and functional aspects of alimentary canal and associated organ development in larval *Leiostomus xanthurus.* Rev. Can. Biol. 39: 69–80.

Govoni, J.J. 1984. Histology. pp. 40–42. *In*: H.G. Moser, W.J. Richards, D.M. Cohen, M.P. Fahay, A.W. Kendall & S.L. Richardson (ed.) Ontogeny and Systematics of Fishes, Amer. Soc. Ichthyol. Herpetol. Spec. Publ. 1, Lawrence.

Govoni, J.J., D.S. Peters & J.V. Merriner. 1982. Carbon assimilation during the larval development of the marine teleost *Leiostomus xanthurus* Lacépède. J. Exp. Mar. Biol. Ecol. 64: 287–299.

Higuchi, S., M. Suga, A.M. Wannenberg & B.H. Schofield. 1979. Histochemical demonstration of enzyme activities in plastic and paraffin embedded tissue sections. Stain Technol. 54: 5–12.

Hjelmeland, K. 1983. Proteinase inhibitors in the muscle and serum of cod (*Gadus morhua*). Isolation and characterization. Comp. Biochem. Physiol. B. Comp. Biochem. 76: 365–372.

Hjelmeland, K., I. Huse, T. Jorgensen, G. Molvik & J. Raa. 1983. Trypsin and trypsinogen as indices of growth and survival potential of cod (*Gadus morhua* L.) larvae. Flodevigen Rapp. 3: 1–17.

Hoagman, W.J. 1974. Vital activity parameters as related to the early life history of larval and post-larval lake whitefish (*Coregonus clupeaformis*). pp. 547–558. *In:* J.H.S. Blaxter (ed.) The Early Life History of Fish, Springer-Verlag, Berlin.

Houde, E.D. & R.C. Schekter. 1980. Feeding by marine fish larvae: developmental and functional responses. Env. Biol. Fish. 5: 315–334.

Houde, E.D. & R.C. Schekter. 1983. Oxygen uptake and comparative energetics among eggs and larvae of three subtropical marine fishes. Mar. Biol. 72: 283–293.

Hughes, G.M. 1984. Measurement of gill area in fishes: practices and problems. J. Mar. Biol. Assoc. U.K. 64: 637–655.

Hulet, W.H. 1978. Structure and functional development of the eel leptocephalus *Ariosoma balearicum* (De La Roche, 1809). Philos. Trans. R. Soc. Lond. Biol. Sci 282: 107–138.

Hunter, J.R. 1977. Behavior and survival of northern anchovy *Engraulis mordax* larvae. Calif. Coop. Oceanic Fish. Invest. Rep. 19: 138–146.

Hunter, J.R. 1981. Feeding ecology and predation of marine fish larvae. pp. 34–77. *In:* R. Lasker (ed.) Marine Fish Larvae: Morphology, Ecology, and Relation to Fisheries, Washington Sea Grant Program, Seattle.

Hunter, J.R. & C.A. Kimbrell. 1980. Early life history of Pacific mackerel, *Scomber japonicus.* U.S. Fish. Bull. 78: 89–101.

Ikeda, A. 1959. Embryological and histochemical studies on the development of the digestive system in a teleost fish, *Oryzias latipes.* Hiroshima J. Med. Sci. 8: 71–88.

Iwai, T. 1964. Feeding and ciliary conveying mechanisms in larvae of salmonid fish *Plecoglossus altivelis* Temminck et Schlegel. Physiol. Ecol. Jpn. 12: 38–44.

Iwai, T. 1967a. The comparative study of the digestive tract of teleost larvae – I. Fine structure of the gut epithelium in larvae of Ayu. Bull. Jpn. Soc. Sci. Fish. 33: 489–496.

Iwai, T. 1967b. The comparative study of the digestive tract of teleost larvae – II. Ciliated cells of the gut epithelium of pond smelt larvae. Bull. Jpn. Soc. Sci. Fish. 33: 1116–1119.

Iwai, T. 1968a. The comparative study of the digestive tract of teleost larvae – V. Fat absorption in the gut epithelium of goldfish larvae. Bull. Jpn. Soc. Sci. Fish. 34: 973–978.

Iwai, T. 1968b. Fine structure and absorption patterns of intestinal epithelial cells in rainbow trout alevins. Z. Zellforsch. Mikrosk. Anat. 91: 366–379.

Iwai, T. 1969. Fine structure of gut epithelial cells of larval and juvenile carp during absorption of fat and protein. Arch. Histol. Jpn. 30: 183–199.

Iwai, T. & M. Tanaka. 1968a. The comparative study of the digestive tract of teleost larvae – III. Epithelial cells in the posterior gut of halfbeak larvae. Bull. Jpn. Soc. Sci. Fish. 34: 44–48.

Iwai, T. & M. Tanaka. 1968b. The comparative study of the digestive tract of teleost larvae- IV. Absorption of fat by the gut of halfbeak larvae. Bull. Jpn. Soc. Sci. Fish. 34: 871–875.

Iwai, T. & H. Rosenthal. 1981. Ciliary movements in guts of early clupeoid and salangid larvae. Mar. Ecol. Prog. Ser 4: 365–367.

Johannes, R.E. & M. Satomi. 1967. Measuring organic matter retained by aquatic invertebrates. J. Fish. Res. Board Can. 24: 2467–2471.

Kapoor, B.G., H. Smit & I.A. Verighina. 1975. The alimentary canal and digestion in teleosts. Adv. Mar. Biol. 13: 109–239.

Kawai, S. & S. Ikeda. 1973a. Studies on digestive enzymes of fishes – III. Development of the digestive enzymes of rainbow trout after hatching and the effect of dietary change on the activities of digestive enzymes in the juvenile stage. Bull. Jpn. Soc. Sci. Fish. 39: 819–823.

Kawai, S. & S. Ikeda. 1973b. Studies on digestive enzymes of fishes – IV. Development of the digestive enzymes of carp and black sea bream after hatching. Bull. Jpn. Soc. Sci. Fish. 39: 877–881.

Kendall, A.W., E.H. Ahlstrom & H.G. Moser. 1984. Early life history stages of fishes and their characters. pp. 11–24. *In:* H.G. Moser, W.J. Richards, D.M. Cohen, M.P. Fahay, A.W. Kendall & S.L. Richardson (ed.). Ontogeny and Systematics of Fishes, Amer. Soc. Ichthyol. Herpetol. Spec. Publ. 1, Lawrence.

Kjelson, M.A. & G.M. Johnson. 1976. Further observations on the feeding ecology of postlarval pinfish, *Lagodon rhomboides,* and spot, *Leiostomus xanthurus.* U.S. Fish. Bull. 74: 423–432.

Kostomarova, A.A. 1962. Effect of starvation on the development of the larvae of bony fishes. Tr. Inst. Morfol. Zhivotn. Severtsova Akad. Nauk. SSSR 40: 4–77 (In Russian).

Kurata, H. 1959. Preliminary report on the rearing of the herring larvae. Bull. Hokkaido Reg. Fish. Res. Lab. 20: 117–138. (In Japanese).

Lauff, M. & R. Hofer. 1984. Proteolytic enzymes in fish development and the importance of dietary enzymes. Aquaculture 37: 335–346.

Laurence, G.C. 1971. Digestion rate of larval largemouth bass. N.Y. Fish. Game J. 18: 52–56.

Laurence, G.C. 1977. A bioenergetic model for the analysis of feeding and survival potential of winter flounder, *Pseudopleuronectes americanus,* larvae during the period from hatching to metamorphosis. U.S. Fish. Bull 75: 529–546.

Mironova, N.V. 1974. The energy balance of *Tilapia mossambica.* J. Ichthyol. 14: 431–438.

Moser, H.G. 1981. Morphological and functional aspects of marine fish larvae. pp. 90–131. *In:* R. Lasker (ed.) Marine Fish Larvae: Morphology, Ecology, and Relation to Fisheries, Washington Sea Grant Program, Seattle.

Noaillac-Depeyre, J. & N. Gas. 1974. Fat absorption by enterocytes of the carp. Cell Tissue Res. 155: 353–365.

Noaillac-Depeyre, J. & N. Gas. 1976. Electron microscopic study on gut epithelium of the tench (*Tinca tinca* L.) with respect to its absorptive functions. Tissue Cell 8: 511–530.

O'Connell, C.P. 1976. Histological criteria for diagnosing the starving condition in early post yolk sac larvae of the northern anchovy, *Engraulis mordax* Girard. J. Exp. Mar. Biol. Ecol. 25: 285–312.

O'Connell, C.P. 1981. Development of organ systems in the northern anchovy, *Engraulis mordax,* and other teleosts. Amer. Zool. 21: 429–446.

Oppenheimer, J.M. 1937. The normal stages of *Fundulus heteroclitus.* Anat. Rec. 68: 1–15.

Pedersen, B.H. 1984. The intestinal evacuation rates of larval herring (*Clupea harengus* L.) predating on wild plankton. Dana Rep. 3: 321–330.

Prakash, A. 1961. Distribution and differentiation of alkaline phosphatase in the gastro-intestinal tract of steelhead trout. J. Exp. Zool. 146: 237–251.

Reifel, C.W. & A.A. Travill. 1977. Structure and carbohydrate histochemistry of the esophagus in ten teleostean species. J. Morphol. 152: 303–313.

Reifel, C.W. & A.A. Travill. 1978. Structure and carbohydrate histochemistry of the stomach in eight species of teleosts. J. Morphol. 158: 155–167.

Reifel, R.W. & A.A. Travill. 1979. Structure and carbohydrate histochemistry of the intestine in ten teleostean species. J. Morphol. 162: 343–359.

Richards, W.J. & G.R. Dove. 1971. Internal development of young tunas of the genera *Katsuwonus, Euthynnus, Auxis,* and *Thunnus* (Pisces, Scombridae). Copeia 1971: 72–78.

Roberts, R.J., M. Bell & H. Young. 1973. Studies on the skin of plaice (*Pleuronectes platessa* L.). II. The development of larval plaice skin. J. Fish Biol. 5: 103–108.

Rosenthal, H. & G. Hempel. 1970. Experimental studies in feeding and food requirements of herring larvae (*Clupea harengus* L.). pp. 344–364. *In:* J.H. Steele (ed.) Marine Food Chains, University of California Press, Berkeley.

Ryer, C.H. & G.W. Boehlert. 1983. Feeding chronology, daily ration, and the effects of temperature upon gastric evacuation in the pipefish, *Syngnathus fuscus.* Env. Biol. Fish. 9: 301–306.

Smith, D.G. 1984. Elopiformes, notacanthiformes, and anguilliformes: relationships. pp. 94–102. *In:* H.G. Moser, W.J. Richards, D.M. Cohen, M.P. Fahay, A.W. Kendall & S.L. Richardson (ed.) Ontogeny and Systematics of Fishes, Amer. Soc. Ichthyol. Herpetol. Spec. Publ. 1, Lawrence.

Sorokin, Ju. I. 1966. Carbon-14 method in the study of the nutrition of aquatic animals. Int. Rev. Gesamten Hydrobiol. 51: 209–224.

Sorokin, Ju. I. & D.A. Panov. 1966. The use of C^{14} for the quantitative study of the nutrition of fish larvae. Int. Rev. Gesamten Hydrobiol. 51: 743–756.

Stroband, H.W.J. 1977. Growth and diet dependent structural adaptations of the digestive tract in juvenile grass carp (*Ctenopharyngodon idella,* Val.). J. Fish Biol. 11: 167–174.

Stroband, H.W.J. & F.M.H. Debets. 1978. The ultrastructure and renewal of the intestinal epithelium of the juvenile grasscarp *Ctenopharyngodon idella* (Val.). Cell Tissue Res. 187: 181–200.

Stroband, H.W.J., H. v.d. Meer & L.P.M. Timmermans. 1979. Regional functional differentiation in the gut of the grasscarp, *Ctenopharyngodon idella* (Val.). Histochemistry 64: 235–249.

Stroband, H.W.J. & K.R. Dąbrowski. 1979. Morphological and physiological aspects of the digestive system and feeding in freshwater fish larvae. pp. 355–376. *In:* M. Fontaine (ed.) La Nutrition des Poisons, CNERNA, Paris.

Stroband, H.W.J. & A.G. Kroon. 1981. The development of the stomach in *Clarias lazera* and the intestinal absorption of protein macromolecules. Cell Tissue Res. 215: 397–415.

Stroband, H.W.J. & F.H. v.d. Veen. 1981. Localization of protein absorption during transport of food in the intestine of the grasscarp, *Ctenopharyngodon idella* (Val.). J. Exp. Zool. 218: 149–156.

Szlaminska, M. 1980. A histochemical study of digestive enzymes in pike larvae. Fish. Manage. 11: 139–140.

Tanaka, M. 1969a. Studies on the structure and function of the digestive system of teleost larvae – I. Development of the digestive system during prelarval stage. Jap. J. Ichthyol. 16: 1–9 (In Japanese).

Tanaka, M. 1969b. Studies on the structure and functions of the digestive systems in teleost larvae – II. Characteristics of the digestive system in larvae at the stage of first feeding. Jpn. J. Ichthyol. 16: 41–49. (In Japanese).

Tanaka, M. 1971. Studies on the structure and function of the digestive system in teleost larvae – III. Development of the digestive system during post larval stage. Jpn. J. Ichthyol. 18: 164–174. (In Japanese).

Tanaka, M. 1972a. Studies on the structure and function of the digestive system in teleost larvae – IV. Changes in the epithelium related to fat absorption in the anteriomedium part of the intestine after feeding. Jpn. J. Ichthyol. 19: 15–25. (In Japanese).

Tanaka, M. 1972b. Studies on the structure and function of the digestive system in teleost larvae – V. Epithelial changes in the posterior gut and protein ingestion. Jpn. J. Ichthyol. 19: 172–180. (In Japanese).

Tanaka, M., S. Kawai & S. Yamamoto. 1972. On the development of the digestive system and changes in activities of digestive enzymes during larval and juvenile stage in Ayu. Bull. Jpn. Soc. Sci. Fish. 38: 1143–1152. (In Japanese).

Theilacker, G.H. 1978. Effect of starvation on the histological and morphological characteristics of jack mackerel, *Trachurus symmetricus,* larvae. U.S. Fish. Bull. 76: 403–414.

Twongo, T.K. & H.R. MacCrimmon. 1977. Histogenesis of the oropharyngeal and oesophageal mucosa as related to early feeding in rainbow trout, *Salmo gairdneri* Richardson. Can. J. Zool. 55: 116–128.

Ugolev, A.M. 1965. Membrane (contact) digestion. Physiol. Rev. 45: 555–595.

Umeda, S. & A. Ochiai. 1973. On the development of the structure and function of the alimentary tract of the yellowtail from the larval to the juvenile stage. Bull. Jpn. Soc. Sci. Fish. 39: 923–930. (In Japanese).

Umeda, S. & A. Ochiai. 1975. On the histological structure and function of digestive organs of the fed and starved larvae of the yellowtail, *Seriola quinqueradiata.* Jpn. J. Ichthyol. 21: 213–219. (In Japanese).

Veith, W.J. 1980. Viviparity and embryonic adaptations in the teleost *Clinus superciliosus.* Can. J. Zool. 58: 1–12.

Vu, T.T. 1976. Étude du développement du tube digestif des larves de bar *Dicentrarchus labrax* (L.). Arch. Zool. Exp. Gén. 117: 493–509.

Vu, T.T. 1983. Etude histoenzymologique des activities proteasiques dans le tube digestif des larves et des adultes de bar, *Dicentrarchus labrax* (L). Aquaculture 32: 57–69.

Ware, D.M. 1975. Growth, metabolism, and optimal swimming speed of a pelagic fish. J. Fish. Res. Board Can. 32: 33–41.

Watanabe, Y. 1981. Ingestion of horseradish peroxidase by the intestinal cells in larvae or juveniles of some teleosts. Bull. Jpn. Soc. Sci. Fish. 47: 1299–1307.

Watanabe, Y. 1982a. Intracellular digestion of horseradish peroxidase by the intestinal cells of teleost larvae and juveniles. Bull. Jpn. Soc. Sci. Fish. 48: 37–42.

Watanabe, Y. 1982b. Ultrastructure of epithelial cells of the anteromedian intestine and the rectum in larval and juvenile teleosts. Bull. Fac. Fish. Hokkaido Univ. 33: 217–228. (In Japanese).

Watanabe, Y. 1984a. An ultrastructural study of intracellular digestion of horseradish peroxidase by the rectal epithelium cells in larvae of a freshwater cottid fish *Cottus nozawae.* Bull. Jpn. Soc. Sci. Fish. 50: 409–416.

Watanabe, Y. 1984b. Morphological and functional changes in rectal epithelium cells of pond smelt during postembryonic development. Bull. Jpn. Soc. Sci. Fish. 50: 805–814.

Watanabe, Y. 1984c. Postembryonic development of intestinal epithelium of masu salmon (*Oncorhynchus masou*). Bull. Tohoku Reg. Fish. Res. Lab. 46: 1–14. (In Japanese).

Watanabe, Y. 1984d. Digestion and absorption in fish larvae. Aquabiology 6: 191–197. (In Japanese).

Watanabe, Y. & N. Sawada. 1985. Larval development of digestive organs and intestinal absorptive functions in the freshwater goby *Chaenogobius annularis.* Bull. Tohoku Reg. Fish. Res. Lab. 47: 1–10.

Werner, R.G. & J.H.S. Blaxter. 1980. Growth and survival of larval herring (*Clupea harengus*) in relation to prey density. Can. J. Fish. Aquat. Sci. 37: 1063–1069.

Weinberg, S. 1976. Morphology of the intestine of the goldfish (*Carassius auratus*). Bijdr. Dierk. 46: 35–46.

Winberg, G.G. 1956. Rate of metabolism and food requirements of fishes. Fish. Res. Board Can. Transl. Ser. 194. 234 pp.

Wourms, J.P. & D. Cohen. 1975. Trophotaeniae, embryonic adaptations in the viviparous ophidioid fish, *Oligopus longhursti:* A study of museum specimens. J. Morphol. 147: 385–401.

Yamada, J. 1968. A study on the structure of surface cell layers in the epidermis of some teleosts. Annot. Zool. Jpn. 41: 1–8.

Yamamoto, T. 1966. An electron microscope study of the columnar epithelial cell in the intestine of fresh water teleosts: goldfish (*Carassius auratus*) and rainbow trout (*Salmo irideus*). Z. Zellforsch. Mikrosk. Anat. 72: 66–87.

Yamamoto, T. 1982. Ultrastructural basis of intestinal absorption. Arch. Histol. Jpn 45: 1–22.

Received 3.1.1985. *Accepted 6.5.1985.*

SEBASTES FASCIATUS, Grd.

Feeding by larvae of *Hypoatherina tropicalis* (Pisces: Atherinidae) and its relation to prey availability in One Tree Lagoon, Great Barrier Reef, Australia

Patricia D. Schmitt
Southwest Fisheries Center, La Jolla Laboratory, National Marine Fisheries Service, NOAA, P.O. Box 271, La Jolla, CA 92038, U.S.A.

Keywords: Microzooplankton, Selectivity, Density-dependent consumption

Synopsis

Feeding of and food availability for larvae of *Hypoatherina tropicalis* were investigated in One Tree Lagoon, Great Barrier Reef, Australia, during November 1981 and January 1982. These surface-dwelling larvae and their microzooplankton prey were sampled as near to simultaneously as possible on 12 occasions during the daytime. Larvae of all sizes (5–17 mm SL) fed successfully over the observed range of mean prey densities (12–235 per liter), and the overall feeding incidence was 98.9%. Larger larvae consumed greater numbers and more categories of prey than did smaller larvae. Larvae selected copepods of all sizes, and nauplii, gastropods, bivalves, and foraminiferans that were greater than 75 μm in width. Tintinnids (mostly 37–74 μm in width) were generally avoided by larvae, but were occasionally important in the diets when they constituted more than 60% of the total available prey, regardless of the density of the selected prey categories. Larvae less than 14 mm SL ingested meroplankton (gastropods, bivalves, foraminiferans, and polychaetes) in direct relation to the densities available, and without regard to the densities of copepods available. However, the largest larvae (14–17 mm SL) ingested meroplankton in inverse relation to the density of copepods available, indicating that larvae consumed more meroplankton when the concentration of copepods was low. Such flexibility and opportunism in feeding behavior may increase the larvae's chances of obtaining adequate nutrition during periods of suboptimal feeding conditions.

Introduction

Very few studies have examined the feeding and food availability of tropical marine fish larvae (Burdick 1969, Jenkins et al. 1984, Johnson 1982, Uotani et al. 1981). A considerable amount of information is available about feeding in various temperate and subtropical fish larvae (see review by Hunter 1980). Several generalizations have resulted from this information, concerning the trophic ecology and feeding behavior of marine fish larvae.

The various life stages of copepods are the main food through much of the larval period of most species (Blaxter & Hunter 1982, Burdick 1969, Cohen & Lough 1983, Dekhnik et al. 1970, Kauffman et al. 1981, Last 1978a, Lebour 1921, Sumida & Moser 1980). Some notable exceptions include the piscivorous fish larvae, such as barracuda, marlin, and many scombroid larvae (Hunter 1980, Jenkins et al. 1984, Uotani et al. 1981), and larvae that feed mainly on appendicularians, such as plaice (Shelbourne 1953).

In the earliest feeding stages, most species studied to date tend to be more euryphagous than in later larval stages, and may consume tintinnids, dinoflagellates, diatoms, and mollusc larvae as well as copepod nauplii (Blaxter & Hunter 1982, Govoni et al. 1983, Hunter 1980, Last 1978a, b).

Selective feeding in relation to available food assemblages has been demonstrated both in the field and in the laboratory, with indications that prey size is the most important factor in selectivity (Hunter 1980, Stepien 1976). Selectivity for certain types of prey has also been reported (Bainbridge & Forsyth 1971, Burdick 1969, Checkley 1982, Stoecker & Govoni 1984). Diet composition or selectivity patterns may vary temporally with the composition of the available plankton (Berner 1959, Cohen & Lough 1983, Laroche 1982, Schnack 1974).

Given the presumed differences in plankton composition, standing stocks, and seasonal production cycles between temperate coastal and tropical near-reef waters (Drake et al. 1978, Revelante & Gilmartin 1982), tropical reef-associated fish larvae may be confronted with qualitatively and quantitatively different feeding conditions than are larvae living in temperate coastal waters. With respect to the above generalizations about feeding in fish larvae, I chose to examine the trophic ecology of larvae of the tropical atherinid *Hypoatherina tropicalls* (Whitley) in One Tree Lagoon (152′E, 23.5′S).

H. tropicalis is a small, schooling fish, which attains a maximum size of about 7 cm SL. It is typically found in surface waters near reefs and around coastal islands, and is distributed from New Guinea to southern Queensland (Ivantsoff 1978). During development, it does not undergo the pelagic to benthic metamorphosis between larval and juvenile stages that is characteristic of most reef fish. The larval period in *H. tropicalis* lasts from first feeding (ca. 5 mm TL) until individuals attain their full complement of fin elements and begin to form scales and adult pigmentation (17–19 mm SL). In this study, I describe the diets of field-collected larvae from 5–17 mm SL, in relation to the available microzooplankton prey.

Methods

Larvae of *H. tropicalis* and microzooplankton were collected from the surface waters at one location inside One Tree Lagoon. Preliminary examination of gut contents indicated that prey items ranged from 37 to 250 μm in width, so the size-range of microzooplankton collected was 37–355 μm. Sampling was done on 6 occasions each month during November 1981 and January 1982, i.e. during high and low tides on each of 3 twenty four hour periods each month.

Sampling gear

Larvae were sampled with a neuston net constructed of 333 μm nylon mesh, with a rectangular mouth opening of 0.03 m^2 and an open area ratio of 8.5 (Tranter & Smith 1968). While under tow at 1 m sec^{-1}, the bottom of the net mouth was submerged to a depth of 10–20 cm, and the top remained above the surface. A General Oceanics flowmeter was suspended across the lower half of the net mouth, to determine the area of sea surface filtered during each tow.

Microzooplankton was sampled by two methods. In January 1982, a 37 μm nylon mesh net was used to sample the entire size range of microzooplankton. The net was attached to a Clarke-Bumpus net frame from which the opening-closing device had been removed. The net mouth was 12.3 cm in diameter, and the net had an open area ratio of 9. A General Oceanics flowmeter was mounted in the center of the cylindrical net frame to measure the volume filtered. The flowmeter was calibrated in the frame, with the net attached. The filtration coefficient was 74.8%, indicating the percentage of water accepted by the sampler (Tranter & Heron 1965). Judging from the appearance and behavior of the net at the end of each tow, clogging was rarely a problem.

During November 1981, a similar Clarke-Bumpus net of 74 μm nylon mesh was used simultaneously with a Jabsco Water Puppy pump to sample the size fraction from 37 to 74 μm. The net had the same dimensions as the 37 μm net. The impeller-type pump delivered water at a mean rate of 18.2 l min^{-1} (s = 0.1, n = 4) into a nested set of sieves of 37 and 265 μm.

Sampling procedure

The neuston net was towed from a boom, which held it 1–1.5 m away from the side of the boat and generally out of the visibly-disturbed water created by the boat's passage. Each tow lasted 10 min and covered an area of 600–900 m^2. Immediately after the net was retrieved, the sample was washed into the cod-end with pumped seawater, and preserved in 10% formalin in seawater buffered with sodium borate.

Immediately before, after or between the duplicate neuston tows, duplicate samples of microzooplankton were collected along the same tow track in the top 0.5 m of the water column, excluding the air-sea interface. All microzooplankton samples were taken from the undisturbed water 1 m abeam of the boat, at a speed of 1 m sec^{-1}. Each 5-min tow sampled a volume of ca. 3–5 m^3. The duplicate tows together covered approximately the same distance as each neuston tow.

Laboratory procedures

Gut contents. – For each sampling time, I examined the gut contents of a maximum of 5 fish in each of four arbitrarily-chosen size classes: 5.1–8.0 mm SL, 8.1–11.0 mm SL, 11.1–14.0 mm SL, 14.1–17.0 mm SL. In samples that contained more than 5 fish in a given size class, the examined fish were randomly chosen. The standard length was measured to the nearest 0.1 mm, and the gape width of the jaws (maximum width with jaws open at an angle of ~60°) was measured to the nearest 40 μm, using an ocular micrometer. The gut was then dissected out in glycerin. Larvae of *H. tropicalis* have a coiled but apparently undifferentiated gut (i.e. the stomach was not discernible from the intestine). Since food passage rates are unknown, only the contents of the descending half of the anterior-most loop were examined. This was done in order to: (1) ensure as much as possible that the gut contents were recently ingested from the same microzooplankton assemblage as was sampled; (2) minimize the digestion of soft-bodied forms, and hence underestimation of their numbers; and (3) minimize the amount of unidentifiable material.

The anterior section of the gut was placed in a drop of glycerin on a flat slide, the contents teased out with fine needles, and the gut wall removed and discarded. The entire slide was then examined under a compound microscope with 63–160× magnification, using phase contrast lighting. All prey items were identified to 18 major categories (e.g. crustacean nauplii, including copepod nauplii; copepodites and adult copepods; gastropod and bivalve veligers; tintinnids; trochophores; foraminiferans), counted, and the maximum width measured to the nearest 20 μm with an ocular micrometer. Since polychaetes were generally recognizable only by the setae, it was not possible to determine the number of individuals in a gut. For purposes of analyses, I assumed that one specimen had been ingested by each larva whose gut contained polychaete setae.

For each larva examined, the mean size of prey consumed and the range of sizes was also determined. For the prey types in which the whole width was not always measureable, such as foraminiferans, polychaetes, and trochophores, the size was estimated as follows. The foraminiferans found in the guts were often broken, and only fragments containing the center were counted. The mean size of all intact foraminiferans found in guts each season was used to estimate the size of fragments. For polychaetes and trochophores, the modal size found in the microzooplankton samples in each season was used to estimate the sizes ingested.

Microzooplankton samples. – The samples were filtered through a 355 μm sieve to remove any large plankters, and then stained with Biebrich Scarlet to aid in distinguishing whole crustaceans from empty exoskeletons, and zooplankters from sediment particles. The stained samples were suspended in a known volume of water (30–500 ml), stirred to homogenize, and two 0.5 ml aliquots drawn with an Eppendorf pipette (mean tip diameter = 780 μm, s = 45 μm, n = 5; mean volume = 0.499 ml, s = 0.002, n = 10). Each aliquot was placed in a small Bogorov tray, and zooplankters were counted and measured under a compound microscope at 63–160×. Organisms were identified to the same categories as the gut contents. Measurements were made of the maximum width excluding appendages. The mean coefficient of variation between aliquot pairs for each major food taxon was 19.5% (range: 16.9–21.7%).

Data presentation and analyses

For each sampling time and size-class of larvae, data on gut contents were expressed in two ways, for each of the following arbitrarily-chosen classes of prey width: 37–74 μm, 75–125 μm, 126–175 μm, 176–225 μm, and >225 μm. First, the frequency of occurrence (% FO) was determined as the percentage of larvae which had consumed that category of prey. Second, the gut contents of all fish examined in each size class at each sampling time were pooled, and the percentage that each category of prey represented of the total number of items consumed was calculated (%N). Also, the %N and the % FO for each category of prey consumed by overall larvae at each sampling time was calculated as the mean over all size classes of larvae present at that time.

The change in the composition and size-structure of diet with size of larvae was examined graphically using the mean %N over all sampling times, and by 1-factor ANOVA using the %N values for each sampling time, for each major category and size-class of prey. Homogeneity of variances was tested using Cochran's test (Winer 1971), and appropriate transformations applied to heteregeneous data (Underwood 1981).

Each category and size of microplankter in each sample was tabulated in two ways: as the concentration or density (number per liter); and as the percentage (%N) of all microzooplankters present.

The percentage of variance in the concentrations of some commonly-ingested microzooplankters due to various temporal and spatial scales was examined with nested analyses of variance (Sokal & Rohlf 1969). The replicate samples represent variation on a spatial scale of 300–600 m and a temporal scale of 15–30 min. Other temporal scales sampled were high and low tides within days, days within seasons, and seasons.

Feeding selectivity was assessed using Wilcoxon's signed-ranks test (Kohler & Ney 1982, Sokal & Rohlf 1969). This non-parametric statistical procedure used the differences between the proportion of each prey category consumed and the proportion available for each sampling time, and assessed whether there was a significant pattern in the direction of these differences over all the sampling times. Selectivity for each prey category was assessed for each size-class of larvae and for overall larvae. Several prey categories were not tested because they were extremely rare (isopods) or unquantifiable in either the gut samples or the microzooplankton samples (polychaetes, appendicularians, algae). Size of prey was also tested for the common prey categories.

The relationships between the concentrations of prey categories available and the numbers ingested by larvae of various size-classes were examined with multiple linear regressions (Snedecor & Cochran 1980, Sokal & Rohlf 1969). The most important prey categories were divided into 3 groups: nauplii and copepods combined, meroplankton (gastropod and bivalve veligers, foraminiferans, and polychaetes), and tintinnids. The number of each prey category consumed at each sampling time was standardized to represent the number consumed by 5 larvae in each size class.

Results

The following results are based on the examination of the gut contents of 182 larvae of *H. tropicalis* (Table 1). A total of 6960 prey items were identified and measured.

Feeding incidence

The overall incidence of larvae with food in their guts was 98.9%. Only 2 of the larvae examined, both 5.6 mm SL, had no food in the anterior gut.

The mean number of food items per individual increased with size of larvae, from 16.9 in 5–8 mm larvae to 98.5 in 14–17 mm larvae (Table 1). The variances of the number of food items per larva remained heterogeneous ($p = 0.01$) after a logarithmic transformation, precluding a statistical comparison of the means between size classes.

Composition of diet

H. tropicalis larvae ingested 16 categories of prey (Table 2). The number of categories of prey eaten showed a tendency to increase with size of fish, and was highest in 11–14 mm larvae. Six prey categories were considered important components of the diet because they either constituted more than 5% of the total prey consumed (%N) or occurred in more than 25% of larvae (% FO) in one or more size classes of larvae: (1) nauplii; (2) copepodites and adult copepods; (3) gastropod veligers; (4) bivalve veligers; (5) foraminiferans; and (6) tintinnids. Nauplii and copepods were the most important prey types for all sizes of larvae, constituting about half (48–59%) of all prey items and ingested by approximately 90% of all larvae. Gastropods and bivalve veligers, foraminiferans, and tintinnids were ingested by 32–58% of total larvae, and each constituted 3.0–11.3% of prey items (Table 2). Trochophores and polychaetes were occasionally important. Trochophores were ingested by all sizes of larvae in January, but not in November, and overall constituted only 2.8% of the diets and were ingested by 14% of larvae. Polychaetes were ingested by larger larvae (13–14% of 11–17 mm SL), and were more important in November than in January. Prey types eaten rarely (% FO) <15% for all sizes of larvae) and in low numbers (%N <2%) included dinoflagellates; appendicularians; radiolarians, ostracods; isopods; unidentified spheres, which were probably dinoflagellate cysts (G.M. Hallegraeff, personal communication); unidentified ovals; and algae (including diatoms).

The proportions of nauplii, copepods, meroplankton, tintinnids, and trochophores consumed were not significantly different between sizes of

Table 1. Number of *Hypoatherina tropicalis* larval guts examined and the number of prey items found, by larval size class and sampling time. In the sample designation, L indicates low tide, H indicates high tide.

	Standard length (mm) 5.1–8.0		8.1–11.0		11.1–14.0		14.1–17.0	
Sample	Number of fish	prey	fish	prey	fish	prey	fish	prey
1 Nov L	0		4	53	6	98	2	22
2 Nov H	5	15	5	96	5	97	2	11
6 Nov L	3	43	5	253	5	217	5	367
6 Nov H	5	39	5	90	5	122	5	125
22 Nov L	5	236	5	275	5	338	4	516
22 Nov H	5	47	9	136	7	210	0	
Subtotal	23	380	33	903	33	1082	18	1041
18 Jan L	2	96	5	224	5	246	4	1364
18 Jan H	2	42	4	83	3	155	2	111
25 Jan L	5	54	5	148	4	151	0	
25 Jan H	5	54	5	88	3	80	0	
31 Jan L	0		5	129	5	153	3	144
31 Jan H	0		3	61	5	171	0	
Subtotal	14	246	27	733	25	956	9	1619
Total	37	626	60	1636	58	2038	27	2660
$\bar{X}$ No. of prey/larva	16.9		27.3		35.1		98.5	
s	17.0		18.7		22.4		122.6	

Table 2. Composition of diets of *H. tropicalis* larvae of various sizes and overall larvae, expressed as the percentage that each prey type represented of the total diet (%N) and as the frequency of occurrence (% FO). Each value shown is the mean over all sampling times: n = 9 for 5.1–8.0 mm SL larvae; n = 12 for 8.1–11.0 and 11.1–14.0 mm SL larvae; n = 8 for 14.1–17.0 mm SL larvae; and n = 12 for overall larvae. The standard deviation is shown beneath each mean in parentheses.

Prey type	Standard length (mm)									
	5.1–8.0		8.1–11.0		11.1–14.0		14.1–17.0		Overall	
	%N	% FO	%N	% FO	%N	% FO	%N	% FO	%N	% FO
Nauplii	28.4	74	32.2	90	29.7	93	26.1	94	31.2	90
	(16.6)	(26)	(17.8)	(24)	(16.2)	(20)	(17.4)	(18)	(15.0)	(16)
Copepods	23.8	74	21.4	89	29.1	93	21.4	89	24.1	87
	(27.0)	(24)	(21.6)	(18)	(21.0)	(20)	(21.5)	(21)	(20.5)	(18)
Gastropods	4.4	34	8.3	52	11.0	69	11.6	66	9.3	58
	(5.3)	(29)	(8.0)	(35)	(12.0)	(32)	(12.8)	(44)	(8.5)	(28)
Bivalves	1.7	17	3.4	36	2.6	39	3.1	50	3.0	36
	(3.2)	(23)	(5.3)	(32)	(3.1)	(26)	(4.6)	(44)	(3.9)	(24)
Foraminiferans	3.2	18	4.8	34	6.5	47	8.3	64	6.2	43
	(5.4)	(29)	(8.3)	(43)	(8.0)	(35)	(10.8)	(35)	(6.6)	(34)
Tintinnids	21.6	40	12.1	30	6.9	28	14.8	42	11.3	32
	(25.3)	(45)	(21.8)	(41)	(12.6)	(39)	(23.2)	(49)	(18.3)	(41)
Trochophores	3.5	22	3.7	15	3.1	14	2.5	19	2.8	14
	(8.1)	(44)	(10.1)	(34)	(7.3)	(33)	(6.6)	(37)	(7.2)	(34)
Polychaetes	0		0.4	7	0.7	13	0.7	14	0.5	8
			(1.1)	(15)	(1.3)	(22)	(1.6)	(21)	(1.2)	(16)
Dinoflagellates	0		<0.1	2	0.3	8	0.4	8	0.2	6
			(0.2)	(7)	(0.5)	(13)	(1.2)	(24)	(0.5)	(11)
Appendicularians	0.2	2	0.2	2	0.3	4	0		0.2	3
	(0.6)	(7)	(0.6)	(7)	(0.9)	(14)			(0.5)	(7)
Radiolarians	0		0		0.1	3	0		<0.1	1
					(0.3)	(8)			(0.1)	(3)
Ostracods	0		0		0.1	3	<0.1	2	<0.1	1
					(0.3)	(7)	(<0.1)	(7)	(<0.1)	(3)
Isopods	0		0		<0.1	2	<0.1	2	<0.1	1
					(0.1)	(7)	(0.1)	(7)	(<0.1)	(3)
Spheres	1.3	6	0.5	7	0.4	10	0.2	14	0.4	11
	(2.6)	(12)	(1.3)	(20)	(0.7)	(18)	(0.3)	(26)	(0.8)	(18)
Ovals	0		0		0.3	6	0.1	2	0.1	2
					(0.6)	(11)	(0.3)	(7)	(0.2)	(4)
Algae (inc. diatoms)	0		0		0.2	5	0.6	8	0.2	2
					(0.6)	(12)	(1.6)	(25)	(0.6)	(4)
Unknowns	5.4	49	5.0	54	2.9	47	3.5	45	3.8	46
	(5.0)	(38)	(7.6)	(35)	(3.6)	(34)	(4.3)	(42)	(4.3)	(31)
Unid. crustaceans	6.1	49	6.1	58	5.7	71	5.4	69	5.9	63
(Damaged)	(5.0)	(36)	(4.7)	(31)	(4.3)	(38)	(5.1)	(35)	(3.8)	(27)
Total 37–74 μm	43.7		29.2		22.1		27.6		27.6	
	(24.0)		(22.8)		(18.2		(22.0)		(19.6)	
75–125 μm	36.8		45.2		44.1		39.4		43.5	
	(15.0)		(16.8)		(12.4)		(17.0)		(12.9)	
126–175 μm	11.9		15.7		22.7		20.6		18.8	
	(10.8)		(12.4)		(13.3)		(15.5)		(11.7)	
176–225 μm	0.7		1.5		2.2		2.4		1.8	
	(1.5)		(1.9)		(1.6)		(3.0)		(1.5)	
>225 μm	3.3		4.3		5.2		6.8		5.2	
	(4.9)		(7.4)		(6.7)		(7.9)		(5.9)	
Number of guts	37		60		58		27		182	
Number of prey items	626		1636		2038		2660		6960	

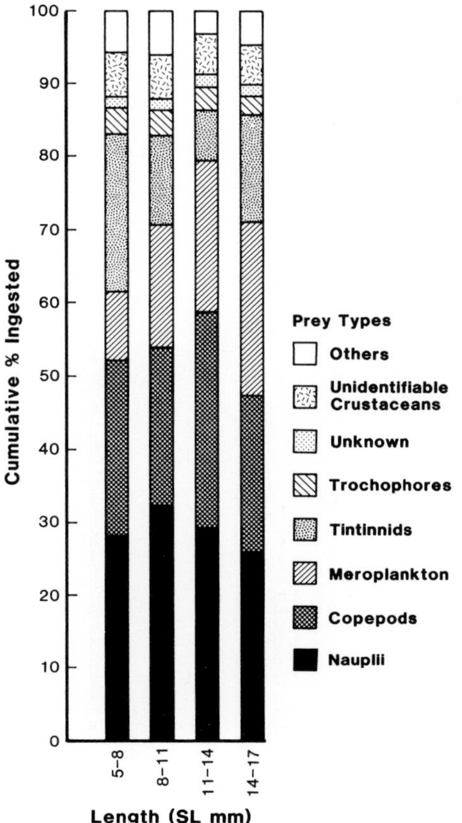

Fig. 1. Composition of the diet of *H. tropicalis* for each size class of larvae.

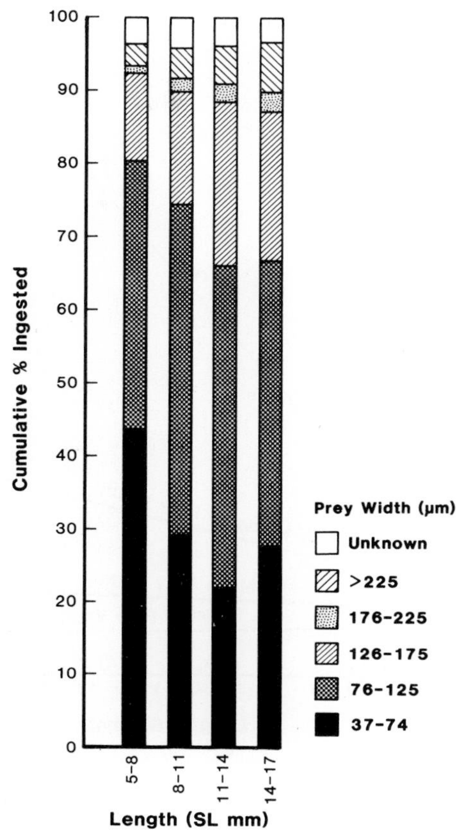

Fig. 2. Size-structure of the diet of *H. tropicalis* for each size class of larvae.

larvae (ANOVA, $p>0.10$). There was a tendency for the mean proportion of meroplankton consumed to increase with larval length, and for the mean proportion of tintinnids eaten to decrease with larval length up to 14 mm SL, and then to increase again in the largest larvae (Fig. 1).

The differences between size-classes of larvae in proportions of various prey sizes consumed were also not significant (ANOVA, $p>0.10$ for all prey sizes). In all size-classes of larvae and in both seasons, more than 80% of prey items consumed were less than 175 μm in width. Larvae tended to consume increasing proportions of prey greater than 75 μm in width, as larval size increased to 14 mm SL. Larvae of 14–17 mm SL consumed a slightly greater proportion of small prey (less than 75 μm in width) than did larvae of 11–14 mm (Fig. 2).

Mean and maximum prey widths increased significantly with fish length, but only 4.2 and 14.6% of the variation in mean and maximum prey widths was explained by differences in length of larvae (Fig. 3). There was no relationship between minimum prey width and larval length, and the range of minimum prey widths observed was 43–86 μm.

Gape width was highly correlated with larval length, and increased from ~250 μm at 5 mm SL to ~850 μm at 17 mm SL (Fig. 3). It appears that *H. tropicalis* larvae were capable of swallowing larger prey than they consumed in this study. Analysis of covariance indicated significant differences between gape size and maximum prey width for various lengths of larvae ($F_{slopes} = 123.05$; df = 1,300; $P<0.001$). The ratio of gape size to maximum prey width increased from 1.8 at 5 mm SL to 4.3 at 17 mm SL, indicating that the discrepancy, between mouth size and prey width increased as larvae grew (Fig. 3).

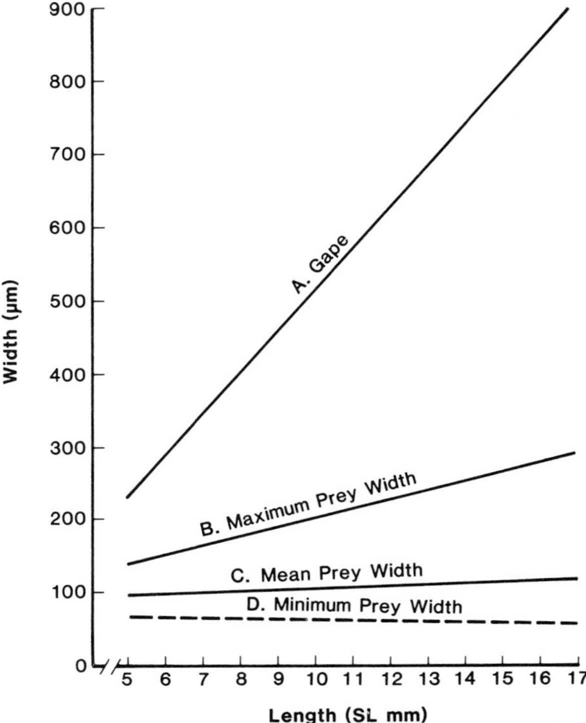

Fig. 3. Regressions of gape width of *H. tropicalis* and the mean, maximum, and minimum prey widths against standard length. Solid lines indicate the regression was significant at $p<0.05$; dashed line indicates regression not significant. Regression parameters are as follows: A. gape = 50.6 (SL) + 2.9, $r^2 = 0.78$, n = 124; B. maximum prey width = 12.7 (SL) + 73.5, $r^2 = 0.15$, n = 180; C. mean prey width = 2.3 (SL) + 82.8, $r^2 = 0.04$, n = 180; D. minimum prey width = −0.5 (SL) + 67.6, $r^2 <0.01$, n = 180.

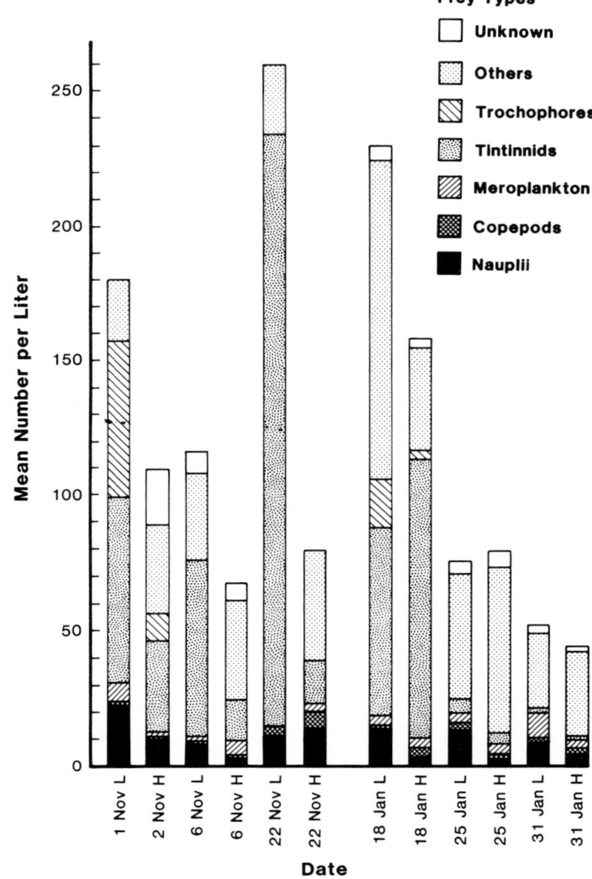

Fig. 4. The mean concentrations of various microzooplankton categories in replicate samples taken in One Tree Lagoon on 12 sampling times in November 1981 and January 1982. L = low tide, H = high tide.

Abundance and composition of available microzooplankton

The mean concentration of microzooplankton categories that were commonly ingested by larvae (naupiii, copepods, gastropods, bivalves, foraminiferans, polychaetes, tintinnids, and trochophores) showed a 20× difference over the 12 sampling times, and ranged from 12 to 235 per liter (Fig. 4). Other microzooplankton categories that were eaten only rarely or in very low numbers (prey type 'others' in Figure 4) ranged from 23 to 118 per liter. Thus, the concentrations of commonly-ingested microzooplankters were of the same order of magnitude as the concentrations of the rarely-ingested prey categories. The rarely-ingested prey available consisted mainly of dinoflagellates, radiolarians, and spheres. Invertebrate taxa that occurred in the microzooplankton samples but were never ingested by larvae included rotifers, chaetognaths, nematodes, and zoeae. The total density of uningested microzooplankters was always less than 3 per liter.

The majority of variance in potential prey availability was between days within seasons and tides within days (Table 3). For all taxa except bivalves, the percentage of the variance due to replicates and seasons was relatively small compared to the variance due to either tides or days. The composition of the prey assemblage also varied over sampling times, from assemblages dominated by tintinnids

(22 Nov Low Tide and 18 Jan High Tide) to assemblages dominated by crustaceans and meroplankton (31 Jan Low and High Tides) (Fig. 4). Thus, it appears that fish larvae in One Tree Lagoon were faced with a food supply that varied at least 20× in density over time, and markedly in the proportions of various prey types available.

Comparison of ingested and available microzooplankton

Nauplii, gastropods, and bivalves greater than 75 μm width were selected by overall larvae, as were copepods of all widths (Table 4). There were differences in selectivity by larvae of different sizes. Larger larvae tended to select wider nauplii and gastropods. The smallest larvae (5–8 mm SL) only selected copepods from 75–175 μm width, while larger larvae selected all widths of copepods from 37–175 μm. The smallest and largest larvae did not select bivalves of any size. Foraminiferans were selected only by 11–14 mm larvae. The rest of the prey categories, including tintinnids, trochophores, and all the rare forms, were either fed upon randomly or were avoided by all sizes of larvae. These results indicate that there were some differences in selectivity among sizes of larvae, but that these differences mainly involved the sizes of prey selected, rather than the categories. Overall, *H. tropicalls* larvae selected prey >75 μm width.

Copepods in the guts showed roughly the same size-distribution as copepods in the plankton, except that animals larger than 175 μm in width were slightly underrepresented in the guts (Fig. 5). Larvae tended to ingest larger sizes of nauplii, gastropods, and bivalves among those available. Tintinnids were almost all between 37 and 74 μm for both available and ingested prey. Foraminiferans found in the guts were consistently and disproportionately the largest ones available. Trochophores and polychaetes in the plankton were mainly 37–125 μm in width. The size distribution of other microzooplankters which were rarely eaten was dominated by 37–74 μm organisms.

For each sampling time, the percentages of each prey type in the diets of overall larvae compared to the percentages of only the commonly-ingested prey types available indicated that copepods were consistently consumed in higher proportions than they were available (Fig. 6). Nauplii were consumed in higher proportions than available on 5 of 12 occasions. Tintinnids were important in the diets of larvae only when they constituted more than 60% of the commonly-ingested prey available. Trochophores were available on 3 dates, 1–2 Nov, 18 Jan, and 25 Jan, although they were consumed only on 18 Jan.

Multiple linear regressions of the concentrations of prey available and the numbers ingested by larvae indicated that the number of nauplii and copepods consumed was not linearly related to either the concentration of nauplii and copepods or of meroplankton, in any size-class of larvae (Table 5A). The number of meroplankton consumed was positively related to their concentration, and not dependent on the concentrations of nauplii and copepods available, in larvae from 8–14 mm SL (Table 5B). In larvae smaller than 8 mm SL, the number of meroplankton ingested was not related to concentrations of either nauplii and copepods,

Table 3. Percentages of total variance in densities of major food categories due to various temporal and spatial scales and to subsampling, determined from nested ANOVA of microzooplankton densities from November 1981 and January 1982 in One Tree Lagoon.

Taxon	Seasons	Days	Tides	Replicates	Aliquots
Nauplii	6.4%	0	67.3%	16.0%	10.3%
Copepods	0	63.5%	7.1	23.2	6.1
Gastropods	0.7	0	58.7	32.3	8.3
Bivalves	9.0	8.0	0	34.3	48.8
Tintinnids	16.4	50.0	26.6	5.5	1.4
Total prey	11.7	36.0	45.1	5.5	1.6

Table 4. Results of Wilcoxon's signed-ranks test to assess selectivity, using proportions of prey types and sizes found in guts of *H. tropicalis* larvae of various sizes, and proportions of prey available in corresponding microzooplankton samples. + indicates selection at p<0.05, ++ indicates selection at p<0.01, ns indicates random feeding (p>0.05), – indicates avoidance at p<0.05, –– indicates avoidance at p<0.01, ut indicates too few pairs available to test (n <6).

Prey type and width	Standard length (mm)				
	5.1–8.0	8.1–11.0	11.1–14.0	14.1–17.0	Overall
Nauplii					
37–74	++	ns	ns	ns	ns
75–125	++	++	++	++	++
126–175	ns	+	++	+	++
Total	++	++	++	++	++
Copepods					
37–74	ns	+	++	ut	++
75–125	++	++	++	+	++
126–175	+	++	++	+	++
176–225	–	ns	ns	ns	+
Total	++	++	++	+	++
Gastropods					
37–74	–	––	ns	ns	ns
75–125	+	++	++	ns	++
126–175	ns	++	++	+	++
Total	ns	++	++	+	++
Bivalves					
37–74	–	–	ns	–	–
75–125	ns	ns	+	ns	++
126–175	ns	ns	++	ns	++
Total	ns	+	+	ns	++
Foraminiferans					
Total	ns	ns	++	ns	++
Tintinnids					
Total	ns	––	––	–	––
Trochophores					
Total	ns	ns	ns	ns	ns
Dinoflagellates					
Total	––	––	––	––	––
Radiolarians					
Total	––	––	––	––	––
Ostracods					
Total	––	––	ns	ut	ns
Spheres					
Total	––	––	––	––	––
Ovals					
Total	––	––	––	––	––
Total microzooplankton					
37–74 μm	––	––	––	––	––
75–125 μm	++	++	++	++	++
126–175 μm	++	++	++	++	++
176–225 μm	ns	ns	++	+	++
>225 μm	ns	+	++	++	++

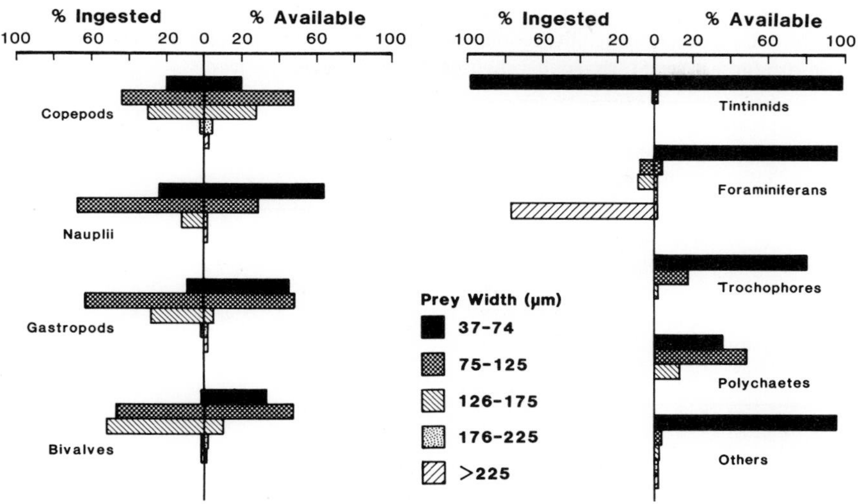

Fig. 5. Comparison of the sizes of various categories of prey ingested by overall larvae of *H. tropicalis* and the sizes available.

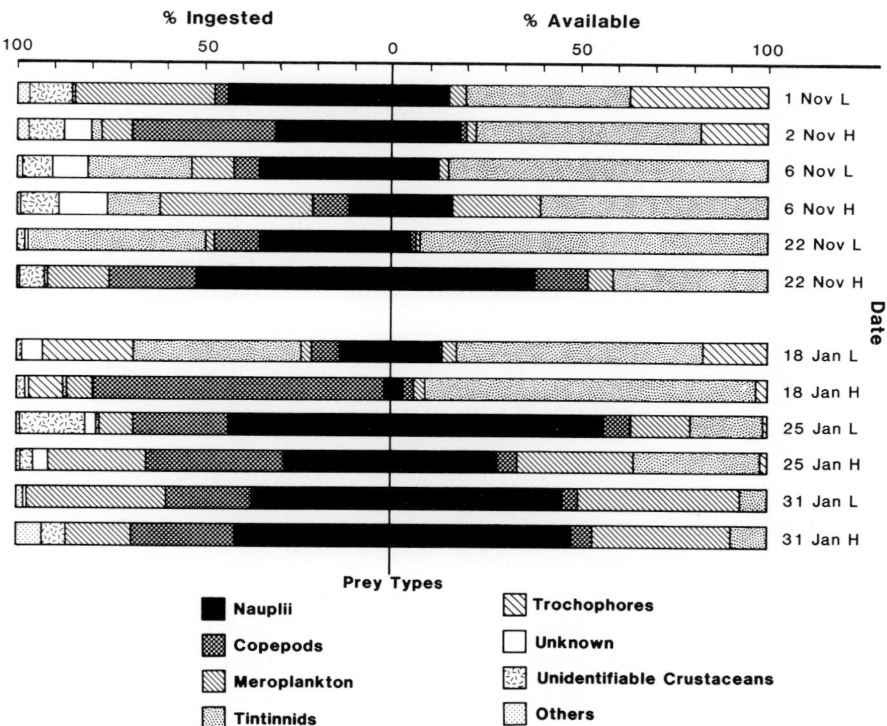

Fig. 6. Comparison of the composition of the diets of overall larvae of *H. tropicalis* and the composition of the commonly-ingested prey available, for each sampling time.

Table 5. Results of multiple linear regressions of the numbers of common prey consumed by each size class of *H. tropicalis* larvae and the densities of common prey available. For all significant regressions, the partial regression coefficients, and the partial and multiple coefficients of determination are shown. The significance of the correlation coefficients, r, are indicated for each r^2: ns indicates $p>0.05$; * $p<0.05$; ** $p<0.01$. n = 9 for 5–8 mm SL larvae; n = 12 for 8–11 and 11–14 mm SL larvae; n = 8 for 14–17 mm SL larvae.

A. Number of nauplii + copepods consumed.
X_1 = Density of nauplii + copepods available.
X_2 = Density of meroplankton available.

Size class	$r^2_{X_1}$	$r^2_{X_2}$	$r^2_{X_1+X_2}$
5–8 mm	ns	ns	ns
8–11 mm	ns	ns	ns
11–14 mm	ns	ns	ns
14–17 mm	ns	ns	ns

B. Number of meroplankton consumed.
X_1 = Density of nauplii + copepods available.
X_2 = Density of meroplankton available.

Size class	b_1	$r^2_{X_1}$	b_2	$r^2_{X_2}$	$r^2_{X_1+X_2}$
5–8 mm		ns		ns	ns
8–11 mm		ns	3.56	0.45*	0.47*
11–14 mm		ns	6.29	0.67**	0.68**
14–17 mm	−3.76	0.48*		ns	0.66**

C. Number of tintinnids consumed.
X_1 = Density of nauplii + copepds + meroplankton available.
X_2 = Density of tintinnids available.

Size class	$r^2_{X_1}$	b_2	$r^2_{X_2}$	$r^2_{X_1+X_2}$
5–8 mm	ns	0.67	0.69**	0.71**
8–11 mm	ns	0.74	0.66**	0.67**
11–14 mm	ns	0.41	0.52**	0.52**
14–17 mm	ns		ns	ns

or of meroplankton. In larvae larger than 14 mm SL, the numbers of meroplankton consumed was inversely related to the concentration of nauplii and copepods available, and was unrelated to the concentration of meroplankton available. The numbers of tintinnids consumed was positively related to the concentrations of tintinnids available for all larvae less than 14 mm SL, and was not dependent on the concentrations of other prey available (Table 5C). In the largest larvae, the consumption of tintinnids was not related to their availability or to the availability of other prey.

Discussion

Types of prey consumed

Larvae of *Hypoatherina tropicalis* in One Tree Lagoon consumed a wider range of prey categories than has been reported for other tropical fish larvae or for other atherinid larvae. The diet of the Hawaiian anchovy, *Stolephorus purpureus,* from 2–12 mm SL, has been documented to consist almost exclusively of copepod nauplii and adults; other prey, including dinoflagellates, foraminiferans, diatoms, and tintinnids, were only found in larvae <7 mm SL (Burdick 1969). In first-feeding larvae (<6 mm SL) of the Hawaiian anchovy, about 80% of the diet was copepod nauplii, and the remainder was copepodites, appendicularian pellets, and rarely gastropod and bivalve larvae (Johnson 1982). Larvae of various scombroid species in the waters off northwestern Australia ate primarily copepods, cladocerans, and appendicularians (Uotani et al. 1981), and in shelf waters of the Great Barrier Reef ate primarily appendicularians and fish larvae (Jenkins et al. 1984).

The diets determined for *H. tropicalis* in One Tree Lagoon are also the most diverse yet reported among atherinids. Atherinid larvae from a southern California embayment ate copepod nauplii, tintinnids, and diatoms until 9 mm SL. The diets of larger fish were almost exclusively (>90%) nauplii and copepods (Kauffman et al. 1981). Larvae of 11–18 mm SL tidewater silversides, *Menidia beryllina,* from an estuarine seagrass bed in Florida ate veliger larvae, detritus, copepods, and cypris larvae (Carr & Adams 1973). In a similar habitat in Florida, Harrington & Harrington (1961) found that copepods accounted for 38% of the stomach contents of tidewater silversides 5–15 mm SL, and other items included protozoa, small gastropods, cladocerans, and insects.

The change in types of prey consumed with increasing size of *H. tropicalis* larvae was slight. There was no tendency for larger larvae to be more stenophagous than smaller larvae. This was in contrast to the larvae of clupeoids (Arthur 1976, Blaxter 1965, Burdick 1969, Checkley 1982, de Mendiola 1974, Duka 1969, Lasker 1975) and gadoids (Last 1978b). Little change in the composition of the diet with increasing size has been reported for cottids (Laroche 1982) and hake (Sumida & Moser 1980).

Size of prey consumed

H. tropicalis consumed prey that were mostly 37–175 μm in width, which is similar to the prey widths reported for many clupeoid larvae (Burdick 1969, Cohen & Lough 1983, Hunter 1980). It is somewhat smaller than the prey widths reported for the larvae of scombroids (Hunter 1980), cottids (Laroche 1982), pleuronectiforms (Last 1978a), gadoids (Last 1978b, Sumida & Moser 1980), and sea bream (Stepien 1976).

Larvae of *H. tropicalis* were capable of swallowing much larger prey than they were observed to consume, and the discrepancy between mouth size and prey size increased as larvae grew. Several other studies have found that prey size tends to level off at the some point in larval life, even though mouth size continues to increase (Arthur 1976, Blaxter & Hunter 1982, Checkley 1982, Cohen & Lough 1983, de Mendiola 1974, Detwyler & Houde 1970, Last 1978a & b). Others have found little ontogenetic increase in prey size for larvae that start feeding on prey >100 μm (Laroche 1982, Sumida & Moser 1980). The mouth size establishes the upper limit to prey size, but the optimal and minimum sizes must be a function of energy costs in capture in relation to prey concentration (Blaxter & Hunter 1982).

Possible explanations for larvae consuming smaller prey than they are capable of ingesting are: (1) larger prey are generally scarce; and (2) the larvae are unable to capture them efficiently. Prey larger than 175 μm in width comprised less than 0.2% of the available prey between 37 and 355 μm in One Tree Lagoon in November 1981 and January 1982. The rapid decline in abundance with increasing size is a common feature of the size distribution of particles in the sea (Hunter 1980).

Most (76%) of the prey larger than 175 μm which were consumed by *H. tropicalis* were foraminiferans. Copepods larger than 175 μm were fairly rare in the diets, even when they were available. Since planktonic foraminiferans are not very mobile compared to copepods, it suggests that larvae were better able to capture large foraminiferans but not large copepods, at least at the concentrations observed in One Tree Lagoon.

Selectivity and density-dependent consumption

Copepod stages are the most frequently selected type of food by a variety of species of larvae (Bainbridge & Forsyth 1971, Burdick 1969, Checkley 1982, Cohen & Lough 1983, Detwyler & Houde 1970). For the larvae of *H. tropicalis* in One Tree Lagoon, the number of categories of prey selected was larger than has generally been reported. Larvae showed selectivity for nauplii, copepods, gastropods, bivalves, and foraminiferans, and for prey larger than 75 μm in all these categories except copepods. Prey types that were randomly consumed or were avoided were predominantly <75 μm in width. It appeared that selection for categories was related to the size of the organisms, and that size was the most important factor in prey selectivity in my study.

It was interesting that tintinnids were important in the diets of *H. tropicalis* larvae only when they constituted more than 60% of the total available prey, regardless of the density of other prey such as nauplii, copepods, and meroplankton. Possibly the larvae incidentally consumed tintinnids when they were present in high concentrations. Alternatively, it was possible that the larvae detected the relatively small tintinnids only when they reach some threshold abundance relative to the more preferred prey types (O'Brien et al. 1976). First-feeding larvae of the northern anchovy fed on a dinoflagellate only when its concentration exceeded a threshold level (Lasker 1975).

The largest larvae of *H. tropicalis* consumed meroplankton in numbers inversely related to the availability of nauplii and copepods. Schnack (1974) found that gastropods and bivalves were selected by herring larvae only when the densities of larger copepods was low. Mollusc larvae have been found to be of low nutritional value compared to copepods (Checkley 1982), and to be largely undigested by herring larvae (Schnack 1974). Mollusc veligers were highly preferred by a variety of sub-tropical fish larvae (Houde & Lovdal 1984), although it was possible that such apparent preference was an artifact of slow digestion of these shelled prey. Tintinnids and other ciliates were consumed by lined sole and sea bream larvae only when nauplii concentrations were below 100 per liter (Houde & Schekter 1980). The consumption of tintinnids and meroplankton may have been less energetically beneficial to the larvae than the consumption of nauplii and copepods, although the tintinnids and meroplankton may have been less evasive and therefore less energetically expensive to catch.

Densities of nauplii and copepods appears to be lower in One Tree Lagoon (4–22 l^{-1}) than in other tropical and subtropical locations, such as Kaneche Bay, Hawaii, (36–201 l^{-1}) (Burdick 1969, Johnson 1982); and Biscayne Bay, Florida (132.1 l^{-1}) (Houde & Lovdal 1984). The concentrations in One Tree Lagoon were also lower than most values reported for the open sea (15–58 l^{-1}) or partially-enclosed water masses in temperate areas (95–223 l^{-1}) (Hunter 1980). These apparent differences may be partially due to differences in sampling techniques or the portion of the water column sampled. Nevertheless, the wide range of prey categories consumed by *H. tropicalis* larvae may have been the result of relatively poor feeding conditions in the surface-waters of One Tree Lagoon.

Acknowledgements

This research was part of my dissertation completed at the University of Sydney. I would like to thank my supervisor, P.F. Sale, for his excellent support and advice; A.J. Underwood and K. McGuinness for statistical advice; P. Hunt, S. Kennelly, and K. McGuinness for help with field work; and J. Hunter, G. Moser, and B. Sumida for reviewing early drafts of the manuscript. Thanks also to two anonymous reviewers for their helpful comments. Financial support was provided by a University of Sydney Postgraduate Research Studentship, the Great Barrier Reef Marine Park Authority, and the Australian Coral Reef Society. The recent support of a post-doctoral fellowship from the National Research Council, Washington D.C., is also gratefully acknowledged. Finally, I'd like to thank G. Cailliet and C. Simenstad for their editorial assistance and for organizing an excellent workshop.

References cited

Arthur, D.K. 1976. Food and feeding of larvae of three fishes occuring in the California Current, *Sardinops sagax, Engraulis mordax,* and *Trachurus symmetricus.* U.S. Fish. Bull. 74: 517–530.

Bainbridge, V. & D.C.T. Forsyth. 1971. The feeding of herring larvae in the Clyde. Rapp. et P.-v. Reun. Cons. Int. Explor. Mer. 160: 104–113.

Berner, L. 1959. The food of the larvae of the northern anchovy. Bull. Inter-Am. Trop. Tuna Comm. 4: 3–22.

Blaxter, J.H.S. 1965. The feeding of herring larvae and their ecology in relation to feeding. Calif. Coop. Oc. Fish. Inv. Rep. 10: 79–88.

Blaxter, J.H.S. & J.R. Hunter. 1982. The biology of the clupeoid fishes. Adv. Mar. Biol. 20: 1–223.

Burdick, J.E. 1969. The feeding habits of nehu (Hawaiian anchovy) larvae. M.Sc. Thesis, University of Hawaii, Honolulu. 54 pp.

Carr, W.E.S. & C.A. Adams. 1973. Food habits of juvenile marine fishes occupying seagrass beds in the estuarine zone near Crystal River, Florida. Trans. Amer. Fish. Soc. 102: 511–540.

Checkley, D.M. Jr. 1982. Selective feeding by Atlantic herring (*Clupea harengus*) larvae on zooplankton in natural assemblages. Mar. Ecol. Prog. Ser. 9: 245–253.

Cohen, R.E. & R.G. Lough. 1983. Prey field of larval herring *Clupea harengus* on a continental shelf spawning area. Mar. Ecol. Prog. Ser. 10: 211–222.

Dekhnik, T.V., L.A. Duka & V.I. Sinyukova. 1970. Food supply and the causes of mortality among the larvae of some common Black Sea fishes. J. Ichthyol. 10: 304–310.

de Mendiola, B.R. 1974. Food of the larval anchoveta *Engraulis ringens* J. pp. 277–285. *In*: J.H.S. Blaxter (ed.) The Early Life History of Fish, Springer-Verlag, Berlin.

Detwyler, R. & E.D. Houde. 1970. Food selection by laboratory-reared larvae of the scaled sardine *Harengula pensacolae* (Pisces, Clupeidae) and the bay anchovy *Anchoa mitchilli* (Pisces, Engraulidae). Mar. Biol. 7: 214–222.

Drake, C., J. Imbrie, J.A. Knauss & K. Turekian. 1978. Oceanography. Holt, Rinehart and Winston, New York. 447 pp.

Duka, L.A. 1969. Feeding of larvae of the anchovy *Engraulis encrasicholus maeoticus* Pusanov in the Azov Sea. J. Ichthyol. 9: 223–230.

Govoni, J.J., D.E. Hoss & A.J. Chester. 1983. Comparative feeding of three species of larval fishes in the Northern Gulf of Mexico: *Brevortia patronus, Leiostomus xanthurus,* and *Micropogonias undulatus*. Mar. Ecol. Prog. Ser. 13: 189–199.

Harrington, R.W. & E.S. Harrington. 1961. Food selection among fishes invading a high subtropical salt marsh: from onset of flooding through progress of a mosquito brood. Ecology 42: 646–666.

Houde, E.D. & J.A. Lovdal. 1984. Seasonality of occurrence, foods and food preferences of ichthyoplankton in Biscayne Bay, Florida. Est. Coast. Shelf Sci. 18: 403–419.

Houde, E.D. & R.C. Schekter. 1980. Feeding by marine fish larvae: developmental and functional responses. Env. Biol. Fish. 5: 315–334.

Hunter, J.R. 1980. The feeding behavior and ecology of marine fish larvae. pp. 287–330. *In*: J.E. Bardach, J.J. Magnuson, R.C. May & J.M. Reinhart (ed.) Fish Behavior and its Use in the Capture and Culture of Fishes, ICLARM Conf. Proc. 5, Manila.

Ivantsoff, V. 1978. Taxonomic and systematic review of the Australian fish species of the family Atherinidae with references to related species of the old world. Ph.D. Dissertation, Macquarie University, North Ryde. 701 pp.

Jenkins, G.P., N.E. Milward & R.F. Hartwick. 1984. Food of larvae of spanish mackerels genus *Scomberomorus* (Teleostei: Scombridae) in shelf waters of the Great Barrier Reef. Aust. J. Mar. Freshw. Res. 35: 477–482.

Johnson, L.R. 1982. Feeding chronology and daily ration of first-feeding larval Hawaiian anchovy, *Stolephorus purpureus*. M.Sc. Thesis, University of Hawaii, Honolulu. 78 pp.

Kauffman, T.A., J. Lindsay & R. Leithiser. 1981. Vertical distribution and food selection of larval atherinids. Rapp. P.-v. Reun. Cons. Int. Explor. Mer 178: 342–343.

Kohler, C.C. & J.J. Ney. 1982. A comparison of methods for quantitative analysis of feeding selection of fishes. Env. Biol. Fish. 7: 363–368.

Laroche, J.L. 1982. Trophic patterns among larvae of five species of sculpins (Family: Cottidae) in a Maine estuary. U.S. Fish. Bull. 80: 827–840.

Lasker, R. 1975. Field criteria for survival of anchovy larvae: the relationship between inshore chlorophyll maximum layers and successful first feeding. U.S. Fish. Bull. 73: 453–462.

Last, J.M. 1978a. The food of four species of pleuronectiform larvae in the Eastern English Channel and Southern North Sea. Mar. Biol. 45: 359–368.

Last, J.M. 1978b. The food of three species of gadoid larvae in the Eastern English Channel and Southern North Sea. Mar. Biol. 48: 377–386.

Lebour, M.V. 1921. The food of young clupeoids. J. Mar. Biol. Ass. U.K. 12: 458–467.

O'Brien, W.J., N.A. Slade & G.L. Vinyard. 1976. Apparent size as the determinant of prey selection by bluegill sunfish (*Lepomis macrochirus*). Ecology 57: 1304–1310.

Revelante, N. & M. Gilmartin. 1982. Dynamics of phytoplankton in the Great Barrier Reef lagoon. J. Plank. Res. 4: 47–76.

Schnack, D. 1974. On the biology of herring larvae in the Schlei Fjord, Western Baltic. Rapp. P.-v. Reun. Cons. Int. Explor. Mer 166: 114–123.

Shelbourne, J.E. 1953. The feeding habits of plaice post-larvae in the Southern Bight. J. Mar. Biol. Ass. U.K. 32: 149–59.

Snedecor, G.W. & W.G. Cochran. 1980. Statistical methods, 7th Edition. Iowa State University Press, Ames. 507 pp.

Sokal, R.R.& F.J. Rohlf. 1969. Biometry. W.H. Freeman and Co, San Francisco. 776 pp.

Stepien, W.P. Jr. 1976. Feeding of laboratory-reared larvae of the sea bream *Archosargus rhomboidalis* (Sparidae). Mar. Biol. 38: 1–16.

Stoecker, D.K. & J.J. Govoni. 1984. Food selection by young larval gulf menhaden (*Brevoortia patronus*). Mar. Biol. 80: 299–306.

Sumida, B.Y. & H.G. Moser. 1980. Food and feeding of Pacific hake larvae, *Merluccius productus,* off southern California and northern Baja California. Calif. Coop. Fish. Inv. Rep. 21: 161–165.

Tranter, D.J. & H.C. Heron. 1965. Filtration characteristics of

Clarke-Bumpus samplers. Aust. J. Mar. Freshw. Res. 16: 281–291.

Tranter, D.J. & P.E. Smith. 1968. Filtration performance. pp. 27–56. *In*: Zooplankton Sampling, Unesco Press, Paris.

Uotani, I., K. Matsuzaki, Y. Sakino, K. Noda, O. Inamura & M. Horikawa. 1981. Food habits of larvae of tuna and their related species in the area northwest of Australia. Bull. Jap. Soc. Sci. Fish. 47: 1165–1172.

Underwood, A.J. 1981. Techniques of analysis of variance in experimental marine biology and ecology. Oceanogr. Mar. Biol. Ann. Rev. 19: 513–605.

Winer, B.J. 1971. Statistical principles in experimental design. McGraw-Hill Kogakusha, Tokyo. 907 pp.

Received 15.12.1984 *Accepted 20.5.1985*

When to feed: decision making in sticklebacks, *Gasterosteus aculeatus*

David L.G. Noakes
Animal Behaviour Research Group, Department of Zoology, South Parks Road, Oxford University, Oxford OX1 3PS, England
Present address: Department of Zoology, Group for the Advancement of Fish Studies, University of Guelph, Guelph, Ontario N1G 2W1, Canada

Keywords: Behaviour, Causation, Choise, Displacement activities, Ethology, Function, Reproduction, Territory, Fish

Synopsis

The functional analysis of feeding, territory defense and nest-directed activities by reproductively mature male three-spined sticklebacks, *Gasterosteus aculeatus,* was carried out under controlled laboratory conditions. Territorial males were placed in motivational conflict between feeding and nest-directed activities by making performance of the two necessary activities spatially incompatible. During a limited time each day they had to choose between staying close to their nests to carry out nest-directed activities, and moving into a separate compartment of the observation tank to feed. Males were trapped for varying lengths of time within a central, neutral compartment as they moved between the nest compartment at one end of the tank and the feeding compartment at the other. Their choice of either compartment following such trapping was recorded as indicative of their dominant motivation state. Males tended to be nest dominant in these observations, but motivational dominance could be altered. Males deprived of food for longer periods of time, or presented with an empty food cup during the observation were more likely to be food dominant. If they had not been previously deprived of food, or if a sexually mature conspecific had just been presented in the territory, males were more likely to be nest dominant. The length of time males were trapped in a neutral compartment while in transit between the nest and food also significantly influenced their subsequent behavior. Longer interruptions were more likely to result in a change in the direction in which the male was moving, but only when he was moving towards the subdominant activity. This effect supports the hypothesis of time sharing as the mechanism regulating motivation in these fish.

Introduction

Few fish species have been so intensively studied, and yet still hold so much potential for ethological study as the three-spined stickleback, *Gasterosteus aculeatus* (e.g. Wootton 1976, Coad 1981). Previously, there was a preponderence of observational and experimental data with few hypotheses to account for them, but this balance has now shifted with a rapid growth in theorizing, so that now hypotheses exist which have yet to be tested by experimental observations. Earlier work on analysis of motivational mechanisms in this species (McFarland 1974a, Cohen & McFarland 1979) has dealt with some aspects of time sharing in courtship and nest-directed activities in territorial males. I investigated the possible motivational mechanisms involved in feeding and nest-directed activities in these males.

Motivation has been a long-standing topic of ethology (Tinbergen 1963) and, in addition to earlier questions of causation, more recently has become a cornerstone of functional considerations (McFarland 1974b). It has long been recognized that the behaviour of animals at times when they appear to have two or more mutually incompatible drives or tendencies is of particular interest and importance. How animals resolve such conflicts, i.e. which of the mutually incompatible activities they perform, or in which sequence they alternate between the alternatives, should provide an insight into their motivational mechanisms. Displacement activities have been a prime example of this type of consideration. These apparently irrelevant activities, performed when animals are thought to be in motivational conflict, have had a number of interpretations. Originally, they were suggested as truly irrelevant, at least with respect to the conflicting behavioural tendencies most in evidence (Tinbergen & van Iersel 1947, Tinbergen 1952). Subsequent investigators have variously proposed that displacement activities serve to allow the animal to re-adjust its motivational state (Wilz 1970a, 1970b), or that displacement activities may have evolved to serve a signal function (Ridley & Rechten 1981), or that they may in fact be functional activities (McFarland 1969, 1974b). The intention of this study was not to resolve the broad controversy of displacement activities, but to specifically address the question arising from a well-known example of a displacement activity, nest-directed activities by territorial sticklebacks. That question concerns the means by which the male apportions his time between these nest-directed activities and the mutually incompatible activity of feeding.

Behavioural activities

The male stickleback performs a number of nest-directed activities during the reproductive cycle (Tinbergen & van Iersel 1947, van Iersel 1953, Sevenster 1961, van den Assem 1967). Fanning is one of the most frequent, conspicuous and readily quantified of these. The function of fanning during the parental phase of the reproductive cycle has been established as enhancing the survival of embryos developing within the nest (Sevenster 1961, van Iersel 1970). Fanning also occurs during the sexual (= courtship) phase of the reproductive cycle, but there is still controversy as to its function in this context (Ridley & Rechten 1981). The form of fanning is essentially indistinguishable in the two different phases, and it has been argued that the motivational mechanisms may be the same (or similar) in the two cases as well (Sevenster 1961). Sexual fanning apparently results from a motivational conflict between sexual and aggressive systems. However, the mechanism whereby these conflicting motivations produce the temporal sequence and temporal patterning of fanning is still in debate (Wilz 1970a, 1970b, McFarland 1974a, Cohen & McFarland 1979).

Materials and methods

Animals and pre-treatment

Adult sticklebacks were captured in unbaited traps from ponds and canals near Oxford, England in January 1981. They were held collectively in aquaria (45 l or 100 l), with a layer of fine sand on the bottom and clumps of aquatic plants for cover. Each aquarium had a compressed air diffuser to aerate the water, and an ad libitum supply of live tubificid worms for food in a floating plastic cone. Live enchytraeid worms and dried tropical fish food were also provided from time to time. Water temperatures varied between 17° and 20° C, with overhead incandescent bulbs providing light on a 16:8 hour light:dark photoperiod. As individual fish began to show signs of reproductive maturity, males and females were segregated into separate holding tanks (45 l), with increased amounts of plants for cover. Males showing signs of nest building activity were chosen as needed for experimental observations. No male used in this study had the opportunity to spawn with a female during the current breeding season.

Residential territorial males with nests in one compartment of an observation tank were required to travel to a distant compartment in the tank to feed during a limited time alloted each day. The technique of trapping fish in transit through a neutral, middle compartment between the nest and feeding compartments and then judging their motivation by their choice of compartment when released (McFarland 1979, Cohen & McFarland 1979) was used.

Three kinds of behavioural activities were selected for study from among those essential for the successful reproductive behaviour of male sticklebacks: feeding, territorial defense and nest-directed activities. These three were defined as mutually incompatible activities, in terms of competing for the behavioural final common pathway (McFarland & Sibly 1975). All three have been studied and described in detail by previous investigators (van Iersel 1953, Sevenster 1961, van dem Assem 1967).

Observations were carried out in two large aquaria (260 by 50 by 40 cm deep). Each aquarium was divided into a series of three equal-sized compartments (60 by 50 by 40 cm deep) by opaque (white) plastic partitions (Fig. 1). Each partition had a vertical, sliding door that was normally kept closed, but that could be opened by remote control (similar to that described by McFarland 1974a). A layer of fine sand 5 cm deep covered the bottom of each tank, and clumps of artificial (plastic) plants were situated in the rear corners of each compartment to provide cover. Overhead incandescent lights provided the same 16:8 hour light:dark photoperiod as in the holding tanks. Thermostatically controlled heaters maintained water temperature at 20 ± 1° C.

The compartment at one end of each tank was the home compartment for a territorial male, the other two sections were the neutral and feeding compartments. Males showing nest building activity in the holding tanks were transferred as required to home compartments of observation aquaria. Any male that did not construct a nest in the home compartment within 12 h was replaced by another male from the holding tank. By placement of artificial plants (van dem Assem 1967, Jenni 1972) and, if necessary, gradually moving the nest as the male constructed it, the nest of each male ended up near the middle of the home compartment, about two-thirds of the way towards the end wall away from the partition into the adjacent (neutral) aquarium compartment (Fig. 1).

During a 30 minute period each day, while the

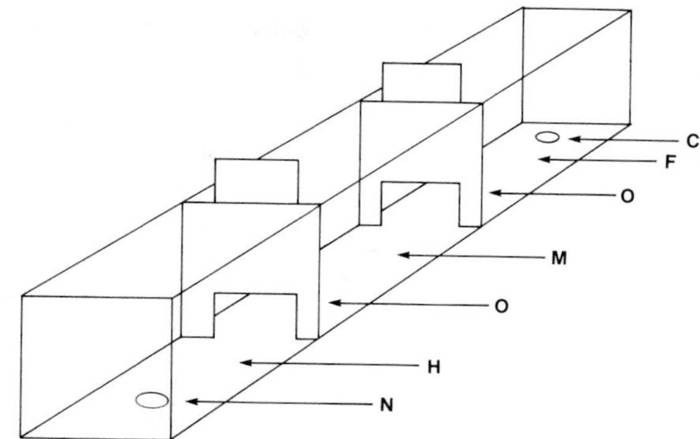

Fig. 1. Diagram of observation aquarium, showing division into sections by opaque plastic partitions (not to scale). N = nest in nest compartment of resident male, M = middle compartment, F = feeding compartment, O = opaque plastic partition (with vertical sliding door), C = plastic food cup, H = location where intruder presented in clear glass flask in nest compartment.

male was establishing his nest in his home compartment, he was fed live tubifex worms in a small plastic cup placed on the substrate near the partition. After about 3 or 4 days, the male was required to pass daily through the opened door in the partition, into the neutral compartment, to feed on the worms during that daily 30 minute period. No other food was provided to the males, and they quickly learned to respond to the presence of the cup by food-seeking and ingesting the worms. Each male was confined to his home compartment by closing the door in the partition at all other times.

After about an additional 3 or 4 days, males were required to pass through the neutral (middle) compartment and into the feeding (distant) compartment of the aquarium during the daily feeding period. The plastic cup with the live tubifex was now presented only in the feeding compartment during the daily feeding period. Males quickly adjusted to the procedure of moving through the open doors of the two partitions to reach the feeding compartment. They were then accustomed to the additional procedure of opening and closing the sliding doors in these partitions during the daily feeding period. Again, they quickly adjusted to this, and showed no signs of alarm or alteration in behaviour as a result of movement of the doors. When males were fully accustomed to moving to the feeding compartment each day, and to having doors opened and closed during that period, experimental observations were begun.

About once every second day while becoming accustomed to the testing procedure, each male was exposed to another male in breeding condition, confined in a clear glass flask in the home (nest) compartment for a period of about 10 minutes. A series of regularly-spaced holes around the bottom of the flask allowed for exchange of water between the flask and the aquarium. This exposure to another male was carried out during the time each day when the doors were not open. Resident males reacted quickly and vigorously to the intruding male in the flask (several different individual males were used as intruders during the experiment). Biting at the glass wall of the flask near the intruding male was the most frequent response of residents, but they also showed nest-directed activities on some occasions. On a few occasions, resident males were exposed to a reproductively mature female in the flask, instead of a rival male. Again, resident males responded immediately and vigorously, but now they showed mostly courtship (zig zagging, leading) and nest-directed (fanning, etc.) activities.

For each experimental male, a series of experimental observations was carried out once he was accustomed to feeding in the food compartment, to the opening and closing of doors with trapping in the neutral compartment, and to exposure to an intruder in a flask in the home compartment.

Preliminary observations established the number of live tubifex worms eaten by individual male sticklebacks each day under ad libitum conditions, and the minimum time period required to eat that number of worms. Live worms were supplied (about twice the number eaten daily under ad libitum conditions) in a layer of gravel in the bottom of a small, white plastic cup. This allowed more food than required for the male, but required him to feed on worms individually or at most only a few at a time, since he had to pull them out of the gravel. If worms were presented in a plain cup clustered into a tight mass, the males took one or two large mouthfuls and quickly satisfied their feeding requirements. Worms dispersed into the gravel and projected up out of the gravel individually a few minutes after being placed into the cup. This procedure (feeding worms in a layer of gravel) tended to equalize the feeding rate of males over a longer period of time. Presenting worms in the plastic cup, rather than on the substrate of the tank, prevented worms from becoming dispersed throughout the aquarium, and encouraged males to feed only from the cup. In fact, with experience, males quickly came to show food-searching only in or around the cup even if no worms were present inside the cup. As more than the typical daily ration of worms was present in the cup, this lessened the likelihood of males seriously depleting the food source during the course of the observation.

The basic experimental design followed that previously described by McFarland (1974a) and Cohen & McFarland (1979). While males were in the middle compartment, passing to and from the feeding and home compartments, they were trapped for varying period of time by closing the doors in the two partitions. The durations for which males were trapped in the middle compartment during observations were chosen according to criteria similar to those used by Cohen & McFarland (1979), to provide short, medium and long interruptions (5, 15 and 30 seconds, respectively). Each male was trapped twice for each duration, in each direction (nest to food, and food to nest) in each observation, with the sequence and duration of traps determined for each observation in advance by random lot. The male was allowed at least one uninterrupted passage in each direction through the middle compartment between traps.

During each observation, a spoken commentary of the position and behaviour of the male was made on a tape recorder. Males were trapped in the middle compartment according to the predetermined schedule, the duration of which was determined using a stopwatch, to the nearest second. The doors in the partitions were controlled remotely from behind an opaque black curtain, where I sat to make observations through a narrow viewing slot. The times that the doors were closed and opened for each interruption were recorded on the commentary of each observation, to be verified when the tape was transcribed. The record included: (1) the times at which the male passed through the doors between compartments; (2) the direction in which he was travelling; (3) the times at which he picked up individual tubifex worms from the feeding cup; (4) the times at which he performed any nest-directed activities (including durations, where applicable); and (5) any additional activities other than hovering in position or swimming from one compartment to another. Nest-directed activities were defined as fanning, pushing, boring, glueing, creeping through, adding plant material (filamentous algae) to the nest, and digging (see Sevenster 1961, van den Assem 1967, van Iersel 1970 for more detailed descriptions of these activities). Tape recorded observations were transcribed onto prepared data record sheets, including the times of all trap intervals in the middle compartment and the frequency and duration for all behavioural activities noted above.

In the course of these observations, males were variously exposed to an intruder in a clear glass flask in the home (nest) compartment for 10 minutes immediately prior to opening the doors to initiate the feeding period, or to no intruder prior to the daily observation. Males were normally allowed a 20 minute feeding period each day, whether or not there was a behavioural observation recorded. To test the effects of increased food deprivation, in some cases males were not given food for a period of 48 h prior to recording a behavioural observation. In some other cases, the food cup was placed in the food compartment as usual, but the cup contained only gravel and no tubifex worms.

Once established, males performed reliably over a period of about 3 weeks, with no indication of change in feeding or nest-directed behaviour. Males were removed from experimental observations when they ceased to respond aggressively to an intruder male in the flask in the home compartment. Invariably, this failure to respond to an intruder was followed within a day or so by the resident male losing his bright colouration and territorial behaviour, apparently reverting to non-reproductive condition.

Results

Males tended to move directly from one compartment to the other, with little hesitation, and with only sufficient time spent in the middle compartment for the transit between the feeding and nest compartments. While in the food compartment, the males' activities were focussed on the feeding cup, and consisted mainly of ingesting worms, or apparently searching for food in the cup. While in the nest compartment, males spent most of their time either close to the nest and showing one or more nest-directed activities, or swimming slowly or hovering elsewhere in the compartment. When trapped in the middle compartment, males usually held position by hovering or swimming slowly until the doors were re-opened. In some cases they moved back and forth close to one of the doors, usually moving through that door as soon as it was opened.

Although there was variability in the responses of males, both between trials for the same male and between males for the same condition, patterns emerging in the results seem clear. Males performed a number of activities, including several different nest-directed activities, but only feeding responses and fanning occurred with sufficient frequency and consistency across different observations to allow analyses. As expected, the general behaviour of males, including their territorial and nest-directed activities as well as their feeding behaviour, were as described by previous investigators. Consequently, only the quantitative aspects of their activities will be analysed and discussed.

It was necessary to pool results for analysis so an initial assessment of dominance had to be made for each observation. A male was judged to be food dominant if he returned to the food compartment following a majority of interruptions during that observation, and similarly nest dominant if he returned to the nest compartment. I used the specific criterion of 5 or 6 returns (out of 6 interruptions) to determine dominance, similar to the criterion established by Cohen & McFarland (1979).

Applying this criterion to a total of 76 observations on 6 different males, dominance could be assessed in 29 observations (38% of total) (Table

Table 1. Motivational dominance shown by male sticklebacks during observations with interruptions. Dominance decided on basis of 5 or 6 returns to particular compartment following 6 interruptions.

Observation condition	Motivational dominance during observation
No intruder, no food	2 Food dominant, 12 no dominance
No intruder, food present	4 Nest, 2 Food, 6 no dominance
Male intruder, no food	1 Food, 11 no dominance
Female intruder, food present	4 Nest, 4 no dominance
Male intruder, food present	16 Nest, 14 no dominance

1). Of the dominant observations, 24 (83%) were nest dominant, and 5 (17%) were food dominant. (Using a more liberal criterion of 4 or more returns, 78% of the 76 observations were dominant, with 81% of those nest dominant and 19% food dominant). The conditions of the observation appeared to influence dominance, so the relative numbers of nest and food dominant observations should be interpreted in that light (see below).

Initial analysis assessed the effect of length of the interruptions on the behaviour of the males, for which only dominant observations were used (n = 29). The behaviour of each male following each interruption was scored as moving towards the dominant or subdominant compartment (without regard to whether a particular observation was nest or food dominant). The results (Table 2) show a significant (χ^2, $p = 0.02$) effect of the length of interruptions of subdominant behaviour, in which short interruptions were less likely to cause the male to change direction than were either medium or long interruptions.

On the other hand, interruptions of dominant behaviour (Table 3) had no effect, regardless of the length of the interruption. The probability of returning to the subdominant compartment (i.e. changing direction following the interruption) was not significantly different (χ^2, $p = 0.30$) (and not significantly different from zero) for interruptions of any length.

Comparisons of individual behavioural measures between nest and food dominant observations (Table 4), between food dominant and no dominance observations (Table 5), and between nest dominant and no dominance observations (Table 6) showed a number of significant differences. In all cases these differences were significant (t-test, $p<0.05$) in favour of the dominant activity.

Nest dominant observations had longer mean nest visits, more fanning bouts, longer total fanning duration, more nest activities (other than fanning), and more time spent in the nest compartment than did food dominant observations (Table 4). Nest dominant observations had more fanning bouts, a greater total duration of fanning, a greater percentage of time spent in the nest compartment, and a longer mean duration of nest visits than did observations with no dominance (Table 6). Observations with no dominance had a greater percentage of time spent in the food compartment, and a longer mean duration of visits in the food compartment than did the nest dominant observations. There was no significant difference between these two

Table 2. Returns to 'dominant' compartment, following interruptions of varying lengths, by male sticklebacks when moving from 'dominant' towards 'subdominant' compartment, i.e. interruption of 'subdominant' activity (values expressed as percentage of totals). Total of 29 observations from six males.

Return to:	Following interruption of: 5 seconds	15 seconds	30 seconds
Dominant	69%	90%	93%
Subdominant	31%	10%	7%
Total interruption	29	29	29

Table 3. Returns to 'subdominant' compartment, following interruptions of varying lengths, by male sticklebacks moving from 'subdominant' towards 'dominant' compartment, i.e. interruption of 'dominant' activity. Total of 29 observations from six males.

Return to:	Following interruption of: 5 seconds	15 seconds	30 seconds
Dominant	100%	96%	93%
Subdominant	0%	4%	7%
Total interruptions	29	29	29

sets of observations in terms of the total nest activities other than fanning.

Food dominant observations, conversely, had significantly longer visits to the food compartment and a greater percentage of time spent in the food compartment than did nest dominant observations. Food dominant observations had a greater percentage of time spent in the food compartment, and greater mean duration of visits to the food compartment than did observations with no dominance (Table 5). The observations with no dominance had a greater percentage of time spent in the nest compartment, and longer mean duration of visits to the nest compartment.

An overview of the conditions of each observation, as related to dominance shows some apparent patterns (Table 1). Nest dominance occurred only in observations in which food was present during the observations. There were also several observations with no dominance under this condition, so

Table 4. Comparisons of 'nest dominant' (ND) and 'food dominant' (FD) observations of individual behavioural measures of male sticklebacks (for ND, n = 24; for FD, n = 5). (Diff. SIG indicates p<0.05; NON indicates p>0.05, t-test).

Behavioural measure	ND mean	FD mean	Diff.
Nest visits, mean duration	33.7	9.2	SIG
Food visits, mean duration	30.8	75.0	SIG
Fanning bouts	6.3	0.0	SIG
Fanning duration	48.2	0.0	SIG
Total activities directed to nest, other than fanning	32.6	2.8	SIG
% Time in nest compartment	43.5	5.6	SIG
% Time in food compartment	28.5	70.0	SIG

Table 5. Comparisons of differences between 'food dominant' (FD) and observations with no dominance (O) for individual behaviour measures of male sticklebacks (for FD, n = 5, for O, n = 47). (Diff. SIG indicates p<0.05, NON indicates p<0.05, t-test).

Behaviour measure	FD mean	O mean	Diff.
% Time in nest compartment	5.6	25.0	SIG
% Time in food compartment	70.0	44.5	SIG
Nest visits, mean duration	9.2	21.0	SIG
Food visits, mean duration	75.8	47.4	SIG

Table 6. Comparisons of differences between 'nest dominant' (ND) and observations with no dominance (O) on individual behaviour measures of male sticklebacks (for ND, n = 24, for O, n = 47). (Diff. SIG indicates p<0.05, NON indicates p>0.05, t-test).

Behaviour measure	ND mean	O mean	Diff.
Fanning bouts	6.2	2.2	SIG
Fanning, total duration	48.5	17.9	SIG
Nest activities, total other than fanning	31.0	22.8	NON
% Time in nest compartment	43.5	25.6	SIG
% Time in food compartment	28.5	44.6	SIG
Nest visits, mean duration	33.7	21.8	SIG
Food visits, mean duration	30.8	47.1	SIG

the relationship is not a necessary one. In fact observations with no dominance occurred on some occasions under every condition tested. Food dominance occurred under either one of two conditions. If no food was present during the observation (i.e. no worms in the cup in the food compartment), some fish were food dominant. The other condition under which fish were sometimes food dominant was if food was present but there had not been any conspecific intruder presented to the resident male prior to the observation.

Discussion and conclusions

The predictions from a time sharing motivational hypothesis (Cohen & McFarland 1979) include two related specifically to the effects of interruptions of ongoing behaviour (Fig. 2). Interrupting dominant behaviour should have no effect on the subsequent behaviour, regardless of the length of the interruption. That is precisely the result seen in this study (Table 3). Interrupting subdominant behaviour, however, should result in the animal continuing to the subdominant compartment only if the interruption is of short duration. Again, the results agree precisely with this prediction (Table 2). The fact that these predictions are fulfilled can be taken as support for the time sharing hypothesis. This conclusion is not unexpected, given previous evidence of this mechanism in other studies (McFarland &

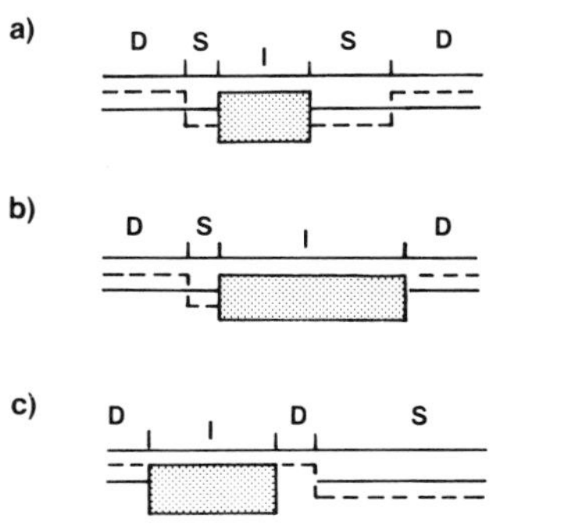

D = performance of dominant activity, S = performance of subdominant activity, I = interruption (trap)
——— = level of motivational factors for subdominant activity
- - - - = level of motivational factors for dominant activity
▭ = interruption (trap)

Fig. 2. Diagram of trapping procedure, to indicate assumed effects of traps of different lengths on motivational state of test fish (from McFarland 1974b). Time progresses from left to right.

Lloyd 1973, McFarland 1974a, 1983a, b, 1985, Sibly & McFarland 1976), including courtship behaviour of sticklebacks (Cohen & McFarland 1979).

Dominance in observations in this study also correlated with other behavioural measures of motivational state. Nest dominant observations had significantly more (and/or longer) nest visits and nest-related activities than did either food dominant or no dominance observations. The same was true for feeding activities in food dominant observations, as compared to nest dominant or no dominance observations. This agrees with similar findings on stickleback courtship (Cohen & McFarland 1979), and suggests that the performance of nest-related activities during periods of feeding arises from factors associated with nest maintenance, and is not an example of irrelevant or motivationally displaced behaviour.

The effects of observation conditions on the probability of motivational dominance are predictable, given the hypothesis that both kinds of behaviour (nest activities, feeding) are functionally significant. With only one exception, exposure of a male to an intruder near his nest increased the likelihood that the male would become nest dominant. The sex of the intruder appeared to have no effect, similar to the results of other observations of courtship fanning (Noakes, unpublished). With no prior exposure to an intruder, resident males were not subsequently nest dominant, but were more likely to be food dominant. If food was withheld during an observation, that also increased the likelihood of food dominance (through its effect on hunger, presumably).

These manipulations would be expected to alter the respective (feeding, nest maintenance) motivational state of the resident male. The fact that the manipulations alter the motivational dominance in subsequent observations confirms the functional significance of the two kinds of behaviour in the observations.

The numbers of food and nest dominant observations sessions should be interpreted in light of this, since the numbers of observations under different conditions were not equal (Table 1). There were only 26 observations in which no intruder was presented, compared to 42 with an intruder. Similarly, there were only 25 observations with no food present, but 51 with food present. Knowing the relationship between these treatments and motivational dominance, we would expect there to be more nest dominant than food dominant observations (about 51 nest dominant and 25 food dominant observations). The actual numbers, 24 nest dominant and 5 food dominant, suggest that nest dominance was relatively, as well as absolutely, more common than food dominance.

In light of current hypotheses concerning the optimality of behaviour, it is worth noting the distinction made by McFarland (1985). The behaviour we observe in an animal can be thought of as optimal in two different ways. It could be optimal in terms of the best way for the animal to accomplish a particular objective, i.e. optimal in the functional sense. It could also be though of as optimal in the causal sense, i.e. as the best compromise for the motivational state of the animal. For a variety of reasons, elaborated at length by a number of authors (e.g. Dawkins 1982, McFarland 1985) it is unlikely that any particular animal will be truly optimal at any point in time. However, the concept of optimality is useful as an idealized standard we can use for postulating expected values toward which an animal might be moving.

Thus we can assume that the causal mechanism(s) evolved to regulate the feeding and nest maintenance behaviour of territorial male sticklebacks, and the functional consequences of the performance of those two types of behaviour in particular sequences and for particular durations, should at least approximate optimal solutions to the problems faced by the fish. All fish must feed at least some time, in order to survive as vehicles, but just as surely all fish must breed to survive as replicators (Dawkins 1982).

Time sharing appears to be the mechanism regulating the expression of conflicting behavioural activities in a variety of circumstances for a number of animal species, although the reasons why this should be so are not yet clear (McFarland 1985).

Acknowledgements

Thanks to David McFarland for providing space and facilities in the A.B.R.G. to carry out this work, and for his discussion and suggestions concerning the work and analysis of results, to Mr. Hardy, Biology Master at Radley College for permission to collect sticklebacks on the College grounds; to Nick Styles for assistance in capturing and maintaining the fish; to Catie Rechten and Mark Ridley for advice on collecting, maintaining and observing sticklebacks; to Karen Hollis, Cathy Kennedy and Bori Olla for discussions of time sharing; to Gene Helfman, an anonymous referee, Greg Cailliet and Charles (Si) Simenstad for constructive criticisms and comments on the manuscript, to Pat and Jeff for their understanding; and to Tim Beardsley, Dennis Lendrem, Meg McVey and Catie Rechten for tolerating my presence with grace and good humour, making my stay a memorable and enjoyable experience. Financial support was provided by a Faculty Research Leave Award from the University of Guelph, and an operating grant from N.S.E.R.C., Canada.

References cited

Coad, B.W. 1981. A bibliography of the sticklebacks (Gasterosteidae: Osteichthyes). Syllogeus 35: 1–142.

Cohen, S. & D.J. McFarland. 1979. Time-sharing as a mechanism for the control of behaviour sequences during courtship of the threespined stickleback (*Gasterosteus aculeatus*). Anim. Behav. 27: 270–283.

Dawkins, R. 1982. The extended phenotype. W.H. Freeman, San Francisco. 382 pp.

Jenni, D.A. 1972. Effects of conspecifics and vegetation on nest site selection in *Gasterosteus aculeatus* L. Behaviour 42: 97–118.

McFarland, D.J. 1969. Mechanisms of behavioural disinhibition. Anim. Behav. 17: 238–242.

McFarland, D.J. 1974a. Time-sharing as a behavioural phenomenon. pp. 251–282. *In:* D. Lehrman et al. (ed.) Advances in the Study of Behavior, Vol. 5, Academic Press, New York.

McFarland, D.J. 1974b. Motivational Control Systems Analysis. Academic Press, New York. 425 pp.

McFarland, D.J. 1983a. Behavioural transitions: a reply to Roper and Crossland (1982). Anim. Behav. 31: 305–307.

McFarland, D.J. 1983b. Time sharing: a reply to Houston (1982). Anim. Behav. 31: 307–308.

McFarland, D.J. 1985. Animal behaviour. Pitman Press, New York. 576 pp.

McFarland, D.J. & I.H. Lloyd. 1973. Time-shared feeding and drinking. Quart. J. Exp. Psychol. 25: 48–61.

McFarland, D.J. & R.M. Sibly. 1975. The behavioural final common path. Phil. Trans. Roy. Soc. B, 270: 265–293.

Ridley, M. & C. Rechten. 1981. Female sticklebacks prefer to spawn with males whose nests contain eggs. Behaviour 76: 152–161.

Sevenster, P. 1961. A causal analysis of a displacement activity (fanning in *Gasterosteus aculeatus* L.). Behaviour Suppl. 9: 1–170.

Sibly, R.M. & D.J. McFarland. 1976. On the fitness of behaviour sequences. Amer. Nat. 110: 610–617.

Siegel, S. 1956. Nonparametric statistics for behavioural sciences. McGraw-Hill, New York. 276 pp.

Tinbergen, N. 1952. 'Derived' activities; their causation, biological significance, origin and emancipation during evolution. Quart. Rev. Biol. 27: 1–32.

Tinbergen, N. 1963. On the aims and methods of ethology. Z. Tierpsychol. 20: 410–433.

Tinbergen, N. & J.J.A. van Iersel. 1947. 'Displacement reactions' in the three-spined stickleback. Behaviour 1: 56–63.

van dem Assem, J. 1967. Territory in the three-spined stickleback, *Gasterosteus aculeatus* L. An experimental study in intra-specific competition. Behaviour Suppl. 16: 1–164.

van Iersel, J.J.A. 1953. An analysis of the parental behaviour of the three-spined stickleback (*Gasterosteus aculeatus* L.). Behaviour Suppl. 3: 1–159.

Wilz, K.J. 1970a. Causal and functional analysis of dorsal pricking and nest activity in the courtship of the three-spined stickleback, *Gasterosteus aculeatus*. Anim. Behav. 18: 115–124.

Wilz, K.J. 1970b. The disinhibition interpretation of the 'displacement' activities during courtship in the three-spined stickleback, *Gasterosteus aculeatus*. Anim. Behav. 18: 682–687.

Wootton, R.J. 1976. The biology of sticklebacks. Academic Press, London. 386 pp.

Received 21.12.1984 *Accepted 18.7.1985*

The relationship of diet to growth and ammonium excretion in salt marsh fish

David S. White[1]*, Charlene D'Avanzo[2], Ivan Valiela[1], Carlos Lasta[3] & Miguel Pascual[4]
[1] *Boston University Marine Program, Marine Biological Laboratory, Woods Hole, MA 02543, U.S.A.*
[2] *School of Natural Science, Hampshire College, Amherst, MA 01002, U.S.A.*
[3] *Instituto Nacional de Investigacion y Desarrollo Pesquero, C.C. 175, 7600 – Mar del Plata, Argentina*
[4] *Instituto de Biologia Marina y Pesquera 'Alte. Storni', Av. Costanera s/n, 8520 – San Antonio Oeste, Argentina*

Keywords: Detritus, Epibiota, Nutrition, Assimilation, Mortality, Adaptations, Maintenance, Food chain dynamics, Feeding ecology

Synopsis

Detritus is an abundant but poor quality food source for consumers in salt marsh ecosystems. Here we present results of feeding experiments to determine the ability of *Fundulus heteroclitus, Cyprinodon variegatus,* and *Mugil cephalus,* three major detritivores in Great Sippewissett Marsh, Massachussetts, to assimilate detritus and use it for growth. *C. variegatus,* the sheepshead minnow, gained weight on a detrital diet, but *F. heteroclitus,* the marsh killifish, and juvenile *M. cephalus,* the striped mullet, lost weight and suffered high mortality on detrital diets. *C. variegatus* is a herbivore with morphological adaptations for ingesting plant material. *F. heteroclitus* is a carnivore poorly suited to effectively assimilate detritus from the diet. Although adult *M. cephalus* are adapted for ingesting and assimilating detrital material, the young may lack these adaptations and thus do not assimilate detritus. *C. variegatus* excretes ammonium at a lower rate than *F. heteroclitus* when fed animal food, or when starved for short periods of time. This protein sparing effect could be crucial for survival when detritus is the only food available at certain times of the year. While both *C. variegatus* and *M. cephalus* are visitors in the marsh and may leave to exploit food sources elsewhere, *F. heteroclitus* is a year-round resident and is the most abundant species of fish in the marsh. Yet when high quality food sources become scarce in late summer and fall, detritus, although plentiful, is apparently not a suitable alternative.

Introduction

Detritus is an abundant but poor quality food source for consumers in coastal marine ecosystems (Odum et al. 1979, Rice 1982, Tenore et al. 1982). In highly productive salt marshes of the eastern United States, average aboveground plant production ranges from 1–2 kg dry wt m^{-2} yr^{-1} (Reimold et al. 1975, Valiela et al. 1982). Less than ten percent of the *Spartina alterniflora* production is consumed directly by herbivores. The remainder forms the basis of a complex detrital food chain (Teal 1962). This litter is composed mainly of refractory organic compounds, since the potentially more nutritious labile dissolved organics such as proteins and amino acids quickly leach out of the newly dead matter (Tenore & Rice 1980, Marinucci et al. 1983, Valiela et al. 1985).

Several of the fish species that occur in the salt marsh ingest large amounts of *Spartina* detritus (Field 1906, Darnell 1964, Odum 1970, Jeffries 1972, Clymer 1978, Peters & Schaaf 1981, Werme 1981). Three of these species, the killifish *Fundulus heteroclitus,* the sheepshead minnow *Cyprinodon*

* To whom all correspondences should be addressed.

variegatus, and the striped mullet *Mugil cephalus* represent a gradient in the amount of detritus food in their guts. *F. heteroclitus* has a diet that ranges from 20–40% detritus, 5–80% algae, and 5–40% animals (Werme 1981). This species is a common year-round resident of the marsh, comprising approximately ninety per cent of the fish present (Werme 1981). *C. variegatus* is a herbivore whose diet may contain as much as 90% detritus (Werme 1981). It is not a permanent resident of the marsh, being present throughout the summer and early fall. *M. cephalus* spawns in offshore waters during the fall and winter, and the young enter the marsh, using it as a nursery (Anderson 1958, Arnold 1958, McHugh 1967). In salt marshes, its diet consists of approximately 50% benthic algae and vascular plant detritus, and 50% inorganic sediment (Odum 1970).

These three species of fish show a range of differing adaptations for feeding on detritus. *F. heteroclitus* seems ill-suited for detritivory. It has no teeth for shredding plant material, and a relatively short gut (equal to its body length), a characteristic typical of carnivorous cyprinid fishes (Kapoor et al. 1975). In contrast, *C. variegatus* has a gut length approximately three times its body length, and stout pharyngeal and large tricuspid teeth, characteristics that seem better suited for herbivory or detritivory (Bigelow & Schroeder 1953, Harrington & Harrington 1961, Warlen 1964, Dahlberg 1972, Derickson & Price 1973, Kapoor et al. 1975, Clymer 1978, Humphries & Miller 1981). Likewise, *M. cephalus* has several adaptations suited to detritivory, including a gut length to body length ratio of from 3–5 (larger fish tend to have bigger ratios), a highly modified gizzard-like pyloric stomach for grinding plant material (Odum 1970, Kapoor et al. 1975), and an alkaline hindgut which may function to chemically release and absorb proteins bound up in refractory compounds common in vascular plant detritus (Payne 1978, Berenbaum 1980, Martin et al. 1980, Bernays & Woodhead 1982). These three species of fish provide us with a gradient of detritivory from the carnivorous *F. heteroclitus* to the more detritivorous *C. variegatus*.

Several field and laboratory studies have been conducted to estimate growth rates and to assess the ability of the detrital component of the diet to promote growth in salt marsh fish. *F. heteroclitus* and *M. cephalus* in the field grow significantly more slowly late in the summer than during the spring or early summer, regardless of age (Cech & Wohlschlag 1975, Valiela et al. 1977). This lowered growth rate may be related to high water temperatures or to seasonal changes in food supply (Cech & Wohlschlag 1975). *F. heteroclitus* populations seem closely coupled to the abundance of their food. Unlike what seems to be the case for pelagic fish (Valiela 1984), the growth and mortality of coastal shallow-water species such as *F. heteroclitus* may thus be density dependent. In experiments where densities of fish were manipulated (Kneib 1981, S.B. Weisberg & B.A. Lotrich, unpublished data), with no simultaneous alteration of food supply, there were decreases in growth and increases in mortality where densities were higher. These density dependent effects occurred within the range of densities of natural populations of *F. heteroclitus*. This fish thus lives at densities that can tax available food resources.

Werme (1981) found that the proportion of animal food in the gut contents of *F. heteroclitus* and eight other species of fish in a salt marsh is sharply reduced during late summer relative to early summer. The number of macroinvertebrates available to fish in creek bottoms declined as the number of fish increased through the summer months. This pattern suggests that predation pressure by the fish may limit their prey and may therefore be the cause of the decrease in animal food in their diets over the course of the summer. Data of Wiltse et al. (1984) support this conclusion. In a series of caging experiments designed to exclude predators from portions of the bare creek bottoms, macroinvertebrates within cages increased in abundance while those not protected by cages showed a seasonal decline similar to that found by Werme (1981). Predation clearly has an important effect on the benthic invertebrates in salt marsh creek bottoms.

We have calculated, from earlier data on growth and production (Prinslow et al. 1974, Valiela et al. 1977), that the consumption of benthos by *F. heteroclitus* alone is sufficient to turn over the entire population of macrobenthos within weeks to a few months. The gut analysis by Werme (1981) supports this conclusion. While the gut of *F. heteroclitus* contained over 50% animal food in the spring, algae and detritus were by far more common in the summer.

If predation by *F. heteroclitus* and other species of fish decimates the benthic invertebrates, they may be forced to consume lower quality foods such as detritus late in the summer. Consequently their growth rates may decline during this time. Prinslow et al. (1974) found that near-adult *Fundulus heteroclitus* grew no better on animal food diets to which detritus was added than on the animal food diets alone.

Fish can use ingested protein for growth or as a substrate for oxidative metabolism. When a food contains protein in excess of dietary requirements for growth, fish can use this excess as an energy source. Amino acids are deaminated and the nitrogen is excreted primarily in the form of ammonium. The amount of ammonium excreted is directly proportional to the amount of protein in the diet (Cowey & Sargent 1972, 1979, Brett & Groves 1979, Walton & Cowey 1982). Animal food diets should thus yield greater growth and higher rates of ammonium excretion than algal and detrital diets.

In this paper we have investigated the ability of three species of salt marsh fish, adult *F. heteroclitus, C. variegatus,* and juvenile *M. cephalus* to use animal, algal, and detrital diets by measuring growth and rates of ammonium excretion in fish fed each of these diets.

Materials and methods

Individuals of all three species were collected from the creeks of Great Sippewissett Marsh, Falmouth, Massachussetts using a 5 mm mesh seine. Individuals (0.3–0.7 g wet wt) from each species were divided into groups to be assigned to one of five dietary regimes: (1) animal food, consisting of pieces of the ribbed mussel *Geukensia demissa* (*F. heteroclitus* and *C. variegatus* only), or a standard commercial fish food (*M. cephalus* only), (2) *Spartina* detritus obtained from the bottom of tidal creeks, (3) standing dead *Spartina* litter, (4) a green filamentous alga *Enteromorpha clathrata,* or (5) starved controls. All fish were kept in continuously bubbled 1 gallon jars at a temperature of approximately 19–20° C. Groups were fed ad libitum, and water and food were changed every two days.

For the growth study, three groups of ten fish each were assigned to each of the above diets (*M. cephalus* was not placed on the detrital diet and there were no starved controls for *F. heteroclitus* or *C. variegatus*). Wet weights were recorded before assignment and again after 30 d on a particular diet (12 d for *M. cephalus*). Growth was expressed as per cent initial weight per day. Mortalities for each group were recorded after 50 d (12 d for *M. cephalus*).

To determine rates of ammonium excretion, three groups of three fish each (one group for *M. cephalus*) were starved for 48 h and then placed on each of four diets (the detritus diet was not included in this experiment and only *M. cephalus* was placed on the algal diet). After a 2 h feeding period, each group was placed in a darkened, sealed 500 ml screw-cap glass jar filled with unfiltered sea water which had been bubbled at room temperature for 12 h to prevent oxygen depletion. Seawater controls without fish were also prepared in a similar manner to account for ammonium excretion by microscopic heterotrophs. Water was gently mixed with a magnetic stir bar which was separated from the fish by nylon mesh. Aliquots of 8 ml were removed every 5 min for 15 min. The concentration of ammonium in these samples was immediately analyzed by means of a Technicon Autoanalyzer II, using the phenol-hypochlorite method of Solorzano (1969). The rate of ammonium excretion was calculated by fitting linear regression lines to the increase in concentration of ammonium over time for fish on each of the four diets, and for the seawater controls. Net rates of ammonium excretion were obtained by subtracting the rates of the control jars from those with fish. Weight-specific rates of ammonium excretion were then determined by dividing the net rate by the wet weight of the fish in each of the jars.

Results

All three species gained weight on the animal diets (Table 1). *C. variegatus* gained weight on the detritus diet, while *F. heteroclitus* lost weight (*M. cephalus* was not fed detritus). *C. variegatus* maintained its weight on the algal diet, while *F. heteroclitus* lost weight. *M. cephalus* lost weight at a rate almost equivalent to that of starved controls. All three species lost weight on the litter diet, but *C. variegatus* lost weight more slowly than *F. heteroclitus*. *M. cephalus* lost weight at a rate equivalent to that of the starved controls.

All three species survived well on the animal diets (Table 2). The ten percent mortality in *M. cephalus* was probably caused by a fungal infection noted at the beginning of the experiment. *F. heteroclitus* experienced very high mortality on all three of the other diets while *C. variegatus* suffered very little or no mortality on these same diets. *M. cephalus* showed higher mortality than *C. variegatus* but significantly lower than *F. heteroclitus*.

Rates of ammonium excretion (Table 3) for all three species were significantly higher ($p<0.05$) on animal food diets than on either litter or algal (*M. cephalus* only) diets. Excretion rates of fish fed litter or algae (*M. cephalus* only) were not significantly different ($p<0.05$) from those of starved controls, with the exception of *F. heteroclitus*, whose rate of excretion for starved controls was higher than for those on the litter diet.

Table 1. Weight change, expressed as percent initial weight per day $\bar{x} \pm$ s.e., for *Fundulus heteroclitus*, *Cyprinodon variegatus*, and *Mugil cephalus* on different diets.

Diet	*F. heteroclitus*	*C. variegatus*	*M. cephalus*
Animal food	+0.5 ± 0.15	+0.68 ± 0.1	+5.1 ± 0.3
Detritus	−1.2 ± 0.12	+0.28 ± 0.06	
Litter	−13.5 ± 1.0	−8.4 ± 1.2	−1.9 ± 0.4
Algae	−8.2 ± 0.68	−0.01 ± 0.3	−1.5 ± 0.2
Starved			−2 ± 0.1

Table 2. Percent mortality on various diets for *Fundulus heteroclitus* and *Cyprinodon variegatus* after 50 d, and for *Mugil cephalus* after 12 d.

Diet	*F. heteroclitus*	*C. variegatus*	*M. cephalus*
Animal food	0	0	10
Detritus	77	0	
Litter	100	5	37.5
Algae	54	5	30
Starved			52

Table 3. Rates of ammonium excretion* (u-moles NH_4-N g wet wt^{-1} h^{-1}, $\bar{x} \pm$ s.e.) of *Fundulus heteroclitus*, *Cyprinodon variegatus*, and *Mugil cephalus* on various diets.

Diet	*F. heteroclitus*	*C. variegatus*	*M. cephalus*
Animal food	7.4 ± 2.2	2 ± 0.2	5.8
Litter	0.2 ± 0.2	0.2 ± 0.2	3.7
Algae			3.6
Starved	1.5 ± 0.2[a]	0.4 ± 0.2[a]	3.1
	1.4 ± 0.2[b]	0.6 ± 0.07[b]	

[a] 1983 data.
[b] 1984 data.
* A multiple comparisons test was made of slopes of regression lines to determine significant differences in rates of ammonium excretion for each species on each of the above diets.

Discussion

F. heteroclitus is not able to use detritus, litter, or algae for growth or maintenance. It suffered high weight loss and mortality on these diets. The low rate of ammonium excretion on the litter diet shows that it did not use litter protein as an energy source. *F. heteroclitus* must therefore rely on animal food as its primary source of protein for growth and maintenance, a conclusion consistent with its known adaptations for carnivory. To a small extent, it may be able to supplement its diet with creek detritus colonized by microorganisms. Several studies have shown that although plant detritus itself is a poor quality food, associated epibiota are a potentially better source of nutrition (Fenchel 1970, Hargrave 1970, Cammen 1980, Tenore & Findlay 1982, Stuart et al. 1982, Newell & Field 1983, Seiderer et al. 1984). Weight loss of *F. heteroclitus* on the detrital diet was not as great as it was on the algal and litter diets. Preliminary results of another study using the stable isotope tracer ^{15}N

(data of C. D'Avanzo, to be published elsewhere) indicate that *F. heteroclitus* was not able to assimilate either ^{15}N labelled detrital plant matter or its associated epibiota. Detritus alone is not sufficient for maintenance and growth as evidenced by the 77% mortality rate after 50 d. The high rate of ammonium excretion of the starved controls relative to that of fish on the litter diet could be due to catabolism of body proteins. Carnivores have the enzymes capable of breaking down these body proteins (Brett & Groves 1979), and during periods of starvation, *F. heteroclitus* may slough off and oxidize them, thereby increasing ammonium excretion.

C. variegatus is better able than *F. heteroclitus* to survive on plant diets. It suffered no weight loss and very low mortality (5%) on the algal diet, although it could not grow on it. It did lose weight on the litter diet and the rate of ammonium excretion was not significantly different from starved controls, but the rate of weight loss and the mortality were less than for *F. heteroclitus*. *C. variegatus* gained weight and suffered no mortality on the detrital diet. The epibiota associated with the detritus may provide enough protein for growth. Preliminary data of C. D'Avanzo (to be published elsewhere) indicate that, unlike *F. heteroclitus, C. variegatus* assimilated ^{15}N labelled epibiota associated with detritus, but not the labelled plant matter itself. In addition to its morphological adaptations for herbivory already discussed, *C. variegatus* may be able to retain assimilated protein for growth more efficiently than *F. heteroclitus*. This physiological adaptation may explain why ammonium excretion by *C. variegatus* when fed the animal diet was significantly less than for *F. heteroclitus*. When the amount of protein in the diet is small, such as with detritus or algae, this protein sparing quality in *C. variegatus* may account for the lower weight loss and mortality on the detrital, algal and litter diets.

Young *M. cephalus* fed algal and litter diets are not able to use them for growth and maintenance. Growth rates on these diets were not significantly different from those of starved controls. Mortality was significant, although not as high as that for *F. heteroclitus* on the same diet. Cech & Wohlschlag (1975) found reduced growth rates in *M. cephalus* from Texas coastal waters during the summer. They suggested that the high water temperatures, rather than a decline in food quality or abundance, caused the reduced growth rates due to a temperature-dependent shift to less efficient metabolic pathways. Results of the present study do not discount this possibility, but do point to the importance of food quality as a determinant of growth in *M. cephalus*. They must depend on at least some animal food in their diet. Previous dietary studies of *M. cephalus* (Odum 1970) were based on gut analysis where detritus was an important part of the digestive contents. The fish studied by Odum averaged 200 mm in length, had developed a gizzard-like buccal sac for grinding plant matter, and had a long gut length to body length ratio for digesting it. Perhaps the smaller, younger mullet in the present study had not yet developed these adaptations and consequently relied more heavily on animal food for growth and maintenance than the larger ones. Alternatively, the detritus in the gut may not be efficiently assimilated. If so, then the nutritional importance of the detritus may be due to the epibiota associated with it. Moriarty (1976) showed that bacteria comprise 15–30% of the carbon intake of *M. cephalus* feeding on sea grass flats.

Microorganisms associated with detritus may be an important dietary component for other fish species in the marsh as well. For instance, based on calculated energy intake of various dietary components, Peters & Schaaf (1981) concluded that detritus was an important determinant of growth in juvenile Atlantic menhaden (*Brevoortia tyrannus*) inhabiting coastal estuaries. Peters & Lewis (1984) have shown that half of the organic material associated with *Spartina* detritus fed to juvenile menhaden was protein from bacteria, ruptured algae, and reaggregated dissolved amino acids. Since the results of the present study show that the *Spartina* detritus itself cannot promote growth in the three species investigated, but that associated epibiota can in at least one, it is possible that *B. tyrannus* may also use microorganisms associated with the detritus for growth.

In the salt marsh, when animal food becomes scarce during the latter part of the summer and fall, the three species included in this study may employ different strategies for coping with this problem. *C. variegatus* may rely more heavily on algal and detrital material (especially the associated epibiota) than benthic invertebrates as food sources, and since it is not a permanent resident of the marsh it may leave the marsh to seek prey elsewhere. *M. cephalus,* which comes into Great Sippewissett Marsh only as a juvenile, may exploit the benthic fauna available in the spring and early summer, and then leave the marsh to exploit other habitats. *F. heteroclitus* is the only one of these three species which is a year-round resident of the marsh. It can exploit the benthic fauna in the spring and early summer, but it may become severely food-limited during the late summer and fall because of its inability to use algae or detritus as an alternative food source. This conclusion is somewhat paradoxical because *F. heteroclitus* is by far the dominant fish species in Great Sippewissett Marsh. More research on the feeding ecology and behavior of this fish is necessary to more fully understand its role in the food chain dynamics and energy flow of the salt marsh ecosystem.

Acknowledgements

We thank Dorothea Gifford and Salt Pond Inc. for the generous use of their land for research, Melissa Morrison for helping to collect the data, Sarah Allen for data analysis, and Ken Foreman for helpful comments on the manuscript. Support for C. Lasta and M. Pascual was provided by the Tinker Foundation and the Marine Biological Laboratory, Woods Hole.

References cited

Anderson, W.W. 1958. Larval development, growth, and spawning of striped mullet (*Mugil cephalus* L.) along the south Atlantic coast of the United States. Fishery Bull. Fish Wildl. Serv. 58: 501–525.

Arnold, E.L. & J.R. Thompson. 1958. Offshore spawning of the striped mullet (*Mugil cephalus*) in the Gulf of Mexico. Copeia 1958: 84.

Berenbaum, M. 1980. Adaptive significance of midgut pH in larval Lepidoptera. Amer. Nat. 115: 138–146.

Bernays, E.A. & S. Woodhead. 1982. Plant phenols utilized as nutrients by a phytophagus insect. Science 216: 201–203.

Bigelow, H.B. & W.C. Schroeder. 1953. Fishes of the Gulf of Maine. Fishery Bull. Fish Wildl. Serv. 53: 1–577.

Brett, J.R. & T.D.D. Groves. 1979. Physiological energetics. pp. 279–352. *In:* W.S. Hoar, D.J. Randall & J.R. Brett (ed.) Fish Physiology, Vol. 8, Bioenergetics and Growth, Academic Press, New York.

Cammen, L.M. 1980. The significance of microbial carbon in the nutrition of the deposit feeding polychaete *Nereis succinea.* Mar. Biol. 61: 9–20.

Cech, J.J. Jr. & D.E. Wohlschlag. 1975. Summer growth depression in the striped mullet *Mugil cephalus* L. Contr. Mar. Sci. 19: 91–100.

Clymer, J.P. 1978. The distribution, trophic activities, and competitive interactions of three salt marsh killifishes (Pisces: Cyprinodontidae). Ph.D. Thesis, Lehigh University, Easton. 281 pp.

Cowey, C.B. & J.R. Sargent. 1972. Fish nutrition. Adv. Mar. Biol. 10: 383–492.

Cowey, C.B. & J.R. Sargent. 1979. Nutrition. pp. 1–69. *In:* W.S. Hoar, D.J. Randall & J.R. Brett (ed.) Fish Physiology, Vol. 8, Bioenergetics and Growth, Academic Press, New York.

Dahlberg, M.D. 1972. An ecological study of Georgia coastal fishes. Fishery Bull. Fish Wildl. Serv. 70: 323–345.

Darnell, R.M. 1964. Organic detritus in relation to secondary production in aquatic communities. Verh. int. Verein. theor. angew. Limnol. 15: 462–470.

Derickson, W.K. & K.S. Price Jr. 1973. The fishes of the shore zone of Rehoboth and Indian River Bays, Delaware. Trans. Amer. Fish. Soc. 102: 552–563.

Fenchel, T. 1970. Studies on the decomposition of organic detritus derived from the turtle grass, *Thalassia testudinum.* Limnol. Oceanogr. 15: 14–20.

Field, I.A. 1906. Unutilized fishes and their relation to the fishing industries. *In:* Report of the Commission of Fisheries 1906 and Special Papers, Bureau of Fisheries Document No. 622, Government Printing Office, Washington, D.C. 50 pp.

Hargrave, B. 1970. The utilization of benthic microflora by *Hyalella azteca* (Amphipoda). J. Anim. Ecol. 39: 427–437.

Harrington, R.W. & E.S. Harrington. 1961. Food selection among fishes invading a high tropical salt marsh. Ecol. 42: 646–665.

Humphries, J.M. & R.R. Miller. 1981. A remarkable species flock of pupfishes, genus *Cyprinodon,* from Yucatan, Mexico. Copeia 1981: 52–64.

Jeffries, H.P. 1972. Fatty-acid ecology of a tidal marsh. Limnol. Oceanogr. 17: 433–440.

Kapoor, B.G., H. Smit & I.A. Verighina. 1975. The alimentary canal and digestion in teleosts. Adv. Mar. Biol. 13: 109–239.

Kneib, R.T. 1981. Size-specific effects of density on the growth, fecundity, and mortality of the fish *Fundulus heteroclitus* in an intertidal salt marsh. Mar. Ecol. Prog. Ser. 6: 203–212.

Marinucci, A.C., J.E. Hobbie & J.V.K. Helfrich. 1983. Effect of litter nitrogen on decomposition and microbial biomass in *Spartina alterniflora*. Microb. Ecol. 9: 27–40.

Martin, M.M., J.S. Martin, J.J. Kukor & R.W. Merritt. 1980. The digestion of protein and carbohydrates by the stream detritivore *Tipula abdominalis* (Diptera: Tipulidae). Oecologia 46: 360–364.

McHugh, J.L. 1967. Estuarine nekton. pp. 581–620. *In:* G.H. Lauff (ed.) Estuaries, AAAS, Washington, D.C.

Moriarty, D.J.W. 1976. Quantitative studies on bacteria and algae in the food of the mullet *Mugil cephalus* L. and the prawn *Metapanaeus bennettae* (Racek and Dall). J. Exp. Mar. Biol. Ecol. 22: 131–143.

Newell, R.C. & J.G. Field. 1983. The contribution of bacteria and detritus to carbon and nitrogen flow in a benthic community. Mar. Biol. Lett. 4: 23–36.

Odum, W.E. 1970. Utilization of the direct grazing and plant detritus food chains by the striped mullet *Mugil cephalus*. pp. 207–222. *In:* J.H. Steele (ed.) Marine Food Chains, University of California Press, Berkeley.

Odum, W.E., P.W. Kirk & J.C. Zieman. 1979. Non-nitrogen compounds associated with particles of vascular plant detritus. Oikos 32: 363–367.

Payne, A.I. 1978. Gut pH and digestive strategies in estuarine grey mullet (Mugilidae) and tilapia (Cichlidae). J. Fish. Biol. 13: 627–629.

Peters, D.S. & W.E. Schaaf. 1981. Food requirements and sources for the juvenile Atlantic menhaden. Trans. Amer. Fish. Soc. 110: 317–324.

Peters, D.S. & V.P. Lewis. 1984. Estuarine productivity: relating trophic ecology to fisheries. pp. 255–264. *In:* B.J. Copeland, K. Hart, N. Davis & S. Friday (ed.) Research for Managing the Nation's Estuaries: Proceedings of a Conference in Raleigh, North Carolina, UNC Sea Grant Publication 84-08, Raleigh.

Prinslow, T.E., I. Valiela & J.M. Teal. 1974. The effect of detritus and ration size on the growth of *Fundulus heteroclitus* (L.). J. Exp. Mar. Biol. Ecol. 16: 1–10.

Reimold, R.J., J.L. Gallagher, R.A. Linthurst & W.J. Pfeiffer. 1975. Detritus production in coastal Georgia salt marshes. pp. 217–228. *In:* L.E. Cronin (ed.) Estuarine Research, Academic Press, New York.

Rice, D.L. 1982. The detritus nitrogen problem: new observations and perspectives from organic geochemistry. Mar. Ecol. Prog. Ser. 9: 153–162.

Seiderer, L.J.,, C.L. Davis, F.T. Robb & R.C. Newell. 1984. Utilization of bacteria as nitrogen resource by kelp-bed mussel *Choromytilus meridionalis*. Mar. Ecol. Prog. Ser. 15: 109–116.

Solorzano, L. 1969. Determination of ammonia in natural waters by the phenolhypochlorite method. Limnol. Oceanogr. 14: 799–801.

Stuart, V., J.G. Field & R.C. Newell. 1982. Evidence for absorption of kelp detritus by the ribbed mussel *Aulacomya ater* using a new ^{51}Cr-labelled microsphere technique. Mar. Ecol. Prog. Ser. 9: 263–271.

Teal, J.M. 1962. Energy flow in the salt marsh ecosystem of Georgia. Ecology 43: 614–624.

Tenore, K.R. & D.L. Rice. 1980. Trophic factors affecting secondary production of deposit feeders. pp. 325–340. *In:* K.R. Tenore & B.C. Coull (ed.) Marine Benthic Dynamics, University of S. Carolina Press, Columbia.

Tenore, K.R. & S. Findlay. 1982. Nitrogen source for a detritivore: detritus substrate versus associated microbes. Science 218: 371–373.

Tenore, K.R., L. Cammen, S.E.G. Findlay & N. Phillips. 1982. Perspectives of research on detritus: do factors controlling the availability of detritus to macroconsumers depend on its source? J. Mar. Res. 40:473–490.

Valiela, I. 1984. Marine ecological processes. Springer-Verlag, New York. 546 pp.

Valiela, I., J.E. Wright, J.M. Teal & S.G. Volkmann. 1977. Growth production, and energy transformation in the salt marsh killifish *Fundulus heteroclitus*. Mar. Biol. 40: 135–144.

Valiela, I., B. Howes, R. Howarth, A. Giblin, K. Foreman, J.M. Teal & J.E. Hobbie. 1982. Regulation of primary production in a salt marsh ecosystem. pp. 151–168. *In:* B. Gopal, R.E. Turner, R.G. Wetzel & D.F. Whigham (ed.) Wetlands: Ecology and Management, Natl. Inst. Ecol., Jaipur, and Intl. Sci. Pub.

Valiela, I., J.M. Teal, S. Volkmann, R. van Etten & S. Allen. 1985. Decomposition in salt marsh ecosystems: the phases and major factors affecting the disappearance of organic matter above ground. J. Exp. Mar. Biol. Ecol. 89: 29–54.

Walton, M.J. & C.B. Cowey. 1982. Aspects of intermediary metabolism in salmonid fish. Comp. Bioch. Physiol. 73B: 59–79.

Warlen, S.M. 1964. Some aspects of the life history of *Cyprinodon variegatus,* Lacepède 1803, in southern Delaware. M.S. Thesis, University of Delaware, Newark. 40 pp.

Werme, C. 1981. Resource partitioning in a salt marsh fish community. Ph.D. Thesis, Boston University, Boston. 198 pp.

Wiltse, W.I., K.H. Foreman, J.M. Teal & I. Valiela. 1984. Effects of predators and food resources on the macrobenthos of salt marsh creeks. J. Mar. Res. 42: 923–942.

Received 6.12.1984 *Accepted. 8.7.1985*

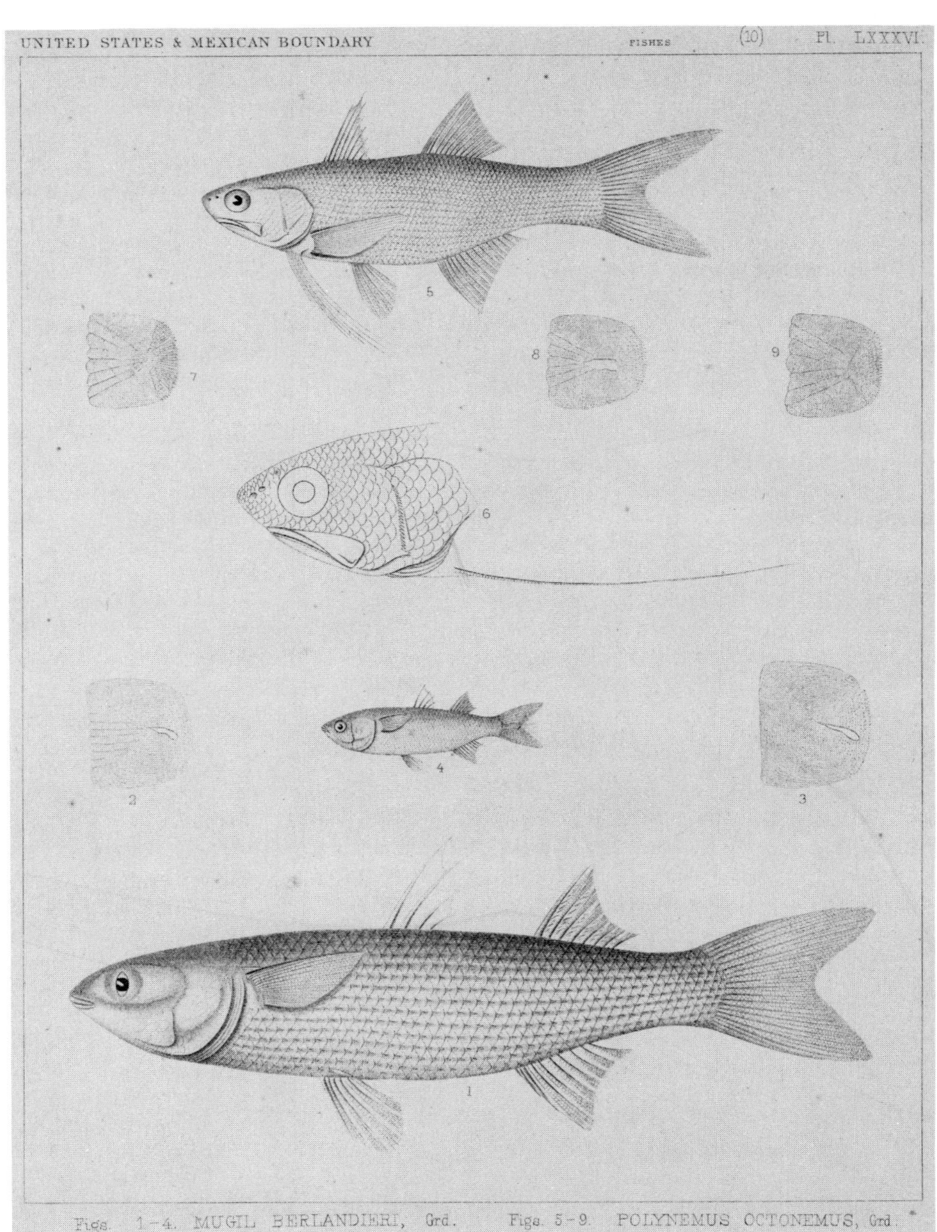

Figs. 1-4. MUGIL BERLANDIERI, Grd. Figs. 5-9. POLYNEMUS OCTONEMUS, Grd.

Feeding rate and attack specialization: the roles of predator experience and energetic tradeoffs

James R. Bence
Department of Biological Sciences, The University of California, Santa Barbara, CA 93106, U.S.A.

Keywords: Foraging, Optimal diet, Learning, Behavior, Freshwater, Fish, *Gambusia*

Synopsis

Individual mosquitofish, *Gambusia affinis,* can adopt a broad range of attack selectivities. In part, this variation can be explained by the past experiences of a fish. Individuals selected the more profitable *Ceriodaphnia dubia* (Cladocera) over less profitable cyclopoid copepods to a greater degree after being exposed to both prey types than did individuals experienced with only one of the prey types. Feeding rate (biomass ingested per unit time) declined with increased attack specialization on the profitable prey (*Ceriodaphnia*) when such prey were scarce, a result in agreement with assumptions of optimal diet theory. When profitable prey were abundant feeding rate was a bimodal function of the intensity of specialization on profitable prey; fish that specialized on cyclopoid copepods (the less profitable prey type) fed at higher rates than did generalists. This may be the result of antagonistic learning that precluded feeding efficiently on more than one type of prey at a time. The data are consistent with the hypothesis that rejection of unsuitable prey involves a time cost. The two preceeding aspects of foraging behavior, which are absent from most optimal diet models, could lead to failure in predicting the attack specialization of some predators. An additional aspect of the results was the generally weak relationship between feeding efficiency and specialization behavior. This suggests that feeding rate may not have been as tightly linked to the specialization behavior a predator adopts as is assumed by current foraging theory.

Introduction

Optimal diet theory has provided insight into the fitness implications of selective predation (Werner & Hall 1974, Pulliam 1974, Pyke 1984). Selective predation is of further interest because of its role in determining community structure and population dynamics (e.g. Brooks & Dodson 1965, Paine 1966, Hall et al. 1970, Connell 1975, Murdoch & Oaten 1975, McCauley & Briand 1979, Zaret 1980, Murdoch et al. 1984, Fulton 1985). Optimal diet theory can make general predictions about patterns of prey selection, and therefore it potentially has great utility for population and community ecologists (e.g. Werner & Hall 1979, Werner et al. 1981, Hubbard et al. 1982, Holt 1983).

Optimal diet predictions focus on the extent to which a predator will specialize on profitable prey (profitability is defined as the ratio of the net energy content of a prey item to the predator's handling time for it). A general qualitative prediction of the theory is that the degree of such specialization will increase when profitable prey are more abundant. The premise is that when profitable prey are abundant specializing on them will maximize feeding efficiency (the net rate of energy aquisition), while generalizing is the best strategy when profitable prey are scarce. In part because of the focus provided by optimal diet theory, we have learned that fish can actively select among sizes and taxa of planktonic prey, and that the degree of selectivity for a prey type may (e.g. Werner et al. 1981, 1983, Dill 1983), or may not (e.g. Wright & O'Brien 1984) vary with prey abundance.

Few optimal diet studies have gone much beyond testing whether the degree of specialization on profitable prey increases when such prey are abundant. Since all models will fail to some extent, we are frequently left with post hoc speculation on why the model 'failed' or explanations of how the results 'really' fit the model. Here optimal diet theory is evaluated at a more basic level. My goal is to determine whether feeding efficiency and the degree of specialization on profitable prey are related in the manner assumed by optimal diet theory. I do this by attempting to explain variation in feeding efficiency by variation in behavioral specialization by the predator.

I evaluate the relationship between feeding efficiency and the degree of specialization on profitable prey by using experimental data for the mosquitofish, *Gambusia affinis,* feeding upon zooplankton. This fish can adopt a wide range of predatory behaviors under identical conditions; as shown below, the behavior of a particular fish depends in part on its past experience. It possesses flexible behavior (a prerequisite for this study) and variation in behavior can be obtained by appropriate training. Thus, mosquitofish are well suited for the purposes of this work.

Methods

Definitions and indices

Working definitions of the terms 'feeding efficiency', 'degree of specialization upon profitable prey', and 'prey profitability' are given below.

Feeding efficiency here is equated to the rate at which biomass is ingested (feeding rate). This measure of foraging success is frequently used in optimal diet studies because it is likely to be tightly correlated with the net rate of energy aquisition (e.g. Pulliam 1974, Werner et al. 1981, Pyke 1984) and because it is relatively easy to measure. In calculations of feeding rates, prey biomasses were estimated from length-weight regressions (McCauley 1983, Bence, unpublished data).

The degree of specialization upon profitable prey is measured by the index Sp which ranges from zero to one, with larger values indicating stronger specialization on profitable prey. This index is based upon Manly's (1974) selection index, defined by

$$P_i = \frac{S_i N_i}{\sum_{j=1}^{k} S_j N_j}, \quad i = 1, 2, \ldots, k,$$

where P_i is the probability that the next selection will be type i, N_j is the density of type j and S_j is the selectivity for type j. Here, P_i, refers to the probability of an attack (rather than the probability of consuming a prey item) because optimal diet theory predictions are concerned with how attacks are distributed among prey types. Methods of estimating the S_j's are described in detail by Chesson (1983) and by Bence & Murdoch (1986).

In all but one of the experiments reported here only two prey types (the cladoceran *Ceriodaphnia dubia* and the copepod *Acanthocyclops vernalis*) were used. In this situation Sp can unambiguously be equated to the selectivity (S_j) for the more profitable prey type. When there are more than two prey types, relative selectivity values among the less profitable prey types needs to be taken into account. To do this, I define the following general formula: $Sp = S_k + ((k\text{-}2)/(k\text{-}1))S_{k-1} + ((k\text{-}3)/(k\text{-}1)) S_{k-2} + \ldots + (1/(k\text{-}1))S_2$, where S_k, S_{k-1}, ..., S_1 are the selectivities for the k prey types ranked from the most profitable (k) to the least profitable (1). This index retains the same range as S_j (zero to one) and is composed of a weighted sum of the S_j's, with the S_j's for the more profitable prey types being weighted most heavily.

The *profitabilities* of the prey types used in this study were ranked as: (1) *Ceriodaphnia dubia* > (2) *Acanthocyclops vernalis* > (3) *Diaptomus pallidus* > (4) unidentified ostracod spp. Information (from Bence 1985) on both the ratio of prey biomass (B) to handling time (T_h) and on the growth rates of fish fed ad libitum on a diet of each prey type singly were considered to determine these rankings. Both measures of profitability rank the prey types in the same order with the exception of ostracods. Ostracods produced the lowest growth rate even though its B to T_h ratio was the second highest (Bence 1985). Ostracods were rarely attacked in the single experiment in which they were used; thus, the estimates of Sp were only slightly influenced by the ranking of this prey type.

Experimental methods

All the data come from laboratory experiments done at 25° C using young female mosquitofish (mean standard length ± one standard deviation = 18.5 ± 1.0) observed singly in a two liter aquarium. In each trial the appropriate numbers of each prey type were added to the aquarium prior to introduction of a fish.

The relationship between feeding rate and specialization on profitable prey was analyzed for three sets of data. The first used fish immediately after they were captured from the field. Eighteen collections of mosquitofish were made from California rice fields within a one month period (Bence 1985). Fish were observed in laboratory feeding trials 24 to 48 hs after capture. In these trials equal numbers of four prey types (total prey density of $50\,l^{-1}$) were presented to single fish, which were observed until at least 15 prey items were eaten. The average results for each field collection (five to 10 fish each) are treated as a single observation (i.e. N = 18 rather than 118).

The second experiment evaluated the influence of prey abundance on the relationship between feeding rate and attack specialization. To do this fish were trained to yield a wide range of Sp values. Fish were initially given a diet of either pure *A. vernalis* ($50\,l^{-1}$) or a mixture of *A. vernalis* ($50\,l^{-1}$) plus *C. dubia* ($50\,l^{-1}$) for seven consecutive days during a ten minute feeding period each day. For the experiment, fish initially trained on pure cyclopoid copepods were observed at three different densities (2, 10, or $50\,l^{-1}$) of *C. dubia* (>5 replicate fish at each density). Fish trained on the mixture of both prey types were observed only at the lowest ($2\,l^{-1}$) *Ceriodaphnia* density (3 replicate fish). During the experimental period, as during the initial period, fish were again presented with prey only for 10 minutes each day over a seven day period. The less profitable prey (*A. vernalis*) is treated as part of the (constant) environment in this experiment and was present at the same density as it was during training ($50\,l^{-1}$) in all treatments. This was the only experiment in which prey were replaced; whenever either prey type was depleted by 25% its numbers were replenished by gently pippetting precounted groups of prey into the aquarium.

Experimental data consisted of attack behavior during the first two minutes of each trial. This minimized satiation effects (Bence & Murdoch 1986). The mean length of prey were 0.8 mm and 1.0 mm for *C. dubia* and *A. vernalis* respectively and the c.v. in length was <15% for both prey species.

Polynomial regressions were fit to the data from the second experiment to explore potential relationships between feeding rate and Sp. These regression models were restricted to third order or less due to limitations of the data. The highest order model for which all slope parameters were significant ($P<0.05$) was retained. A problem with this method of analysis is that random scatter might be mistaken for a complicated polynomial relationship. Consequently, high order polynomial models were used here only to suggest hypotheses rather than as rigorous tests of predictions.

The last experiment was designed to test for the existence of a bimodal relationship between feeding rate and Sp that was suggested by polynomial regression techniques. This was done by comparing the feeding rates of fish at three crucial points along the Sp axis. Fish were initially trained: (1) on *A. vernalis* to obtain fish that specialized on the less profitable *A. vernalis* (low value of Sp), (2) on *C. dubia* to obtain generalists that attacked both prey species (intermediate values of Sp), and (3) on a mixture of *A vernalis* and *C. dubia* to obtain specialists on the more profitable prey, *C. dubia* (high values of Sp). The training period for this experiment lasted five days, during which fish were housed individually in 500 ml containers and fed ad libitum on the appropriate diet for a one hour period each day. Following a 24 h starvation period at the end of the training period, fish from all three training treatments were observed feeding singly on a mixture of prey composed of equal numbers of both prey types (total prey density $20\,l^{-1}$).

Theoretical feeding rate functions

The basic premise of the original, simplest and most widely used optimal diet model (hereafter optimal diet theory) is that predators that acquire energy at higher rates will have higher fitness (Pyke 1984). Net rate of energy acquisition is almost always equated to the rate at which biomass is ingested (the feeding rate). Predictions of the theory stem from how the relationship between feeding rate and the degree to which a predator specializes (concentrates its attacks) on profitable prey (Sp) changes with changes in prey density.

Consider a hypothetical consumer choosing between two prey types, A and B, where A is the more profitable. When prey are scarce optimal diet theory predicts that the predator will attack both prey types whenever it encounters them. The assumption is that an intermediate Sp maximizes feeding rate (Fig. 1). As the profitable prey type becomes more abundant, the cost of rejecting low quality items becomes less; the chances increase that a high quality item will be found soon after a low quality item is rejected. Thus, although attacking all prey items may still be the optimal strategy, the cost of 'overspecializing' is less as prey density increases (Fig. 1).

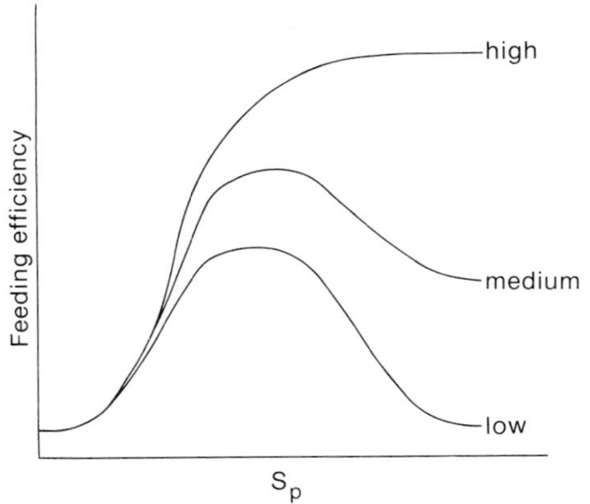

Fig. 1. The (qualitative) relationship between feeding efficiency and degree of specialization upon profitable prey (Sp) assumed by optimal diet theory. The relationship is depicted for three densities (low, medium and high) of the most profitable prey type.

When profitable prey are abundant enough, feeding rate no longer peaks in at intermediate degree of specialization but instead increases monotonically with specialization. At this prey density the predicted optimal strategy switches sharply from always attacking both prey types to attacking only type A. The sharp change between generalizing and specializing is not a robust prediction of the theory (e.g. Emlen 1984, Pyke 1984, Bence & Murdoch 1986). Thus, it is probably more reasonable to interpret the prediction qualitatively: as the abundance of the profitable prey type (or prey density in general) increases, the degree of specialization on profitable prey should increase. The thrust of this paper is to examine empirically the relationship between feeding efficiency and attack specialization behavior that underlies this basic optimal diet prediction.

Results

Feeding rate versus specialization in fish from the field. – Fish collected from the field showed a wide range of variation in their degree of specialization on profitable prey (Sp) (Fig. 2). Much of this variation can be explained by variation in characteristics of the prey communities where different groups of fish were collected (Bence 1985). Overall, feeding rate (micrograms of prey eaten per minute) was related to Sp for these fish. Specifically, there was a significant increase in mean feeding rate as a function of mean Sp when fish were exposed to high densities of prey. This result is qualitatively consistent with the basic premises of optimal diet theory (Fig. 1). However, Sp only explained 32% of the variation in feeding rate despite removal of within group variation (i.e. average feeding rates and Sp values for each group of fish were used in the regression).

Temporal trends in specialization for laboratory

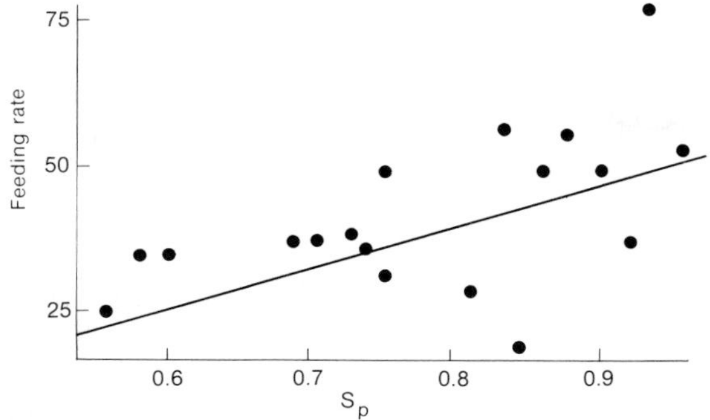

Fig. 2. The relationship between mean feeding rate (micrograms per minute) and mean degree of specialization on profitable prey (Sp) for 18 collections of fish from California rice fields. The regression line is given by $Y = -12.7 + 63.0X$, $R^2 = 0.32$, $P<0.01$.

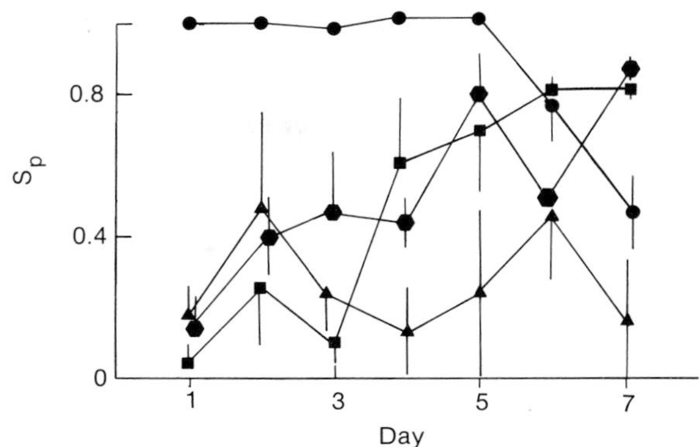

Fig. 3. The degree of specialization on profitable prey (Sp) during the seven day experimental period (methods) for laboratory-trained fish feeding upon a mixture of cyclopoid copepods and *Ceriodaphnia.* Fish which were given only copepods during the prior training period were given two (▲), 10 (⬢), or 50 (■) *Cerodaphnia* l^{-1} during the observational period. Fish which were given both cyclopoid copepods and *Ceriodaphnia* during the prior training period (●) were given two *Ceriodaphnia* l^{-1} during the observation period. Vertical bars indicate standard errors. Sp changed significantly ($P<0.01$) for all treatments except for fish initially trained on cyclopoid copepods and observed at the $2 l^{-1}$ *Ceriodaphnia* density ($P>0.10$) (one way analysis of variance).

trained fish. – Fish previously trained only on copepods started the experimental observation period with a low Sp. After these individuals were exposed to the more profitable *Ceriodaphnia* Sp increased significantly ($P<0.01$) through time for the two higher prey density treatments (Fig. 3). There was no clear trend for the group of fish given the lowest density of *Ceriodaphnia* (Fig. 3). Fish that were initially trained on mixture of copepods and *Ceriodaphnia* and then given the lowest density of *Ceriodaphnia* during the experiment displayed high Sp that decreased significantly ($P<0.01$) by the seventh day of the observation period (Fig. 3).

Feeding rate versus specialization – the effect of prey density. – The temporal patterns noted above produced a wide range of Sp values for each of the three prey densities. These data were then used to evaluate the relationship between feeding rate and Sp for each density of the profitable prey (*C. dubia*). Feeding rate and Sp were most tightly related when *Ceriodaphnia* were scarce ($2 l^{-1}$); 60% of the variation in feeding rate was explained. In this case, there was a clear cost (in terms of feeding rate) to individuals that 'over-specialized' on the profitable prey (Fig. 4A). The 'break' in the data between Sp of 0 and 0.3 is a consequence of the relative scarcity of *Ceriodaphnia*; cyclopoid copepods were 25 times as abundant as *Ceriodaphnia,* so if any *Ceriodaphnia* were eaten, the resulting Sp was estimated as substantially greater than zero. The theoretical decline in feeding rate with decreasing Sp below some threshold Sp (Fig. 1) is only weakly supported by these data. One explanation for this result is that, since *Ceriodaphnia* were relatively scarce, the cost associated with ignoring them was too small to detect.

There was no obvious relationship between feeding rate and Sp when *Ceriodaphnia* were moderately abundant ($10 l^{-1}$) (Fig. 4B). However, when *Ceriodaphnia* were abundant ($50 l^{-1}$) there was a significant but weak cubic relationship (Fig. 4C). The cubic regression suggests that as Sp increased, feeding rate first decreased, then increased from a low value over a broad range of Sp values before declining again at very high Sp. If real, this bimodal pattern contrasts sharply with the theoretical relationship for high prey densities, which is a monotonic increase (Fig. 1).

Experimental test for bimodality of the feeding rate function. – A final experiment explored whether feeding rate was a bimodal function of Sp as sug-

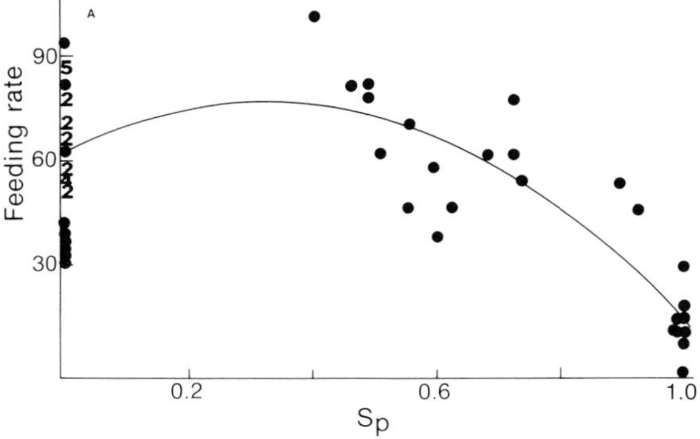

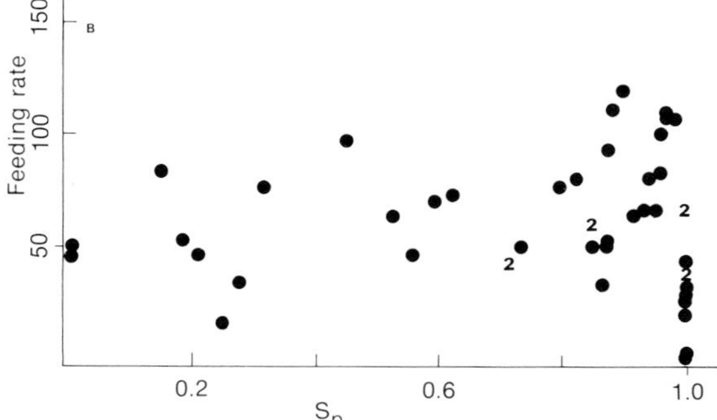

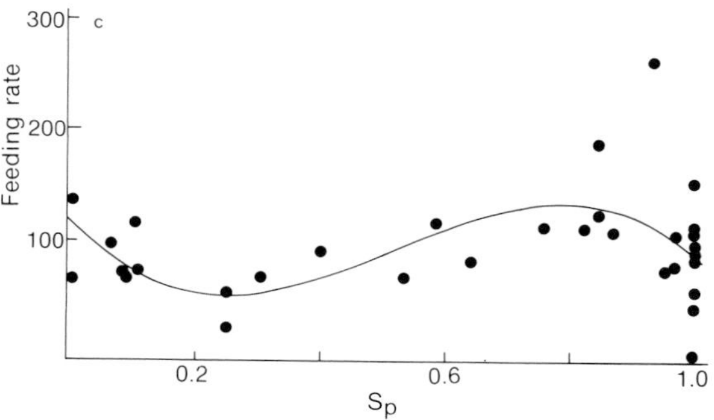

Fig. 4. The relationship between feeding rate (micrograms per minute) and the degree of specialization on profitable prey (Sp) for individual laboratory-trained fish when *Ceriodaphnia* were at (A) 2, (B) 10, and (C) 50 l^{-1}. The polynomial curves are given by (A) $Y = 62.6 + 87.5X - 134.7X^2$, $R^2 = 0.60$, $P < 0.0001$ and (C) $Y = 115.4 - 584.2X + 157.4X^2 - 1010.2X^3$, $R^2 = 0.24$, $P < 0.05$.

gested by the regression analysis. The results strongly indicated that feeding rate can have a bimodal relationship to Sp: feeding rate was lower for the generalists than for fish that specialized on the less or more profitable prey type (Table 1). In addition, the feeding rate of specialists on the less profitable type (cyclopoid copepods) was nearly as high as the feeding rate of specialists on the profitable prey type (*C. dubia*). These results, from three strategically chosen points along the Sp axis, show clearly that the relationship can have at least two local maxima.

Table 1. Comparison of the feeding rates (micrograms per minute) of cyclopoid copepod specialists, generalists, and *Ceriodaphnia dubia* specialists. Sp is a measure of the intensity of specialization on the more profitable *Ceriodaphnia*. Both Sp and feeding rates vary significantly among treatments (one way analysis of variance, $P < 0.01$).

Behavior	Sp	Feeding rate
Cyclopoid copepod specialist	0.12 ± 0.12	36.0 ± 3.0
generalist	0.52 ± 0.07	17.4 ± 1.8
Ceriodaphnia dubia specialist	0.92 ± 0.03	43.8 ± 9.0

Discussion

The relationship between feeding rate and the degree of specialization on profitable prey (Sp) by the mosquitofish was consistent with optimal diet theory to some extent. The positive relationship between feeding rate and Sp for field collected fish (Fig. 2) was in agreement with the theoretical expectation when prey are abundant (Fig. 1). When profitable prey (*Ceriodaphnia dubia*) were scarce, the feeding rate of fish trained in the laboratory appeared to peak at an intermediate Sp (Fig. 3a), and this indicates that a generalist strategy produced the highest feeding rate. This result also is consistent with optimal diet theory (Fig. 1). There was one further agreement with expectations under optimal diet theory. The theory assumes that the relationship between feeding rate and Sp will change with changes in the abundance of profitable prey. This was evident in my laboratory experiments.

Perhaps of more interest were the deviations from the theoretical optimal diet relationships. The feeding rate of laboratory trained fish did not increase monotonically with Sp when the profitable prey type was abundant (Fig. 4c). This result contrasts sharply with the theoretical optimal diet relationship (Fig. 1). In fact, the relationship between feeding rate and Sp appears to have been bimodal with specialists on either the low or high ranked prey feeding at higher rates than generalists (Table 1). This bimodal relationship seems to conflict with the positive relationship between feeding rate and Sp for field collected fish (Fig. 2). This difference may reflect, however, the fact that groups of field collected fish did not show extreme specialization on either cyclopoid copepods or *Ceriodaphnia* whereas individuals trained in the laboratory did. In fact if the data in Figure 4c were censored by excluding extreme Sp values ($Sp<0.10$ or $Sp>0.90$) feeding rate would then be positively related to Sp ($R^2 = 0.54$, $P<0.01$).

Multiple peaks in the feeding rate- attack specialization function suggest that there are energetic trade-offs in feeding efficiently on one prey type or another. Specialization seems likely to be a better strategy than generalization when particular physiological or behavioral traits are required to feed efficiently on each prey type (Murdoch 1969, Murdoch et al. 1975, Cornell 1976, Werner et al. 1981, Glasser 1982, Dill 1983, Holt 1983). Such tradeoffs may be more common when prey types are of different taxa rather than of different sizes. If this is true, optimal diet theory, in its most widely applied form, might fail to correctly predict selection among taxa more often than it correctly predicts selection among sizes of prey. Optimal diet theory also might be more likely to fail when prey are classified by taxa rather than size for other reasons. For example, energetic costs per unit time might not be the same for attacks on prey of different taxa if attacks require specific tactics for each prey taxa.

Among freshwater macroinvertebrates sedentary and mobile predators feed most efficiently on different types of prey (Cooper et al. 1985). Sedentary predators, which accelerated rapidly during their attacks, successfully captured evasive copepods. Mobile predators, on the other hand, were not successful in capturing copepods. Instead they encountered and ate Cladocera at higher rates than did sedentary predators. Raptorially feeding fish can adopt either strategy but antagonistic learning (sensu Werner et al. 1981) may prevent them from feeding efficiently on both prey types simultaneously. Rapid acceleration from a stationary position has frequently been reported for fish experienced with evasive prey (e.g. Vinyard 1980, Mathias & Li 1982, Winfield 1983). In addition, some fish increase their attack rates with Cladocera as they gain experience on them (e.g. Werner et al. 1981), learn progressively to reject more evasive prey such as copepods (Vinyard 1980, Bohl 1982, Bence 1985), but then have low capture rates when they attack copepods (Bence 1985).

In contrast with the predictions of optimal diet theory, populations composed of individuals that specialize on different prey types may be common (Allan 1981, Werner et al. 1981, Bryan & Larkin 1972, Rosenthal & Hempel 1970, Dill 1983, Schmitt & Coyer 1982). This sort of strong but variable selectivity has been taken to indicate a propensity towards switching behavior (see Murdoch et al. 1975, Murdoch & Oaten 1975). It is possible that strong but variable preferences are symptomatic of a predator that does better specializing than generalizing, and that switching occurs at the population level because a greater proportion of the predators tend to specialize on a prey type when it is abundant.

Optimal diet theory predicts absolute specialization on the most profitable prey type at high prey densities. This prediction is rarely met in practice (e.g. Emlen & Emlen 1975, Kislalioglu & Gibson 1976, Bence & Murdoch 1986). This prediction may fail because the time it takes to reject prey items makes absolute specialization a poor strategy (Hughes 1979). My qualitative observations suggest that sometimes fish would take a substantial period of time evaluating prey items, sometimes fixating upon a given item several times at different angles. The quantitative data suggest that pure specialization may have lowered feeding rates in comparison with a 'less perfect' strategy at intermediate and high prey densities (Fig. 4b, 4c). Comparison of pure specialists on the profitable *Ceriodaphnia* ($Sp = 1.0$) and specialists ($0.9<Sp<1.0$) shows that pure specialists fed at significantly lower rates than specialists at both prey densities (Table 2). This analysis is post hoc; however it does suggest that the theoretical idea of rejection time warrants more empirical attention.

Table 2. Comparison of feeding rates of pure specialists on *Ceriodaphnia dubia* (Sp = 1.0) with specialists on *C. dubia* (0.9<Sp<1.0). Data are for laboratory trained fish presented with *C. dubia* at 10 or 50 l^{-1}. Cyclopoid copepods were available at a density of 50 l^{-1} to all fish.

A. Mean ± 1 S.E. feeding rate (micrograms per minute).

	Specialist	Pure specialist
C. dubia density		
10	90.3 ± 6.5 (10)	32.0 ± 5.8 (13)
50	120.9 ± 36.0 (5)	82.5 ± 12.6 (11)

B. Two-way analysis of variance. ** indicates P<0.01, NS indicates a nonsignificant result (P>0.10).

Source	Sum of squares	d.f.	F
C. dubia density	16162	1	9.0**
Specialization type	22131	1	17.6**
Specialization* *C. dubia* density interaction	173	1	0.1 NS
Error	50117	34	–

An additional deviation from the predictions of conventional optimal diet theory may be common; namely a time lag in predator response to changes in prey density. Mosquitofish took more than seven days of laboratory training to stabilize their behavior following a change in prey density. Similar lags have been documented in other laboratory studies of fish feeding behavior (Beukema 1968, Murdoch et al. 1975, Werner et al. 1981) and many other predator taxa (e.g. Murdoch 1969, Landenberger 1968, Dawkins 1971, Curio 1976, Bayliss 1982). Training bias for familiar food may reflect this same phenomenon and has been reported frequently for fish (e.g. LeBrasseur 1969, Bryan 1973, Milinski & Loewenstein 1980). In addition, the mosquitofish appeared to respond with a lag to changes in the density of prey in the field (Bence 1985), and in the field bluegill sunfish lagged approximately a week behind the time an optimality model predicted in their shift of habitats (Werner et al. 1981). The marked training effects are also strong evidence for active choice in contrast to the idea that prey selection results simply from 'hardwired' behavior (e.g. Wright & O'Brien 1984).

Finally, a striking aspect of the results was that at high prey densities, specialization behavior explained a relatively small proportion of the variation in feeding rate. One interpretation is that individual performance is only weakly related to individual specialization behavior. An alternative is that individual fish vary from one another in their average feeding rates. Thus, it might be that the feeding rate of an individual is highly dependent upon its prey selection behavior. The hypothesis here is that among individual variation in average feeding rate is a covariate that obscures the important within individual relationship. Future work should distinguish between these two alternatives.

Acknowledgements

S. Bence, G. Cailliet, S. Holbrook, G. Mittelbach, R. Schmitt, C. Simenstad and one anonymous reviewer provided helpful comments on earlier versions of this paper. S. Cooper suggested some pertinent literature citations, W. Murdoch provided valuable intellectual input, and S. Swarbrick and S. Bence helped draw the figures. This research was supported in part by a grant from the University of California's program for appropriate technology.

References cited

Allan, J.D. 1981. Determinants of diet of brook trout (*Salvelinus fontinalis*) in a mountain stream. Can. J. Fish. Aquat. Sci. 38: 184–192.

Bayliss, D.E. 1982. Switching by *Lepsiella vinosa* (Gastropoda) in South Australian mangroves. Oecologia 54: 212–226.

Bence, J.R. 1985. Selection of prey by the mosquitofish and its predatory impact on invertebrates. Ph. D. Thesis, University of California, Santa Barbara. 272 pp.

Bence, J.R. & W.W. Murdoch. 1986. Prey size selection by the mosquitofish and its relation to optimal diet theory. Ecology (in press).

Beukema, J.J. 1968. Predation by the three spined stickleback (*Gasterosteus aculeatus* L.): the influence of hunger and experience. Behavior 30: 1–126.

Bohl, E. 1982. Food supply and prey selection in planktivorous cyprinidae. Oecologia 53: 134–138.

Brooks, J.L. & S.I. Dodson. 1965. Predation, body size, and composition of plankton. Science 150: 28–35.

Bryan, J.E. 1973. Feeding history, parental stock, and food selection in rainbow trout. Behavior 45: 123–153.

Bryan, J.E. & P.A. Larkin. 1972. Food specialization by individual trout. J. Fish. Res. Board Can. 29: 1615–1624.

Chesson, J. 1983. The estimation and analysis of preference and its relationship to foraging models. Ecology 59: 211–215.

Connell, J.H. 1975. Some mechanisms producing structure in natural communities. pp. 460–491. *In:* M.L. Cody & J.M. Diamond (ed.) Ecology and Evolution of Communities, Harvard University Press, Boston.

Cooper, S.D., D.W. Smith & J.R. Bence. 1985. Prey preferences of freshwater predators with different foraging strategies. Can. J. Fish. Aquat. Sci. 42: 1720–1732.

Cornell, H. 1976. Search strategies and the adaptive significance of switching in some general predators. Amer. Nat. 110: 317–320.

Curio, E. 1976. The ethology of predation. Springer-Verlag, New York. 250 pp.

Dawkins, M. 1971. Shifts of 'attention' in chicks during feeding. Animal Behavior 19: 575–582.

Dill, L.M. 1983. Adaptive flexibility in the foraging behavior of fishes. Can. J. Fish. Aquat. Sci. 40: 398–408.

Emlen, J.M. & M.G.R. Emlen. 1975. Optimal choice in diet: test of a hypothesis. Amer. Nat. 109: 427–435.

Emlen, J.M. 1984. Population biology – the coevolution of population dynamics and behavior. MacMillan Publishing Company, New York. 547 pp.

Glasser, J.W. 1982. A theory of trophic strategies: the evolution of facultative specialists. Amer. Nat. 119: 250–262.

Fulton, R.S. 1985. Predator-prey relationships in an estuarine littoral copepod community. Ecology 66: 21–29.

Hall, D.J., W.E. Cooper & E.E. Werner. 1970. An experimental approach to the production dynamics and structure of freshwater communities. Limnol. Ocean. 15: 839–928.

Holt, R.D. 1983. Optimal foraging and the form of the predator isocline. Amer. Nat. 122: 521–541.

Hubbard, S.F., R.M. Cook, J.G. Glover & J.J.D. Greenwood. 1982. Apostatic selection as an optimal foraging strategy. J. Anim. Ecol. 51: 625–634.

Hughes, R. 1979. Optimal diets under the energy maximization premise: the effects of recognition time and learning. Amer. Nat. 113: 209–221.

Kislalioglu, M. & R.W. Gibson. 1975. Field and laboratory observations on prey size selection in *Spinachia spinachia* (L.). pp. 29–41. *In:* H. Barnes (ed.) Proceedings of the 19th European Marine Biology Symposium, The Aberdeen University Press, Aberdeen.

Landenberger, D.E. 1968. Studies on selective feeding in the pacific starfish *Pisaster* in Southern California. Ecology 49: 1062–1075.

LeBrasseur, R.J. 1969. Growth of juvenile chum salmon (*Oncorhynchus keta*) under different feeding regimes. J. Fish. Res. Board Can. 26: 1631–1645.

Manly, B.F. 1974. A model for certain types of selection experiments. Biometrics 30: 281–294.

Mathias, J.A. & S. Li. 1982. Feeding habits of walleye larvae and juveniles: comparative laboratory and field studies. Trans. Amer. Fish. Soc. 111: 722–735.

McCauley, E. 1983. The estimation of the abundance and biomass of zooplankton in samples. pp. 228–262. *In:* J.A.D. Downing & F.H. Rigler (ed.) The Assessment of Secondary Productivity in Freshwaters, Blackwell Scientic Publications, Oxford.

McCauley, E. & F. Briand. 1979. Zooplankton grazing and phytoplankton species richness: field tests of the predation hypothesis. Limnol. Ocean. 24: 243–252.

Milinski, M. & C. Loewenstein. 1980. On predator selection against abnormalities of movement: a test of an hypothesis. Zeit. für Tier. 53: 323–340.

Murdoch, W.W. 1969. Switching in general predators: experiments on predator specificity and stabilitv of prey populations. Ecol. Monogr. 39: 336–353.

Murdoch, W.W., S. Avery & M.E. Smythe. 1975. Switching in a predatory fish. Ecology 56: 1094–1105.

Murdoch, W.W., M.A. Scott & P. Ebsworth. 1984. The effects of the general predator, *Notonecta* (Hemiptera) upon a freshwater community. J. Anim. Ecol. 53: 791–808.

Murdoch, W.W. & A. Oaten. 1975. Predation and population stability. Adv. Ecol. Res. 9: 1–131.

Paine, R.T. 1966. Food web complexity and species diversity. Amer. Nat. 100: 65–74.

Pulliam, H.R. 1974. On the theory of optimal diets. Amer. Nat. 108: 59–75.

Pyke, G.H. 1984. Optimal foraging theory: a critical review. Ann. Rev. Ecol. System. 15: 523–571.

Rosenthal, H. & G. Hempel. 1970. Experimental studies in feeding and food requirements of herring larvae (*Clupea harengus* L.). pp. 344–364. *In:* J.H. Steele (ed.) Marine Food Chains, Oliver & Boyd, Edinburgh.

Schmitt, R.J. & J.A. Coyer. 1982. The foraging ecology of sympatric marine fish in the genus *Embiotoca* (Embiotocidae): importance of foraging behavior in prey size selection. Oecologia 55: 369–378.

Vinyard, G.L. 1980. Differential prey vulnerability and predator selectivity: the effects of evasive prey on sunfish (*Lepomis*) predation. Can. J. Fish. Aquat. Sci. 37: 2294–2299.

Werner, E.E. & D.J. Hall. 1974. Optimal foraging and size selection of prey by the bluegill sunfish *Lepomis macrochirus*. Ecology 55: 1042–1052.

Werner, E.E. & D.J. Hall. 1979. Foraging efficiency and habitat switching in competing sunfishes. Ecology 60: 256–264.

Werner, E.E., G.G. Mittelbach & D.J. Hall. 1981. Foraging profitability and the role of experience in habitat use by bluegill sunfish. Ecology 62: 116–125.

Werner, E.E., G.G. Mittelbach, D.J. Hall & J.F. Gilliam. 1983. Experimental tests of optimal habitat use in fish: the role of relative habitat profitability. Ecology 64: 1525–1539.

Winfield, I.J., G. Peirson, M. Cryer & C. Townsend. 1983. The behavioral basis of prey selection by underyearling bream *Abramis brama* (L.) and roach *Rutilus rutilus* (L.). Freshwater Biol. 13: 139–149.

Wright, D.I. & W.J. O'Brien. The development and field test of a tactical model of the planktivorous feeding of white crappie (*Pomoxis annularis*). Ecol. Monogr. 54: 65–98.

Zaret, T.M. 1980. Predation and freshwater communities. Yale University Press, New Haven. 187 pp.

Received 6.12.1984 *Accepted 9.9.1985*

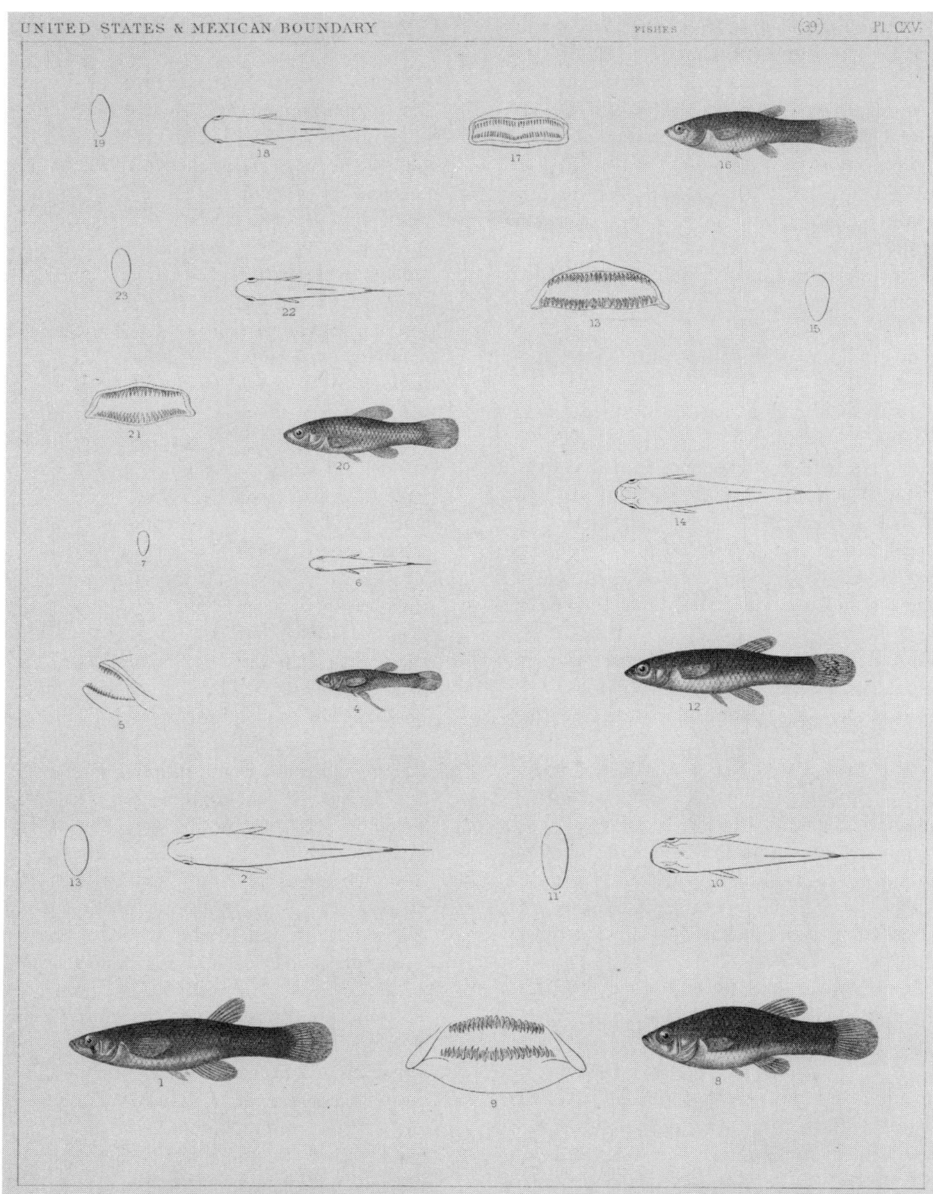

Figs. 1 - 7. GAMBUSIA PATRUELIS, Grd. Figs. 8 - 11. GAMBUSIA NOBILIS, Grd.
Figs. 12 - 15. GAMBUSIA AFFINIS, Grd. Figs. 16 - 19. GIRARDINUS OCCIDENTALIS, Grd.
Figs. 20 - 23. LUCANIA VENUSTA, Grd.

Foraging in surfperches: resource partitioning or individualistic responses?

Alfred W. Ebeling & David R. Laur
Department of Biological Sciences and Marine Science Institute, University of California, Santa Barbara, CA 93106, U.S.A.

Keywords: California, Embiotocidae, Feeding behavior, Kelp forest, Microhabitat, Niche overlap, Randomization model, Reef

Synopsis

On Naples Reef off southern California, five sympatric species of surfperch (Embiotocidae) eat small animal prey associated with algal turf and other benthos. Although they tend to select different foods in different microhabitats, their different foraging traits are probably 'individualistic responses' (non-interactive or autecological species responses to resources) because the species' pairwise resource overlaps are neither uniformly small nor generally complementary. Mean overlap in both food and foraging microhabitat was significantly greater than 'random' as most species converged in resource use. Only two exploiters of extremes in the available spectra of prey types and microhabitats had significantly narrow overlap. Thus, resource overlaps may simply reflect similarities and differences in the species' independently evolved foraging traits, which were rendered more or less inflexible as different morphological specializations for picking, crunching, or winnowing prey. Therefore, any one species may occur about the reef to its best advantage independently of the others. Its distribution and abundance may simply reflect its tolerance of the local environment. The evidence indicates that the surfperches are not concertedly partitioning resources on Naples Reef, i.e., that the five species had not all coevolved to avoid interspecific competition in the past nor have they all mutually 'shifted their niches' to minimize it in the present.

Introduction

A feeding guild of sympatric surfperches (Embiotocidae) on Naples Reef off Santa Barbara, southern California differ in their microhabitat distributions (Alevizon 1975a) and in the ways they exploit food and foraging space (Laur & Ebeling 1983). This is not surprising, because they are morphologically distinct species and differences could be due to many factors, including interspecific competition, historical quirks, physiological tolerances, physical disturbances, environmental change, fluctuations in resource levels, and predation risks (e.g. Connor & Simberloff 1983, Grossman 1985, Ross 1986, Toft 1985). The question more relevant to community theory is whether the species' differences in resource use are synecological or autecological: whether the differences reflect present or past interactions among the species, or independent lines of specialization (Strong 1983). For argument's sake we call the synecological differences 'resource partitioning' [in the strict sense of Roughgarden (1976), but not in the broad sense of Schoener (1968), Ross (1986), or Toft (1985)].

In this sense, resource partitioning is a concerted process of dividing up limited supplies to mitigate interspecific competition (cf. Schoener 1974). It may come about through primary coevolution of competing species or by selective invasion of a community by species at niche positions where competition is low (Roughgarden 1976, 1983). For the coevolutionary alternative, resource partitioning is due to character displacement allowing closely-related competitors to come together and live in sympatry; for the invasive alternative, it is due to adaptive niche shifting to accommodate a species that is sufficiently dissimilar from the others to enter the system and exploit underutilized resources (Roughgarden 1976).

Gleason's (1926) 'individualistic concept' of how a local assemblage of species is made up obviates the requirement that member populations must have coevolved or undergone niche shifting to accommodate each other. Instead, this concept emphasizes the primary importance of each species' individualism – its autecology, migratory capacity, and resource requirements – rather than its relations with others – its synecology, competitive interactions, and resource sharing (Strong 1983). Since a species must tolerate a distribution of suboptimal to optimal environments throughout its total geographic range, furthermore, its position within a local assemblage is viewed in biogeographic perspective (reviewed in Brown 1984): a given locality may provide conditions more suitable for some species than others because, for purely historical reasons, adaptations of no two species are alike. Each species' '... individual peculiarities of migration and environmental requirements' (Gleason 1926, p. 26), and not necessarily its acquisition of resource-partitioning mechanisms, determines its presence and abundance at the site. Hence, the species' differences in resource use may simply reflect the species' separate autecologies and evolutionary histories.

We present evidence that the surfperches' preferences for food and foraging space are due to their autecologies. Compared with overlaps generated at random, their observed pairwise overlaps in food and space were not uniformly small as predicted for species that had coevolved or undergone mutual niche shifting (reviewed in Ebeling & Laur 1982). Nor were overlaps in food generally complementary with those in space (see Schoener 1974, Werner 1977, Yoshiyama 1980). These results are compatible with species showing little more than individualistic (non-interactive or autecological) responses to their resources.

The system

Five species of medium to large-sized viviparous surfperch forage over the bottom of Naples Reef, an area of reef and kelp located about 1.6 km offshore near Santa Barbara, southern California (34° 25′ N, 119° 57′ W). Naples Reef is a semi-isolated system of shale ridges and rills covering approximately 2.2 ha and averaging 8–12 m in depth (see maps in Ebeling et al. 1980a, b, 1985). During the study period in 1973, the reef bore a thick kelp forest of three canopies: (1) a surface layer of giant kelp, *Macrocystis pyrifera,* (2) an understory of low kelps, mostly *Pterygophora californica,* or bushy red algae, *Gelidium robustum,* and (3) a carpet of algal turf over much of the reef bottom.

The five surfperches comprise a feeding guild within the Naples fish assemblage, in that they form a distinct cluster of species separated from other such groups by lower average overlap in food and space use (Ebeling, unpublished, see Pianka et al. 1979). All eat small benthic animals living in and about the algal turf, and their feeding mechanisms are specialized for harvesting prey from the complex cover of reef bottom, e.g., for handling and crushing larger prey or for winnowing smaller soft-shelled prey from the turf (Laur & Ebeling 1983). Consequently, the fish rarely if ever switch to alternative food such as plankton or nekton, and seldom forage away from the reef. Adults of four species (*Embiotoca jacksoni, E. lateralis, Rhacochilus toxotes, Damalichthys vacca*) are year-round residents of the reef (personal observations, Hixon 1979 & personal communication). The fifth (*Hypsurus caryi*) is a seasonal species on the reef, arriving in May and leaving near the end of fall (Ebeling & Laur 1985). Nonetheless, individuals of all species may spend a large part of their lives together on the reef.

The five species comprise more than 20% of total fish numbers and biomass at Naples (Ebeling et al. 1980a, b) and are major exploiters of the turf food source. Since they seldom seek shelter during the day or night (Ebeling & Bray 1976), their turf-associated food and foraging space would seem to be their most critical resources. Adults measure 15–30 cm in SL, and usually wander about fully exposed showing little fear of divers. Comparatively large and robust newborn young hide in understory kelp and other algae, but sally forth within a few months (Ebeling & Laur 1985, Schmitt & Holbrook 1985).

Methods

Sampling

To compute dietary overlap, we analyzed gut contents of 149 adult surfperch (about 30 individuals of each species) with food in their foreguts. Equivalent numbers (about 10 of each species) were sampled during April–May, June–July, and August–September, 1973. Fish were speared during daylight hours (0800–1800 h) haphazardly as they were encountered. Frozen specimens were thawed, sexed and measured before their entire guts were excised. 'Foregut' was defined as the first quarter length of their long, simple, tubular digestive tract (Ebeling & Bray 1976). We estimated foregut fullness as the percent of its length distended by food, before we squeezed out the contents for sorting into ten prey categories ('food items'). We measured the collective volume of each item by water displacement in a syringe. The 10-item dietary array excluded those of doubtful food value that are probably ingested incidentally (plants, chitinous amphipod tubes, crustose debris, and branched hydroids + bryozoans) (Laur & Ebeling 1983). Since rank concordance among the corresponding three dietary arrays was high for each species (Kendall's $W = 0.70–0.85$, $P<0.01$), percentage values per item were based on total volumes summed over all fish pooled among the three bimonthly periods.

We estimated availability of food items from benthic samples of algal turf and other unconsolidated materials. Twelve 0.25 m^2 quadrats were placed randomly about midreef, across all microhabitats (see below). All attached material along with its resident prey were scraped from the quadrats and collected by an air-lift sampler. We preserved samples in 5% buffered formalin, separated out organic material, subsampled in aliquots, and measured water-displacement volumes of food items. The high among-sample concordance of ranked food item abundances justified pooling as one array of available food (Laur & Ebeling 1983). Hence, percentage values per item were of total volumes summed over the 12 samples.

To compute spatial overlap, we observed the fish's foraging effort by counting the number of bites each of 590 haphazardly encountered adult individuals (86–142 per species) took during 5 min among five microhabitats: (1) reef crest, supporting dense stands of bushy red algae and thick algal turf; (2) reef slope, with rich patches of turf; (3) reef flat, often partly overgrazed by sea urchins and supporting sparse turf; (4) reef cobble, with scattered but thick turf and burrowing prey in soft conglomerate deposits in rills; and, (5) deep cobble, with sparse turf in deep water (13 m) about the reef base (Fig. 1). To conserve diving effort, observations were limited to the western third of the reef where landmarks were well known, fish were abundant, and microhabitats were distinct yet close together (Laur & Ebeling 1983).

Availability of foraging space was estimated from reef topography. We measured areas of microhabitats drawn on the west section of a detailed contour map of the reef (Fig. 1, insert). We adjusted the two-dimensional areas to total surface using functions of topographic complexity determined in situ by lengths of weighted lines (Laur & Ebeling 1983).

Breadth, selectivity, and overlap

Resource breadth and overlap were estimated for food items and foraging microhabitat as percent similarity (*PS*) between arrays consumed by species relative to arrays available in the environment: $PS = 100\ (S)$, where, for example,

$$S_1 = \sum_{j=1}^{r} \min\ (p_{1j},\ q_j)$$

for species 1's breadth of usage of the available array of $j = 1, \ldots, r = 10$ food items or $1, \ldots, r = 5$

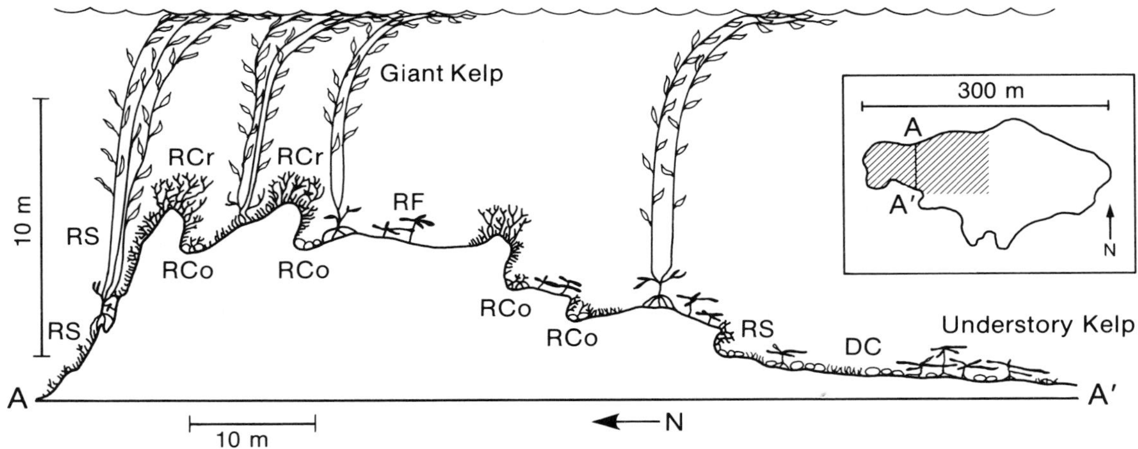

Fig. 1. Section of Naples Reef (A-A′) within the area (hatched) where surfperch feeding bites were counted. Microhabitats are: DC – deep cobble, RCo – bottoms of rills with algal turf and kelp plants growing on cobble, RCr – reef crest with dense stands of *Gelidium* algae and scattered understory kelp, RF – reef flat, RS – reef slope with turf, bushy algae, and scattered kelp plants.

microhabitats; p_{1j} is species 1's proportionate dietary volume of item *j* or number of foraging bites taken in microhabitat *j*; q_j is the proportionate volume of item *j* in the benthic samples or area of microhabitat *j* in the study site (Feinsinger et al. 1981); and

$$S_{12} = \sum_{j=1}^{r} \min (p_{1j}, p_{2j})$$

for overlap of usage between species 1 and 2 (Schoener 1968). Using computer simulations to compare different overlap indices, Linton et al. (1981) showed that *S* was the most consistent and unbiased estimator of the true population overlap if 16 or more observations per resource state are made for each species. Our observations numbered 13–17 per food item and 20–39 per foraging microhabitat.

Selectivity (*PE*) of species 1 for resource *j* was estimated as the difference between resource use and availability: $PE = 100\ (E)$, where, for example, $E_{1j} = p_{1j} - q_j$, and $-100 < PE < +100$ (Strauss 1979).

We used Sale's (1974) randomization model to compute mean pairwise 'competition free' overlap for 'null guilds' of hypothetical species. The model assumes that each species has its resource-usage peak at some position along the observed spectrum of available resources but the position is randomly chosen. Thus, the method preserves the observed pattern of both members' resource use for each species pair, leaving only the positions of these patterns on the spectrum – their degree of overlap – to randomize (procedure illustrated in Ebeling & Laur 1982, Fig. 1). The advantage of Sale's method is that overlaps depend on the usage of each item in the spectrum if all items were equally available (Schoener 1974), i.e., on the fish's preference for certain items over others, independent of the item's abundance in the environment. For the complete feeding guild of five species and incomplete guilds of four, three, and two species, mean observed overlap was compared with the average of 100 means of null guilds by *t*-tests between a single observation and a sample mean (Sale 1974).

Results

Most fish foraged almost continually during the day in and about rich patches of algal turf. Excepting *R. toxotes*, 90% of all fish examined had food in their foreguts, which averaged 64 ± 36% (S.D.) full. Unlike the others, *R. toxotes* feeds intermittently during the day and also at night (Ebeling & Bray 1976), but its diurnal and nocturnal diets were virtually identical. Only *D. vacca* selected large and common hardshelled items such as gastropods, crabs, and brittle stars (Table 1). The other species preferred abundant but small items with soft shells, especially gammarid amphipods. However, *R. toxotes* favored crabs as well, and *E. jacksoni* was the only species to select polychaete worms.

Fish concentrated their foraging in microhabitats – reef slope, crest, and cobble (Table 2) – having largest volumes of food items (Laur & Ebeling 1983). *E. lateralis* was most selective, taking some 90% of its foraging bites from among the bushy algae (mostly *Gelidium robustum*) at reef crest (Table 2). *Hypsurus caryi* often selected the reef slope, while *R. toxotes* and *D. vacca* preferred slope and cobble microhabitats. *Embiotoca jacksoni* and *D. vacca* had the greatest spatial as well as dietary breadths (Tables 1, 2, *PS*), extending their foraging ambits to the less productive microhabitats of reef flat and (with *R. toxotes*) deep cobble.

For both diet and foraging space, average observed overlap was significantly greater than random (Table 3), as also indicated by its position beyond 95% of the means of 100 null guilds (Fig. 2). This was generally true even of such comparisons for the artificial incomplete guilds of fewer than five species (Table 3). In fact, only when incomplete guilds included the most divergent species –

Table 1. Percent dietary volumes (P), selectivities (PE), usage breadth (PS), and availabilities (Q) of surfperch food items. Availabilities were measured by benthic samples from Naples Reef.

Food item	*E. jacksoni*		*E. lateralis*		*H. caryi*		*R. toxotes*		*D. vacca*		Benthic samples
	P	PE	P	PE	P	PE	P	PE	P	PE	Q
Polychaete worms	23.5	10.7	0.9	−11.9	5.0	−7.8	0.6	−12.2	0.2	−12.6	12.8
Bivalves	11.4	−20.4	2.1	−29.7	2.0	−29.8	0.2	−31.6	15.9	−15.9	31.8
Gastropods	0.4	−7.2	0.2	−7.4	0.3	−7.3	0.2	−7.4	10.3	2.7	7.6
Isopods	1.2	−0.1	3.0	1.7	2.1	0.8	0.7	−0.6	0.1	−1.2	1.3
Large (>3 mm) gammarid amphipods	24.7	18.7	21.5	15.5	32.9	26.9	34.6	28.6	1.0	−5.0	6.0
Small (<3 mm) gammarid amphipods	14.3	4.5	9.8	0.0	41.7	31.9	7.5	−2.3	1.2	−8.6	9.8
Caprellid amphipods	17.9	11.8	37.5	31.4	3.5	−2.6	2.2	−3.9	0.1	−6.0	6.1
Shrimps	0.5	−0.5	23.2	22.2	4.2	3.2	10.1	9.1	0.4	−0.6	1.0
Crabs	4.4	−5.6	1.8	−8.2	4.7	−5.3	43.3	33.3	27.8	17.8	10.0
Brittle stars	1.7	−12.1	0.0	−13.8	3.5	−10.3	0.7	−13.1	43.1	29.3	13.8
PS	54.3		29.2		37.1		29.1		50.3		

Table 2. Percent feeding bites taken in (P), selectivities of (PE), usage breadth of (PS), and availabilities of (Q) microhabitats comprising the surfperches' foraging space. Availabilities were measured by estimating areas on Naples Reef. Numbers in parentheses are total fish watched; total bites counted.

Microhabitat	*E. jacksoni* (24;98)		*E. lateralis* (32;161)		*H. caryi* (37;186)		*R. toxotes* (31;147)		*D. vacca* (42;155)		Areal estimates
	P	PE	P	PE	P	PE	P	PE	P	PE	Q
Reef crest	24.5	15.9	89.4	80.8	15.0	6.4	2.7	−5.9	3.2	−5.4	8.6
Reef slope	43.9	15.9	9.3	−18.7	60.2	32.2	35.4	7.4	41.9	13.9	28.0
Reef cobble	4.1	−9.0	0.0	−13.1	22.6	9.5	49.0	35.9	36.8	23.7	13.1
Reef flat	8.2	−6.9	0.6	−14.5	2.2	−12.9	2.0	−13.1	11.0	−4.1	15.1
Deep cobble	19.4	−15.8	0.6	−34.6	0.0	−35.2	10.9	−24.3	7.1	−28.1	35.2
PS	68.3		19.1		51.9		56.7		62.4		

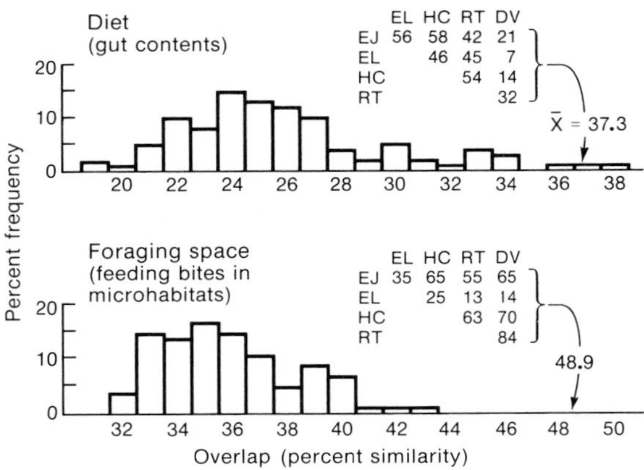

Fig. 2. Frequency distributions of means of 100 randomized-guild overlaps compared with observed means (arrows) of overlaps (insets) in diet and foraging space between pairs of species comprising the surfperch guild at Naples Reef. Species abbreviations: EJ – *E. jacksoni,* EL – *E. lateralis,* HC – *H. caryi,* RT – *R. toxotes,* and DV – *D. vacca.*

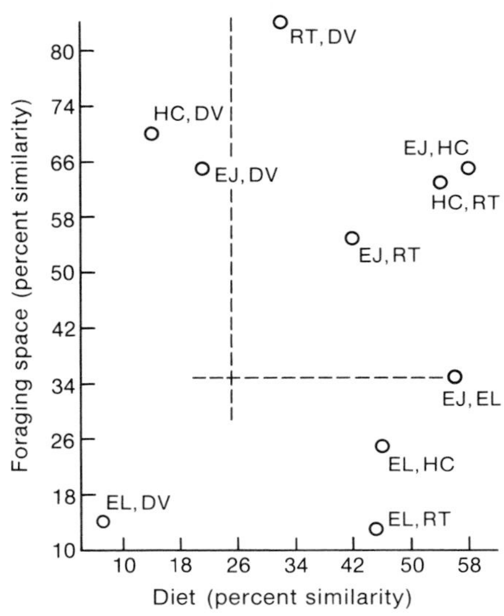

Fig. 3. Relation between overlap in diet and foraging space among pairs of species (coordinates) comprising the surfperch guild. Dashed lines indicate random levels (averages for randomized guilds in Table 3). Species abbreviations in Figure 2.

either *D. vacca* for diet or *E. lateralis* for foraging space – did observed overlaps approach random. Only these two species had a dietary overlap significantly less than random, i.e., they exploited opposite ends of the available-food spectrum.

The relation between pairwise dietary and spatial overlaps revealed no general pattern of complementarity (Fig. 3). Half the species pairs had greater than random overlap in both diet and foraging space. Exceptions included *D. vacca,* which had a diet unlike the others but a broad foraging range. Therefore, its overlaps with other broad-ranged species (*E. jacksoni, H. caryi*) were narrow in diet but wide in space because microhabitats are closely packed within the limited reef area. Conversely, *E. lateralis* ate many of the same prey as

Table 3. Comparison of mean observed resource overlap with 100 means of random overlaps for the complete surfperch guild and for guilds with species arbitrarily deleted. Observed means are significantly greater than (>), less than (<) or equal to (=) random by *t*-test. The standard deviation (S.D.) of random means estimates the standard error of the observed mean (Ricklefs & Lau 1980). Species abbreviations in Figure 2.

Species combination	Overlap in diet			Overlap in foraging space		
	Observed	Random ± S.D.	Conclusion	Observed	Random ± S.D.	Conclusion
Complete guild	37.3	25.9 ± 3.9	>	48.9	35.9 ± 2.5	>
EJ, EL, RT, DV	33.6	25.3 ± 3.8	>	44.3	34.9 ± 3.1	>
EJ, EL, HC, RT	50.0	23.0 ± 4.8	>	42.7	31.7 ± 3.3	>
EJ, HC, RT, DV	36.0	29.2 ± 4.7	=	66.9	47.9 ± 1.3	>
EJ, HC, RT	51.1	25.0 ± 6.3	>	61.0	45.9 ± 4.7	>
EJ, EL	55.8	22.5 ± 8.9	>	35.0	18.5 ± 5.4	>
EJ, HC	57.7	30.7 ± 9.9	>	65.2	47.1 ± 10.6	>
RT, DV	31.8	24.6 ± 9.6	=	84.0	44.9 ± 9.5	>
EL, DV	7.0	23.8 ± 10.8	<	14.0	18.1 ± 6.8	=
EL, RT	44.8	15.8 ± 13.6	>	13.3	18.4 ± 8.6	=

did others, but its few individuals seemed best adapted to forage in the more turbulent zone of reef crest.

Discussion

Even though our results indicate that the surfperches select different arrays of prey from different suites of microhabitats, they provide little evidence that all five species do so in response to past or present interspecific competition. The species' pairwise overlaps in diet and foraging space were neither uniformly small nor generally complementary, as predicted for species whose coexistence is a product of coevolution in the past or niche shifting in the present (e.g., Sale 1974, Schoener 1974, Werner 1977, Yoshiyama 1980, Grossman 1985, but see Abrams 1981, Colwell & Winkler 1984). On the contrary, mean pairwise overlap significantly exceeded random, as though most guild members actually converged both in diet and foraging space. Only the most divergent species pairs had overlaps measuring significantly less than random (Table 3, EL & DV).

Ebeling & Laur (1982) discussed problems in interpreting the randomization model as applied to the Naples Reef surfperch guild. The test was apparently not sensitive enough to distinguish mean overlap that was decreased by equalizing distances between the species' resource-utilization peaks from mean overlap that was determined at random (Fig. 4). Only when 'specialization' (narrowing of peaks) augmented dispersion (moving of peaks) was overlap significantly less than random. Thus, there is inherent bias toward accepting the 'null hypothesis' of no concerted partitioning (Abrams 1981), though the same model has been used to reject this hypothesis for a guild of tropical reef fishes (Gladfelter & Johnson 1983). Furthermore, since mean pairwise overlap – but obviously not the sum of interspecific interactions – increases when the most divergent member is excluded from a guild, the amount of this bias may depend arbitrarily on how the guild is delimited (Thomson & Rusterholz 1982). This is shown by excluding divergent members from arbitrarily reduced surfperch

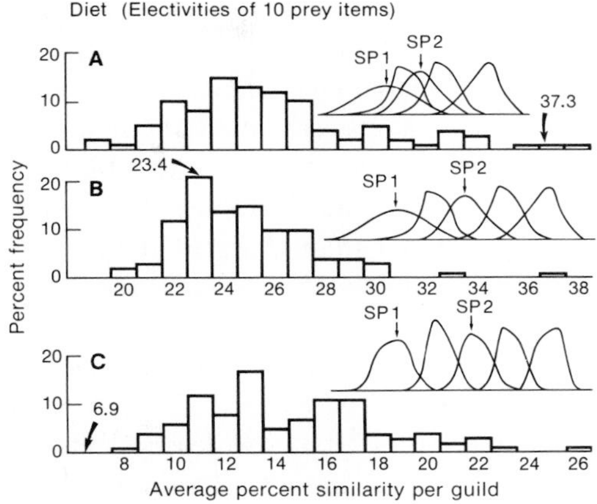

Fig. 4. Frequency distributions of means of 100 randomized-guild overlaps in diet, compared with the observed mean (large arrow) of overlaps between all real pairs of surfperches in the guild (A), between hypothetical pairs whose dietary spectra are altered to represent food partitioning by dispersion of dietary peaks (B) or by dispersion and narrowing of peaks (C). Inset curves illustrate how the observed food utilization peaks (A: SP 1 = *E. jacksoni*, SP 2 = *E. lateralis*) are adjusted to produce partitioning (B, C), as detailed in Ebeling & Laur (1982, Fig. 2).

guilds (Table 3). For example, when *D. vacca* is excluded, the ratio of 'observed' to 'random' overlap increases markedly from 37.3/25.9 (complete guild) to 50.0/23.0 (incomplete guild of the four remaining species). Because both ratios exceed unity, however, the results of this analysis as applied to Naples Reef surfperch are robust.

The conclusion that Naples Reef surfperches tend to converge in their exploitation of available food and foraging space has a sound biological basis. Most species selected similar prey – amphipods and other small crustaceans – from the richest microhabitats (Laur & Ebeling 1983). Only *D. vacca* preferred larger, hard-shelled prey such as brittle stars, crabs, gastropods, and bivalves. Since all prey items occur in all microhabitats (Laur & Ebeling 1983), the species may seek out similar microhabitats whose food is most available. For example, individuals of *E. jacksoni* and *E. lateralis* that forage in the shallower microhabitats obtain the greater biomass of prey (Schmitt & Holbrook, this volume). Thus, surfperch may choose a microhabitat in which to forage before they select food.

Hence, overlaps may simply reflect a common

phylogenetic origin in the species independently evolved foraging traits. Perhaps the species' remarkable pharyngeal mechanism (Liem & Greenwood 1981) was the basis for evolving skills at picking, crunching, or winnowing prey to optimize food gathering by a particular technique (e.g. Maynard Smith 1978), rather than character displacements to minimize interspecific competition (Roughgarden 1976). If the traits are not very flexible, the species will have separate autecologies (Schlosser & Toth 1984, Ross 1985).

Thus, the Naples Reef surfperches may occur about the reef to their best advantage independently of one another. For example, *R. toxotes* and *H. caryi* converge in diet and foraging space (Fig. 2) because both happen to be specialized winnowers that can efficiently separate out food from unconsolidated substrates in reef slope and cobble microhabitats. More generalized in its feeding mode (Laur & Ebeling 1983), *E. jacksoni* overlaps broadly with both, as well as frequenting the best places for winnowing and picking prey from bushy algae at reef crest. The best microhabitats for winnowing are also probably the best for unearthing hard-shelled prey. Therefore, *D. vacca* forages in microhabitats favored by winnowers, but uses its heavy pharyngeal teeth for crunching quite different prey, including crabs. Although *R. toxotes*'s thick lips may function primarily as a gasket seal for sucking up turf, etc. to be winnowed, they also serve well for moving small rocks and plucking small crabs (Laur & Ebeling 1983). Hence, *R. toxotes* forages in the same places as, and overlaps in diet with, *D. vacca* because it is adept at both winnowing amphipods and uncovering crabs.

Indeed, diets of most of the species are more or less invariant in both space and time, although foraging space may vary as food availability fluctuates. For instance, four of the species eat about the same prey as they do at Naples Reef at localities near Los Angeles (Ellison et al. 1979) and San Diego (Quast 1968b) where the fifth species *E. lateralis* is absent. At Naples Reef during a 3-year period when the kelp forest was eliminated and the food supply was severely depressed (Ebeling et al. 1985), species diets varied remarkably little, even though all fish converged in microhabitat as they foraged together in remaining rich patches of turf and prey (Stouder 1983). This is just the opposite of the interactive response predicted from niche theory, i.e., microhabitat partitioning to share limited food supplies (e.g. Werner et al. 1977).

Perhaps competition and migration played more important roles in population regulation during this period of 'ecological crunch' (see Wiens 1977). Even during less stringent times, all individuals of *H. caryi* as well as many subadults and some adults of resident species join the surfperch guild only during the spring–fall seasons when food is abundant (Ebeling & Laur 1985). Predictably, *H. caryi*'s sojourns became much briefer as their abundance on Naples Reef declined during the sustained crunch. *H. caryi* has broad dietary overlap with the other amphipod-eaters (Fig. 2), yet is slightly smaller than and aggressively subordinate to *E. jacksoni* and *E. lateralis* (M. Hixon, personal communication). Rusterholz (1981) pointed out that subordinate species should be among the first to drop out of a feeding guild as resources dwindle.

Yet, interspecific competition between the *Embiotoca* congeners undoubtedly causes resource partitioning wherever both species abound. There is excellent experimental evidence that *E. lateralis* influences the diet and distribution of *E. jacksoni* in the few localities where both congeners live together in abundance. Reef environments along Santa Cruz Island, located some 30 km across a deep channel from Naples Reef, are favorable for both species and support large stocks of each (Alevizon 1975b, Ebeling et al. 1980a). Here, *E. lateralis* interferes with *E. jacksoni*'s access to shallow stands of bushy *Gelidium* algae harboring dense concentrations of larger prey (Hixon 1980). This induces *E. jacksoni* to restrict both its microhabitat distribution and diet, even though both species retain their different foraging behaviors (Hixon 1980, Schmitt & Coyer 1982, Schmitt & Holbrook, this volume). Furthermore, at other, more southerly Channel Islands where *E. lateralis* is absent, *E. jacksoni* expands its diet to include larger prey which are favored by *E. lateralis* elsewhere (Schmitt & Coyer 1983).

Nonetheless, we suggest that the surfperches' geographic ecology accounts for their relative abundance, microhabitat distribution, and diet at Naples Reef, where *E. lateralis* is comparatively rare. At any particular locality, the range of physical conditions and resources are more favorable for growth and reproduction of some species than for others (e.g. James et al. 1984). Thus, locally rare species may represent tails of distributions of species that are more abundant elsewhere (Brown 1984). At Naples Reef, for example, *E. jacksoni* is the numerical dominant (327 individuals ha^{-1}), *E. lateralis* is rarest (40 ha^{-1}) and the others are at intermediate densities (60–92 ha^{-1}) (Ebeling et al. 1980a, b, Ebeling & Laur 1985). Surfperch densities have also been estimated farther south off Los Angeles (Stephens & Zerba 1980) and San Diego (Quast 1968a), off Santa Cruz Island (Ebeling et al. 1980a), and off Monterey, central California (Miller & Giebel 1973) north of a faunal barrier near Point Conception (e.g. Horn & Allen 1978). If the highest of the density estimates per species is assigned 100%, densities at Naples Reef are 100% for *E. jacksoni,* 90% (of Redondo maximum) for *R. toxotes,* 33% (of Monterey maximum) for *H. caryi* and *D. vacca,* but only 20% (of Monterey maximum) for *E. lateralis*. Naples Reef is near the geographic center of ranges of *E. jacksoni, H. caryi,* and *R. toxotes,* but near the southern extremes of more northerly ranging *D. vacca* and *E. lateralis* (Ho 1983). Why *H. caryi* was not more abundant at Naples Reef is unknown; but it is the only seasonal migrant of the five and may be difficult to assess, especially in a variable environment. *Embiotoca lateralis* is comparatively rare at Naples Reef because it endures a suboptimal part of its range south of the thermal discontinuity at Point Conception (see Hixon 1979, 1980). To the north, *E. lateralis* is not only more abundant, but has a broader diet and microhabitat distribution (Haldorson & Moser 1979). The few scattered individuals at Naples Reef limit their foraging to reef crest (Table 2), and are so widely spaced that they seldom interfere with *E. jacksoni*'s access to this microhabitat (personal observation). Hence, interspecific competition between the two congeners, one of which is rare (see Schoener 1983), may play but a minor role in population structuring and resource partitioning at this locality.

Using only our descriptive study, we cannot outrightly reject the hypothesis of competition-derived resource partitioning among Naples Reef surfperches. This would require enormous effort in replicated, long-term field experiments (Connell 1980, 1983). Even so, our evidence indicates that the fishes' foraging behavior and guild structure at Naples Reef are not determined in major part by concerted partitioning of limited resources among all five species. Without evidence to the contrary, the individualistic explanation that, for the most part, the species have independently evolved, noninteractive responses to resources is the more parsimonious (Connor & Simberloff 1983, Strong 1983).

Acknowledgements

We thank our colleagues for criticizing and improving (but not necessarily affirming) the manuscript: earlier drafts by L. Crowder, R. Larson, M. Hixon, P. Sale, and R. Warner; later drafts by D. Breitburg, P. Dayton, S. Holbrook, R. Schmitt, and editors and referees of this symposium. Several former students helped with benthic sampling, gut-content analysis, and fish watching: especially L. Asakawa, A. Bjordal, R. Cowen, S. Edwards, L. Leum, M. Rode, D. Sarver, and S. Swartz. Steve Edwards directed the mapping of Naples Reef. Norm Lammer assisted with equipment, diving, and boating operations. This work was sponsored by DOC-NOAA Sea Grants 2-35208-6 and 04-3-158-22 (Project R-FA-14); and by NSF Grants GA 38588, OCE76-23301, and OCE79-08452. The Marine Science Institute provided supplementary funding and secretarial services.

References cited

Abrams, P.A. 1981. Comparing randomly constructed and real communities: a comment. Amer. Nat. 118: 776–782.

Alevizon, W.S. 1975a. Spatial overlap and competition in congeneric surfperches (Embiotocidae) off Santa Barbara, California. Copeia 1975: 352–356.

Alevizon, W.S. 1975b. Comparative feeding ecology of a kelp-bed embiotocid (*Embiotoca lateralis*). Copeia 1975: 608–615.

Brown, J.H. 1984. On the relationship between abundance and distribution of species. Amer. Nat. 124: 255–279.

Colwell, R.K. & D.S. Winkler. 1984. A null model for null models in biogeography. pp. 344–359. *In*: D.R. Strong Jr., D. Simberloff, L.G. Abele & A.B. Thistle (ed.) Ecological Communities: Conceptual Issues and the Evidence, Princeton University Press, Princeton.

Connell, J.H. 1980. Diversity and the coevolution of competitors, or the ghost of competition past. Oikos 35: 131–138.

Connell, J.H. 1983. On the prevalence and relative importance of interspecific competition: evidence from field experiments. Amer. Nat. 122: 661–696.

Connor, E.F. & D. Simberloff. 1983. Interspecific competition and species co-occurrence patterns on islands: null models and the evaluation of evidence. Oikos 41: 455–465.

Ebeling, A.W. & R.N. Bray. 1976. Day versus night activity of reef fishes in a kelp forest off Santa Barbara, California. U.S. Fish. Bull. 74: 703–717.

Ebeling, A.W., R.J. Larson & W.S. Alevizon. 1980a. Habitat groups and island-mainland distribution of kelp-bed fishes off Santa Barbara, California. pp. 403–431. *In*: D.M. Power (ed.) Multidisciplinary Symposium on the California Islands, Santa Barbara Museum of Natural History, Santa Barbara.

Ebeling, A.W., R.J. Larson, W.S. Alevizon & R.N. Bray. 1980b. Annual variability of reef-fish assemblages in kelp forests off Santa Barbara, California. U.S. Fish. Bull. 78: 361–377.

Ebeling, A.W. & D.R. Laur. 1982. Does resource partitioning have a descriptive null hypothesis? pp. 158–165. *In*: G.M. Cailliet & C.A. Simenstad (ed.) Proc. Third Pacific Workshop Fish Food Habits Studies, University of Washington, Seattle.

Ebeling, A.W. & D.R. Laur. 1985. The influence of plant cover on surfperch abundance at an offshore temperate reef. Env. Biol. Fish. 12: 169–179.

Ebeling, A.W., D.R. Laur & R.J. Rowley. 1985. Severe storm disturbances and reversal of community structure in a Southern California kelp forest. Mar. Biol. 84: 287–294.

Ellison, J.P., C. Terry & J.S. Stephens Jr. 1979. Food resource utilization among five species of embiotocids at King Harbor, California with preliminary estimates of caloric intake. Mar. Biol. 52: 161–169.

Feinsinger, P., E.E. Spears & R.W. Poole. 1981. A simple measure of niche breadth. Ecology 62: 27–32.

Gladfelter, W.B. & W.S. Johnson. 1983. Feeding niche separation in a guild of tropical reef fishes (Holocentridae). Ecology 64: 552–563.

Gleason, H.A. 1926. The individualistic concept of the plant association. Bull. Torrey Botanical Club (N.Y.) 53: 7–26.

Grossman, G.D. 1985. Division of food resources among fishes of the rocky intertidal zone. J. Zool. (Lond.) (in press).

Haldorson, L. & M. Moser. 1979. Geographic patterns of prey utilization in two species of surfperch (Embiotocidae). Copeia 1979: 567–572.

Hixon, M.A. 1979. Competitive interactions and spatiotemporal patterns among California reef fishes of the genus *Embiotoca*. Ph.D. Dissertation, University of California, Santa Barbara. 213 pp.

Hixon, M.A. 1980. Competitive interactions between California reef fishes of the genus *Embiotoca*. Ecology 61: 918–931.

Ho, J. 1983. Copepod parasites of Japanese surfperches: their inference on the phylogeny and biogeography of Embiotocidae in the Far East. Ann. Rep. Sado Mar. Biol. Sta. Niigata Univ. 13: 31–62.

Holbrook, S.J. & R.J. Schmitt. 1986. Food acquisition by competing surfperch on a patchy environmental gradient. Env. Biol. Fish. 16: 135–146 (this volume).

Horn, M.H. & L.G. Allen. 1978. A distributional analysis of California coastal marine fishes. J. Biogeogr. 5: 23–42.

James, F.C., R.F. Johnston, N.O. Wamer, G.J. Niemi & W.J. Boecklen. 1984. The Grinnellian niche of the wood thrush. Amer. Nat. 124: 17–30.

Liem, K.F. & P.H. Greenwood. 1981. A functional approach to the phylogeny of the pharyngognath teleosts. Amer. Zool. 21: 83–101.

Linton, L.R., R.W. Davies & F.J. Wrona. 1981. Resource utilization indices: an assessment. J. Animal Ecol. 50: 283–292.

Laur, D.R. & A.W. Ebeling. 1983. Predator-prey relationships in surfperches. Env. Biol. Fish. 8: 217–229.

Maynard Smith, J. 1978. Optimization theory in evolution. Ann. Rev. Ecol. Systematics 9: 31–56.

Miller, D.J. & J.J. Geibel. 1973. Summary of blue rockfish and lingcod life histories: a reef ecology study; and giant kelp, *Macrocystis pyrifera,* experiments in Monterey Bay, California. Calif. Dep. Fish Game, Fish Bull. 158: 1–137.

Pianka, E.R., R.B. Huey & L.R. Lawlor. 1979. Niche segregation in desert lizards. pp. 67–115. *In*: D.J. Horn, G.R. Stairs & R.D. Mitchell (ed.) Analysis of Ecological Systems, Ohio State University Press, Columbus.

Quast, J.C. 1968a. Estimates of the populations and the standing crop of fishes. pp. 57–79. *In*: W.J. North & C.L. Hubbs (ed.) Utilization of Kelp-bed Resources in Southern California, Calif. Dep. Fish Game, Fish Bull. 139.

Quast, J.C. 1968b. Observations on the food of the kelp-bed fishes. pp. 109–147. *In*: W.J. North & C.L. Hubbs (ed.) Utilization of Kelp-bed Resources in Southern California, Calif. Dep. Fish Game, Fish Bull. 139.

Ricklefs, R.E. & M. Lau. 1980. Bias and dispersion of overlap indices: results of some Monte Carlo simulations. Ecology 61: 1019–1024.

Ross, S.T. 1986. Resource partitioning in fish assemblages: a review of field studies. Copeia (in press).

Roughgarden, J. 1976. Resource partitioning among competing species – a coevolutionary approach. Theoret. Pop. Biol. 9: 388–424.

Roughgarden, J. 1983. Competition and theory in community ecology. Amer. Nat. 122: 583–601.

Rusterholz, K.A. 1981. Competition and structure of an avian foraging guild. Amer. Nat. 118: 173–190.

Sale, P.F. 1974. Overlap in resource use, and interspecific competition. Oecologia 17: 245–256.

Schlosser, I.J. & L.A. Toth. 1984. Niche relationships and population ecology of rainbow (*Etheostoma caeruleum*) and fantail (*E. flabellare*) darters in a temporally variable environment. Oikos 42: 229–238.

Schmitt, R.J. & J.A. Coyer. 1982. The foraging ecology of sympatric marine fish in the genus *Embiotoca* (Embiotocidae): importance of foraging behavior in prey size selection. Oecologia 55: 369–378.

Schmitt, R.J. & J.A. Coyer. 1983. Variation in surfperch diets between allopatry and sympatry: circumstantial evidence for competition. Oecologia 58: 402–410.

Schmitt, R.J. & S.J. Holbrook. 1985. Patch selection by juvenile black surfperch (Embiotocidae) under variable risk: interactive influence of food quality and structural complexity. J. Exp. Mar. Biol. (in press).

Schoener, T.W. 1968. The *Anolis* lizards of Bimini: resource partitioning in a complex fauna. Ecology 49: 704–726.

Schoener, T.W. 1983. Field experiments on interspecific competition. Amer. Nat. 122: 240–285.

Schoener, T.W. 1974. Resource partitioning in ecological communities. Science 185: 27–39.

Stephens, J.S. & K.E. Zerba. 1981. Factors affecting fish diversity on a temperate reef. Env. Biol. Fish. 6: 111–121.

Stouder, D.J. 1983. Effects of a severe weather disturbance on foraging patterns of a guild of surfperches. M.A. Thesis, University of California, Santa Barbara. 48 pp.

Strauss, R.E. 1979. Reliability estimates for Ivlev's electivity index, the forage ratio, and a proposed linear index of food selection. Trans. Amer. Fish. Soc. 108: 344–352.

Strong, D.R. Jr. 1983. Natural variability and the manifold mechanisms of ecological communities. Amer. Nat. 122: 636–660.

Thomson, J.D. & K.A. Rusterholz. 1982. Overlap summary indices and the detection of community structure. Ecology 63: 274–277.

Toft, C.A. 1985. Resource partitioning in amphibians and reptiles. Copeia 1985: 1–21.

Werner, E.E. 1977. Species packing and niche complementarity in three sunfishes. Amer. Nat. 111: 553–575.

Werner, E.E., D.J. Hall, D.R. Laughlin, D.J. Wagner, L.A. Wilsmann & F.C. Funk. 1977. Habitat partitioning in a freshwater fish community. J. Fish. Res. Board Can. 34: 360–370.

Wiens, J.A. 1977. On competition and variable environments. Amer. Sci. 65: 590–597.

Yoshiyama, R.M. 1980. Food habits of three species of rocky intertidal sculpins (Cottidae) in Central California. Copeia 1980: 515–525.

Received 15.12.1984 *Accepted 15.5.1985*

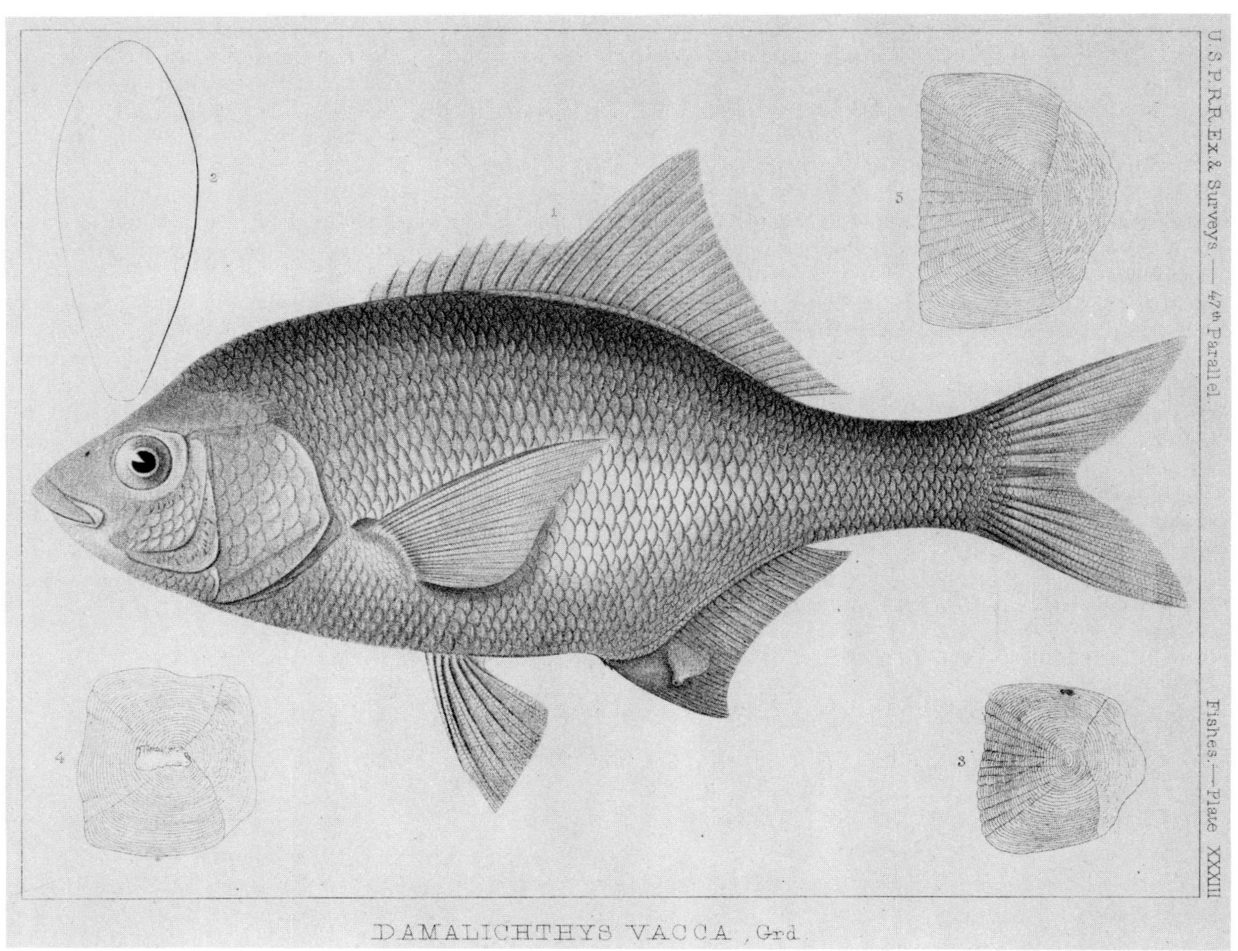

DAMALICHTHYS VACCA, Grd.

Food acquisition by competing surfperch on a patchy environmental gradient

Sally J. Holbrook & Russell J. Schmitt
Department of Biological Sciences and Marine Science Institute, University of California, Santa Barbara, Santa Barbara, CA 93106, U.S.A

Keywords: Foraging behavior, Foraging success, Competitive interactions, Marine fish, Among-individual variation

Synopsis

Black surfperch, *Embiotoca jacksoni,* and striped surfperch, *Embiotoca lateralis,* coexisted along steep sloping rocky habitats at Santa Cruz Island, California. The range of depths occupied (to 15 m) was characterized by a strong gradient in abundance of prey and a changing mosaic of substrate types from which surfperch harvested food. Availability of prey and diversity of benthic substrates were greatest in shallowest areas and both declined with increasing depth. Individuals of both surfperch species were residential within a narrow range of depths, with the result that different segments of their populations were consistently exposed to different foraging environments. These two phenomena (residential behavior combined with a gradient in availability of resources) resulted in variation in foraging behaviors and diets among individuals that resided at different depths. The pattern of within-population variation differed between the surfperch species. Black surfperch individuals achieved similar taxonomic diets and expended similar foraging effort at all depths, but deep-water foragers captured much less prey biomass per unit effort. The taxonomic composition of striped surfperch diets differed among depths, and although similar amounts of prey biomass were captured everywhere, individuals in deep areas expended much greater effort to obtain that level of food return. For both species, habitat profitability (food return to foraging effort) declined with depth. The difference in habitat profitability appeared to influence fitness components of both surfperches. Individuals occupying deep habitats were about 5% shorter in standard length than conspecifics of the same chronological age living in shallow areas; the disparity in body size resulted in an estimated difference in clutch size of 10–18%.

Introduction

Interspecific competition can lead to divergence in use of resources by co-occurring species (Schoener 1974, Connell 1978, 1980). For that reason, patterns of resource use have been examined for evidence that competition is an important structuring process (Diamond 1978, Connor & Simberloff 1979, 1983, Connell 1980). However, it is difficult to determine the actual processes underlying resource partitioning without corroborating evidence (e.g. Ebeling & Laur 1986). Diets of sympatric fishes have often been compared within the context of resource partitioning (e.g. Nilsson 1963, Keast 1965, 1978, Alevizon 1975a, Ross 1977, Targett 1978, MacPherson 1979, Werner 1979, Schmitt & Coyer 1982, 1983, Mittelbach 1984). Such studies have tended to focus on populations as a whole, comparing the average or composite diet between species. Another approach has contrasted diets of populations of the same species in different geographical regions (e.g. Alevizon 1975b, Haldorson & Moser 1979, Schmitt & Coyer 1983). To ensure that these comparisons are meaningful, it is necessary to consider the extent and significance of variation in resource use within a single population.

For the most part, investigations that consider within-population variation in resource use of fishes have addressed patterns related to ontogeny of morphology or behavior (e.g. Keast 1965, 1968, 1977a, b, Carr & Adams 1973, Coyer 1979, Grossman 1980, Grossman et al. 1980, Lemly & Dimmick 1982, Mittelbach 1984, Schmitt & Holbrook 1984a, b). Relatively few studies have explored mechanisms that produce variation among adults (but see Bryan & Larkin 1972, Milinski 1982, Ringler 1983), and fewer still have attempted to link such variation to differences in components of individual fitness. Populations of many species of fish dwell in spatially heterogeneous environments, and if patch use is not uniform among individuals, substantial differences in resource acquisition can result. Our goals here are to quantify the extent of within-population variation in resource use by a pair of fish species known to compete (Hixon 1980), to establish the mechanisms underlying these patterns, and to determine whether measured variation results in differences in fitness components of individual fish.

We examined two locally sympatric reef fishes, black surfperch, *Embiotoca jacksoni,* and striped surfperch, *Embiotoca lateralis.* General attributes of the diet of these species have been described with regard to local populations (Alevizon 1975a, Ellison et al. 1979, Laur & Ebeling 1983), geographical trends (Alevizon 1975b, Haldorson & Moser 1979, Schmitt & Coyer 1983), possible resource partitioning (Schmitt & Coyer 1982, 1983), and ontogenetic patterns (Schmitt & Holbrook 1984a, b). Hixon (1980) demonstrated that the two species competed for space over the vertical range of their depth distribution at Santa Cruz Island, where the present study was done. As described here, the range of depths over which local populations (stocks) were distributed represented a steep gradient in availability of crucial resources. We examine the consequences of acquiring food along this patchy environmental gradient.

Methods

Both black and striped surfperch were continuously distributed across a range of depths (to about 15 m) over sloping, rocky habitat at Santa Cruz Island, California (34°5′N:119°40′W). To facilitate comparisons of fish diets and activity along the depth gradient, we delineated three depth zones [shallow (3 m), middle (6 m) and deep (9 m)] and sampled resource availability and use within each zone. The data presented here were obtained from three similar sites on the north (lee) side of the island; unless otherwise noted, data are presented as the mean of site averages. For a particular depth stratum, MANOVA of substrate cover and ANOVA of prey density revealed no statistical differences among the sites. Hence, each site represented a replicate of the environmental variability of interest, variation among-depths within a stock.

For each site, availability of substrate types and density of invertebrate food were estimated within each of the three depth strata. Percent cover of various substrates was assessed by sampling along four line transects, each 50 m in length, within each depth. The type of substrate (e.g., species of algae, turf, rock, sand) under 100 randomly selected points along each transect line was determined (N = 400 points per depth stratum). The mean distance between each pairwise combination of substrate types was estimated by measuring their linear separation (in meters) along the transect lines. A total of 25 measurements of each combination were made at each depth.

Density of available prey at each site was estimated from 12 random samples per depth of benthic substrates; all benthic materials in 0.1 m^2 samples were removed by scraping and placed in plastic bags underwater, sealed, and returned to the laboratory for preservation in 10% buffered formalin and later processing. Invertebrates in each sample were removed from substrates, identified, counted, measured, and converted to biomass equivalence using techniques reported previously (Schmitt & Coyer 1982, 1983).

Short-term fidelity to a depth by a foraging fish was examined by following 200 actively feeding fish (101 *E. jacksoni*; 99 *E. lateralis*) for 5 min periods.

For each fish, the number of transitions between depth zones, inter- and intra-specific chases, and other activity were noted. Longer-term depth fidelity was investigated by marking fish (25 *E. jacksoni*; 10 *E. lateralis)*) with individually color-coded Floy tags. Individuals were resighted during weekly visits spread over a six month period; upon resighting, the tag color, exact depth, and activity of the individual were recorded. For each species, a pooled variance was calculated for deviation around the mean depth occupied by each individual as:

$$\sigma^2 = \sum_{i=1}^{K} \sum_{j=1}^{n_i} \frac{(X_{ij} - \bar{X}_{i.})^2}{n_i - 1},$$

where n_i is the number of observations for fish i, K is the number of fish observed, and X_{ij} is the depth that the ith fish was observed at time j. A 95% prediction interval was estimated as:

$$\text{P.I.} = \bar{X} \pm t_{[0.975,\ (\sum_{i=1}^{k}(n_i - 1))]} \cdot \sqrt{\sigma^2 \left(1 + \frac{1}{\sum_{i=1}^{k} n_i}\right)}$$

Dietary patterns were established by collecting (with pole spear) adult fish of both species within each depth zone for analysis of stomach contents. The sample size for dietary analyses was 20 individuals per species per depth. Upon collection, fish were immediately injected with 10% buffered formalin and kept at −20° C in a laboratory freezer until analyzed. Before guts were removed, morphological measurements of each fish were made and otoliths were removed. Taxonomic analysis of gut contents was done following techniques reported in Schmitt & Coyer (1982, 1983). Here we report the contribution to the diets of the five most important taxonomic components (gammarid amphipods, caprellid amphipods, shrimp, crabs, and isopods), which together comprise about 90% of the diet of either surfperch species. To age each specimen, otoliths were cleared in glycerol and read to the nearest 0.25 yr under 20 × of a dissecting microscope. Repeated counts verified that estimates were always within 0.25 yr. Otolith rings are annual (Schmitt & Holbrook, unpublished); our otolith-age estimates for fish of various sizes corresponded closely to age estimates based on scale annuli (e.g. Hixon 1979).

Foraging activities of individual fish were documented by SCUBA divers with stopwatches who monitored time, depth, and foraging substrate of all bites taken for up to 5 min. For each surfperch species, approximately 300 individuals were observed to take about 2200 bites; these observations were evenly distributed among the three depth zones. The information allowed us to determine depth-specific substrate use, foraging rates (bites per min), probabilities of repeated bites from the same patch, and transitions between substrate types. Using the transition probabilities among patches (Schmitt & Holbrook, unpublished) and distances separating substrate pairs, we estimated the average distance travelled between successive bites by black and striped surfperch in each depth stratum. We calculated this as follows:

$$E(D) = \sum_{i=1}^{m} \sum_{j=1}^{m} P_{ij} \cdot D_{ij},$$

where P_{ij} is the transition probability between substrate types i and j, D_{ij} is the average linear distance (in meters) between substrates i and j, and m is the number of substrate types. The surfperch species principally used five major substrate types (turf, *Gelidium robustum,* laminarian algae, *Rhodymenia* sp., and articulating coralline algae); consequently, these constituted the set from which pairwise transition probabilities and intervening distances were estimated.

Results

Characteristics of habitats and availability of prey

Black and striped surfperch harvested invertebrate prey from benthic substrates. The kinds and diversity of such substrates varied considerably with depth. Shallow water areas had relatively equitable cover of the major substrate types, although turf, *Gelidium robustum,* and several species of articulating coralline algae were the most common (Fig. 1). The shallowest depth was also characterized by a mosaic of patches of various sizes and mor-

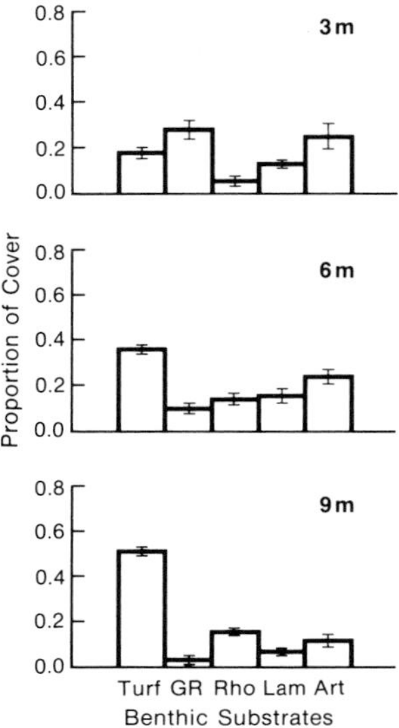

Fig. 1. Availability of benthic substrates in three depth zones at Santa Cruz Island. Mean percent cover (±1SE) of each substrate is presented; availabilities differed significantly among depths (MANOVA; $F_{10,32}$ = 15.05; P<0.0001). GR = *Gelidium robustum*, Rho = *Rhodymenia* – like algae, LAM = *Eisenia arborea* + *Laminaria* spp., ART = articulated coralline algae.

phological structures (e.g., growth form, height above bottom). By contrast, deep water areas were dominated by a single substrate type, turf (Fig. 1), and these regions were open and relatively two-dimensional. All other substrates occurred in deep areas, but were relatively rare. A major consequence of this depth-related pattern was a difference in absolute distance between kinds of foraging patches (Table 1). In general, the average distance separating various combinations of substrate types favored as foraging locations (see below) was considerably shorter in shallow areas. As shown below, this had important implications regarding the effort expended by fish while foraging.

In addition to the changing mosaic of foraging patches, there was a strong depth gradient in characteristics of the prey assemblage available to the surfperches. The shallowest depth had considerably higher prey densities than deeper areas (Table 2). This gradient in abundance of food also reflected prey biomass, because the mean size of prey did not vary much with depth (Table 2). The taxonomic representation of the invertebrate prey assemblage changed significantly with depth (Fig. 2). Caprellid amphipods dominated the shallow-water prey assemblage, but were relatively rare in deeper areas. Gammarid amphipods displayed an opposite pattern, being proportionately more abundant in deep areas.

The patterns of substrate and prey characteristics described here represent continuous variation along the depth gradient. For our purposes, we recognized three distinct 'habitats' (i.e. depth

Table 1. Mean distance between benthic substrates in the shallowest (3 m) and deepest (9 m) habitats examined at Santa Cruz Island. Data are the mean (±1 SE) linear separation (in meters) between adjacent patches of each pairwise combination of substrates; means based on average of three sites. Five substrate types were the major patches used by surfperch, and together comprised more than 80% of the total benthic cover (see Figure 1 for cover and name of each substrate).

Depth zone		Substrate type				
		TURF	GR	RHO	LAM	ART
3 m	TURF	2.6 (0.5)	3.2 (0.6)	5.6 (2.1)	3.9 (0.1)	3.1 (0.5)
	GR	–	1.7 (0.6)	5.9 (2.1)	4.4 (0.9)	2.6 (0.2)
	RHO	–	–	4.7 (0.3)	5.5 (1.6)	5.2 (0.9)
	LAM	–	–	–	3.4 (0.6)	2.8 (0.1)
	ART	–	–	–	–	2.1 (0.6)
9 m	TURF	0.9 (0.1)	5.8 (0.2)	2.2 (0.1)	4.3 (0.3)	4.7 (0.9)
	GR	–	8.5 (1.9)	7.6 (2.4)	10.0 (1.8)	9.4 (1.1)
	RHO	–	–	2.9 (0.3)	5.2 (0.7)	5.4 (1.1)
	LAM	–	–	–	6.1 (0.4)	5.8 (1.2)
	ART	–	–	–	–	5.8 (2.3)

Table 2. Abundance and size of available prey in each of three depth zones at Santa Cruz Island. Prey density is the mean number (±1SE) 0.1 m^{-2} benthic sample (N = 36 samples depth −1); data were log-transformed for ANOVA. Prey size is the mean biomass (±1SE) of an item in benthic samples. For both analyses, means with same letter do not differ by SNK test at P<0.05 level.

Density of prey (No. m^{-2})				
Depth zone				
3 m	6 m	9 m		
5461.2 (1554.9)[a]	784.8 (133.4)[b]	586.2 (76.2)[b]		
ANOVA				
Source	*df*	*MS*	F_s	*Significance*
Among depths	2	2.773	9.77	P<0.001
Error	105	0.284		
Mean size of prey (mg)				
Depth zone				
3 m	6 m	9 m		
1.84 (0.19)[c,d]	1.98 (0.14)[c]	1.51 (0.07)[d]		
ANOVA				
Source	*df*	*MS*	F_s	*Significance*
Among depths	2	2.294	3.12	P = 0.048
Error	105	0.735		

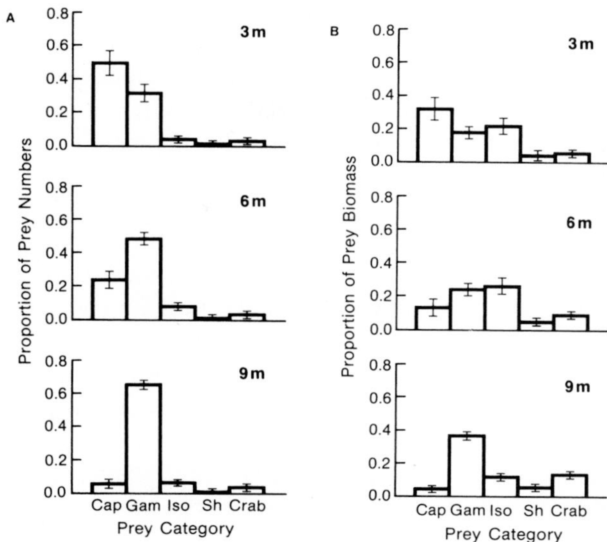

Fig. 2. Taxonomic composition of prey on benthic substrates in each depth zone. Presented are mean (±1SE) proportions of items in the five major taxonomic groups eaten by the surfperch. CAP = caprellid amphipods, GAM = gammarid amphipods, ISO = isopods, SH = shrimp. (A) Proportion of prey types by number. Availability of prey types differed among depths (MANOVA; $F_{10,128}$ = 5.56; P<0.0001). (B) Proportion of prey types by biomass. Composition of prey assemblage by biomass differed among depths (MANOVA; $F_{10,128}$ = 4.16; P<0.0001).

zones): the shallow and deep ends of the continuum and a middle zone having intermediate characteristics.

Depth-related patterns of food acquisition

Stocks of both surfperch species were characterized by individuals that displayed strong fidelity to a depth. While feeding, individuals moved horizontally along depth contours, but did not move much between depths. Of the 200 foragers observed for the five minute periods, only three individuals (one of 101 *E. jacksoni*; two of 99 *E. lateralis*) left the depth zone in which the foraging bout was initiated. Fully a third of these fish (N = 31 black surfperch, N = 35 striped surfperch) were interrupted during foraging by an intraspecific or interspecific chase; these individuals were driven away from their foraging location, but returned to the same depth zone to resume feeding. We took this as evidence that short-term fidelity to a depth zone was great and equal for both surfperch species. Individuals also displayed a strong tendency to remain at a depth over long periods (months). Repeated sightings of individually tagged fish indicated that animals were residential within narrow depth ranges. A 95% prediction interval was calculated for black surfperch based on a total of 141 resightings of 25 tagged individuals over a six month period. For fish observed more than once (N = 20), the prediction interval was ±1.67 m of the individual's mean depth of residence. Although based on fewer observations (20 resightings of 7 fish), *E. lateralis* had a similar prediction interval (±1.70 m). These mean that although individuals might occasionally make substantial changes in depth, we would expect 95% of sightings of an individual to fall within the calculated prediction interval. This vertical range in which individual fish concentrated activity (<3.5 m) was considerably smaller than the range of each designated depth zone (~5 m). Given such a high degree of depth fidelity, different segments of each fish stock were exposed for prolonged periods to different environmental conditions associated with the particular depth zone inhabited.

We characterized several aspects of foraging by surfperches that might vary due to environmental differences among habitats. These included foraging behavior, substrate selection, effort expended, and diets achieved by fish within each depth zone. These are described for the two species in turn.

The use of substrate types as feeding locations by black surfperch varied among depths (Table 3). In shallow water, black surfperch mainly harvested prey from turf and *Gelidium robustum,* whereas in deep habitats most bites were taken from turf. Despite differences in substrate use, several indicators of foraging effort did not differ substantially among the depths. First, feeding rates (bites per min) were uniform among depths [black surfperch foraging rates ($\bar{X}$ bites per min): 3 m = 2.10, 6 m = 2.12, 9 m = 2.06, $F_{2,198} = 0.07$, NS]. The probability that a subsequent bite would be taken from the same plant also did not differ among depths (Table 4a). Finally, there was no depth-related difference in the average distance travelled between bites (Table 4b). These data suggested that the cost of foraging to black surfperch did not differ along the depth gradient. However, there were certain aspects of the diet of *E. jacksoni* that did vary among depths. The taxonomic composition of prey in diets was quite similar at all depths (Figs. 3, 4). Gammarid amphipods and shrimp were the major biomass components, with gammarids always the numerically dominant taxon. Despite the high degree of taxonomic similarity, there were substantial differences in the total amount of prey obtained by individual fish feeding in the various depth zones. Black surfperch in deep water contained significantly less prey biomass than individuals foraging

Table 4a. Probability a successive bite was taken from the same individual patch by black surfperch or striped surfperch in shallow and deep habitats. Data are the mean (±1SE) proportion of repeated bites on the same plant; data were analyzed by Student's t-test (N = 10 fish species^{-1} depth^{-1}).

Depth zone	3 m	9 m	t-value	Significance
E. jacksoni	0.37 (.06)	0.39 (.07)	0.33	NS
E. lateralis	0.58 (.05)	0.31 (.04)	4.65	P<0.001

Table 4b. Distance travelled between successive bites by surfperch as a function of depth. Data are the mean (±1SE) linear distance (in meters) an individual fish swam between bites (for details, see Methods); means are based on average of 3 sites.

Depth zone	3 m	9 m	t-value	Significance
E. jacksoni	2.9 (0.5)	2.0 (0.1)	1.75	NS
E. lateralis	2.1 (0.3)	6.1 (0.4)	5.63	P<0.01

Table 3. Use of benthic substrates as feeding locations by black surfperch and striped surfperch in each of three depth strata. Data are mean (±1 SE) proportion of bites taken (See Figure 1 for names of substrates); means based on average of three sites.

Depth zone			*Substrate type*		
	TURF	GR	RHO	LAM	ART
3 m					
E jacksoni	0.43 (.08)	0.40 (.05)	0.06 (.01)	0.02 (.01)	0.10 (.05)
E. lateralis	0.03 (.01)	0.83 (.05)	0.02 (.02)	0.03 (.01)	0.08 (.05)
6 m					
E. jacksoni	0.53 (.07)	0.24 (.07)	0.03 (.01)	0.08 (.02)	0.12 (.03)
E. lateralis	0.07 (.01)	0.57 (.06)	0.08 (.05)	0.17 (.10)	0.11 (.05)
9 m					
E. jacksoni	0.73 (.04)	0.03 (.01)	0.12 (.01)	0.04 (.01)	0.08 (.03)
E. lateralis	0.10 (.04)	0.17 (.03)	0.18 (.04)	0.42 (.06)	0.13 (.03)

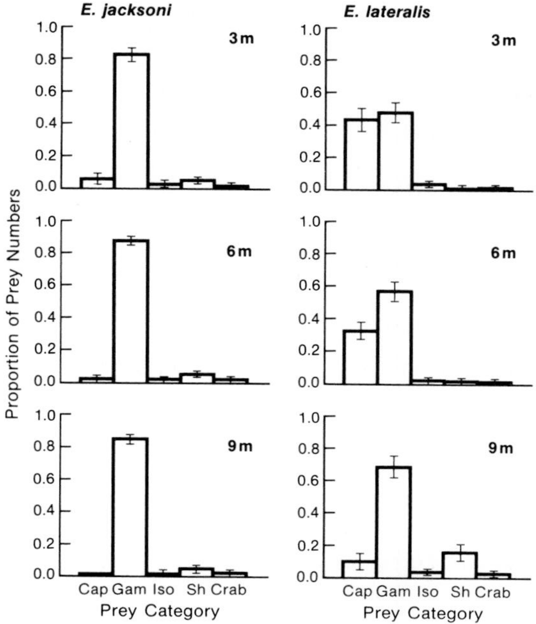

Fig. 3. Taxonomic composition (by number) of prey in diets of black and striped surfperch in three depth zones. Data are mean (±1SE) proportions of items by number. (For taxonomic groups, see Figure 2.) N = 20 fish species^{-1} depth^{-1}. Diets of black surfperch did not vary among depths (MANOVA; $F_{10,104}$ = 0.90; p>0.50); diets of striped surfperch were different (MANOVA; $F_{10,104}$ = 4.46; P<0.0001).

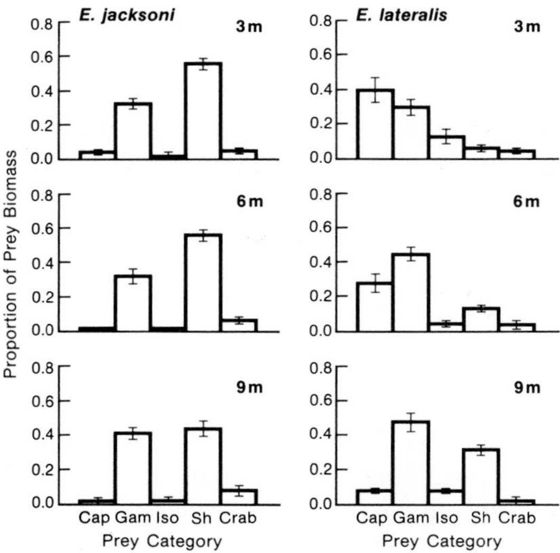

Fig. 4. Taxonomic composition (by biomass) of prey in diets of black surfperch and striped surfperch in three depth zones. Data are mean (±1SE) proportion of types by biomass. (See Figure 2 for taxonomic groups.) Diets of black surfperch were not different among depths (MANOVA; $F_{10,104}$ = 1.21; P>0.25); diets of striped surfperch did vary (MANOVA; $F_{10,104}$ = 4.02; P<0.0001).

in shallower zones (Fig. 5). The disparity was linked to the fact that fewer – not smaller – prey items were harvested by deep-water animals [black surfperch mean prey size (mg): 3 m = 2.01, 6 m = 2.15, 9 m = 1.76, $F_{2,57}$ = 0.75, NS]. Taken together, the analyses of diets and foraging behavior indicated that, although black surfperch in all habitats invested equal amounts of effort into foraging, individuals feeding in deep habitats obtained a significantly lower return of food.

Compared with *E. jacksoni,* striped surfperch showed quite different patterns in diets and foraging behavior as a function of depth. In shallow habitats, bites were concentrated on *Gelidium robustum.* When foraging deeper, other substrates such as laminarian algae and *Rhodymenia* were more frequently visited (Table 3). Although feeding rates did not vary among zones [striped surf-

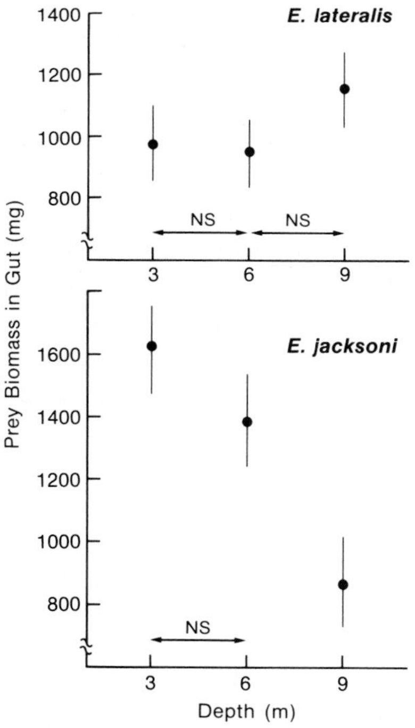

Fig. 5. Total prey biomass captured by individual black and striped surfperch in each of three depth zones. The mean mg (±1SE) of prey in fish guts standardized for body length is given for each depth (N = 20 fish species^{-1} depth^{-1}). ANCOVA with standard length as a covariate revealed no depth difference for striped surfperch ($F_{2,56}$ = 1.70; P>0.20) but significant differences for black surfperch ($F_{2,56}$ = 7.31; P<0.002). Arrows along abscissa connect means that do not differ at P<0.05 level.

perch foraging rates ($\bar{X}$ bites per min): 3 m = 2.81, 6 m = 2.32, 9 m = 2.74, $F_{2, 151} = 2.80$, NS], foragers in deep water travelled more than three times farther between bites (Table 4b). In shallow zones, the probability that a second bite would be taken from the same plant was significantly greater than in deep zones (Table 4a). These results indicated that striped surfperch expended more effort when feeding in deep habitats. The increase stemmed directly from the much wider spacing of substrate types preferred by striped surfperch (Tables 1, 3). Like black surfperch, the environmental gradient resulted in depth-related variation in the diet achieved by *E. lateralis*. Unlike black surfperch, the taxonomic composition of striped surfperch diets differed significantly among depths. *E. lateralis* in shallow areas ate mostly caprellid and gammarid amphipods, whereas individuals in deeper habitats consumed a larger fraction of shrimp (Fig. 3, 4). However, neither the total biomass of prey captured by individual striped surfperch (Fig. 5) nor the mean size of prey [striped surfperch mean prey size (mg): 3 m = 1.24, 6 m = 1.38, 9 m = 1.25, $F_{2,57} = 0.11$, NS] varied with depth. Taken together, the dietary and foraging patterns of striped surfperch indicated that foragers in deep zones expended more effort to obtain the same food return compared with conspecifics in shallow habitats.

For both species of surfperch we found evidence that the ratio of food return to effort expended declined with increasing depth. If, as our evidence suggested, individual fish retained fidelity to a depth for long periods of time, growth rates could be expected to decline with increasing depth. Data obtained from otolith analyses supported this contention. We found that both black and striped surfperch collected from deep habitats were significantly smaller than conspecifics of the same chronological age in shallow zones (Fig. 6). For the average-aged individual (about three years in our Santa Cruz Island stocks), this involved a reduction in standard length of nearly 10 mm or about 5%. Assuming that length-fecundity relationships for *E. jacksoni* and *E. lateralis* in Baltz (1984) hold, we estimated the probable consequence of these body size differences on clutch size. For the average aged black surfperch, predicted fecundity differed by about 10% (deep individuals = 9.3 young per clutch; shallow = 10.4 per clutch) across the depth gradient. The impact was even greater for striped surfperch, where predicted fecundity differed by 18% (deep = 7.6 young per clutch; shallow = 9.4 per clutch).

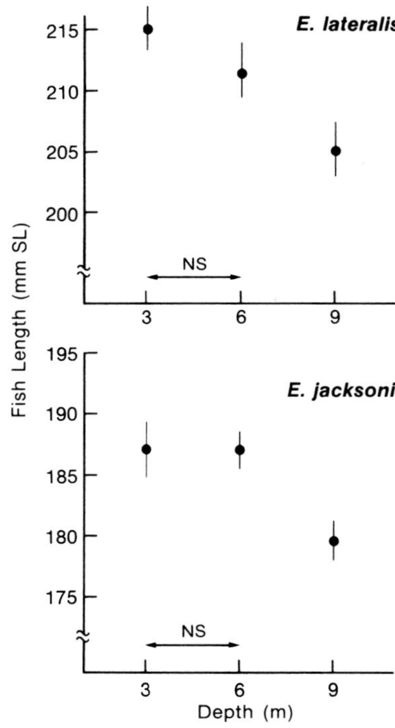

Fig. 6. Body length of striped (top) and black (bottom) surfperch of standardized age (3 yr) in three depth zones. Given are mean (±1SE) standard length (mm) of fish of standardized age. ANCOVA with age as a covariate indicated that the sizes of each species varied among depths (*E. jacksoni:* $F_{2,238} = 5.24$; $P<0.01$; *E. lateralis:* $F_{2,207} = 5.69$; $P<0.05$). Arrows along abscissa connect means that do not differ at $P<0.05$ level.

Discussion

Studies of intraspecific variation in diet of predatory fishes frequently have compared trends among distinctly separate populations. When within-population (stock) variation has been addressed, the focus has been primarily on ontogenetically derived patterns (reviewed in Schmitt & Holbrook 1984a, b). With few exceptions (e.g. Allen 1941, Bryan & Larkin 1972, Werner et al. 1981, Milinski 1982, Ringler 1983), the existence, mechanisms, and consequences of stock variation in diet of fishes have largely gone unexplored. The data presented here indicate that variation in the foraging environment within the scale of a stock can lead to persistent differences in the foraging behavior, diets, and growth performance among adults. These results suggest that comparisons of diet among populations be interpreted with caution.

The between-individual variation in foraging observed for both black and striped surfperch arose for two reasons. First, the depth gradient in availability of substrate types and invertebrate prey provided dramatically different habitats within a very short distance. Second, individuals of both species were residential within fairly limited depth ranges, with the result that segments of each stock were consistently exposed to a particular array of prey and foraging substrates associated with the depth zone inhabited. The immediate consequence of fidelity to a depth zone was that animals living just a few meters apart exhibited different foraging behavior and achieved different diets. In the long-term, these variations in foraging success appeared to result in differences in growth of individuals. Fish living in the deeper, prey-poor end of the gradient were smaller in length than individuals of the same chronological age residing in shallow, food-rich areas. This body size difference (~5%) quite probably reflected a greater disparity in reproductive output; based on length-fecundity relationships in Baltz (1984), we estimated the difference in clutch size for the average aged female was 10% for *E. jacksoni* and 18% for *E. lateralis.*

The variation in success of both black and striped surfperch stemmed from differences in 'profitability' (benefit to cost ratio) of various depth zones. However, the precise mechanism differed for the two fishes. For *E. jacksoni,* individuals invested equal effort foraging but obtained different food return among depths (i.e. only benefits differed among habitats). By contrast, *E. lateralis* in different depths exerted different effort feeding but achieved the same prey return (i.e. only costs differed). It appeared that the different mechanistic reason why profitability varied among depths was directly related to species-specific foraging adaptations of the surfperch.

Black surfperch can separate unwanted food items and debris from edible prey, swallow the desired material, and eject the remainder from the mouth (Laur & Ebeling 1983). This winnowing allows black surfperch to feed from debris-laden substrates such as turf and heavily encrusted algal plants, but it is a relatively indiscriminate process (Holbrook & Schmitt 1984). The variation in foraging return experienced by black surfperch may stem from their use of a winnowing feeding mode, because, given the disparity in prey density, an indiscriminate bite in deeper habitats would yield fewer prey than one in shallower areas. By contrast, striped surfperch forage visually, locate prey items on the surfaces of foliose algae and harvest them one or a few at a time (Laur & Ebeling 1983). When prey are abundant on a particular patch, an individual can take several successive bites before moving to a new plant. However, a visual mode of feeding does not allow for effective use of turf as a foraging substrate (Schmitt & Holbrook 1984a, b). Hence, striped surfperch in deep areas, where turf predominated, expended considerable effort travelling between the widely spaced foliose algal plants that could be used. Deep areas had an additional impact on striped surfperch; because prey densities were lower, very few repeated bites were taken from the same plant, which resulted in more transitions between patches. For *E. lateralis,* deep habitats were poor not only because prey densities were lower, but also because appropriate foraging substrates were rarer and more dispersed.

Because the growth rate consequences of using different foraging habitats by fishes have received little empirical attention (but see Werner et al. 1983a, b), the general results presented here have important implications. For example, Werner et al. (1981) found that bluegill sunfish (*Lepomis macrochirus*) in experimental ponds appeared to select habitats over time on the basis of profitability (relative benefit to cost ratio). For the most part, fish were habitat 'specialists' (i.e. >80% of diet obtained from one habitat type), although a few individuals displayed a relatively mixed or generalized diet. Werner et al. (1981) were able to demonstrate that, compared with habitat specialists, bluegill with a mixed diet contained less prey biomass but did not have smaller body sizes. Werner et al. (1981) attributed the lack of difference in individual growth on the transient nature of habitat and/or dietary specialization in the bluegill system. By contrast, Bryan & Larkin (1972) documented dietary differences among individuals of three salmonid species (*Salvelinus fontinalis, Salmo clarki,* and *Salmo gairdneri*) that persisted for relatively long periods (months). No explanation was offered as to why specialization persisted, nor were the relative costs and benefits of attaining different diets estimated. Bryan & Larkin (1972) did not find any correlation between degree of dietary specialization and either body size or growth rate.

In the surfperch system, fitness may vary among individuals because fish at each depth exploit the environment in a different manner, which in turn yields different costs and benefits. This disparity in short-term success appears to have a cumulative effect, because individuals display strong fidelity to a particular habitat over time. Intra- or interspecific aggression could result in relegation of inherently slower-growing individuals to deeper habitats and could contribute to the patterns we have documented. Nevertheless, we have shown that individuals in deeper habitats encounter lower availability of food and achieve lower foraging success than those in shallower areas. This is a parsimonious explanation for the observed depth-related pattern of fish size at age.

Finally, black and striped surfperch are known to compete with one another for foraging space at Santa Cruz Island (Hixon 1980, Schmitt & Holbrook in preparation), allowing us to speculate on the impact of competition on fish diets. Interspecific competition curtails access to feeding areas on two spatial scales, which in turn might impact the surfperches quite differently. On one level, competition from striped surfperch limits abundance of black surfperch in shallow, prey-rich habitats (Hixon 1980); increased density of *E. jacksoni* in shallow water upon experimental removal of striped surfperch is initially achieved by movement of fish from deeper habitats, not by immigration from other sites (Hixon 1980). Relocated black surfperch have access to richer foraging habitats and would presumably obtain more food per effort and therefore experience enhanced growth rates. This effect of interspecific competition is not reciprocal; the depth distribution of striped surfperch is unaffected by *E. jacksoni* (Hixon 1980). However, interspecific competition also influences the choice of substrates used within a depth zone and, at this finer spatial scale, both surfperch species are affected (Schmitt & Holbrook in preparation). Because we do not know whether characteristics of prey vary among patch types, it is not possible to speculate what impact competition for foraging substrates within depth zones may have on surfperch diets. Nevertheless, the intensity and consequences of competition in the surfperch system will likely vary in space because individuals acquire food along different segments of a patchy environmental gradient.

Acknowledgements

We thank J. Bence, A. Ebeling, D. Laur, and E. McCauley for critical discussion and comments on the manuscript. A. Breyer, M. Carr, M. Casey, K. Collins, W. Douros, and C. Gottschalk assisted in the field. A. Breyer, P. Chiu, K. Collins, J. Crisp, K. Iverson, J. Jolly, and S. Swarbrick helped with laboratory analyses. The California Department of Fish and Game provided permits for field work. This work was supported by National Science Foundation grants (OCE-8110150 and OCE-8314832), and logistic support was provided by the University of California, Santa Barbara.

References cited

Alevizon, W.S. 1975a. Comparative feeding ecology of a kelp-bed embiotocid (*Embiotoca lateralis*). Copeia 1975: 608–615.

Alevizon, W.S. 1975b. Spatial overlap and competition in congeneric surfperches (Embiotocidae) off Santa Barbara, California. Copeia 1975: 352–356.

Allen, K.R. 1941. Studies on the biology of the early stages of the salmon (*Salmo salar*). 2. Feeding habits. J. Anim. Ecol. 10: 47–76.

Baltz, D.M. 1984. Life history variation among female surfperches (Perciformes: Embiotocidae). Env. Biol. Fish. 10: 159–171.

Bryan, J.E. & P.A. Larkin. 1972. Food specialization by individual trout. J. Fish. Res. Board Can. 29: 1615–1624.

Carr, W.E.S. & C.A. Adams. 1973. Food habits of juvenile marine fishes occupying seagrass beds in the estuarine zone near Crystal River, Florida. Trans. Amer. Fish. Soc. 102: 511–540.

Connell, J.H. 1978. Diversity in tropical rain forests and coral reefs. Science 199: 1302–1310.

Connell, J.H. 1980. Diversity and the coevolution of competitors, or the ghost of competition past. Oikos 35: 131–138.

Connor, E.F. & D. Simberloff. 1979. The assembly of species communities: chance or competition? Ecology 60: 1132–1140.

Connor, E.F. & D. Simberloff. 1983. Neutral models of species' co-occurrence patterns. pp. 316–331. *In:* D.R. Strong, Jr., D. Simberloff, L.G. Abele & A.B. Thistle (ed.) Ecological Communities: Conceptual Issues and the Evidence, Princeton University Press, Princeton.

Coyer, J.A. 1979. The invertebrate assemblage associated with *Macrocystis pyrifera* and its utilization as a food source by kelp forest fishes. PhD Dissertation, University of Southern California, Los Angeles. 364 pp.

Diamond, J.M. 1978. Niche shifts and the rediscovery of interspecific competition. Amer. Sci. 66: 322–331.

Ebeling, A.W. & D.R. Laur. 1986. Foraging in surfperches: resource partitioning or individualistic responses? Env. Biol. Fish. 16: 123–133 (this issue).

Ellison, J.P., C. Terry & J.S. Stephens, Jr. 1979. Food resource utilization among five species of embiotocids at King Harbor, California, with preliminary estimates of caloric intake. Mar. Biol. 52: 161–169.

Grossman, G.D. 1980. Ecological aspects of ontogenetic shifts in prey size utilization in the bay goby (Pisces: Gobiidae). Oecologia 47: 233–238.

Grossman, G.D., R. Coffin & P.B. Moyle. 1980. Feeding ecology of the bay goby (Pisces: Gobiidae). Effects of behavioral, ontogenetic, and temporal variation on diet. J. Exp. Mar. Biol. Ecol. 44: 47–59.

Haldorson, L. & M. Moser. 1979. Geographic patterns of prey utilization in two species of surfperch (Embiotocidae). Copeia 1979: 567–572.

Hixon, M.A. 1979. Competitive interactions and spatiotemporal patterns among California reef fishes of the genus *Embiotoca*. Ph.D. Dissertation, University of California, Santa Barbara. 213 pp.

Hixon, M.A. 1980. Competitive interactions between California reef fishes of the genus *Embiotoca*. Ecology 61: 918–931.

Holbrook, S.J. & R.J. Schmitt. 1984. Experimental analysis of patch selection by foraging black surfperch (*Embiotoca jacksoni*). J. Exp. Mar. Biol. Ecol. 79: 39–64.

Keast, A. 1965. Resource division amongst cohabiting fish species in a bay, Lake Opinicon, Ontario. pp. 106–132. *In:* Proc. 8th Conf. Gr. Lakes Res., University of Michigan, Ann Arbor.

Keast, A. 1968. Feeding of some Great Lake fishes at low temperatures. J. Fish. Res. Board Can. 25: 1199–1218.

Keast, A. 1977a. Mechanisms expanding niche width and minimizing intraspecific competition in the rock bass and bluegill sunfish (*Centrarchidae*). Evol. Biol. 10: 333–395.

Keast, A. 1977b. Feeding and food overlaps between the year classes relative to the resource base, in the yellow perch, *Perca flavescens*. Env. Biol. Fish. 2: 55–70.

Keast, A. 1978. Trophic and spatial interrelationships in the fish species of an Ontario temperate lake. Env. Biol. Fish. 3: 7–31.

Laur, D.F. & A.W. Ebeling. 1983. Predator-prey relationships in a guild of surfperches. Env. Biol. Fish. 8: 217–229.

Lemly, A.D. & J.F. Dimmick. 1982. Growth of young-of-the-year and yearling Centrarchids in relation to zooplankton in the littoral zone of lakes. Copeia 1982: 305–321.

MacPherson, E. 1979. Ecological overlap between macrourids in the western Mediterranean Sea. Mar. Biol. 53: 149–159.

Milinski, M. 1982. Optimal foraging: the influence of intraspecific competition on diet selection. Behav. Ecol. and Sociobiol. 11: 109–115.

Mittelbach, G.C. 1984. Predation and resource partitioning in two sunfishes (Centrarchidae). Ecology 65: 499–513.

Nilsson, N. 1963. Interaction between trout and char in Scandinavia. Trans. Amer. Fish. Soc. 92: 276–285.

Ringler, N.H. 1983. Variation in foraging tactics of fishes. pp. 159–171. *In:* D.L.G. Noakes, D.G. Lindquist, G.S. Helfman & J.A. Ward (eds.) Predators and Prey in Fishes, Dev. Env. Biol. Fish. 2, Dr W. Junk Publishers, The Hague.

Ross, S.T. 1977. Patterns of resource partitioning in searobins (Pisces: Triglidae). Copeia 1977: 561–571.

Schmitt, R.J. & J.A. Coyer. 1982. The foraging ecology of sympatric marine fish in the genus *Embiotoca* (Embiotocidae): importance of foraging behavior in prey size selection. Oecologia 55: 369–378.

Schmitt, R.J. & J.A. Coyer. 1983. Variation in surfperch diets between allopatry and sympatry: circumstantial evidence for competition. Oecologia 58: 402–410.

Schmitt, R.J. & S.J. Holbrook. 1984a. Gape-limitation, foraging tactics, and prey size selectivity of two microcarnivorous species of fish. Oecologia 63: 6–12.

Schmitt, R.J. & S.J. Holbrook. 1984b. Ontogeny of prey selection by black surfperch, *Embiotoca jacksoni* (Pisces: Embiotocidae): the roles of fish morphology, foraging behavior, and patch selection. Mar. Ecol. Prog. Ser. 18: 225–239.

Schoener, T.W. 1974. Resource partitioning in ecological communities. Science 185: 27–39.

Targett, T.E. 1978. Food resource partitioning by the pufferfishes *Sphoeroides spengleri* and *S. testudineus* from Biscayne Bay, Florida. Mar. Biol. 49: 83–91.

Werner, E.E. 1979. Niche partitioning by food size in fish communities. pp. 311–322. *In:* H. Clepper (ed.) Predator-prey Systems in Fisheries Management, Sport Fish Inst., Washington, D.C.

Werner, E.E., G.G. Mittelbach & D.J. Hall. 1981. The role of foraging profitability and experience in habitat use by the bluegill sunfish. Ecology 62: 116–125.

Werner, E.E., G.G. Mittelbach, D.J. Hall & J.F. Gilliam. 1983a. Experimental tests of optimal habitat use in fish: the role of relative habitat profitability. Ecology 64: 1525–1539.

Werner, E.E., J.F. Gilliam, D.J. Hall & G.G. Mittelbach. 1983b. An experimental test of the effects of predation risk on habitat use in fish. Ecology 64: 1540–1548.

Received 12.12.1984 *Accepted 21.5.1985*

Ecological and morphological shifts in Lake Michigan fishes: glimpses of the ghost of competition past

Larry B. Crowder
Department of Zoology, North Carolina State University, Box 7617, Raleigh, NC 27695, U.S.A.

Keywords: *Alosa pseudoharengus,* Character displacement, Competitive bottlenecks, *Coregonus hoyi,* Diet shifts, Habitat shifts, Resource partitioning

Synopsis

Both historical patterns and recent evidence of resource partitioning and complementarity within the Lake Michigan fish community provide circumstantial evidence for interspecific competition. But competition is difficult to document in the field without controlled experimentation. In Lake Michigan, controlled experiments on competition within the fish community are nearly impossible, but we still need to understand the interactions among the dominant fishes. For this purpose, I have relied upon hypothesis-based field observation, ' natural experiments' in the field and designed laboratory experiments to evaluate competitive interactions. Resource use patterns and trophic morphology of the bloater, *Coregonus hoyi*, a native cisco, from samples taken before alewife, *Alosa pseudoharengus,* became abundant (1960) were compared to more recent data (1979–80). After the alewife density increase, bloaters had significantly fewer and shorter gill rakers. This suggests a morphological shift toward greater benthic foraging efficiency in response to high abundances of an efficient pelagic planktivore, alewife. Resource use comparisons suggested that bloaters now shift from pelagic zooplanktivory to benthic habitats and diets at least two years earlier in their life history than they did before alewife became abundant. This evidence, albeit not experimental, provides strong support for the importance of competition in the structure of the current Lake Michigan fish community.

In Lake Michigan, seasonal thermal habitat compression can pack fish into a narrow thermal zone across the lake bottom, leading to increased habitat overlap, reduced prey availability and fish diets containing fewer and smaller prey. Thermal habitat compression, which can occur intermittently through the season, may create competitive bottlenecks which help maintain the observed resource partitioning among these fishes.

Introduction

Competition has long been assumed to be an important factor structuring animal communities. Evolutionary ecology and niche theory have often implicitly adopted competition as an important factor and though this may not be unreasonable, the assumption has often gone untested. Competition has come under severe fire in recent years (e.g. Wiens 1977, Conner & Simberloff 1979, Connell 1980), mostly based on the lack of firm field evidence to support the idea that competition acts to create or regulate resource partitioning. Darwin (1859) considered competition to be an important factor and in some ways anticipated the development of both competition-based theory and the current controversy: 'It is good thus to try in imagination to give to any one species an advantage over another. Probably in no single instance should we know what to do. This ought to convince us of our ignorance on the mutual relations of all organic beings; a conviction as necessary as it is difficult to acquire.'

In the 125 years since Darwin, we seem to have made little progress. Our theoretical 'imagination' produced logical, appealing generalizations about the expected outcomes of competition (e.g. resource partitioning, complementarity, limiting similarity, body size ratios), but many of these were accepted somewhat uncritically and without careful testing. Recent criticisms of the competition 'paradigm' (Wiens 1977, Connell 1980, and Simberloff and his collegues) have argued that we have depended too much on 'imagination' and too little on critical observation or experimentation. This critique has again pointed out our 'ignorance', particularly when we depend on descriptive patterns of resource use alone. Recent analyses of the controversy (Salt 1983, Schoener 1983, Connell 1983, Strong et al. 1984) have noted the strengths and weaknesses of the earlier critiques, emphasized the need for a more experimental approach and sought balance between theory and test. In the final analysis, the surest approach requires a detailed knowledge of the natural history of the system to interpret either descriptive or experimental results.

In Lake Michigan, controlled experiments on competition within the fish community are nearly impossible, both because of ongoing sport and commercial fisheries and because of the sheer size of the task. But we still need to understand the interactions among the dominant fishes and to be able to predict the outcome of various management manipulations. For this purpose, I have relied upon hypothesis-based field observations, 'natural' experiments and designed laboratory experiments, which I will summarize here.

Historical dynamics of the Lake Michigan fish community

Both the fish communities and fisheries of the Laurentian Great Lakes have been highly variable over the past half century. This variability has complicated the use and management of fishery resources. Much of this variability (population extinctions, exponential growth of exotic fishes, the collapse and reopening or genesis of commercial or sport fisheries) has occurred through unintentional disturbances as well as planned manipulations by man. These large-scale variations are best known in Lake Michigan and can be considered to be experiments (though uncontrolled) on fish community structure and species interactions.

The historical dynamics of the Lake Michigan fish community were both dramatic and well documented (Smith 1970, Wells & McLain 1973, Christie 1974, Kitchell & Crowder 1986). The initial changes derive from the invasion of sea lamprey, *Petromyzon marinus,* which together with a size selective fishery eliminated the larger commercial fishes including lake charr, *Salvelinus namaycush,* by the mid-1950s. Rainbow smelt, *Osmerus mordax,* was introduced into the Lake Michigan drainage in 1912 and increased rapidly in Lake Michigan in the 1930s. Alewife, *Alosa pseudoharengus,* was first noted in Lake Michigan in 1949 and increased rapidly in the early 1960s, presumably due to the absence of large piscivores. During the increase of alewife and smelt densities, a number of native fishes declined or became locally extinct. Native planktivores originally included seven ciscoes; six of which went extinct during the period when alewife and smelt were increasing.

Because native and exotic fishes ate similar zooplankton prey (Wells & Beeton 1963, Morsell & Norden 1968, Janssen & Brandt 1978, Wells 1980, Crowder et al. 1981) and large zooplankton became rare during the increase of alewife (Wells 1960, 1970), competition may well have been intense. This should not be surprising, since predators did not constrain the increase of alewife or other planktivores during this period. Available data on the feeding behavior of adult and juvenile alewife and bloater, *Coregonus hoyi,* support the idea that alewife is more efficient at feeding on small prey, based on its ability to switch feeding modes from particulate feeding to filtering (Janssen 1976, Crowder & Binkowski 1983, Crowder 1985). But the adult bloater appears to be more efficient than adult alewife when foraging on benthic prey. The zooplankton feeding interval in juvenile bloater, thus becomes a potential 'bottleneck' (sensu Werner & Gilliam 1984) for recruitment of larger bloaters to the benthos . If bloaters make it through the zooplankton feeding interval, it appears that they can succeed in the benthos relative to alewife.

By the mid 1960s, Lake Michigan was yielding little of its natural productivity to commercial or sport fisheries. Alewife went through a population explosion and massive die-off in 1967. In response to the alewife 'pest' and in an effort to improve fishing, management measures were imposed to reduce sea lamprey densities (Lawrie 1970) which permitted re-introduction of lake charr to Lake Michigan in 1965. To further rehabilitate the lake and to reduce the alewife population, several non-native salmonid predators (Pacific salmon, *Oncorhynchus* spp.) were also stocked. But this valuable fishery is strongly management dependent; lamprey control and salmonid stocking must be continuously maintained. Furthermore, the former 'pest', alewife, now occupies a critical role in the system because it supports the entire salmonid fishery as the major forage item for lake charr and Pacific salmon. Though smelt and sculpins account for a small part of the lake trout diet, they and bloaters are relatively unimportant in the diets of Pacific salmon (Stewart et al. 1981).

Descriptive resource partitioning

Competition for food may be minimized by spatial segregation of potential competitiors. In pelagic systems, temperature may provide a 'template' for spatial segregation of species with similar food habits (Magnuson et al. 1979). Brandt et al. (1980) provided evidence for segregation by thermal habitat of major planktivores in Lake Michigan. Complementarity in the use of thermal habitat and food has also been demonstrated in Lake Michigan fishes (Crowder et al. 1981); native fishes that overlap thermal habitats of alewife and smelt use different foods. Direct overlaps between native juvenile fishes and adult alewife appear limited, especially when the lake is thermally stratified. Complementarity in resource use often has been attributed to competition (Schoener 1974), though other interactions and environmental changes will certainly influence resource use. The observed patterns provide, at best, circumstantial evidence for competition.

The available data on the relative efficiency of benthic feeding by bloater (Janssen 1976, Crowder & Binkowski 1983) may also explain the habitat shift in alewife which we observed as bloater increased in abundance in the late 1970s (Crowder & Magnuson 1982). When we began our studies of species interactions in Lake Michigan in 1977, bloaters were extremely rare. The commercial fishery for bloaters had been closed and our catch of bloaters amounted to 0.3% of the total fish caught. Alewife were abundant in day bottom trawls and occurred at temperatures preferred in the laboratory. We repeated the 1977 sampling plan in fall 1979. Between 1977 and 1979, bloater had increased in abundance and dominated the thermal habitat that had been occupied by alewife in 1977; adult alewife near the bottom occupied water about 10° C colder than preferred. Since total catch of alewife was not reduced in 1979, and since predation intensity by salmonids on alewife did not change significantly between 1977 and 1979 (Stewart et al. 1981), we inferred that the altered thermal distribution of alewife resulted from competitive interactions with bloater (Crowder & Magnuson 1982).

The observation that bloater appeared to displace alewife suggested that the interaction leading to the initial declines of many native fishes, including bloater, was more complex than simple competitive dominance by alewife. The ability to filter smaller zooplankton confers a competitive advantage on alewife when small zooplankton predominate as they did in the mid-1960s in Lake Michigan. If zooplankton are larger, as they were in the late 1970s-early 1980s (Gitter 1982), bloater young-of-year do not suffer this relative disadvantage (Crowder & Binkowski 1983). After lake charr and salmon were introduced into Lake Michigan, forage fish stocks (especially alewife) were reduced and zooplankton increased in size relative to those at the peak of alewife abundance. It also appears that bloaters now grow rapidly during the zooplankton feeding interval (Rice 1985).

Thus, the available field data on both the historical dynamics of the Lake Michigan fish community and the recent patterns of resource use in these fishes are consistent with the idea that competition has been an important force structuring this fish community. But this evidence is still circumstantial and provides what Strong et al. (1984) have called 'soft corroboration' for the competition hypothesis. Given that in situ experiments with pelagic Lake Michigan fishes would be difficult, costly, or impossible, what further evidence can be used to support or reject the competition hypothesis?

Ecological and morphological shifts in bloater

Character displacement (Brown & Wilson 1956, Grant 1972) and niche shifts (Pianka 1976, Sale 1979) both provide evidence in support of competition, but few studies of purported character displacement either document the rate of the character change or relate the observed morphological shifts directly to shifts in resource use. I am unaware of any examples from fishes for which even the minimum rate of character displacement can be estimated or for which the morphological shift can be related directly to shift in resource use.

During the increase of alewives in the 1960s, bloater and several other native species declined. Alewives have long, fine gill rakers and can efficiently filter the small zooplankton that were dominant in the mid-1960s. Bloaters have coarser, shorter rakers than alewife and apparently cannot filter (Crowder & Binkowski 1983). Gill raker number is a reasonably stable, heritable character in coregonine fishes and is of great taxonomic importance (Svärdson 1952, Bailey & Smith 1981). Raker morphology is influenced by changes in the environment of developing fishes (Lindsey 1981, Todd et al. 1981), but a change in gill raker number of two or more in nature has often been attributed to artifical or natural selection (Svärdson 1952, Berg 1970, Bailey & Smith 1981).

The only native cisco which remained in Lake Michigan after the increase of alewife was the bloater. An unknown amount of genetic introgression occurred among the seven cisco stocks in the 1950s and 1960s, and may have caused changes in the raker morphology of bloater. Since introgression was possible with a variety of ciscoes – some with fewer and some with more rakers – the net effect of introgression on raker number in bloater is uncertain, but one would expect variance in raker number to be increased by introgression.

During the alewife density increase, bloater numbers were drastically reduced (Brown 1970) and thus competition may have been intense. This competition may have resulted in a diet or habitat shift (Sale 1979) in bloater. Lindsey (1981) suggested the hypothesis that some coregonine fishes in the Great Lakes may have shifted to fewer gill rakers in response to the advent of alewife. I have examined the evidence for shifts in gill raker morphology relative to introgression and competition using gill raker metrics from bloaters collected from southeastern Lake Michigan before and after alewife became abundant (Crowder 1984). I have also assessed the possibility of a habitat shift of bloater in southeastern Lake Michigan using habitat and diet data before and after the alewife density increase (Crowder & Crawford 1984).

Bloaters collected from southeastern Lake Michigan in 1979 had significantly fewer (by 2.1) and significantly shorter (by 15%) gill rakers than bloaters collected there in 1960 (Fig. 1). Because the length of the lower gill arch did not change, the reduction in raker number might be expected to lead to wider raker spacing as well, though this was not measured. The mechanism underlying this shift in raker morphology is difficult to document unequivocally. Introgression could well have contributed to the changes in raker morphology. Other ciscoes were rare by 1960 and introgression would have tended to both reduce mean raker number and increase variance in raker number. Though the mean did decline significantly, the variance remained unchanged (Crowder 1984). Introgression cannot easily explain, however, the apparent loss of individuals with high raker numbers from the population (Fig. 1). No fish examined from 1979 had more than 45 rakers; 17% of the fish examined

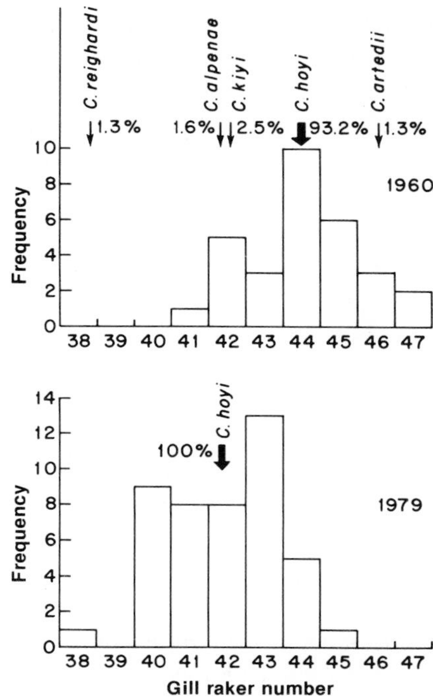

Fig. 1. Histogram of gill raker number for *C. hoyi* collected from fishery statistical district MM7 in southeastern Lake Michigan in 1960 and near Grand Haven, Michigan in 1979. Mean raker number (arrow) and percent of total ciscoes are indicated for *C. hoyi and for rare species of Coregoninae in 1960–61 (from Crowder 1984).*

from 1960 had more than 45 rakers. Introgression may well have contributed to the observed decrease in mean number of rakers, but it probably could not account for the observed reduction in high raker numbers.

In addition to introgression, environmental change may influence raker morphology directly, but we have little evidence to suggest that bloater larvae and young-of-year now occupy different habitats than they did in 1960. Although I cannot eliminate introgression or changes in the environment of early life stages of bloater as contributors to the observed shifts, I suggest that natural selection was likely a major contributor. In particular, selection provides the best explanation for reduced frequency of individuals with high numbers of rakers. Regardless of the underlying mechanism, the observed character displacement results in raker morphology that tends to be more efficient at sifting through benthos than the raker morphology before alewife became abundant.

Our 1979–1980 bottom trawl data (Crowder & Crawford 1984) indicate that bloaters become benthic some time after the first summer of life and before the end of the second summer. The previous evidence based on trawling and diets (Wells & Beeton 1963, Wells 1968) agrees with our observations that large adults occupy the bottom habitat and feed on benthic prey. But smaller bloaters (<161 mm FL – fork length) were rarely caught on the bottom in Wells' extensive trawling in the 1950s and early 1960s (Wells & Beeton 1963, Wells 1968). A few juveniles were occasionally caught in midwater trawls and they had eaten zooplankton exclusively (Wells & Beeton 1963). In 114 small bloaters (<138 mm FL) caught in bottom trawls in 6 years of sampling, an average of 74% of diet biomass was zooplankton. The biomass of zooplankton in diets declined in larger fish, but accounted for less than 1% of the diet biomass only in bloaters larger than 230 mm FL.

By contrast, juvenile bloaters (100–145 mm FL) were common in bottom trawls in both 1979 and 1980 (Crowder & Crawford 1984) and had consumed primarily benthic prey (99% by mass) rather than zooplankton (Table 1). Adult bloaters (145–240 mm FL) consumed almost entirely benthic prey based on biomass in both years (>99.8%). Our evidence thus suggests that juvenile bloaters now shift to the benthic habitat as much as two years earlier than bloaters caught before alewife became abundant. Bloaters as small as 100 mm FL are now common on the bottom in late summer – before alewives became abundant, bloaters as much as 60% longer and two years older were still pelagic (Wells & Beeton 1963). Before the alewife density increase, only bloaters greater than 230 mm FL (about age 5 or more) ate greater than 99% benthic prey; yearling fish about half as long do so now.

This shift in bloater life history and functional morphology toward earlier utilization of the benthic habitat in the presence of alewife provides strong support for the hypothesis that alewife competition has been an important influence on the structure of the current Lake Michigan fish community. Though experimental validation of this

Table 1. Diet biomass of three size (age) classes of bloaters caught from Lake Michigan near Grand Haven, August-September 1979 and 1980. Values are converted prey weights (percent) using biomass conversions from Wells & Beeton 1963 (adapted from Crowder & Crawford 1984).

Prey and fish characteristics	1979		1980		1980 young of year	
	Juv.	Adult	Juv.	Adult	Surface	Midwater
Zooplankton	1	+	1	+	100	100
Benthos						
Mysis relicta	+	20	45	40	0	0
Pontoporeia hoyi	99	80	51	60	0	0
Other	+	+	3	0	0	0
Number of fish examined (with prey)	83 (76)	86 (69)	57 (52)	29 (21)	46 (45)	31 (31)
Size range (mm fork length)	100–145	157–218	80–140	155–240	18–56	28–74
Mean number of prey per stomach	73	53	43	16	169	26

hypothesis is difficult or impossible, the combination of a strong morphological shift and dramatic changes in the ontogeny of habitat and diet shifts is reasonably convincing.

Is competition still occurring among fishes in Lake Michigan?

One of the problems with the 'ghost of competition past' is the question of whether or how competition which has acted in the past may or may not continue to regulate resource use (Connell 1980). Wiens' (1977) chief argument is that in a variable environment one may not see competition as it may only occur intermittently. And if it only occurs intermittently or unpredictably, can it regulate community structure?

Given the reintroduction of piscivorous fishes beginning in 1965, and the fact that alewife in the 1970s were fluctuating around a biomass less than one third their peak level, one might expect that competition for zooplankton in Lake Michigan would be reduced relative to the alewife density peak in the mid-1960s. Further evidence in support of this notion is that mean zooplankton sizes are now larger (Wells 1970, Gitter 1982). So, does competition still occur or are the present resource partitioning patterns simply remnants of the ghost? I think competition still occurs seasonally and may help regulate the established patterns of resource use.

We have examined thermal habitat use by Lake Michigan fishes in the zone where the thermocline intersected the lake bottom in late summer (August-September) during both 1979 and 1980. The thermocline in 1980 was narrower than that encountered in 1979. The mean distance across the lake bottom from the 8 to 17° C isotherm (approximately equivalent to the metalimnion) in 1979 was 2.6 km; in 1980 this distance was significantly reduced (1.8 km, t-test, $p<0.05$). Thus, the total area of the metalimnion may have been reduced by about 30% in 1980 relative to 1979. The overall fish catch per unit effort was up by a factor of three in the metalimnion. Common metalimnion fishes (adult bloater, alewife and rainbow smelt) tended to be significantly more abundant in the metalimnion. Juvenile bloater (<145 mm FL) were more abundant in 1980 catches from 7–18° C (Mann-Whitney U test, $p<0.05$); adult bloaters also increased, but not significantly. Alewife were significantly more abundant in 1980 in the 15–18° C stratum (Mann-Whitney U test, $p<0.01$); smelt were significantly more abundant in 1980 in the 7–14° C strata (Mann-Whitney U test, $p<0.01$).

Associated with this thermal habitat compression and locally increased abundance of metalimnion fishes was an obvious diet shift (Fig. 2) All species of fishes had broader diets in 1980 than in 1979. Juvenile bloater ate primarily *Daphnia* in 1979 though *Pontoporeia* dominated the diet biomass (Crowder & Crawford 1984); in 1980, juvenile bloaters ate primarily *Pontoporeia* but also consumed very small calanoid copepodites and diaptomids. Adult bloater concentrated on *Pontoporeia* and *Mysis* during both years, but consumed fewer, smaller prey on average in 1980 than in 1979 (Table 1, Crowder & Crawford 1984). Both smelt and alewife ate almost exclusively *Daphnia* in 1979, but diets broadened to include much smaller prey in 1980. The most common prey of alewife in 1980 was copepodites. This correlates with a decline in available zooplankton of about 75% between our 1979 and 1980 plankton net samples (Crowder, unpublished data). *Daphnia* declined in availability within the metalimnion by about an order of magnitude.

If zooplankton become scarce and shift to a smaller size, even locally as they did in 1980, the foraging advantage falls to alewife (Crowder & Binkowski 1983). On average, 45% more juvenile bloaters contained prey, and fish that contained prey had 38% more prey by numbers in 1979 than in 1980. Threefold as many adult bloaters contained prey, and each adult had eaten 3 × as many prey, in 1979 than in 1980 (Table 1, Crowder & Crawford 1984). Alewife and smelt ate about the same number of total prey in 1980 as in 1979, but the prey were much smaller on average in all fishes. Calanoid copepodites were the most common prey available in the metalimnion in 1980 and are of a size likely filtered by alewife but unavailable to bloater (Crowder & Binkowski 1983).

The observed thermal habitat compression may occur intermittently in response to storms, though the 1980 data suggest the narrowing was prolonged. Seasonally, the metalimnion narrows as the lake cools in the fall and this may create a seasonal 'bottleneck'. During periods of thermal habitat compression, intense foraging could locally deplete the available foods and because alewives are more efficient harvesting small zooplankton, they would tend to do better during these periods. Although transient, these intermittent periods of compe-

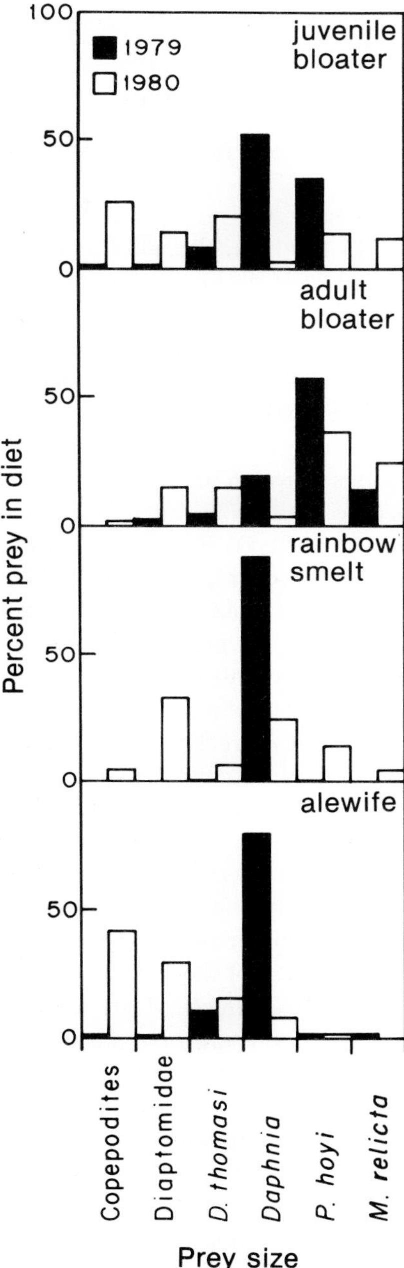

Fig. 2. Percent by numbers diet composition of alewife, smelt, and bloater in August-September 1979 and 1980. Prey catagories are listed in order of size and account for all prey taxa >5% by numbers in the diets.

tition may act to sustain the observed resource use patterns.

Community responses to a recent alewife decline

Alewife biomass has recently declined in Lake Michigan. In fall 1984, the adult alewife biomass available to trawls declined to 12600 metric tons (Wells & Hatch 1985), the lowest level since the early outbreak of alewives (1962 or so). If competition is still important in this system, one might expect to see some response among the other plankton feeding fishes. Bloaters, which are planktivorous through their first summer, have 'continued to recover at an extraordinary rate' (Wells & Hatch 1985). Based on otolith analysis, growth rates of bloater larvae in Lake Michigan in 1982 and 1983 were even more rapid than could be obtained in the laboratory with abundant food (Rice 1985). So, growth through the first summer appeared extremely strong for bloater. Young-of-the-year smelt increased dramatically in 1983 and yellow perch reproduction may have been 'the best since the 1950's' (Wells & Hatch 1984).

One general response to the alewife decline appears to be the formation of strong year classes of fishes whose early life stages depend on zooplankton. Both bloater and perch switch to benthic prey as late young-of-the-year or yearling fish. Other system-wide effects of the alewife decline and hypotheses about causes of the alewife decline are presented elsewhere (Kitchell & Crowder 1986). The available evidence supports the idea that previous alewife foraging on zooplankton may have constrained the formation of strong year classes of some native fishes; at least we have documented strong year classes in the face of an alewife decline.

Discussion

Connell (1980) outlined what he felt were necessary conditions to invoke coevolution of competitors: (1) that divergence between the species in the use of resources has occurred; (2) that it was caused by competition rather than some other mechanism; and (3) that the divergence has a genetic and not simply a phenotypic basis. Connell then defines a set of rigorously controlled experiments to test these hypotheses. But he also acknowledges that controls are difficult to arrange and that certain sorts of organisms cannot be manipulated easily, e.g., open ocean plankton. Connell (1980) does not address the role of competition in maintaining the coexistence of competitors that evolved separately and later came together.

In the Lake Michigan fish community there has been little opportunity for coevolution. Though the native fauna could well have coevolved, the present community is dominated by non-coevolved exotics which invaded relatively recently. Thus, we are not limited to evaluating the results of past competition, i.e., the 'ghost of competition past'. Rather, we can ask if competition only maintains separately evolved differences in resource use among exotic and native fishes or generates divergence in resource use.

We know that alewife evolved in the sea and is an efficient pelagic planktivore. Bloater evolved in a guild of deepwater ciscoes endemic to the Laurentian Great Lakes. What we know of bloater resource use before the invasion of alewife is that they were pelagic planktivores until age 3+ and then switched to benthic feeding for the remainder of their life history. After the alewife invasion and density increase, bloaters had fewer, shorter gill rakers and switched to feeding on benthos at least two years earlier in their life history. If competition was important in this interaction, it did not act to just maintain separately evolved differences, but a divergence has apparently occurred both in functional morphology of the gill rakers and in habitat use and diets of bloater. Connell's (1980) first criterion for coevolution of competitors is thus established.

Was the divergence due to competition or to some other factor? The decline in abundance and size of zooplankton near the peak abundance of the alewife in the mid-1960s (Wells 1970) suggests a limiting resource for many native fishes, particularly because alewife were the only fish in the lake known to be able to filter the small zooplankton then available. Further, evidence of year class failures among native fishes (Brown 1970, Wells & McLain 1973) suggests that recruitment through the zooplankton feeding interval of native fishes may have been inhibited. Now that the alewife have declined and zooplankton are larger and more abundant, strong year classes of native yellow perch and bloater have formed. These year classes apparently have done well in the zooplankton feeding interval and have switched successfully to benthic prey.

The observed shifts in gill raker morphology, particularly the loss of individuals with high numbers of rakers, cannot be completely explained by introgression or other reasonable mechanisms. Although it is impossible to eliminate all other *possible* explanations for the observed patterns, competition cannot be eliminated as a likely cause. In fact, the data strongly suggest that competition led to the observed ecological and morphological shifts.

Does the divergence have a genetic rather than a phenotypic basis? A shift of greater than two rakers can be obtained in the laboratory (Todd et al. 1981), but this may depend upon a change in the environment of the bloaters during their early life history which can directly alter phenotype. Lindsey (1981) catalogued the shifts observed in raker numbers of coregonines in nature and found that most shifts were less than two rakers; one population shifted as much as 3.7 rakers in several generations in Lago Maggiore, Italy. Nearly all populations that shifted two or more rakers were transplanted; small numbers of fish were stocked and changes in the lake environment from the source lake to the receiving lake could have influenced the observed shift in raker number. The bloater population in Lake Michigan differs from the populations Lindsey (1981) cites in that it was not transplanted and we have little evidence to support the idea that bloaters now occupy a different habitat during their first summer than they did before the alewife density increase. A change in gill raker number of two or more in coregonines has often been attributed to artifical or natural selection (Svärdson 1952, Berg 1970, Bailey & Smith 1981). Thus, the shift in raker number observed in Lake Michigan bloater may well have a genetic basis and is probably not strictly a phenotypic change.

Though the evidence for competition as *the* mechanism underlying the observed habitat and morphological shifts and the evidence for a genetic basis for these shifts would not likely convince the ultimate skeptic, I believe the evidence is reasonably convincing when all the available facts are assembled. We may, in fact, have had a better opportunity to see the 'ghost of competition past' by examining the historical and contemporary data on this non-coevolved community than by observing resource use patterns in a coevolved assemblage.

Why would we expect that competition was so important in the Lake Michigan fish community? Alewife increased dramatically in the lake in the mid-1960s, zooplankton size declined and a number of native fishes became rare or locally extinct. All this occurred in the absence of predation, as the major piscivorous fishes were depleted by sea lamprey and overfishing by the mid-1950s. So, competition was perhaps expressed strongly because no predators regulated the population sizes of the planktivores. If lake charr had remained abundant in Lake Michigan and other planktivores had not been depleted by lamprey predation and overfishing, perhaps alewife would not have successfully invaded or would not have had such devastating effects on the native fishes.

Now that the predators on alewife have been re-stocked, the alewife density has declined and zooplankton are again large. Native fishes such as bloater and perch have again developed strong year classes. Is competition now rare, or does it still occur intermittently and thus regulate resource use. Our evidence from 1979–80 suggests that thermal habitat compression can lead to seasonal competitive bottlenecks which might act to maintain the observed resource partitioning.

Competition has been accepted too uncritically and often without appropriate experimental tests (Connell 1980). But it is perhaps just as inappropriate to assume that competition is unimportant because it is difficult to demonstrate as it is to assume it is important in the absence of field experiments. And while an experimental approach is clearly the most powerful and laudable, some systems are just not amenable to direct field experimentation. One reason we know so much about competition in space-limited systems such as the rocky intertidal or in plant communities is because the resource is both identifiable and consistent and because experiments are relatively easy to do. Competition experiments based on time-varying resources are much more difficult.

What about our need to understand ecological systems which are not amenable to controlled field experiments? If we cannot do these experiments are we to abandon all hope of showing the importance of competition or predation or disturbance? Unfortunately, many systems of critical interest to us with respect to resource management or environmental degradation will be difficult to experiment upon. Lake Michigan is a case in point – I would argue that it is important to know, though imperfectly, that competition is likely to be an important interaction among the planktivorous fishes. One alternative in these systems is to employ the adaptive management approach (Holling 1978), i.e. use simulation models to project system responses to perturbation or manipulation and then to follow up these hypotheses by critically examining the system responses to management manipulations. In this way, management can test hypotheses and learn or adapt to changes in expected system behavior.

Competition has been and remains an important interaction among the planktivorous fishes in Lake Michigan. Though controlled field experiments have not been possible, a combination of hypothesis-based field observations, field natural experiments and designed laboratory experiments have provided strong evidence for the importance of competition in this fish community. Predation on alewife and other planktivores has probably reduced the intensity of competition as predator stocking has increased (Stewart et al. 1981), but even the 1982–1984 decline of alewife correlates with the formation of extremely strong year classes of at least two native fishes. Intermittent competition associated with thermal habitat compression may continue to regulate resource use. Thus, it appears that the current lake Michigan fish community continues to be regulated by an interaction of competition and predation.

Acknowledgements

I thank John J. Magnuson, James F. Kitchell, Fred P. Binkowski, Carolyn Lie, Helen Crawford and Michael Gitter for assistance. Bill, Dan and Harold provided inspiration. Two anonymous reviewers, Charles (Si) Simenstad and Greg Cailliet provided very useful comments on the manuscript. This research was funded by the University of Wisconsin Sea Grant College Program under grants from the Office of Sea Grant, National Oceanic and Atmospheric Administration and the State of Wisconsin, Federal Grant NA800–AAD00086, Project R/LR 30.

References cited

Bailey, R.M. & G.R. Smith. 1981. Origin and geography of the fish fauna of the Laurentian Great Lakes Basin. Can. J. Fish. Aquat. Sci. 38: 1539–1561.

Berg, A. 1970. A comparative study of food and growth, and competition between two species of coregonids introduced into Lake Maggiore, Italy. pp. 311–346. *In*: C.C. Lindsey & C.S. Woods (ed.) Biology of the Coregonid Fishes, University of Manitoba Press, Winnipeg.

Brandt, S.B., J.J. Magnuson & L.B. Crowder. 1980. Thermal habitat partitioning by fishes in Lake Michigan. Can. J. Fish. Aquat. Sci. 37: 1557–1564.

Brown, E.H., Jr. 1970. Extreme female predominance in the bloater (*Coregonus hoyi*) of Lake Michigan in the 1960s. pp. 501–514. *In*: C.C. Lindsey & C.S. Woods (ed.) Biology of the Coregonid Fishes, University of Manitoba Press, Winnipeg.

Brown, W.L., Jr. & E.O. Wilson. 1956. Character displacement. Syst. Zool. 5: 49–64.

Christie, W.J. 1974. Changes in fish species composition of the Great Lakes. J. Fish. Res. Board Can. 31: 827–854.

Connell, J.H. 1980. Diversity and the coevolution of competitors, or the ghost of competition past. Oikos 35: 131–138.

Connell, J.H. 1983. On the prevalence and relative importance of interspecific competition: evidence from field experiments. Amer. Natur. 122: 661–696.

Conner, E.F. & D. Simberloff. 1979. The assembly of species communities: chance or competition? Ecology 60: 1132–1140.

Crowder, L.B. 1984. Character displacement and habitat shift in a native cisco in southeastern Lake Michigan: evidence for competition? Copeia 1984: 878–883.

Crowder, L.B. 1985. Optimal foraging and feeding mode shifts

in fishes. Env. Biol. Fish. 12: 57–62.
Crowder, L.B. & F.P. Binkowski. 1983. Foraging behaviors and the interaction of alewife, *Alosa pseudoharengus,* and bloater, *Coregonus hoyi.* Env. Biol. Fish. 8: 105–113.
Crowder, L.B. & H.L. Crawford. 1984. Ecological shifts in resource use by bloaters in Lake Michigan. Trans. Amer. Fish Soc. 113: 694–700.
Crowder, L.B. & J.J. Magnuson. 1982. Thermal habitat shifts by fishes at the thermocline in Lake Michigan. Can. J. Fish. Aquat. Sci. 39: 1046–1050.
Crowder, L.B., J.J. Magnuson & S.B. Brandt. 1981. Complementarity in the use of food and thermal habitat by Lake Michigan fishes. Can. J. Fish. Aquat. Sci. 38: 662–668.
Darwin, C. 1859. On the origin of species by means of natural selection, or the preservation of favoured races in the struggle for life. John Murray, London.
Gitter, M.J. 1982. Thermal distribution and community structure of Lake Michigan zooplankton with emphasis on interactions with young-of-year fishes. Master's Thesis, University of Wisconsin, Madison. 129 pp.
Grant, P.R. 1972. Convergent and divergent character displacement. Biol. J. Linn. Soc. 4: 39–68.
Holling, C.S. 1978. Adaptive environmental assessment and management. Wiley Inter-Science, New York. 377 pp.
Janssen, J. 1976. Feeding modes and prey size selection in the alewife (*Alosa pseudoharengus*). J. Fish. Res. Board Can. 33: 1972–1975.
Janssen, J. & S.B. Brandt. 1980. Feeding ecology and vertical migration of adult alewife (*Alosa pseudoharengus*) in Lake Michigan. Can J. Fish. Aquat. Sci. 37: 177–184.
Kitchell, J.F. & L.B. Crowder. 1986. Predator-prey interactions in Lake Michigan: model predictions and recent dynamics. Env. Biol. Fish. (in press).
Lawrie, A.H. 1970. The sea lamprey in the Great Lakes. Trans. Amer. Fish. Soc. 99: 766–775.
Lindsey, C.C. 1981. Stocks are chameleons: plasticity in gill rakers of coregonid fishes. Can. J. Fish. Aquat. Sci. 38: 869–872.
Magnuson, J.J., L.B. Crowder & P.A. Medvick. 1979. Temperature as an ecological resource. Amer. Zool. 19: 331–343.
Morsell, J.W. & C.R. Norden. 1968. Food habits of the alewife, *Alosa pseudoharengus,* in Lake Michigan. Proc. 11th Conf. Great Lakes Res.: 103–110.
Pianka, E.R. 1978. Evolutionary ecology. Harper and Row Publishers, New York. 397 pp.
Rice, J.A. 1985. Mechanisms regulating survival of larval bloater *Coregonus hoyi* in Lake Michigan. Ph.D. Dissertation, University of Wisconsin, Madison. 213 pp.
Sale, P.F. 1979. Habitat partitioning and competition in fish communities. pp. 323–331. *In*: H. Clepper (ed.) Predator-Prey System in Fish Communities, Sport Fishing Institute, Washington, D.C.
Salt, G.W. 1983. Roles: their limits and responsibilities in ecological and evolutionary research. Amer. Natur. 122: 697–705.
Schoener, T.W. 1974. Resource partitioning in ecological communities. Science 185: 27–39.
Schoener, T.W. 1983. Field experiments in interspecific competition. Amer. Natur. 122: 240–285.
Stewart, D.J., J.F. Kitchell & L.B. Crowder. 1981. Forage fishes and their salmonid predators in Lake Michigan. Trans. Amer. Fish. Soc. 110: 751–763.
Strong, D.R., Jr., D. Simberloff, L.G. Abele & A.B. Thistle. 1984. Ecological communities: conceptual issues and the evidence. Princeton University Press, Princeton. 613 pp.
Svärdson, G. 1952. The coregonid problem. IV. The significance of scales and gill rakers. Rep. Inst. Freshwater Res. Drottningholm 33: 204–232.
Todd, T.N., G.R. Smith & L.E. Cable. 1981. Environmental and genetic contributions to morphological differentiation in ciscoes (Coregoninae) of the Great Lakes. Can. J. Fish. Aquat. Sci. 38: 59–67.
Wells, L. 1960. Seasonal abundance and vertical movements of planktonic crustacea in Lake Michigan. U.S. Fish. Bull. 60: 343–369.
Wells, L. 1968. Seasonal depth distribution of fish in southeastern Lake Michigan. U.S. Fish. Bull. 67: 1–15.
Wells, L. 1970. Effects of alewife predation on zooplankton populations in Lake Michigan. Limnol. Oceanogr. 14: 556–565.
Wells, L. 1980. Food of alewives, yellow perch, spottail shiners, trout-perch, and slimy and fourhorn sculpins in southeastern Lake Michigan. U.S. Fish and Wildlife Ser. Tech. Pap. 98. 12 pp.
Wells, L. & A.M. Beeton. 1963. Food of the bloater, *Coregonus hoyi,* in Lake Michigan. Trans. Amer. Fish. Soc. 92: 245–255.
Wells, L. & A.L. McLain. 1973. Lake Michigan: man's effects on native fish stocks and other biota. Great Lakes Fish. Comm. Tech. Rept. 20. 55 pp.
Wells, L. & R.W. Hatch. 1984. Status of bloater chubs, alewives, smelt, slimy sculpins, deepwater sculpins and yellow perch in Lake Michigan, 1983. Mimeo report to the Great Lakes Fishery Commission, Lake Michigan Committee, March 19–20, 1984. 8 pp.
Wells, L. & R.W. Hatch. 1985. Status of bloater chubs, alewives, smelt, slimy sculpins, deepwater sculpins and yellow perch in Lake Michigan, 1984. Mimeo report to the Great Lakes Fishery Commission, Lake Michigan Committee, March 19, 1985. 8 pp.
Werner, E.E. & J.F. Gilliam. 1984. The ontogenetic niche and species interactions in size-structured populations. Ann. Rev. Ecol. Syst. 15: 393–425.
Wiens, J. 1977. On competition and variable environments. Amer. Scient. 65: 590–597.

Received 12.12.1984 *Accepted 3.9.1985*

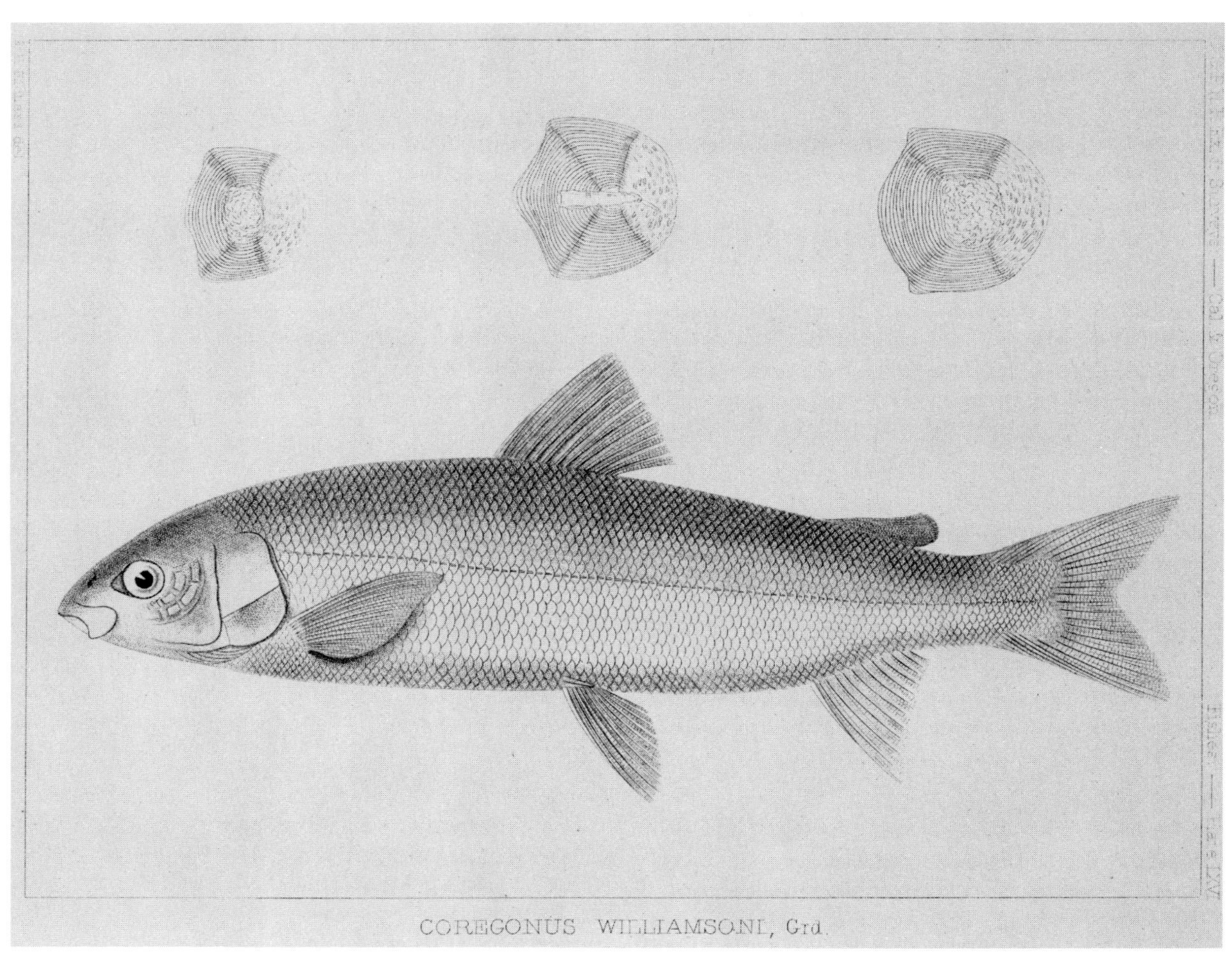

COREGONUS WILLIAMSONI, Grd.

Predator-mediated habitat use: some consequences for species interactions

Gary Mittelbach
Ohio Cooperative Fishery Research Unit and Department of Zoology, Ohio State University, Columbus, OH 43210, U.S.A.

Keywords: Bluegill, Body size, Competition, Foraging, Indirect effects, Predation, Pumpkinseed

Synopsis

Behavioral responses to predators can have a major impact on a fishes' diet and habitat choice. Studies with the bluegill sunfish, *Lepomis macrochirus*, demonstrate that bluegills undergo pronounced shifts in diet and habitat use as they grow in response to changes in their vulnerability to predators. Other species of fish exhibit similar habitat shifts with body size, presumably also in response to changing predation risks and/or foraging gains. An important but little appreciated consequence of this type of predator-mediated habitat use is that predation risk, by structuring size and/or age-specific resource use, may also indirectly affect species interactions. This paper discusses some of the ways in which behavioral responses to predators may affect intra- and interspecific competition in fish. Observational and experimental studies with sunfish (Centrarchidae) provide most of the examples. These studies suggest that the 'nonlethal' effects of predators may be as important as the actual killing of prey.

Introduction

There is a growing awareness among ecologists that the processes of competition and predation often interact in subtle ways, and that the old debate over which of these factors structures communities is largely misdirected. Most studies which have examined the interaction of competition and predation in the same system have generally viewed predation as a factor which reduces interspecific competition, either by lowering demand for resources or removing superior competitors (e.g. Brooks & Dodson 1965, Paine 1966, Dayton 1971, Connell 1975, 1983, Neill 1975, Menge 1976, Yodzis 1976, Caswell 1978, Lubchenco 1978, Morin 1983). However, there are often additional effects of predators besides the simple removal of individuals from a system. For example, prey may respond behaviorally to predators by moving to protective habitats (Stein & Magnuson 1976, Sih 1980, 1982, Schmitt 1982, Cerri & Fraser 1983, Edwards 1983, Werner et al. 1983a, Cooper 1984, Power 1984, Power et al. 1985, Lima et al. 1985), increasing vigilance (Murton et al. 1971, Caraco et al. 1980), reducing foraging distances (Dill & Fraser 1985), and/or limiting feeding time and intake (Maiorana 1976, Stein & Magnuson 1976, Milinski & Heller 1978, Murdoch & Sih 1978, Power 1984, Schmitt & Holbrook 1985). As a consequence of these behaviors, predators can have strong effects on the diet and/or habitat use of a species, and thereby influence both interspecific competition, and/or intraspecific competition between size or age classes. To date, however, such 'indirect' or 'nonlethal' effects of predators on species' interactions have been little studied (the theoretical work of Abrams

1984 on foraging time adjustment is a notable exception).

A shift in habitat use due to the presence of predators is probably the best documented and most pervasive effect of predation risk. In this paper, I discuss how such predator-mediated habitat shifts may affect species and size-class interactions in fish. Most of the examples presented come from small, freshwater lakes because these are the systems I know best. However, the basic ideas about the indirect or nonlethal effects of predators on fish populations should apply to other fish communities as well.

Predation risk and habitat use

Changes in diet and/or habitat during ontogeny are extremely common in fishes (Werner & Gilliam 1984) and are often associated with size-specific predation risks and foraging gains (e.g. Jackson 1961, Bray 1981, Mittelbach 1981, Werner et al. 1983a, Jones 1984, Power 1984, Ebeling & Laur 1985). For example, bluegills in small Michigan lakes show a dramatic shift in diet and habitat with increasing body size (Mittelbach 1984). Small individuals (<75 mm standard length) feed in vegetated habitats, while larger size classes may shift to feeding in the open water on zooplankton (*Daphnia* spp.) when this resource is the most profitable (Fig. 1). Using an optimal diet model, I was able to show that small bluegills can incur as much as a 50% reduction in net energy gain when feeding in the vegetation as opposed to feeding on zooplankton (Mittelbach 1981). Thus, small fish use the vegetation at a significant cost in foraging intake.

Dense vegetation does, however, provide bluegills protection from predators such as the largemouth bass, *Micropterus salmoides* (Glass 1971, Savino & Stein 1982), suggesting that these fish are trading off energy intake for safety. In a recent pond experiment, we were able to demonstrate that when predators are removed, bluegills of all sizes shift to using open-water or bare-bottom habitats if these habitats provide the highest foraging gains (Werner et al. 1983b). However, the presence

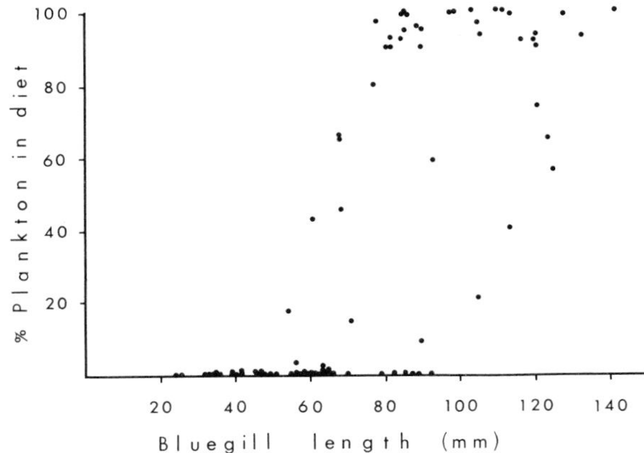

Fig. 1. Percent plankton (dry wt) in the diets of bluegills collected from Lawrence Lake and Three Lakes II, Michigan. Data include all fish collected during the months when bluegills were feeding extensively on plankton. Small bluegills with low percentages of plankton in the diet were feeding on vegetation-dwelling prey (from Mittelbach 1984).

of bass in this same experiment caused small bluegills to increase their use of the vegetation and to have slower growth (Werner et al. 1983a). Larger, less vulnerable, bluegills did not change habitats in the presence of the predator and they utilized habitats based upon maximizing energetic gain (Werner et al. 1983b). Thus, the diets of bluegills are a function of size-specific predation risks and foraging gains.

Predation risk and intraspecific competition

When risk to predation decreases with increased body size, as it does for most fishes (Ware 1975, Werner 1985), small (vulnerable) individuals will often be restricted to protective habitats. If resources are less available in these habitats, intraspecific competition among smaller size classes may be intensified due to the prey's response to risk. No direct tests of this idea exist. The work of Jones (1984), however, provides a likely example. Jones observed that juvenile wrasses, *Pseudolabrus celidotus*, were concentrated in patches of algal cover scattered throughout a coral reef. The prey items eaten by the juvenile fish, however, were more abundant in the surrounding rock flats than in the algal patches. Once *P. celidotus* reached adult size it shifted to foraging on the rock flats, feeding mainly on bivalves and crabs. Thus small wrasses, like small bluegills, are concentrated in protective cover where prey resources are less abundant. Importantly, Jones was able to show that recruitment of *P. celidotus* was correlated with algal-microcrustacean abundance and that both growth and mortality of juvenile fish was density dependent. Thus, the behavioral response of *P. celidotus* to predation risk appears to have a significant impact on juvenile competition and recruitment. Stamps (1983) discusses a similar example in *Anolis* lizards and Cooper (1984) has measured a reduction in the fecundity of water striders due to their avoidance of higher risk habitats.

I have previously suggested that predation risk may also act to reduce competition between juveniles and adults of the same species (Mittelbach 1981, 1983). By restricting small size classes to protective habitats, predators enforce habitat segregation between size classes and make open-water resources exclusively available to large fish. For species such as the bluegill, which show little resource partitioning between size classes on the basis of prey size (Werner 1974), such habitat segregation may be a very important factor preventing the development of 'stunted' populations dominated by very small fish. Simulations with an optimal foraging model show that if bluegill size classes are not segregated by habitat, and all fish sizes are in competition for the same resources, only small fish will be able to maintain positive energy budgets at low resource levels (Mittelbach 1983). There is also abundant empirical evidence showing that bluegill stocked in the absence of predators invariably develop populations of small, 'stunted' individuals (Swingle & Smith 1940, Dillard & Novinger 1975). It will be important in future research to distinquish the indirect effects of predators in promoting resource partitioning from the direct effect of predators in reducing prey numbers. Both factors will tend to reduce competition between prey size classes. From a management perspective, the indirect effect of predation will be most beneficial in systems where open habitats can be managed to provide abundant prey resources for large fish (e.g. productive *Daphnia* populations for bluegill).

Predation risk and interspecific competition

Similar behavioral responses to predators by a number of species may concentrate vulnerable size classes into a common refuge. In sunfish, for example, juveniles of as many as 5–6 coexisting species all occupy the vegetation (Werner et al. 1977, Keast 1978, Mittelbach 1984). Consequently, resource overlap is high early in these species' life history before individuals are able to move into more open areas (Laughlin 1979, Mittelbach 1984). This common response to predation risk by a number of species has the potential to intensify the demand for resources early in their life history and thereby reduce individual growth rates.

Survivorship and growth rate are positively correlated in some fish (see Backiel & LeCren 1967, Ware 1975, Werner 1985 for reviews) and density-dependent growth has long been postulated as the principal factor regulating juvenile mortality in fishes (Ricker & Foerster 1948). Put simply, fish that grow quickly spend less time vulnerable to predators and have higher survival. Gilliam (1982) has further shown mathematically that for those periods of the life history where survivorship is already low (i.e. juvenile), a small reduction in growth can cause a very large reduction in survivorship (see also Werner et al. 1983a). Thus, any density-dependent effects on growth that result from behavioral responses of juvenile fish to predators are likely to have important consequences for survival, recruitment, and overall density.

Below I discuss a specific example of how predation risk may influence size-specific diets and habitat use in two sunfish; the bluegill and the pumpkinseed, *L. gibbosus*. I then present data suggesting that food resources are limiting for small bluegills and pumpkinseeds in the vegetation of natural lakes and that the species are potentially competing for these limited resources.

The bluegill-pumpkinseed example

Bluegill and pumpkinseed are common sunfish in small lakes of the northeastern United States and southern Canada. As adults, these species differ in functional morphology and feeding efficiency

(Keast 1978, Mittelbach 1984), and consequently they feed on different prey types found in separate habitats. In lakes with true limnetic zones, large bluegills feed primarily on open-water zooplankton (*Daphnia*), whereas large pumpkinseeds specialize on snails (Mittelbach 1984, see also Seaburg & Moyle 1964, Keast 1978).

Unlike large fish, the diets and habitat use of small bluegills and pumpkinseeds (≤75 mm) are very similar (Mittelbach 1984). Small fish of both species feed predominantly in the vegetation and 80–90% of their average diet may be vegetation-dwelling prey, excluding snails (Mittelbach 1984). Calculated diet overlaps (using Schoener's 1970 measure) average about 50% for small bluegills and pumpkinseeds, but only 2–8% for large individuals in lakes where *Daphnia* are abundant (Mittelbach 1984). This increased overlap in resource use among small fish is due to two factors. First, piscivorous fish restrict small sunfish to the vegetation where they are less vulnerable to predators (Mittelbach 1981, Werner et al. 1983). Second, although the pumpkinseed's adult prey (snails) occur in the vegetation, juvenile pumpkinseeds are able to crush only very tiny, newly-hatched snails and therefore cannot effectively use the adult snail resource.

Small bluegills and pumpkinseeds thus share a common refuge and much of the same prey. However, upon reaching less vulnerable sizes, bluegills and pumpkinseeds shift to using different resources and show very reduced overlap in diet. Other species of fish show similar patterns of high overlap in resource use as juveniles and little or no overlap as adults (e.g. Lindstrom & Nilsson 1962, Laughlin 1979, McCabe et al. 1983). Whether a prey's response to predators significantly affects interspecific competition in these cases will depend on whether juveniles are competing for limited resources while restricted to protective habitats. The available evidence suggests that resources can be limiting for young bluegills and pumpkinseeds in the vegetation.

Growth in fishes is very labile and is highly sensitive to resource availability (Weatherley 1972, Werner & Gilliam 1984). Table 1 lists the growth rates of young (age 1) bluegills and pumpkinseeds from a series of experimental and natural environments located near the Kellogg Biological Station in southwestern Michigan. For both species, growth was highest in experimental ponds that contained no fish prior to the introduction of bluegills and/or pumpkinseeds and in which prey were very abundant. Growth rates in these pond experiments approximate growth rates reported for bluegills fed ad libidum in the laboratory (Kitchell et al. 1974). In the two natural lakes studied (Lawrence Lake and Three Lakes II), bluegill and pumpkinseed

Table 1. Comparison of bluegill and pumpkinseed growth rates (g dry wt day^{-1}) from experimental and natural environments. Initial fish sizes correspond to year 1+ fish (25–35 mm SL) in each case. Bluegills and pumpkinseeds were stocked in the same pond in Exp. 1 and in two separate ponds in Exp. 2. Pond growth rates based on >100 measured fish per species.

Location	Dates	Bluegill		Pumpkinseed	
		Density (#m^{-2})	Growth	Density (#m^{-2})	Growth
Pond Exp. 1	7/4/84–7/26/75	0/6	0.031	0.6	0.030
Pond Exp. 2	6/25/73–8/3/73	1.3	0.017	1.3	0.011
Lawrence Lake	5/22/79–7/20/79	2.5	0.007[a]		
Lawrence Lake	1981	2.5	0.005[b]	1	0.005[b]
Three Lakes II	1981	?	0.004[b]	?	0.004[b]
Experimental Cages	7/17/84–9/6/84	0.4–3.4	0.009–0.002	0.9[c]	0.002

[a] Growth determined by the change in mean length of age 1+ fish over the time period indicated; n = 244 fish.

[b] Age 1+ growth determined from relationships between scale annuli distances and increments in dry weight. A 186-day growing season is assumed for southern Michigan (Gerking 1966); n = 80–150 fish per species per lake.

[c] Bluegill density of 7 fish per cage (2.9 m^{-2}).

growth rates were only 13–45% of those exhibited in the experimental ponds. Fish were feeding from the vegetation in both the lakes and the ponds, and growth rates in the lakes are consistent with those reported from other northern lakes (Carlander 1977). Finally, in a series of density manipulations conducted in Lawrence Lake (described below), growth rates of caged bluegill ranged from above to below those of uncaged bluegills in the lake, as fish densities in the cages were varied from below to above ambient bluegill density in the lake (Table 1). Thus, data from these field growth studies indicate that resources are potentially limiting for the two sunfish species in nature. These comparisons, however, leave open the question of whether the extent of resource limitation in lakes is affected by changes in the density of young bluegills and pumpkinseeds.

To begin to address this question, I conducted an experiment using five 2.4 m² cages (1.5 m tall) constructed of 0.6-cm mesh netting and placed in the vegetated littoral zone of Lawrence Lake, Michigan. All resident fish were removed from the cages prior to the start of the experiment. Small pumpkinseeds were then placed in each cage at approximately natural density (2 fish per cage) and were subject to a range in bluegill densities (1–8 bluegill per cage; natural density in Lawrence Lake is ≈6 bluegills per cage). Initial pumpkinseed and bluegill lengths were 35 ± 0 and 35.4 ± 0.2 mm, respectively, which is very similar to the mean length of age 1+ fish in the lake at this time (Mittelbach, unpublished data). The cages were inspected at weekly intervals and the fish appeared to be feeding normally. There was little accumulation of periphyton on the nets and the cages remained open to water circulation throughout the 52 day experiment (July 17–September 6, 1984).

Bluegills showed pronounced density-dependent growth in the enclosures (Fig. 2). The regression of bluegill growth as a function of density is $Y = 12.22 - 1.18X$ ($r^2 = 0.94$), and is significant ($P<0.05$). The observed range in bluegill growth rates in the cages spanned the average growth rate of year 1+ bluegills in the lake (Table 1). Pumpkinseed growth also appears to be negatively affected by increased bluegill densities (Fig. 2), although

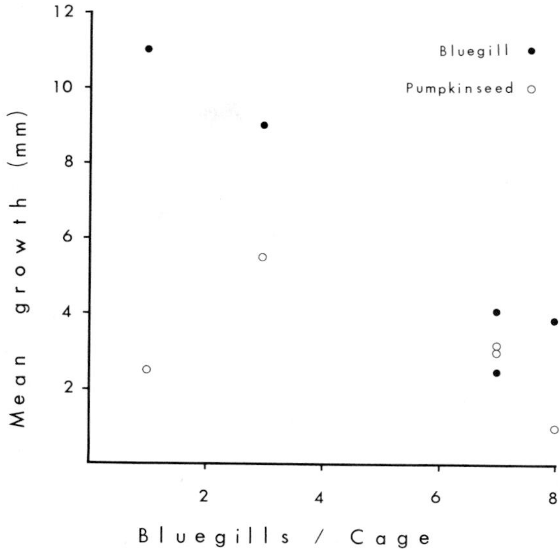

Fig. 2. Mean growth (mm) of bluegills and pumpkinseeds after 52 days in experimental cages in Lawrence Lake, Michigan. Initial bluegill and pumpkinseed lengths averaged 35.0 ± 0 and 35 ± 0.2 mm SL ($\bar{x}$ ± 1 SE) respectively. Mean growth of bluegills and pumpkinseeds was calculated as the average growth of all individuals of a species in a cage. Average coefficients of variation in final length for bluegills and pumpkinseeds are 6.6 and 2.2% respectively.

the form of the relationship for pumpkinseeds is less clear because pumpkinseeds grew so little in the one-bluegill cage. The reason for this is unknown and more experiments are needed to better determine the effect of young bluegills on pumpkinseeds, and vice-versa. However, all the available evidence indicates that young bluegills and pumpkinseeds are competing while restricted to the vegetation refuge.

Because the juvenile is often the most vulnerable and therefore the most likely to be restricted to a refuge, there should be many cases in nature where size-specific predation risk causes species to interact as juveniles or larvae, but not as adults. Under these conditions the outcome of interspecific competition can be complex, with coexistence and the relative abundance of species being affected both by juvenile competitive abilities and adult fecundities (Mittelbach & Chesson 1985). Below I outline a simple two-species model that helps illustrate how predator-induced juvenile competition can affect species abundances. One interesting conclusion form the model is that strong competition in the juvenile period can cause the number of adults of two species to be negetively correlated even though the species do not compete as adults. The model's single species components are similar to Ware's (1980) bioenergetic approach to stock and recruitment.

Juvenile competition and species abundance

In this analysis, each species is considered to have two life intervals, juveniles and adults. Juveniles occupy the same habitat and compete for the same resources, while adults feed on alternate prey and do not compete directly. Juvenile competition is assumed to predominantly affect an individual's probability of surviving to the adult, while adult intraspecific competition affects individual fecundity but not survival. Thus, the two main components of the model are density-dependent adult fecundity and density-dependent juvenile survival (Fig. 3a, b). Density-dependent fecundity is well documented in fish (see Nikol'skii 1962, Schopka & Hempel 1973, McFadden 1977, Bagenal 1978, Ware 1980 for reviews; Parker 1958, Cooper et al. 1979 for examples with bluegills and/or pumpkinseeds), although the shape of the fecundity curve with density is poorly known for most species. Examples of density-dependent juvenile survivorship in fish can be found in (LeCren 1962, 1965, Egglishaw & Shackley 1977, Leggett 1977, Elliott 1984, Beyerle & Williams 1972 for bluegills).

The curves in Figure 3a and Figure 3b define a relationship between the number of adults (X) present at time t and the expected number of recruits (X(t+1)). The equilibrium number of adults of species 1 can therefore be determined from a standard plot of adult numbers versus recruits (Fig. 3c). The intersection of the recruitment curve with the 45° line specifies the point (â) where the adult stock exactly replaces itself and is in equilibrium (Ricker 1954, Ware 1980). In general, intersections on a rising part of the curve are locally stable while intersections on a falling portion of the curve will be stable if the slope is <−1 (May & Oster 1976, Fisher & Goh 1977, Rosenkranz 1983).

Given the processes described in Figure 3, we

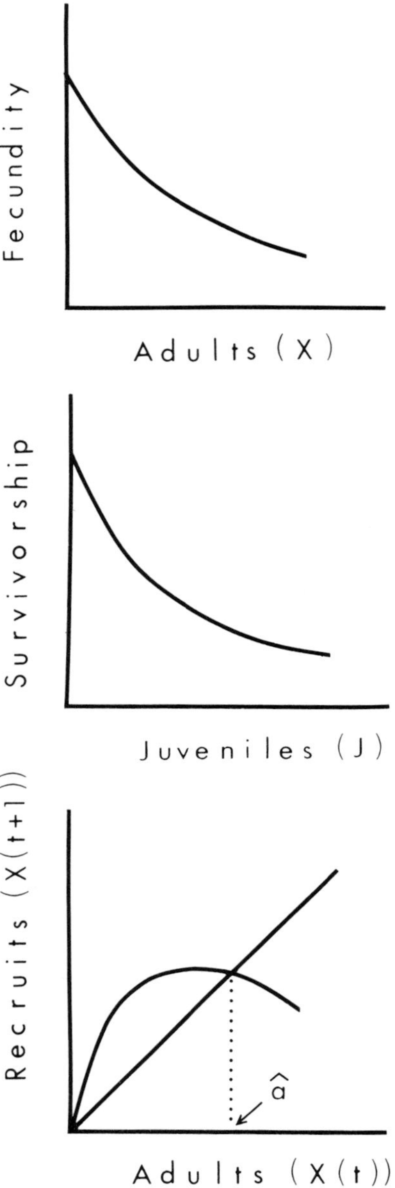

Fig. 3. A graphical model of stock-recruitment for a single species which exhibits (a) density-dependent per-individual fecundity and (b) density-dependent per-individual juvenile survivorship. The equilibrium number of adults in the population occurs at the point where the recruitment curve intersects the 45° replacement line (â in panel c).

can ask what is the effect of adding a second species to the system, when the second species competes for the same juvenile resource as species 1, but uses a completely separate resource as adults. This is the essence of the bluegill-pumpkinseed interaction. As the simplest case, juvenile survivorship is assumed to depend only on the amount of resources each individual can garner, with growth being directly related to an individual's probability of dying (Cushing 1976, Ricker 1979, Lasker 1981). In the specific case of juvenile bluegills and pumpkinseeds, the available evidence indicates that juveniles of the two species are equal competitors for resources (Mittelbach 1984) and that they have similar vulnerabilities to predators (Mittelbach, unpublished data). Thus, I assume for purposes of illustration that their per-capita juvenile survivorship is equally dependent on the total density of juveniles of both species in the environment.

In this case, adding species 2 to the environment will increase the total number of juveniles occupying the refuge and therefore decrease the per-individual probability of survival for species 1 (Fig. 4a). As a result, the number of recruits of species 1 is reduced relative to the situation before species 2 was added (Fig. 4b). The above analysis is only adequate for examining short term changes in abundance. An analysis of equilibrium densities for the two species is more complex. However, for the case where the number of recruits to a species is a monotonically increasing function of the number of juveniles it produces, it can be shown that the above conclusion holds (Mittelbach & Chesson 1985). That is, an increase in the reproductive output of species 2 results in an increase in the equilibrium density of species 2 adults and juveniles, and a decrease in the equilibrium density of species 1 adults. The equilibrium density of species 1 juveniles may either increase or decrease depending on the form of its fecundity curve (Mittelbach & Chesson 1985). Thus, under conditions of juvenile competition, we may observe negative effects between species adults even when they use completely separate resources.

The above analysis is of course a very simplified representation of a number of complex biological processes. Also, it is a deterministic treatment of a system where stochastic forces are well known to modify reproductive success and year class size (Cushing 1982). Consequently, all models of stock and recruitment are perhaps best judged for their usefulness in terms of their heuristic, rather than predictive value (Ware 1980). I believe the above analysis does have heuristic value. First, it demonstrates the need to consider all stages of the life history when interpreting species interactions. Second, it shows how a simple behavioral response to predators may result in the transmission of negative effects between the adults of two species, even when the adults do not interact directly. Finally, it provides further evidence of the potential dangers in attempting to harvest individual fish stocks for maximum sustained yield, if in fact the species compete as juveniles or prerecruits (see also Larkin 1977, Lett 1980, Mercer 1982 for related discussions).

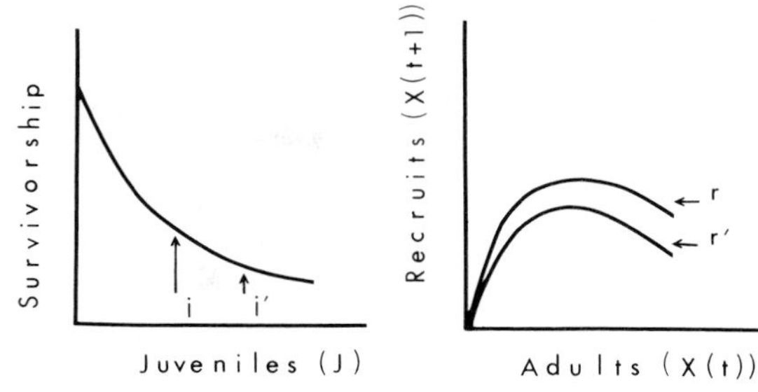

Fig. 4. The short-term effect of adding species 2 to the system depicted in Figure 3 is to increase the total number of juveniles produced (species 1 plus species 2) and to reduce the per-individual juvenile survivorship for species 1 (j to j'). Recruitment of species 1 is thereby lowered from r to r'.

Conclusions and prospectus

A large number of studies now document how prey alter their feeding behavior in the presence of predators (see references in introduction). These studies have developed in consort with theoretical efforts to incorporate both predation risk and foraging gain into models of behavior, and the development and testing of these models is one of the most exciting and prolific areas in foraging ecology. However, far fewer studies have considered how a prey's behavioral response to predators can affect its interactions with other species and its own population dynamics (e.g. Mittelbach 1981, 1984, Sih 1982, Werner et al. 1983, Abrams 1984). Abrams (1984) modeled the case in which a forager must expose itself to a greater risk of predation while foraging, and where it adjusts foraging time so as to maximize fitness. By analogy, Abrams' results can also be extended to the situation where a species uses different habitats in response to predation risk. Abrams found that a reduction in foraging time due to the presence of predators could have strong effects on both predator and prey populations and that these indirect effects could be of the same magnitude or larger than the direct effects between species.

Levine (1976), Holt (1977, 1984), and others have also considered indirect effects caused by predators, but the mechanisms they consider are quite different from those envisioned in this paper. In Levine and Holt's models, predators exert their effects by consuming prey and thus influencing predator and prey densities. The indirect effects considered here and by Abrams (1984) are due to behavioral responses by the prey to the presence of predators. These indirect effects take place much more rapidly than numerical responses and need to be incorporated explicitly into models of population dynamics (Abrams 1984).

In this paper, I have presented a number of examples of size-specific shifts in prey habitat use in response to predators. In some of these studies, there is evidence that resources are limiting in the prey's refuge and that predators may therefore affect competition within and between prey species. However, much more empirical work is needed to better document the existence and consequences of these nonlethal effects of predators. Clearly, one powerful way to proceed is through experimental removals and additions of predators. However, in many systems (e.g. marine pelagic fishes), controlled manipulations of predators are difficult or impossible. The important thing in these cases is to recognize the role predators may play in mediating habitat use and causing size-specific interactions. For example, a change in the number or size distribution of predators present may effect not only the number of prey eaten, but also the nature of the competitive interactions between the prey species. Those charged with regulating the fisheries of interacting species need to be alert for such 'nontraditional' effects of predation.

Acknowledgements

The ideas presented in this paper have evolved from numerous discussions with friends and colleagues. I particularly thank Peter Chesson, Jim Gilliam, Kay Gross, Gene Helfman, Craig Osenberg, and Earl Werner for their insights and suggestions. Jon Bart, Bill DeMott, Rodger Mitchell, Alice Winn, and other members of the Ecology Lunch Group at Ohio State provided comments on the manuscript, for which I am grateful. This work was supported by grant #BSR 81–04697,AO1 from the National Science Foundation.

References cited

Abrams, P.A. 1984. Foraging time optimization and interactions in food webs. Amer. Nat. 124: 80–96.

Backiel, T. & E.D. LeCren. 1967. Some density relationships for fish population parameters. pp. 279–302. *In*: S.D. Gerking (ed.) Ecology of Freshwater Fish Production, J. Wiley and Sons, New York.

Bagenal, T.B. 1978. Aspects of fish fecundity. pp. 75–101. *In*: S.D. Gerking (ed.) Ecology of Freshwater Fish Production, J. Wiley and Sons, New York.

Beyerle, G.B. & J. Williams. 1972. Survival, growth, and production by bluegills subjected to population reduction in ponds. Michigan Department of Natural Resources Development Report 273. 28 pp.

Bray, R.N. 1981. The influence of water currents and zooplankton densities on daily foraging movements of blacksmith, *Chromis punctipinnis*, a planktivorous reef fish. U.S. Fish. Bull. 78: 829–841.

Brooks, J.L. & S.J. Dodson. 1965. Predation, body size and composition of plankton. Science 150: 28–35.

Carlander, K.D. 1977. Handbook of freshwater fishery biology. Volume two. The Iowa State University Press, Ames. 431 pp.

Caraco, T., S. Martindale & H.R. Pulliam. 1980. Avian time budgets and distance to cover. Auk 97: 872–875.

Caswell, H. 1978. Predator-mediated coexistence: a nonequilibrium model. Amer. Nat. 112: 127–154.

Cerri, R.D. & D.F. Fraser. 1983. Predation and risk in foraging minnows: balancing conflicting demands. Amer. Nat. 121: 552–561.

Connell, J.H. 1975.Some mechanisms producing structure in natural communities. A model and some evidence from field experiments. pp. 460–492. *In*: M.L. Cody & J.M Diamond (ed.) Ecology and Evolution of Communities, Belknap Press, Cambridge.

Connell, J.H. 1983. On the prevelance and relative importance of interspecific competition: evidence from field experiments. Amer. Nat. 122: 661–696.

Cooper, E.L., C.C. Wagner & G.E. Krantz. 1971. Bluegills dominate production in a mixed population of fishes. Ecology 52: 280–290.

Cooper, S.D. 1984. The effects of trout on water striders in stream pools. Oecologia 63: 376–379.

Cushing, D.H. 1976. Biology of fishes in the pelagic community. pp. 317–340. *In*: D.H. Cushing & J.J. Walsh (ed.) The Ecology of the Seas, W.B. Saunders Co., Philadelphia.

Cushing, D.H. 1982. Climate and fisheries. Academic Press, New York. 373 pp.

Dayton, P.K. 1971. Competition, disturbance, and community organization: the provision and subsequent utilization of space in a rocky intertidal community. Ecol. Monog. 41: 351–389.

Dill, L.M. & A.H.G. Fraser. 1985. Risk of predation and the feeding behavior of juvenile coho salmon (*Oncorhynchus kisutch*). Behav. Ecol. Sociobiol. 16: 65–72.

Dillard, J.G. & G.D. Novinger. 1975. Stocking largemouth bass in small impoundments. pp. 459–479. *In*: H. Clepper (ed.) Black Bass Biology and Management, Sport Fishing Institute, Washington, D.C.

Ebeling, A.W. & D.R. Laur. 1985. The influence of plant cover on surfperch abundance at an offshore temperate reef. Env. Bio. Fish. 12: 169–179.

Edwards, J. 1983. Diet shifts in moose due to predator avoidance. Oecologia 60: 185–189.

Egglishaw, H.J. & P.E. Shackley. 1977. Growth, survival and production of juvenile salmon and trout in a Scottish stream, 1966–1975. J. Fish Biol. 11: 647–672.

Elliott, J.M. 1984. Numerical changes and population regulation in young migratory trout *Salmo trutta* in a Lake District stream 1966–1983. J. Anim. Ecol. 53: 327–350.

Fisher, M.E. & B.D. Goh. 1977. Stability in a class of discrete time models of interacting populations. J. Math. Biol. 4: 265–274.

Gerking, S.D. 1966. Annual growth cycle, growth potential, and growth compensation in the bluegill sunfish in northern Indiana lakes. J. Fish. Res. Board Can. 23: 1923–1956.

Gilliam, J.F. 1982. Habitat use and competitive bottlenecks in size-structured fish populations. Ph.D. Dissertation, Michigan State University, East Lansing. 107 pp.

Glass, N.R. 1971. Computer analysis of predation energetics in the largemouth bass. pp. 325–363. *In*: B.C. Patten (ed.) Systems Analysis and Simulation in Ecology, Volume 1, Academic Press, New York.

Holt, R.D. 1977. Predation, apparent competition and the structure of prey communities. Theor. Pop. Biol. 12: 197–229.

Holt, R.D. 1984. Spatial heterogeneity, indirect interactions, and the coexistence of prey species. Amer. Nat. 124: 377–406.

Jackson, P.B.N. 1961. The impact of predation especially by the tiger fish (*Hydrocynus vittatus* Cast.) on African freshwater fishes. Proc. Zool. Soc. Lond. 136: 603–622.

Jones, G.P. 1984. The influence of habitat and behavioral interactions on the local distribution of the wrasses, *Pseudolabrus celidotus*. Env. Biol. Fish. 10: 43–58.

Keast, A. 1978. Feeding interrelations between age groups of pumpkinseed sunfish (*Lepomis gibbosus*) and comparisons with the bluegill sunfish (*L. macrochirus*). J. Fish. Res. Board Can. 35: 12–27.

Kitchell, J.F., J.F. Koonce, R.V. O'Neill, H.H. Shugart, Jr., J.T. Magnuson & R.S. Booth. 1974. Model of fish biomass dynamics. Trans. Amer. Fish. Soc. 103: 786–798.

Larkin, P.A. 1977. Fisheries management – an essay for ecologists. Ann. Rev. Ecol. System. 9: 57–73.

Lasker, R. 1981 (ed.). Marine fish larvae: morphology, ecology and relation to fisheries. Washington Sea Grant Program, Seattle. 131 pp.

Laughlin, D.R. 1979. Resource and habitat use patterns in two coexisting sunfish species (*Lepomis gibbosus* and *Lepomis megalotis peltastes*). Ph.D. Dissertation, Michigan State University, East Lansing. 92 pp.

LeCren, E.D. 1962. The efficiency of reproduction and recruitment in freshwater fish. pp. 283–296. *In*: E.D. LeCren & H.W. Holdgate (ed.) The Exploitation of Natural Animal Populations, Blackwell Scientific, Oxford.

LeCren, E.D. 1965. Some factors regulating the size of populations of freshwater fish. Mitt. Internat. Verein. Limnol. 13: 88–105.

Leggett, W.C. 1977. Density dependence, density independence, and recruitment in the American shad (*Alsoa sapidissima*) population in the Conneticut River. pp. 3–17. *In*: W. VanWinkle (ed.) Assessing the Effects of Power-Plant-Induced Mortality in Fish Populations, Pergamon Press, New York.

Lett. P.F. 1980. A comparative study of the recruitment mechanisms of cod and mackerel, their interaction, and its implication for dual stock assessment. Can. Tech. Rep. Fish. Aquat. Sci. 988. vi + 45 pp.

Levine, S.H. 1976. Competition interactions in ecosystems. Amer. Nat. 110: 903–910.

Lima, S.L., T.J. Valone & T. Caraco. 1985. Foraging efficiency-predation risk tradeoff in the grey squirrel. Anim. Behav. 33: 155–165.

Lindstrom, T. & N.-A. Nilsson. 1962. On the competition between whitefish species. pp. 326–340. *In*: E.D. LeCren & H.W. Holdgate (ed.) The Exploitation of Natural Animal Populations, Blackwell Scientific, Oxford.

Lubchenco, J. 1978. Plant species diversity in a marine intertidal community: importance of herbivore preference and algal competitive abilities. Amer. Nat. 112: 23–39.

Maiorana, V.C. 1976. Predation, submergent behavior, and tropical diversity. Evol. Theor. 1: 157–177.

May, R.M. & G.F. Oster. 1976. Bifurcations and dynamic complexity in simple ecological models. Amer. Nat. 110: 573–599.

McCabe, G.T. Jr., W.D. Muir, R.L. Emmett & J.T. Durkin. 1983. Interrelationships between juvenile salmonids and nonsalmonid fish in the Columbia River estuary. U.S. Fish. Bull. 81: 815–826.

McFadden, J.T. 1977. An arguement supporting the reality of compensation in fish and a plea to let them exercise it. pp. 153–183. *In*: W. VanWinkle (ed.) Assessing the Effects of Power-Plant-Induced Mortality in Fish Populations, Pergamon Press, New York.

Menge, B.A. 1976. Organization of the New England rocky intertidal community: role of predation, competition and environmental heterogeneity. Ecol. Monog. 46: 335–369.

Mercer, M.C. (ed.) 1982. Multispecies approaches to fisheries management advice. Can. Spec. Publ. Fish. Aquat. Sci. 59. 169 pp.

Milinski, M. & R. Heller. 1978. Influence of a predator on the optimal foraging behavior of sticklebacks (*Gasterosteus aculeatus L.*). Nature 275: 642–644.

Mittelbach, G.G. 1981. Foraging efficiency and body size: a study of optimal diet and habitat use by bluegills. Ecology 62: 1370–1386.

Mittelbach, G.G. 1983. Optimal foraging and growth in bluegills. Oecologia 59: 157–162.

Mittelbach, G.G. 1984. Predation and resource partitioning in two sunfishes (Centrarchidae). Ecology 65: 499–513.

Mittelbach, G.G. & P.L. Chesson. 1985. Predation risk: indirect effects on fish populations. *In*: W.C. Kerfoot & A. Sih (ed.) Predation: Direct and Indirect Impacts on Aquatic Communities, New England University Press (in press).

Morin, P.J. 1983. Predation, competition, and the composition of larval anuran guilds. Ecol. Monog. 53: 119–138.

Murdoch, W.W. & A. Sih. 1978. Age-dependent interference in a predatory insect. J. Anim. Ecol. 47: 581–592.

Murton, R.K. 1971. Why do some bird species feed in flocks? Ibis 113: 534–536.

Neill, W.E. 1975. Experimental studies of microcrustacean competition, community composition and efficiency of resource utilization. Ecology 56: 809–826.

Nikol'skii, G.V. 1962. On some adaptations to the regulation of population density in fish species with different types of stock structure. pp. 265–282. *In*: E.D. LeCren & H.W. Holdgate (ed.) The Exploitation of Natural Animal Populations, Blackwell Scientific, Oxford.

Paine, R.T. 1966. Food web complexity and species diversity. Amer. Nat. 100: 65–75.

Parker, R.A. 1958. Some effects of thinning on a population of fishes. Ecology 39: 304–317.

Power, M.E. 1984. Depth distributions of armored catfish: predator-induced resource avoidance? Ecology 65: 523–529.

Power, M.E., W.J. Matthews & A.J. Stewart. 1985. Grazing minnows, piscivorous bass and stream algae: dynamics of a strong interaction. Ecology (in press).

Ricker, W.E. 1954. Stock and recruitment. J. Fish. Res. Board Can. 11: 559–623.

Ricker, W.E. 1979. Growth rates and models. pp. 677–743. *In*: W.S. Hoar, D.J. Randall & J.R. Brett (ed.) Fish Physiology, Vol. 8, Academic Press, New York.

Ricker, W.E. & R.E. Foerster. 1948. Computation of fish production. Bulletin Bingham Oceanographic Collection 11: 173–211.

Rosenkranz, G. 1983. On global stability of discrete population models. Math. Biosci. 64: 227–231.

Savino, J.F. & R.A. Stein. 1982. Predator-prey interaction between largemouth bass and bluegills as influenced by simulated, submersed vegetation. Trans. Amer. Fish. Soc. 111: 255–266.

Schmitt, R.J. 1982. Consequences of dissimilar defenses against predation in a subtidal marine community. Ecology 63: 1588–1601.

Schmitt, R.J. & S.J. Holbrook. 1985. Patch selection by juvenile black surfperch (Embiotochidae) under variable risk: interactive influence of food quality and structural complexity. J. Exp. Mar. Biol. Ecol. (in press).

Schoener, T.W. 1970. Nonsynchronous spatial overlap of lizards in patchy habitats. Ecology 51: 408–418.

Schopka, S.A. & G. Hempel. 1973. The spawning potential of populations of herring (*Clupea harengus* L.) and cod (*Gadus marhue* L.) in relation to the rate of exploitation. Rapp. Cons. Int. Expl. Mer. 164: 178–185.

Seaburg, K.G. & J.B. Moyle. 1964. Feeding habits, digestion rates, and growth of some Minnesota warm water fishes. Trans. Amer. Fish. Soc. 93: 269–285.

Sih, A. 1980. Optimal behavior: can foragers balance two conflicting demands? Science 210: 1041–1043.

Sih, A. 1982. Foraging strategies and the avoidance of predation by an aquatic insect, *Notonecta hoffmanni*. Ecology 63: 786–796.

Stamps, J.A. 1983. The relationship between ontogenetic habitat shifts, competition and predator avoidance in a juvenile lizard (*Anolis aeneus*). Behav. Biol. Sociobiol. 12: 19–33.

Stein, R.A. & J.J. Magnuson. 1977. Behavioral response of crayfish to a fish predator. Ecology 58: 571–581.

Swingle, H.S. & E.V. Smith. 1940. Experiments on the stocking of fish ponds. Trans. Amer. Wild. Con. 5: 267–276.

Ware, D.M. 1975. Relation between egg size, growth, and natural mortality of larval fish. J. Fish. Res. Board Can. 32: 2503–2512.

Ware, D.M. 1980. Bioenergetics of stock and recruitment. Can. J. Fish. Aquat. Sci. 37: 1012–1024.

Weatherley, A.H. 1972. Growth and ecology of fish populations. Academic Press, New York. 293 pp.

Wenger, A. 1972. A review of the literature concerning largemouth bass stocking techniques. Technical Series Number 13, Texas Parks and Wildlife Department, Sheldon.

Werner, E.E. 1974. The fish size, prey size, handling time relation in several sunfishes and some implications. J. Fish. Res. Board Can. 31: 1531–1586.

Werner, E.E. 1985. Species interactions in freshwater fish communities. *In*: J.M. Diamond & T. Case (ed.) Ecological Communities, Harper and Row (in press).

Werner, E.E. & J.F. Gilliam. 1984. The ontogenetic niche and species interactions in size-structured populations. Ann. Rev. Ecol. Syst. 15: 393–426.

Werner, E.E., J.F. Gilliam, D.J. Hall & G.G. Mittelbach. 1983a. An experimental test of the effects of predation risk on habitat use in fish. Ecology 64: 1540–1548.

Werner, E.E., G.G. Mittelbach, D.J. Hall & J.F. Gilliam. 1983b. Experimental tests of optimal habitat use in fish: the role of relative habitat profitability. Ecology 64: 1525–1539.

Werner, E.E., D.J. Hall, D.R. Laughlin, D.J. Wagner, L.A. Wilsmann & F.C. Funk. 1977. Habitat partitioning in a freshwater fish community. J. Fish. Res. Board Can. 34: 360–370.

Yodzis, P. 1976. The effects of harvesting on competitive systems. Bull. Math. Biol. 38: 97–109.

Received 15-12-1984 *Accepted 9.7.1985*

Figs. 1-4 POMOTIS HEROS, B. & G. Figs. 5-8. POMOTIS FALLAX, B. & G.

Effect of regurgitation on stomach content data of marine fishes

Ray E. Bowman
National Marine Fisheries Service, Northeast Fisheries Center, Woods Hole Laboratory, Woods Hole, MA 02543, U.S.A.

Keywords: Depth, Food, Feeding, Consumption, Daily ration, Habits, Bias, Silver hake, Spiny dogfish, Red hake

Synopsis

Observations on trawl caught fishes from two bottom depth ranges in Southern New England shelf waters provide evidence that some species regurgitate at different rates when sampled at various depths and further, that fish which regurgitate can't always be detected by external or internal examination. Generally, gadoid fishes are much more prone to regurgitate than flatfish. The consequence of unrecognized regurgitation is discussed in relation to consumption estimates derived by traditional methods.

Introduction

Extreme variability in the types and quantities of food consumed among and within fish species is well documented. Edwards & Bowman (1979) summarized much information on variability in the feeding of Northwest Atlantic fishes. Bowman & Bowman (1980) documented changes in the feeding intensity of silver hake, *Merluccius bilinearis,* by time of day on Georges Bank, and Bowman (1980, 1984) concluded that annual, seasonal, and areal variability are found in the diet of juvenile haddock, *Melanogrammus aeglefinus,* and silver hake. Pennington et al. (1982) examined the weight of the stomach contents of Atlantic cod, *Gadus morhua*, and determined that to estimate the mean stomach content weight during a season within ±10% with 95% certainty, at least 753 fish should be sampled within each 5 cm size class.

Bowman (1981) attempted to list all known and potential causes of variation in fish feeding studies. In that paper it was noted that regurgitation is commonly observed in fishes caught when bottom trawling in deep water (i.e. >100 m). Also mentioned was that some regurgitation may occur and not be detectable when sampling in deep water, thereby biasing stomach content data. The present paper documents observable regurgitation and then addresses the degree of potential bias caused by undetected regurgitation at two depth ranges for several species of marine fish. Silver hake and spiny dogfish, *Squalus acanthias,* are emphasized because they have been identified as major fish predators in the northwest Atlantic ecosystem (Edwards & Bowman 1979, Grosslein et al. 1980). Data were compiled as part of the Marine Resources Monitoring Assessment, and Prediction (MARMAP) program of NOAA's National Marine Fisheries Service (NMFS), Northeast Fisheries Center (NEFC).

Methods and materials

Ship operations

The study was conducted aboard the NEFC research vessel DELAWARE II during a 10 day cruise in December 1981 in continental shelf waters south of Martha's Vineyard, Massachusetts. Two sampling areas were chosen according to bottom water depth, and each represented a square area of 259 km^2 (Fig. 1). Both areas were subdivided into 100 2.6 km^2 squares from which bottom trawling stations were selected at random without replacement. Each sampling area was occupied for at least three 24 h periods, with 0.5 h trawl hauls commencing every 3 h (i.e. 0300, 0600, 0900, etc.). Totals of 27 and 24 hauls were completed at areas A and B, respectively. Bottom water depth ranged 40–53 m at area A and 70–93 m at area B. Waters were essentially isothermal at the two areas (approximately 8.0 °C at both) and cloud cover was almost 100% throughout the study.

Sampling was performed with a standard Yankee No. 36 otter trawl equipped with roller gear and with the cod end and latter section of the upper belly of the trawl lined with 13 mm mesh net to retain small fish. Towing speed was 3.5 knots in the direction of the next random pre-selected station. Catches were processed according to standard NEFC procedures (Grosslein & Azarovitz 1982).

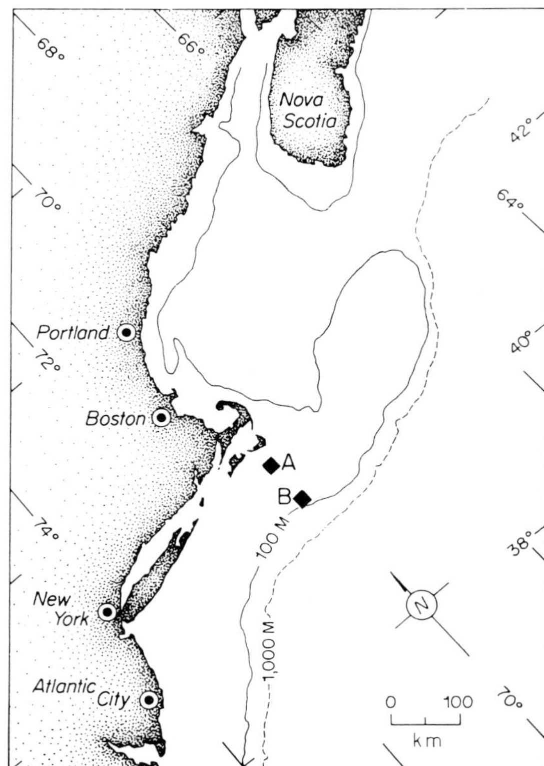

Fig. 1. Locations (A and B) of regurgitation study conducted aboard the R/V DELAWARE II during Cruise 81–08 on 7–17 December 1981. The center of area A (40–53 m) was located at 40° 50′ N, 70° 20′ W and area B (70–93 m) at 40° 25′ N, 70° 20′ W.

Stomach sampling

Observations first involved an examination of each fish for positive evidence of regurgitation (i.e. everted stomach or partially digested food in the mouth). When no evidence of regurgitation was observed the stomach was removed and the total stomach content volume determined. If the stomach contained only water it was considered empty. Prey were identified and the percentage of the contents made up by each particular type of prey evaluated subjectively. Two species in particular, silver hake and spiny dogfish, were intensively sampled (approximately 50 per tow if available). Sampling of other species was based on their relative abundance and the remaining time available between tows.

For comparison purposes stomach content volume was divided by fish weight to obtain percentage body weight (% BW), assuming 1.0 cc equaled 1.0 g, to adjust for differences in fish length among and within stations. Potential differences in stomach content volumes according to time of day were accounted for by using the unweighted overall means (% BW) of directly comparable time periods for each species between areas.

Results

Catches

A total of 46 species was represented in the combined catches at the two study areas (Table 1). The

Table 1. Mean catch (kg) per 30 min trawl haul, percentage composition of total catch, and number of fish examined for species caught at areas A and B. Listing of species is in phyletic sequence.

Species		Area A			Area B		
Scientific name	Common name	(kg)	%	No. exam.	(kg)	%	No. exam.
Petromyzon marinus	Sea lamprey	<0.1	<0.1	0	<0.1	<0.1	0
Squalus acanthias	Spiny dogfish	48.7	30.3	628	86.0	63.4	1050
Torpedo nobiliana	Atlantic torpedo	–	–	–	0.5	0.4	1
Raja erinacea	Little skate	36.2	22.5	149	0.4	0.3	20
Raja laevis	Barndoor skate	0.1	<0.1	1	–	–	–
Raja ocellatus	Winter skate	1.1	0.7	5	–	–	–
Conger oceanicus	Conger eel	–	–	–	<0.1	<0.1	1
Alosa aestivalis	Blueback herring	0.1	<0.1	7	<0.1	<0.1	1
Alosa pseudoharengus	Alewife	4.4	2.7	74	4.0	2.9	95
Alosa sapidissima	American shad	1.0	0.6	39	0.1	0.1	10
Clupea harengus harengus	Atlantic herring	0.1	<0.1	2	<0.1	<0.1	1
(Engraulidae)	Anchovy*	<0.1	<0.1	0	–	–	–
Synodus poeyi	Offshore lizardfish	<0.1	<0.1	0	<0.1	<0.1	1
Lophius americanus	Goosefish	5.5	3.4	20	6.0	4.4	49
Gadus morhua	Atlantic cod	7.9	4.9	45	0.3	0.2	1
Melanogrammus aeglefinus	Haddock	–	–	–	<0.1	<0.1	12
Merluccius bilinearis	Silver hake	3.9	2.4	404	7.5	5.5	571
Pollachius virens	Pollock	–	–	–	0.3	0.2	1
Urophycis chuss	Red hake	4.5	2.8	161	5.7	4.2	495
Urophycis regia	Spotted hake	–	–	–	<0.1	<0.1	1
Urophycis tenuis	White hake	1.2	0.7	72	1.5	1.1	43
Lepophidium cervinum	Fawn cusk-eel	–	–	–	<0.1	<0.1	18
Macrozoarces americanus	Ocean pout	1.4	0.9	44	0.7	0.5	31
Gasterosteus aculeatus	Threespine stickleback	–	–	–	<0.1	<0.1	0
Syngnathus fuscus	Northern pipefish	<0.1	<0.1	0	<0.1	<0.1	0
Centropristis striata	Black sea bass	–	–	–	<0.1	<0.1	2
Ammodytes americanus	American sand lance	<0.1	<0.1	0	–	–	–
Scomber scombrus	Atlantic mackerel	7.3	4.5	24	0.1	0.1	20
Peprilus triacanthus	Butterfish	1.7	1.2	25	4.8	3.5	98
Prionotus carolinus	Northern searobin	–	–	–	0.1	<0.1	1
Hemitripterus americanus	Sea raven	2.4	1.5	77	<0.1	<0.1	2
Myoxocephalus octodecemspinosus	Longhorn sculpin	1.6	1.0	106	<0.1	<0.1	2
Citharichthys arctifrons	Gulf Stream flounder	–	–	–	<0.1	<0.1	16
Paralichthys dentatus	Summer flounder	<0.1	<0.1	2	3.5	2.6	95
Paralichthys oblongus	Fourspot flounder	1.7	1.1	65	8.7	6.4	208
Scophthalmus aquosus	Windowpane	6.3	3.9	156	0.5	0.4	47
Glyptocephalus cynoglossus	Witch flounder	–	–	–	0.2	0.1	5
Limanda ferruginea	Yellowtail flounder	6.8	4.2	180	0.6	0.4	36
Pseudopleuronectes americanus	Winter flounder	14.6	9.1	189	–	–	–
Homarus americanus	Northern lobster*	2.0	1.2	0	1.8	1.3	0
(Caridea)	Caridean shrimp*	0.1	<0.1	0	<0.1	<0.1	0
Cancer irroratus	Rock crab*	<0.1	<0.1	0	–	–	–
Cancer borealis	Jonah crab*	0.1	<0.1	0	0.1	0.1	0
*Rossia sp.**		–	–	–	<0.1	<0.1	0
Loligo pealei	Long-finned squid*	0.1	<0.1	9	2.4	1.8	171
Illex illecebrosus	Short-finned squid*	–	–	–	0.1	<0.1	6
Number of tows			27			24	
Total number of stomachs examined			2484			3111	

* Unidentified species or not listed in AFS Spec. Publ. 12, 1980.

stomach contents of 5595 individuals, representing 36 species of fish and squid, were examined. Silver hake and spiny dogfish accounted for 975 and 1678 samples, respectively.

The major species caught (>5% by weight) at area A (40–53 m) were spiny dogfish, little skate, winter flounder, Atlantic cod, Atlantic mackerel, yellowtail flounder, windowpane and goosefish. At area B (70–93 m) the majority of the catch was made up of spiny dogfish, fourspot flounder, silver hake, goosefish, and red hake.

Observed regurgitation

Positive evidence of regurgitation was seen in 8 of the 36 species examined (Table 2). The best comparisons of the incidence of regurgitation between areas were for spiny dogfish, silver hake, and red hake since large numbers were sampled in both areas A and B. Little indication of regurgitation was seen for these species at area A (totals of 0.5, 2.9, and 3.0%, respectively). In area B no clear evidence of regurgitation was seen for spiny dogfish, but for silver and red hake it was substantial (totals of 24.4 and 49.3%, respectively). A more detailed examination of the silver and red hake data revealed that within area B percentage regurgitation was positively correlated with depth for both species (slopes are >0 at the 95% level) (Fig. 2). No such correlation was noted at area A for either species.

The correlation of regurgitation with depth at area B for the hakes is undoubtedly because hakes

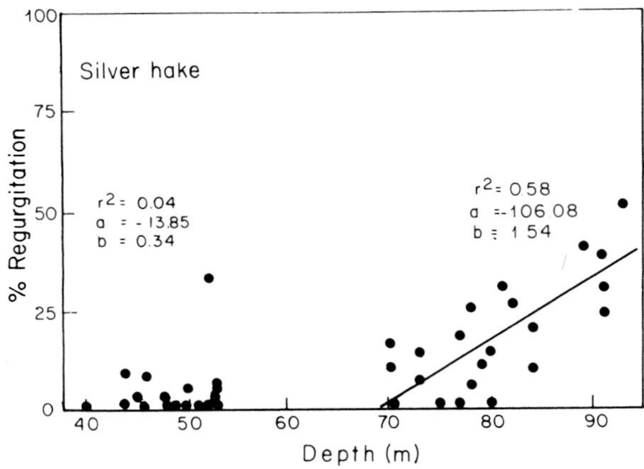

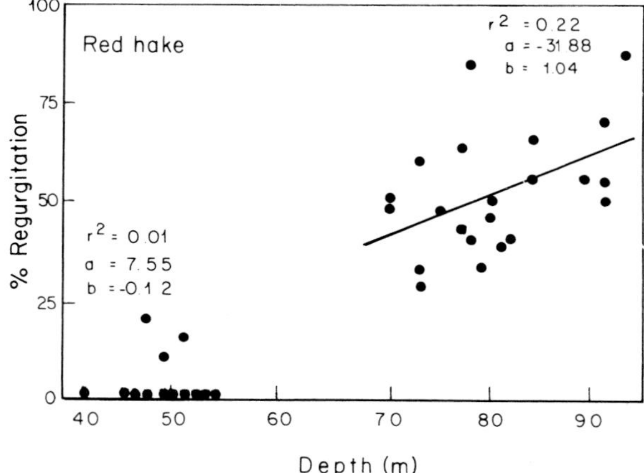

Fig. 2. Percentage detectable regurgitation for silver and red hake sampled at areas A (40–53 m) and B (70–93 m).

Table 2. Percentages of various fish species sampled in areas A and B which were positively observed to regurgitate.

Species	Number examined		Everted stomach		Food in mouth		Total regurg.	
	A	B	A	B	A	B	A	B
Spiny dogfish	628	1050	0.2	0.0	0.3	0.0	0.5	0.0
Little skate	149	20	0.0	0.0	0.0	5.0	0.0	5.0
Goosefish	20	49	5.0	4.1	0.0	0.0	5.0	4.1
Silver hake	404	571	2.2	23.5	0.7	0.9	2.9	24.4
Red hake	161	495	1.2	43.0	1.8	6.3	3.0	49.3
White hake	72	43	0.0	9.3	0.0	4.7	0.0	14.0
Fawn cusk-eel	0	18	–	0.0	–	38.9	–	38.9
Northern sea robin	0	1	–	100.0	–	0.0	–	100.0

have closed gas bladders. However, other forms of stress also cause fish to regurgitate. Bowen (1983) noted capture techniques such as rotenone treatment, electroshocking, gillnetting, and trawling at depth may cause regurgitation. Regurgitation in physoclistous fishes (e.g. silver and red hake) is likely more severe and more easily detected (e.g. everted stomach) than in physostomous fishes or fishes with no gas bladder (e.g. spiny dogfish and yellowtail flounder). Expansion of gas within the bladder, resulting from a decrease in outside pressure as the trawl is rapidly brought to the surface (e.g. haulback times of about 4 and 8 minutes at areas A and B, respectively), enlarges the bladder, or ruptures it and partly fills the body cavity with gas. This is reasonable to assume because the gas within the bladder would expand roughly 6 and 10-fold at A and B, respectively (from bottom to surface according to Boyle's Law). Since the bladder is located in part above and behind the stomach, food in the stomach would probably be expelled, or in extreme cases the stomach would evert, as a result of the increase in pressure within the body cavity. The shape and size of the digestive tract are probably important in this regard since digestive tracts are generally adapted to diet (Lagler et al. 1962). It is recognized that piscivores which eat large prey have large distendable esophaguses and regurgitate more frequently than fishes which feed on small prey and have small esophaguses (Bowen 1983).

Bearing the above in mind, most of the species which had little or no occurrence of observable regurgitation at area B can be grouped into three general categories as follows: (1) cartilaginous fishes – none have gas bladders and diet includes some decapods and fish (e.g. sharks and skates), (2) pelagic fishes – most have gas bladders with various modifications (e.g. in herrings the gas bladder opens to the exterior by a pore near the anus) and mainly feed on small organisms such as copepods, amphipods, and mysids (e.g. herrings, Atlantic mackerel, and butterfish), (3) flatfishes – none have gas bladders when adult, and many species take small prey such as amphipods and polychaetes (e.g. windowpane and yellowtail flounder). Conversely, the gadiform fishes (e.g. silver, red, and white hake) generally had a high incidence of observable regurgitation. They have closed gas bladders and eat large organisms such as fish and decapods. These observations made it apparent that some combination of the presence or absence of a closed gas bladder and digestive tract morphology (as inferred by prey type which is generally a function of mouth, esophagus and stomach size) influenced observable regurgitation.

Therefore, it might be expected that piscivores and physoclistous fishes with full stomachs (i.e. fish with the least space in the body cavity to accomodate the expanded gas) would regurgitate more often, and perhaps more completely, especially when retrieved from deep water. This would increase the proportion of everted and empty stomachs, as well as truncate the upper portion of the frequency distribution of relative stomach content volumes. Since observed regurgitation was substantially higher in area B, it was suspected that there was also a higher incidence of unrecognized regurgitation.

Unrecognized regurgitation

Examination of stomach content volumes (expressed as % BW) of fishes with no visual signs of regurgitation showed 7 of the 8 species for which there was adequate data for analysis had more food in their stomachs at area A than at area B (Fig. 3). Paired t-tests between data for areas A and B, by individual species, showed significantly more food was present in the stomachs of spiny dogfish, silver hake, red hake, and fourspot flounder at area A ($t = 5.75, 7.21, 4.83, 3.23$; D.F. $= 7, 7, 7, 6$, respectively).

A subset of the data for silver hake, red hake, and spiny dogfish (based on strictly comparable length frequencies for each species) was plotted to illustrate the frequency distributions of stomach content volumes in areas A and B (Fig. 4–6). Incidence of visually observed regurgitation is included in the figures for silver and red hake for comparison purposes. In the case of silver hake (Fig. 4) the highest stomach content values (>2% BW) are virtually non-existent from the frequency distribution in area B compared to A, and there was a large

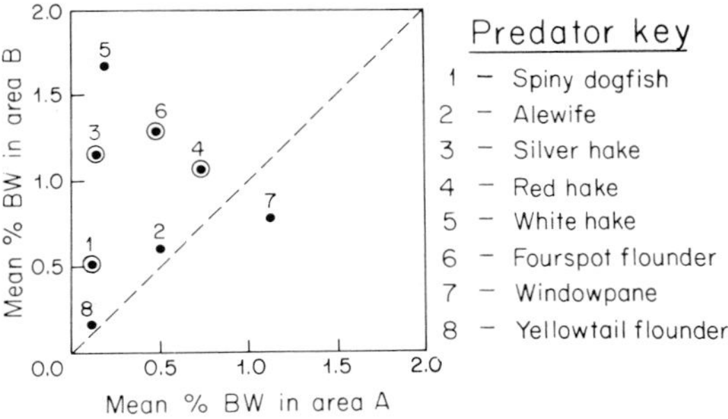

Fig. 3. Scatterplot of the mean % BW values of stomach contents for species sampled at area A versus the same species sampled at area B. Circled data points indicate species for which a significant difference in stomach content volumes was observed between the two areas.

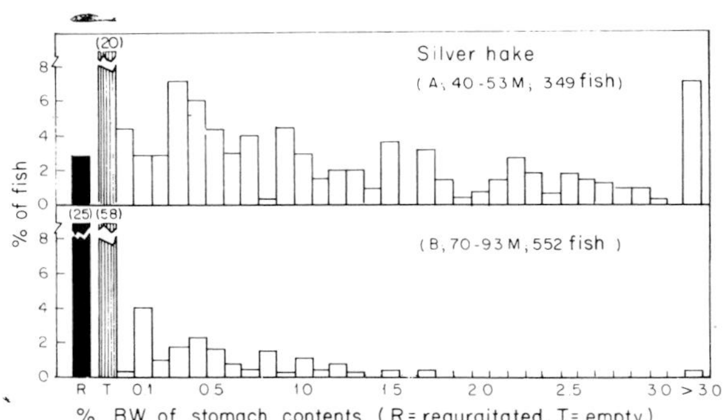

Fig. 4. Percentages of a subset of silver hake sampled at areas A and B which had regurgitated, empty stomachs, and various quantities of food in their stomachs. Number regurgitated was not included in calculation of percentage empty or stomach content values.

increase in the percentage of empty stomachs and detectable regurgitation at B, which is consistent with what was suggested above. The frequency distribution of stomach contents for red hake did not change as much as for silver hake, and there was only a small increase in the percentage empty at area B (Fig. 5). However, the percentage of red hake which regurgitated was much higher in B. This could indicate more complete (i.e. detectable) regurgitation occurred in red hake, possibly because they are more severely affected by a rapid decrease in pressure than silver hake.

For spiny dogfish at B relative to A only a slight decrease was seen in the frequency of stomach contents >2% BW, and only a modest increase was observed in the percentage empty (Fig. 6). The reason spiny dogfish are not often observed with everted stomachs, even though they have large esophaguses, is probably because they have no gas bladder. However, the percentage of fish with stomachs full of water is undoubtedly not a natural phenomenon, and may be an indication of prior regurgitation. Noteworthy in this regard is that spiny dogfish caught in both areas were frequently seen regurgitating as well as gulping down air after the trawl was emptied onto the deck. Apparently spiny dogfish readily regurgitate when caught in twawls. The high percentage of empty stomachs, stomachs full of water, relatively small quantities of food in the stomachs, and visual observations noted above provide evidence that a large number

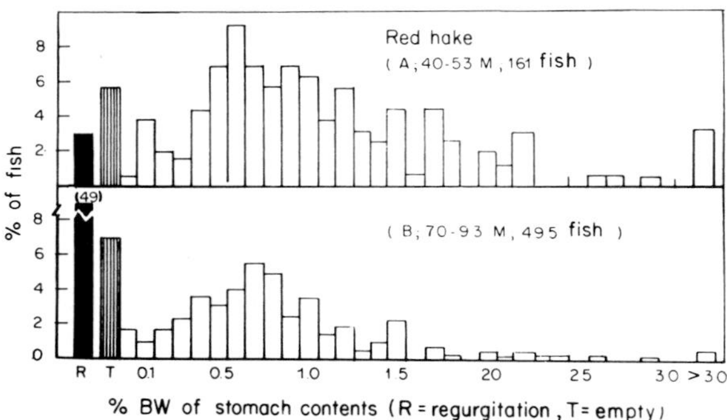

Fig. 5. Percentages of red hake sampled at areas A and B which had regurgitated, empty stomachs, and various quantities of food in their stomachs. Number regurgitated was not included in calculation of percentage empty or stomach content values.

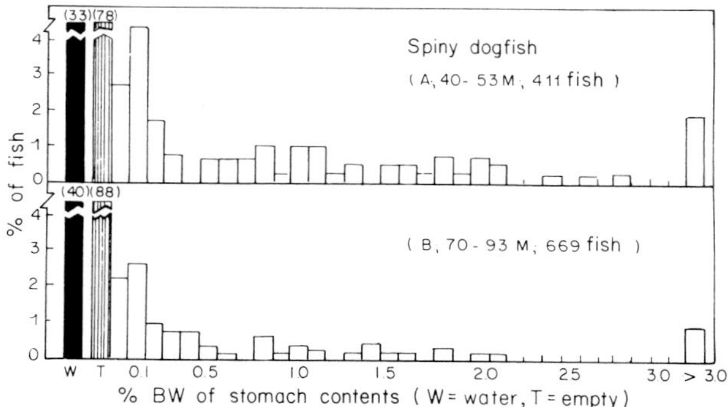

Fig. 6. Percentages of a subset of spiny dogfish sampled at areas A and B which had stomachs full of water, empty stomachs, and various quantities of food in their stomachs. Percentage empty includes stomachs which were filled with water but contained no food.

of dogfish regurgitated in both area A and B.

Without knowing the absolute abundance of preferred prey for each species in each area, it is impossible to sort out the degree to which undetected regurgitation could account for these results. However, as will be shown in the next section, there was no indication of a scarcity of food in B relative to A. This, together with the evidence presented above, suggests that a significant amount of undetected regurgitation probably occurred in area B, especially in silver hake and spiny dogfish, and to a lesser degree in red hake.

Comparison of prey abundance with diet

Relative abundance of prey fish and squid was readily available by calculating the mean catch (number per 30 min trawl haul) of each species at each area (Table 3). Data for documenting the abundance of benthic and pelagic invertebrates were not available. Eight predators were sampled in sufficient numbers at areas A and B to make dietary comparisons (Table 4).

Fish and squid were important prey (>10% of the diet in terms of percentage total volume) of spiny dogfish, silver hake, red hake, white hake, and fourspot flounder. It is readily seen that both the catch per haul indices (Table 3), and the percentages in the diet of various predators (Table 4), of prey such as American sand lance, herrings, Atlantic mackerel, and silver hake (almost all <15 cm fork length), were highest in area A in almost every instance. Conversely, the values for squid and butterfish were highest in area B. Other prey groups such as flatfishes and red hake <15 cm didn't appear to be important prey in a particular area, but both were generally more abundant at area A. The 'other fish' prey category (Table 4) was substantial for several predators (e.g. spiny dogfish and silver hake), but since it was almost exclusively unidentified fish flesh, it could not be considered in the comparison with abundance. In the aggregate, it seems somewhat more prey, especially small silver hake, were available for food in area A. However, their certainly wasn't a scarcity of food at area B, as implied by the fact that the major fish predators (spiny dogfish, silver hake and white hake) were all more abundant at area B (Table 1).

Overall, it doesn't appear the slightly higher abundance of prey in area A would be large enough to account for the drastic differences noted in the % BW values between areas. For example, as seen at the bottom of Table 4, species such as silver hake and spiny dogfish had 8.3 and 4.3 times more food in their stomachs at area A, respectively. Also noteworthy is that no large differences were seen between areas for most predators in terms of major taxonomic food categories. For example, spiny dogfish ate mostly fish and squid (>90% in both A and B), alewife stomachs contained almost totally small invertebrates and decapods (both combined equaled >95% in each area), and the three flounders (i.e. fourspot, windowpane, and yellowtail) consumed roughly the same proportions of some combination of fish and squid, polychaetes and small invertebrates, or decapods depending on the particular predator. Generally there were tradeoffs among the subgroups which resulted in the diet composition being remarkably similar for the major groupings. The obvious exceptions were silver hake and white hake. In the case of silver hake, chaetognaths taken as food in area B (18.1%) resulted in corresponding decreases in the percentages of fish, squid, and decapods in their diet. If

Table 3. Relative abundance of fish and squid taken as prey in areas A and B as determined from trawl catches during study.

Fish and squid	Relative abundance (number per 30 min tow)	
	Area A (40–53 m)	Area B (70–93 m)
Long-finned squid	1.7	94.9
American sand lance	0.8	0.0
Butterfish	13.1	70.0
Herrings	27.0	15.8
Atlantic mackerel	11.1	0.9
Flatfishes	89.4	56.3
Silver hake >15 cm FL	16.4	39.3
Silver hake <15 cm FL	272.3	3.5
Red hake >15 cm TL	18.7	26.5
Red hake <15 cm TL	0.7	0.2

Table 4. Percentage volume of the total stomach contents of dominant prey for predators sampled in both area A and B. The 'other' categories within major prey groupings (e.g. 'other fish') consisted of mainly well digested organisms which couldn't be identified to species. Mean % BW values at bottom of table are only for fish which had no signs of regurgitation.

PREY	Spiny dogfish		Alewife		Silver hake		Red hake		White hake		Fourspot		Windowpane		Yellowtail	
	A	B	A	B	A	B	A	B	A	B	A	B	A	B	A	B
FISH AND SQUID	92.6	92.0	–	–	53.5	44.4	8.7	13.9	21.1	83.3	10.2	10.4	0.5	1.0	–	–
Squid	5.0	31.8	–	–	–	11.0	0.2	1.7	–	26.2	–	8.0	–	–	–	–
American sand lance	1.8	–	–	–	11.8	0.2	1.9	–	1.1	–	–	–	–	–	–	–
Butterfish	0.7	3.5	–	–	–	2.5	–	–	–	–	–	–	–	–	–	–
Herrings	3.8	<0.1	–	–	–	–	–	–	–	–	–	–	–	–	–	–
Atlantic mackerel	33.4	6.5	–	–	–	–	–	–	–	5.9	–	–	–	–	–	–
Flatfishes	7.3	2.6	–	–	–	–	–	0.3	1.4	9.8	0.3	0.8	0.5	–	–	–
Silver hake	1.9	–	–	–	19.9	–	3.8	–	6.3	–	–	1.2	–	–	–	–
Red hake	1.9	2.3	–	–	1.5	–	–	–	–	–	1.2	–	–	1.0	–	–
Other fish	36.8	45.3	–	–	20.5	30.7	2.8	11.9	12.3	41.4	8.7	0.4	–	–	–	–
POLYCHAETES & SMALL INVERTEBRATES	0.1	5.7	85.7	83.4	2.2	20.2	33.9	41.7	17.9	3.0	3.6	2.3	90.7	96.0	96.9	99.2
Polychaeta	0.1	3.1	–	–	0.3	–	8.8	4.1	2.3	1.0	0.1	1.5	0.4	–	36.0	77.9
Amphipoda	<0.1	0.2	38.3	35.8	0.7	1.8	22.6	36.0	14.0	1.0	3.5	0.8	1.6	0.1	60.9	21.3
Mysidacea	–	<0.1	32.3	0.5	1.2	0.2	1.4	0.5	1.5	<0.1	–	–	82.4	4.7	–	–
Other crustaceans	<0.1	0.4	13.7	4.9	<0.1	0.1	0.9	0.5	0.1	–	–	–	0.3	0.1	–	–
Chaetognatha	–	2.0	1.5	42.2	–	18.1	0.2	0.6	–	1.0	–	–	6.0	91.1	–	–
DECAPODS	1.7	0.4	10.2	8.8	41.8	34.0	54.9	37.1	59.8	9.0	85.0	85.1	8.1	3.0	1.3	–
Crangon sp.	<0.1	<0.1	7.4	2.5	24.9	5.3	21.9	2.2	30.2	3.4	28.4	–	7.9	2.1	1.2	–
Pandalidae	<0.1	0.3	2.8	6.3	16.9	28.7	20.5	12.6	19.3	5.6	53.9	48.3	0.2	0.9	–	–
Cancer spp.	0.6	–	–	–	–	–	2.2	6.7	3.6	–	1.3	15.8	–	–	–	–
Other decapod crabs	1.1	0.1	–	<0.1	<0.1	–	10.5	15.6	6.7	–	1.4	21.0	–	–	0.1	–
MISCELLANEOUS	5.6	1.9	4.1	7.8	2.5	1.4	2.5	7.3	1.2	4.7	1.2	2.2	0.7	–	1.8	0.8
Number of fish examined	628	1050	74	95	404	571	161	495	72	43	65	208	156	47	180	36
Number of empty stomachs	463	785	11	32	83	328	9	34	4	13	8	63	48	8	70	9
Mean fish length (cm)	81.2	63.5	24.4	27.0	27.6	30.6	34.3	35.2	34.2	43.2	35.4	33.8	27.1	26.1	33.2	34.1
Mean % BW stom. cont.	0.52	0.12	0.60	0.50	1.16	0.14	1.07	0.73	1.67	0.19	1.28	0.48	0.77	1.12	0.16	0.11

we assume chaetognaths are not a preferred food of silver hake, then the high percentage found for area B may indicate that other food (e.g. fish, squid, and decapods) was not as available as in area A, but this contradicts what was observed for most other species. For white hake it is likely that the difference seen in the types of food eaten between areas A and B was caused by predator length. The mean total lengths of white hake sampled at areas A and B were 34.2 and 43.2 cm, respectively. It is well established that large white hake eat much larger portions of fish and squid than small white hake (Bowman & Michaels 1984).

Discussion

There is little doubt that variability in stomach contents, caused directly or indirectly by sampling depth, is a complex problem. Depending on the particular predator, we have seen different rates of or no detectable regurgitation, different quantities and types of food in the stomachs, and differences in predator and prey abundance. All of these are somehow apparently related to bottom water depth since variables such as water temperature, time of day, season, year or available prey did not (or, in the latter instance, most likely did not) cause the differences. Of particular interest is that most of the species for which no regurgitation was observed had stomach content volumes which were nearly equal between areas (i.e. alewife, windowpane and yellowtail flounder; with an average of only 1.1 times more food in the stomachs at area A). These same predators were more abundant at area A, and their average lengths and major prey groups (polychaetes or small invertebrates) were almost identical between areas. Conversely, for species which were observed to regurgitate (e.g. spiny dogfish, silver hake, red hake, and white hake), the quantities of food in their stomachs were much less in area B (average of about 5.7 times less), all were more abundant at area B, and their major prey was mostly large organisms such as fish, squid, and decapods. We must ask why the abundance of spiny dogfish, silver hake, red hake and white hake would be greater in an area where food was scarce? Also, why were significant differences in the stomach content quantities between areas only found in the species which were observed to regurgitate? I believe the above facts and reasoning provide adequate circumstantial evidence to infer that regurgitation occurs and may go undetected in certain species when they are sampled in deep water.

Food consumption by marine fishes has become a central theme of many large scale fishery research programs initiated in the 1980's. Research conducted in the 1970's and early 1980's has provided evidence that piscivorous fish may not only have a major impact on year class success of species taken as prey, but that these predators may consume larger quantities of species of commercial interest than are harvested by the fisheries (for an extensive review see Sissenwine 1984). Accordingly, more quantitative information must be obtained on predator-prey relationships, and major causes of bias or variability in fish stomach content data must be identified to determine their potential impact on estimates of food consumption. In particular, regurgitation which may occur and go undetected when conducting fish food studies may produce severe underestimation of consumption. Sampling protocols generally address the problem of regurgitation by requesting technicians, at the time samples are collected, to perform the following tasks: (1) inspect the buccal cavity (inside of mouth) for signs of regurgitated food, and the esophageal area (via the body cavity) for eversion. If signs of regurgitation exist discard the fish[1], (2) expanded stomachs which are empty are to be discarded (Daan 1973). The phenomenon known as regurgitation is well known; e.g. 'Many predatory fishes appear to regurgitate large food items from the stomach with great facility. It has been suggested that this is made possible by the pronounced development of striated muscles in the walls of the esophagus extending to the stomach' (Lagler et al. 1962).

The percentage of detectable regurgitation for some species increases considerably with increasing trawl depths. The results presented here docu-

[1] Taken from NEFC sampling protocol.

ment that about 8 times more silver hake and 16 times more red hake regurgitate with an increase of only 40 m bottom water depth (from about 50 to 90 m). Not only is the incidence of detectable regurgitation higher, but even those fish showing no evidence of regurgitation can have measurably lower stomach contents in deep water. In this instance the stomachs of the two species of principal concern, silver hake and spiny dogfish, contained an average of approximately 8 and 4 times more food at the shallow area, respectively. Consumption estimates for these two species could be biased by the same amounts if these differences were caused by undetected regurgitation.

Silver hake and spiny dogfish make up a considerable portion of the total fish biomass and have been identified as the two most significant piscivorous fish in the northwest Atlantic (Edwards & Bowman 1979, Anonymous 1983). Because these two species have been shown to, or are suspect of regurgitating their food, potential severe bias may be inherent in the data one uses to determine the type and quantity of food they consume. Therefore, stomach content data for these two species must be examined critically before attempting to estimate their predatory impact on other fish populations.

When daily ration is estimated according to the method described by Elliott & Persson (1978) or some similar method (e.g. Pennington 1984), and field gathered data is used for the calculations, the results can be of questionable value, especially when the stomach content samples are obtained from different depths. In a recent paper Daan (1984) documented that about 9% of all Atlantic cod he studied from the North Sea had obvious signs of regurgitation. He estimated cod consumption on the remaining samples, as is done traditionally (e.g. Durbin et al. 1983, Livingston 1983). During the study by Durbin et al. on Atlantic cod and silver hake in the northwest Atlantic, it was noted that a large proportion of the silver hake sampled had empty stomachs. Moreover, the average quantity of food present in the stomachs of both silver hake and Atlantic cod was small based on what is known of their energetic requirements. It is probable that most daily ration estimates based on field data are negatively biased, perhaps to a large degree for species with closed gas bladders.

In conclusion, we have seen that detectable regurgitation varies according to species and increases with sampling depth in species with closed gas bladders. Visually undetectable regurgitation was difficult to document but evidence presented suggests that it also occurs for certain species and results in negative bias in average stomach content estimates. The mechanisms of regurgitation, and the relationship of regurgitation with depth and stomach fullness, might be somewhat clarified through experiments on fishes within pressure chambers. However the element of stress and the effects of external pressure (analagous to squeezing of fish inside a full cod end) would be difficult to simulate. Therefore the magnitude of possible bias from undetected regurgitation may be best estimated through further observations such as reported herein, together with analysis of time series data on stomach contents versus depth for selected species.

Acknowledgments

I am especially grateful to the Master of the DELAWARE II, R.E. Adams, and the crew, for their assistance and cooperation throughout the study. Special thanks are also due William Michaels, Thomas Morris, Malcolm Silverman, Alphonze Thrower, Michael Sigler, Richard Ready, Timothy Hughes, Terrance Cianci, Edward Stinchcomb, James Myette, Andrea Swiecicki, Ronald Mack, and Michael Pennington who helped collect stomachs, analyze stomach contents, tabulate data, and provided statistical assistance. I thank Marvin Grosslein for suggestions on data analysis and preparation of the manuscript.

References cited

Anonymous. 1983. Status of the fishery resources off the northeastern United States for 1982. NOAA Tech. Mem. NMFS-F/NEC-22. U.S. Dept. Comm., NOAA, NMFS, Northeast Fisheries Center, Woods Hole, MA 02543. 128 pp.

Bowen, S.H. 1983. Quantitative description of the diet. pp. 325–336. *In*: L.A. Nielsen & D.L. Johnson (ed.) Fisheries Techniques, American Fisheries Society, Bethesda.

Bowman, R.E. 1980. Food of northwest Atlantic juvenile haddock. M.A. Thesis, Bridgewater State College, Bridgewater. 95 pp.

Bowman, R.E. 1981. Examination of known and potential causes of variation in fish feeding studies. U.S. Nat. Fish. Serv., Northeast Fish. Ctr., Woods Hole, MA 02543. Woods Hole Lab. Ref. Doc. No. 81–23. 29 pp.

Bowman, R.E. 1984. Food of silver hake, *Merluccius bilinearis*. U.S. Fish Bull. 82: 21–35.

Bowman, R.E. & E.W. Bowman. 1980. Diurnal variation in the feeding intensity and catchability of silver hake (*Merluccius bilinearis*). Can. J. Fish. Aquat. Sci. 37: 1565–1572.

Bowman, R.E. & W.L. Michaels. 1984. Food of seventeen species of northwest Atlantic fish. NOAA Tech. Mem. NMFS-F/NEC-28. 183 pp.

Daan, N. 1973. A quantitative analysis of the food intake of North Sea cod, *Gadus morhua*. Neth. J. Sea Res. 6: 479–517.

Daan, N. 1984. The ICES stomach sampling project in 1981: Aims, outline and some results. NAFO, 5th Annual Mtg., Sept. 1983, Leningrad, USSR, Cont. A.2.

Durbin, E.G., A.G. Durbin, R.W. Langton & R.E. Bowman. 1983. Stomach contents of silver hake, *Merluccius bilinearis* and Atlantic cod, *Gadus morhua* and estimation of their daily rations. U.S. Fish. Bull. 81: 437–454.

Edwards, R.L. & R.E. Bowman. 1979. Food consumed by continental shelf fishes. pp. 387–406. *In*: H. Clepper (ed.) Predator-Prey Systems in Fisheries Management, Sport Fishing Institute, Washington, D.C.

Elliott, J.M. & L. Persson. 1978. The estimation of daily rates of food consumption for fish. J. Anim. Ecol. 47: 977–991.

Grosslein, M.D., R.W. Langton & M.P. Sissenwine. 1980. Recent fluctuations in pelagic fish stocks of the northwest Atlantic, Georges Bank region, in relation to species interactions. Rapp. P.-v. Reun. Cons. Int. Explor. Mer 177: 374–404.

Grosslein, M.D. & T.R. Azarovitz. 1982. Fish distribution. MESA N.Y. Bight Atlas Monograph No. 15. N.Y. Sea Grant Inst., Albany. 182 pp.

Lagler, K.F., J.E. Bardach & R.R. Miller. 1962. Ichthyology. John Wiley & Sons, New York. 545 pp.

Livingston, P.A. 1983. Food habits of Pacific whiting, *Merluccius productus* off the west coast of North America, 1967 and 1980. U.S. Fish Bull. 81: 629–636.

Pennington, M., R.E. Bowman & R.W. Langton. 1982. Variability of the weight of stomach contents of fish and its implications for food studies. pp. 2–7. *In*: G.M. Cailliet & C.A. Simenstad (ed.) Gutshop '81, Fish Food Habits Studies, Third Pacific Workshop, Washington Sea Grant Publ., University of Washington, Seattle.

Pennington, M. 1984. Estimating the average food consumption by fish in the field from stomach contents data. Int. Con. Explor. Sea, C.M./H: 28. 11 pp.

Sissenwine, M.P. 1984. Why do fish populations vary? pp. 59–94. *In*: R. May (ed.) Exploitation of Marine Communities, Springer-Verlag, Berlin.

Received 15.12.1984 *Accepted 1.7.1985*

Fig.s 1 _ 4. HOMALOPOMUS TROWBRIDGII, Grd.
" 5 _ 8. MORRHUA PROXIMA, Grd.

Recent developments for making gastric evacuation and daily ration determinations

Robert J. Olson & Ashley J. Mullen
Inter-American Tropical Tuna Commission, c/o Scripps Institution of Oceanography, La Jolla, CA 92093, U.S.A.

Keywords: Digestion rates, Food consumption models, Yellowfin tuna, *Thunnus albacares,* Fishes

Synopsis

Gastric evacuation rate estimates often suffer from an important bias caused by fitting experimentally-derived data distributions that are inherently constricted by the X-axis (Y = 0). Monte Carlo simulations were used to evaluate the bias. Truncating constricted distributions prior to curve fitting was suggested as a means to circumvent the problem. A food consumption model developed by D.S. Robson was presented. It employs the integral of the function fit to percentage gastric evacuation data, and does not require an a priori assumption of exponential gastric evacuation. The methods were illustrated using experimental gastric evacuation data and stomach contents data for fishery-caught yellowfin tuna, *Thunnus albacares*.

Introduction

The importance of estimating rates of food consumption in fishes is well recognized (Cohen et al. 1981, Rice & Cochran 1984). With increasing interest in the concept of ecosystem or multispecies management of renewable resources (Mercer 1982, Livingston 1986) the nature and magnitude of species interactions (i.e. predation and competition) must be evaluated. To do this, it is necessary to determine food habits and daily rations of the major predators in the ecosystem. In addition, the top-down approach to estimating productivity at intermediate trophic levels by means of predation rates of apex predators is a practical alternative to the difficult task of directly sampling the prey, especially in the pelagic marine environment. Continued improvement of techniques for determining rates of gastric evacuation and daily ration in fishes is desirable.

There has been considerable discussion regarding models for estimating daily food consumption in fishes (Elliott & Persson 1978, Cochran 1979, Eggers 1979, Elliott 1979, Jobling 1981). Models in common usage today are based on the assumption that gastric evacuation is an exponential process over time (e.g. Elliott & Persson 1978). Although many investigators of gastric evacuation in fishes have fitted their data to the exponential model (Brett & Higgs 1970, Tyler 1970, Elliott 1972, Steigenberger & Larkin 1974, El-Shamy 1976, Thorpe 1977, Heubner & Langton 1982, Macdonald et al. 1982), linear (Hunt 1960, Seaburg & Moyle 1964, Swenson & Smith 1973, Daan 1973, Jones 1974, Bagge 1977, Macdonald et al. 1982), square root (Kariya et al. 1969, Jobling & Davies 1979, Jobling 1981), and other models (Magnuson 1969, Tyler 1970) have also been used. A single method for estimating food consumption which is applicable to a variety of evacuation models is conceptually more simple than developing a method for each evacuation model on an ad-hoc basis.

Since the shape of gastric evacuation data distributions influences estimates of daily ration, it is important to consider factors that affect the fit. One such factor is a bias inherent in many such experimental data sets. The problem becomes important when the duration of gastric evacuation experiments is such that at least some test individuals empty their stomachs (e.g. Magnuson 1969, Steigenberger & Larkin 1974, Macdonald et al. 1982). Empty stomachs must be omitted during data analysis because the exact time they became empty cannot be determined. Prior to the time when the fastest digestors in a sample begin to empty their stomachs, the data include the full range of intraspecific variability expected. But subsequent to that time, an ever-increasing proportion of the sample representing the faster digestors is eliminated from the distribution. Thus, as post-prandial time increases, the distribution becomes constricted by the time axis. If not adjusted for, 'constricted' data distributions can result in biased evacuation rate estimates and a false indication or exaggeration of curvilinearity. This problem has not been addressed by previous workers.

The purpose of this paper is two-fold:

1. To evaluate the importance of this bias when fitting gastric evacuation data. Truncating the data is suggested to circumvent the problem.
2. To present a food consumption model which is advantageous in that it does not require an a priori assumption of exponential gastric evacuation.

Materials and methods

The effect of constricted data distributions on estimating gastric evacuation rates was investigated using the Monte Carlo simulation method. Figure 1 is a diagramatic representation of the simulations. For a straight line equation, $Y = 100 - bX$, 2N values of X were chosen spaced uniformly between 0 and 2(100/b). Y values were generated from a normal distribution of specified standard deviation about the line. Data for which $Y \leq 0$ were omitted and least-squares regression analysis was performed on the remainder (broken line, Fig. 1). Therefore, the expected number of data subjected to the regressions was N. For a given N, standard deviation, and value of b, the process was repeated 500 times. The mean and standard deviation of each set of 500 estimates of b were obtained.

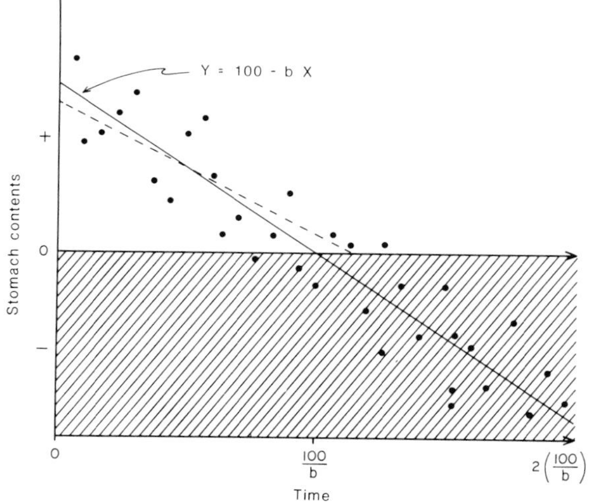

Fig. 1. Schematic diagram of one run of a simulation. The points represent Y values generated from a normal distribution about the solid line. The shaded area (negative values) marks those points which could never be observed in practice. The broken line represents the least squares fit to the positive points.

In 1970 D.S. Robson developed a model for estimating feeding rate in fishes using data on average stomach contents and the integral of the gastric evacuation function. The model is presented here much as it was developed by Robson in his unpublished manuscript[1], except that we expand the model for a predator that eats a variety of prey which are evacuated at different rates.

Gastric evacuation and stomach contents data for yellowfin tuna, *Thunnus albacares,* are used to illustrate the use of the model and the effect of the bias. The laboratory experiments were not designed to test the Robson model. However, the data are included here to demonstrate the utility of the model when a priori assumptions necessary for other models are not met. Detailed descriptions of the experiments and ration calculations are given

[1] Robson, D.S. 1970. On the relation between feeding rate and stomach content in fishes. Biometrics Unit Mimeo Series BU-328-M. Cornell University, Ithaca (unpublished manuscript).

by Olson (1983) and Olson & Boggs (1986). Briefly summarizing, mixed meals of mackerel, *Scomber japonicus,* squid, *Loligo opalescens,* smelt, *Hypomesus pretiosus,* and nehu, *Stolephorus purpureus,* were eaten voluntarily at one feeding time by 69 differentially tagged yellowfin. Each food particle was weighed prior to ingestion and after removal from the stomachs at various post-prandial intervals, and was expressed as a percentage of the amount eaten (Fig. 2). After truncating the data, as described below, to adjust for bias, linear, exponential, and square root functions were fitted using the four data sets. Residual analysis showed that the linear model gave a superior fit. Meal size (% of body mass) did not have a significant effect on the evacuation rates ($\% \cdot h^{-1}$) for three of the four food species; the exception was mackerel, the largest, most digestion resistant test food with the highest lipid content.

Stomach contents data for 3581 yellowfin tuna captured by the eastern Pacific Ocean purse-seine fishery during 1970–1972 were used in conjunction with the laboratory-derived gastric evacuation rates for estimating daily ration using Robson's model. For comparison purposes, rations were re-calculated here using the same stomach contents data and evacuation functions fitted to complete data sets (unedited for biased distributions). Residual analyses were performed after fitting the linear, exponential, and square root models. They were identical to those used by Olson (1983) upon truncated data and included tests for normality (Filliben 1975), constant variance (homoscedasticity), and statistical independence (autocorrelation) of residuals (Wesolowski 1976).

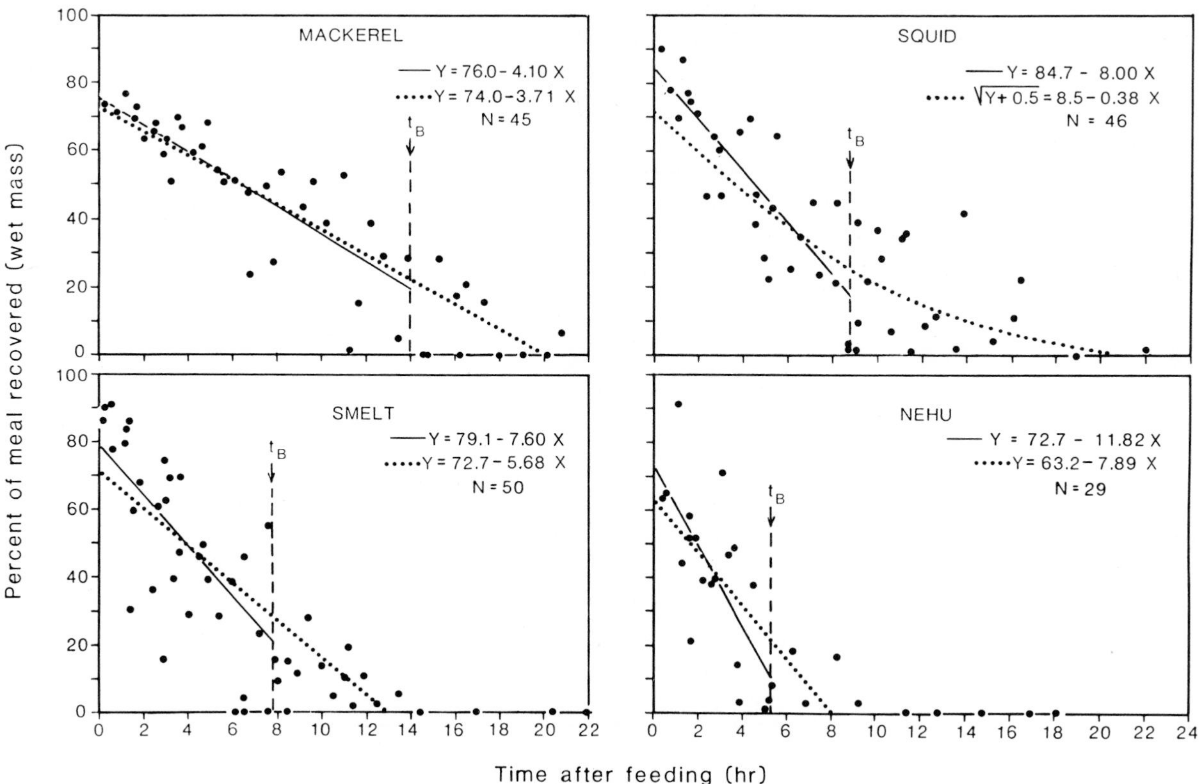

Fig. 2. Percent of initial wet mass of four experimental foods recovered from the stomachs of captive yellowfin versus time after feeding. T_B is the point beyond which data were truncated prior to curve fitting. Solid lines are regression lines fitted to truncated distributions and dotted lines are regression lines fitted to entire distributions, less empty stomachs.

Results

Bias simulations

The Monte Carlo simulations provided calculations of mean gradients and their standard deviation for each group of 500 iterations. Preliminary regressions on selected individual data sets showed that calculated gradients (broken line, Fig. 1) were significantly lower than given gradients (t-tests, $P<0.01$). The simulation confirmed that the number of points in the distribution did not affect the bias, but the variance of the calculated slopes increased as the number of points decreased. Slope had no effect on the bias, because as b varies, the proportion of data for which any given percentage of the distribution overlaps $Y=0$ remains constant. The given standard deviation of generated Y values determines the magnitude of the bias (Fig. 3). For evacuation data with variance comparable to that for yellowfin tuna (Fig. 2), the data constriction causes a $-18\pm9\%$ (mean ±1 SD) bias in evacuation rate estimate.

The bias can be circumvented by, prior to curve fitting, eliminating those data associated with post-prandial times when empty stomachs appear. Regardless of whether a distribution is straight or exponential, a decrease in slope would be expected to occur in the region of the data constriction rela tive to the slope at earlier post-prandial times. The point at which this change begins marks the beginning of the data constriction, and if well-defined, can be used as a guide for choosing a point of truncation. A simple procedure of detecting the beginning of this trend in yellowfin gastric evacuation data (Fig. 2) was adopted by Olson (1983). A series of least-squares linear regressions were calculated on subsets of data commencing with the first three points at earliest post-prandial time and incrementing points one-by-one with increasing time. Separate regressions were calculated as each datum was added (N-2 regressions, N = non-empty stomachs) and regression coefficients were plotted versus N. The point of truncation was determined by estimating visually the post-prandial time (t_B) associated with a systematic decline in slope. Yel-

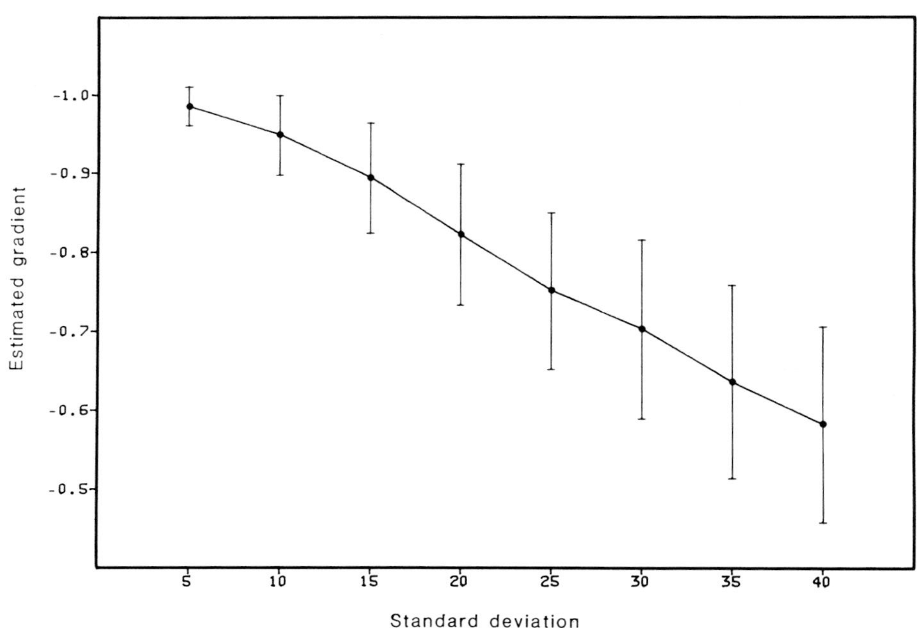

Fig. 3. The results of Monte Carlo simulation to study the effect of constricted gastric evacuation data distributions on estimating gastric evacuation rates. Points are means ±1 standard deviation of 500 regression gradients calculated for generated constricted data distributions with $b=1.0$ and $N=40$.

lowfin gastric evacuation data were truncated at t_B's of 14.0 h for mackerel, 8.8 h for squid, 7.9 h for smelt, and 5.3 h for nehu (Fig. 2). Clearly, this method includes an element of subjectivity which perhaps cannot be eliminated, but could be made constant in the interest of repeatability by using a suitable algorithm

Food consumption model

If the proportion of food, f(t), remaining in the stomach t time units after a single meal is known, then the absolute quantity of food (W) remaining from any single meal (M) is simply

$$W = M\, f(t).$$

But the food found in the stomach of a fish taken from its natural environment may be derived from several meals taken at intervals T_1, T_2, ... before digestion was interrupted (see Fig. 4). Therefore,

$$W = M_1\, f(T_1) + M_2\, f(T_1 + T_2) + \ldots$$

Over a period of K feeding intervals, $T_1 + T_2 + \ldots + T_k$, the mean value of food in the stomach ($\bar{W}_k$) derived from these meals is given by

$$\bar{W}_k = \frac{1}{(T_1 + T_2 + \ldots + T_k)} \int_0^{T_1+T_2+\ldots+T_k} W\, dt$$

$$= \frac{1}{(T_1 + T_2 + \ldots + T_k)} (M_1 \int_0^{T_1} f(t)dt +$$

$$M_2 \int_0^{T_1+T_2} f(t)dt + \ldots + M_k \int_0^{T_1+T_2+\ldots+T_k} f(t)dt)$$

$$= \frac{1}{(T_1 + T_2 + \ldots + T_k)} \times$$

$$((M_1 + M_2 + \ldots + M_k) \int_0^{T_1} f(t)dt +$$

$$(M_2 + M_3 + \ldots + M_k) \int_{T_1}^{T_2} f(t)dt +$$

$$\ldots + M_k \int_0^{T_k} f(t)dt).$$

AS $k \to \infty$ $\bar{W}_k \to \bar{W} =$

$$\frac{(M_1 + M_2 + \ldots + M_k)}{(T_1 + T_2 + \ldots + T_k)} \int_0^{\infty} f(t)dt.$$

$$\bar{W} = \frac{\bar{M}}{\bar{T}} \int_0^{\infty} f(t)dt,$$

where $\bar{M}$ = mean meal size and $\bar{T}$ = mean interval between meals. $r = \bar{M}/\bar{T}$ is the feeding rate, and if we define A to be the integral of the function fit to gastric evacuation data,

$$A = \int_0^{\infty} f(t)dt,$$

then

$$\hat{r} = \frac{\bar{W}}{A}.$$

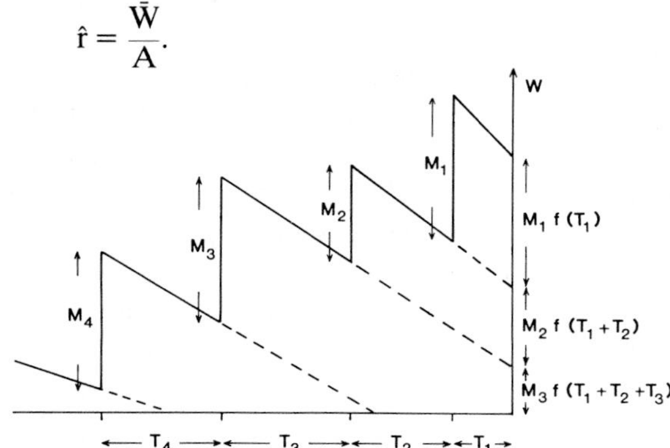

Fig. 4. Diagramatic representation of quantity of food in the stomach as a function of meal sizes and elapsed times.

Reliable estimates of $\bar{W}$ may be obtained by measuring the stomach contents of fish from the field if those fish are sampled at random times, or if it can be assumed that the fish feed at random intervals. Thus, if f(t) is known from controlled experiments, then this information may be combined with field data in a simple manner to obtain estimates of daily ration.

Robson's model is expanded here for a predator that consumes a variety of prey organisms which are evacuated at different rates, assuming mixed meals are evacuated at a rate consistent with the rates for the prey composing the meal, as found by Olson (1983).

$$\hat{r} = \sum_{i=0}^{I} \frac{\bar{W}_i}{A_i},$$

where subscripts i refer to each of the I prey types. If gastric evacuation is measured as an hourly rate, $\hat{r}$ is an estimate of the hourly feeding rate. Daily meal is calculated by multiplying $\hat{r}$ by 24, and daily ration is daily meal expressed as a percent of wet body mass.

Daily rations for yellowfin tuna calculated previously using the Robson model were 2.8, 4.6, 3.6, and 4.5% of body mass per day for four age-classes.

These estimates were corroborated by ration calculations based on bioenergetics measurements of yellowfin and Cs/K ratios in yellowfin and their prey (Olson & Boggs 1986). Refitting linear, exponential, and square root evacuation functions to the unedited data sets (less empty stomachs) resulted in the equations represented by dotted lines in Figure 2. For all four food types, residual mean squares (untransformed) of the three fits were homogeneous ($P>0.05$, F_{max}-test, Sokal & Rohlf 1969). Therefore, best fit was evaluated by residual analysis. The linear function gave the best fit for the mackerel, smelt, and nehu data sets. Squid evacuation data were best described by the square root function. Recalculated daily rations averaged 11.2% lower than those adjusted for the bias.

Discussion

Biased gastric evacuation rates

We have provided unequivocal evidence that, when the duration of gastric evacuation experiments is long enough for some test individuals to empty their stomachs, fitting the resulting constricted data distribution will result in underestimating evacuation rates. For data with variance comparable to that found for yellowfin tuna (Fig. 2) the bias appears to be about −18%. In the context of changes in the integral of the evacuation function, the parameter used in Robson's consumption rate model, refitting constricted distributions resulted in larger integrals, and illustrated a −11% bias in daily ration of yellowfin tuna. Constricted evacuation data distributions can cause the illusion of a tail at later stages of digestion. This alone can explain a better fit to curvilinear functions (e.g. squid, Fig. 2) than to the linear model. This may partly explain why many investigators have fitted gastric evacuation data to the exponential rather than the square root or linear models.

Karpevitch & Bakoff (1937), Swenson & Smith (1973), and Macdonald et al. (1982) described a leveling off in the rate of gastric evacuation after about 90% of natural food is digested. They attributed this to relatively rapid digestion of muscle and other soft tissue followed by a 'residual' phase (Karpevitch & Bakoff 1937) in which hard parts (e.g. chitin and bone) are softened and evacuated more slowly. There is no evidence of this in yellowfin and skipjack tuna (Magnuson 1969), active fishes with high evacuation rates. The illusion of a tail at advanced digestion stages when constricted data distributions are graphed may in part account for these workers' hypothesis.

Ration and evacuation models

Robson's food consumption model is advantageous over other models in that it is flexible and appropriate for a variety of evacuation functions. The Elliott & Persson (1978) model may be viewed as a special case of this method. The Robson model requires, however, that the residual food in the stomach, when expressed as a proportion of the original meal, be independent of meal size. The exponential model of evacuation is attractive because, in addition to the proportion of the remaining meal being dependent only upon post-prandial time and not upon meal size, the percentage of food remaining within the stomach over any period is constant. Thus, the Elliott & Persson (1978) model can be used to obtain diel consumption rates, but the Robson model is inadequate for that purpose.

For models other than the exponential, some confusion arises in comparing different studies because some authors fit functions to absolute food quantities and others used percentages of the amounts eaten. The linear model assumes that a constant absolute amount is evacuated per unit time. If the evacuation rate ($g\ h^{-1}$) is independent of meal size for any given food type, the time for total evacuation will be dependent upon meal size, i.e. small meals will be eliminated from the stomach sooner than larger meals (Barrington 1957, Steigenberger & Larkin 1974, Jones 1974). An increasing body of evidence shows that linear gastric evacuation rates ($g\,h^{-1}$) often increase in proportion to the *initial* meal size (Hunt 1960, Kitchell & Windell 1968, Windell et al. 1969, Bagge 1977, Olson & Boggs 1986, see review by Windell 1978). As with the exponential model, the time for total evacuation of different sized meals is constant. In that case, and for a binomial model fit to evacuation data for skipjack tuna (Magnuson 1969), a constant proportion of the meal is evacuated per unit time regardless of meal size, and Robson's model is applicable.

A problem which has not been addressed is that of one meal affecting the rate of evacuation of another which has already been consumed but not completely evacuated. Persson (1984) suggested that consecutive meals can indeed affect rates of evacuation, but if one considered only total stomach contents the exponential model continued to be adequate. It is difficult, however, to envisage that the model would work for the total stomach contents but not for the constituents. It is important that experiments similar to those of Persson (1984) are performed on other fishes.

Windell et al. (1969) hypothesized that the gastric motility that is initiated at the onset of evacuation stays relatively constant as long as food remains in the stomach. Gastric distension stimulates secretion of gastric juices (Smit 1967, Norris et al. 1973) and leads to an increase in the amplitude of gastric peristaltic contractions (Jobling et al. 1977). Thus, in lieu of increases in the frequency of contractions (Hunt & Knox 1968), increasing volumes of stomach contents could be pumped per peristaltic cycle as meal size increases (Windell et al. 1969). It is not clear, however, why gastric motility should remain constant throughout the time course of evacuation. In the Appendix, we present a speculative conceptual digestion model which is consistent with linear gastric evacuation rates, and is dependent on the surface area of the food bolus and limited secretion of gastric juices. We have no empirical evidence for such a model, and offer it for those who dismiss the linear gastric evacuation model outright as lacking a biological basis.

Acknowledgements

We wish to thank Izadore Barrett and Richard Shomura of the National Marine Fisheries Service for providing facilities for laboratory work with live tunas at the Kewalo Research Facility, Honolulu, Hawaii. James Joseph of the Inter-American Tropical Tuna Commission made this work possible. We greatly appreciate the assistance and advice of Christofer Boggs, Lo-chai Chen, David Farris, Patrick Tomlinson, and Martin Hall.

References cited

Bagge, O. 1977. Meal size and digestion in cod (*Gadus morhua* L.) and the sea scorpion (*Myoxocephalus scorpius* L.). Meddr. Danm. Fisk.-og Havunders. 7: 437–446.

Brett, J.R. & D.A. Higgs. 1970. Effect of temperature on the rate of gastric digestion in fingerling sockeye salmon, *Oncorhynchus nerka*. J. Fish. Res. Board Can. 27: 1767–1779.

Cochran, P.A. 1979. Comments on some recent methods for estimating food consumption by fish. J. Fish. Res. Board Can. 36: 1018.

Cohen, E., M. Grosslein, M. Sissenwine, F. Serchuk & R. Bowman. 1981. Stomach contents studies in relation to multispecies fisheries analysis and modeling for the northwest Atlantic. Internat. Coun. Explor. Sea. ICES C.M. 1981/G: 66. 18 pp.

Daan, N. 1973. A quantitative analysis of the food intake of North Sea cod, *Gadus morhua*. Neth. J. Sea Res. 6: 479–517.

Eggers, D.M. 1979. Comments on some recent methods for estimating food consumption by fish. J. Fish. Res. Board Can. 36: 1018–1019.

Elliott, J.M. 1972. Rates of gastric evacuation in brown trout, *Salmo trutta* L. Freshwat. Biol. 2: 1–18.

Elliott, J.M. 1979. Comments on some recent methods for estimating food consumption by fish. J. Fish. Res. Board Can. 36: 1020.

Elliott, J.M. & L. Persson. 1978. The estimation of daily rates of food consumption for fish. J. Anim. Ecol. 47: 977–991.

El-Shamy, F.M. 1976. Analyses of gastric emptying in bluegill (*Lepomis macrochirus*). J. Fish. Res. Board Can. 33: 1630–1633.

Filliben, J.J. 1975. The probability plot correlation coefficient test for normality. Technometrics 17: 111–117.

Huebner, J.D & R.W. Langton. 1982. Rate of gastric evacuation for winter flounder, *Pseudopleuronectes americanus*. Can. J. Fish. Aquat. Sci. 39: 356–360.

Hunt, B.P. 1960. Digestion rate and food consumption of Florida gar, warmouth, and largemouth bass. Trans. Amer. Fish. Soc. 89: 206–211.

Hunt, J.N. & M.T. Knox. 1968. Control of gastric emptying. Amer. J. Digest. Diseases. 13: 372.

Jobling, M. 1981. Mathematical models of gastric emptying and the estimation of daily rates of food consumption of fish. J. Fish Biol. 19: 245–257.

Jobling, M. & P.S. Davies. 1979. Gastric evacuation in plaice,

Pleuronectes platessa L.: effects of temperature and meal size. J. Fish Biol. 14: 539–546.

Jobling, M., D. Gwyther & D.J. Grove. 1977. Some effects of temperature, meal size and body weight on gastric evacuation time in the dab, *Limanda limanda* (L.). J. Fish Biol. 10: 291–298.

Jones, R. 1974. The rate of elimination of food from the stomachs of haddock *Melanogrammus aeglefinus*, cod *Gadus morhua* and whiting, *Merlangius merlangus*. J. Cons. Internat. Explor. Mer 35: 225–243.

Kariya, T., H. Hotta & M. Takahashi. 1969. Relation between the condition of the stomach mucous folds and the stomach content in the mackerel. Bull. Jap. Soc. Sci. Fish. 35: 441–445.

Karpevitch, A. & E. Bokoff. 1937. The rate of digestion in marine fishes. Zool. Zh. 16: 28–44. (In Russian)

Kitchell, J.F. & J.T. Windell. 1968. Rate of gastric digestion in pumpkinseed sunfish, *Lepomis gibbosus*. Trans. Amer. Fish. Soc. 97: 489–492.

Livingston, P.A. 1986. Incorporating fish food habits data into fish population assessment models. Dev. in Env. Biol. Fish. 7: 225–234.

Macdonald, J.S., K.G. Waiwood & R.H. Green. 1982. Rates of digestion of different prey in Atlantic cod (*Gadus morhua*), ocean pout (*Macrozoarces americanus*), winter flounder (*Pseudopleuronectes americanus*), and American plaice (*Hippoglossoides platessoides*). Can. J. Fish. Aquat. Sci. 39: 651–659.

Magnuson, J.J. 1969. Digestion and food consumption by skipjack tuna (*Katsuwonus pelamis*). Trans. Amer. Fish. Soc. 98: 379–392.

Mercer, M.C. (ed.) 1982. Multispecies approaches to fisheries management advice. Can. Spec. Publ. Fish. Aquat. Sci. 59. 169 pp.

Norris, J.S., D.O. Norris & J.T. Windell. 1973. Effect of simulated meal size on gastric acid and pepsin secretory rates in bluegill (*Lepomis macrochirus*). J. Fish. Res. Board Can. 30: 201–204.

Olson, R.J. 1983. Gastric evacuation and daily ration in yellowfin tuna, *Thunnus albacares*. M.A. Thesis, San Diego State University, San Diego. 94 pp.

Olson, R.J. & C.H. Boggs. 1986. Apex predation by yellowfin tuna: independent estimates of prey turnover from feeding studies, bioenergetics, and cesium concentrations. Can. J. Fish. Aquat. Sci. (in press).

Persson, L. 1984. Food evacuation and models for multiple meals in fishes. Env. Biol. Fish. 10: 305–309.

Rice, J.A. & P.A. Cochran. 1984. Independent evaluation of a bioenergetics model for largemouth bass. Ecology 65: 732–739.

Seaburg, K.G. & J.B. Moyle. 1964. Feeding habits, digestive rates, and growth of some Minnesota warmwater fishes. Trans. Amer. Fish. Soc. 93: 269–285.

Smit, H. 1967. Influence of temperature on the rate of gastric juice secretion in the brown bullhead, *Ictalurus nebulosus*. Comp. Biochem. Physiol. 21: 125–132.

Sokal, R.R. & F.J. Rohlf. 1969. Biometry: the principles and practice of statistics in biological research. W.H. Freeman, San Francisco. 776 pp.

Steigenberger, L.W. & P.A. Larkin. 1974. Feeding activity and rates of digestion of northern squawfish (*Ptychocheilus oregonensis*). J. Fish. Res. Board Can. 31: 411–420.

Swenson, W.A. & L.L. Smith, Jr. 1973. Gastric digestion, food consumption, feeding periodicity, and food conversion efficiency in walleye (*Stizostedion vitreum vitreum*). J. Fish. Res. Board Can. 30: 1327–1336.

Thorpe, J.E. 1977. Daily ration of adult perch, *Perca fluviatilis* L. during summer in Loch Leven, Scotland. J. Fish Biol. 11: 55–68.

Tyler, A.V. 1970. Rates of gastic emptying in young cod. J. Fish.

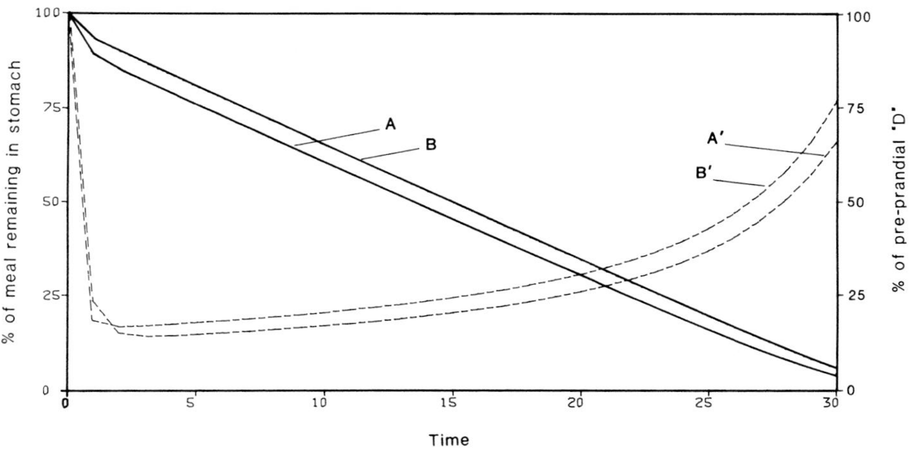

Fig. 5. Output from the simulation described in Appendix. The solid lines (A & B) show the proportion of food remaining in the stomach, and dotted lines (A′ & B′) show the active component of digestive fluid. The parameters for run A were: $S = 0.25$, $D_R = 1.0$, $a = 2.0$, $b = 0.25$, and the meal size was 1.0. For run B the meal size was 2.0 and $S = 0.5$, otherwise parameters were identical.

Res. Board Can. 27: 1177–1189.

Wesolowski, G.O. 1976. Multiple regression and analysis of variance: an introduction for computer users in management and economics. John Wiley & Sons, New York. 292 pp.

Windell, J.T. 1966. Rate of digestion in the bluegill sunfish. Invest. Indiana Lakes and Streams 7: 185–214.

Windell, J.T. 1978. Digestion and daily ration of fishes, pp. 159–183. *In:* S.D. Gerking (ed.) Ecology of Freshwater Fish Production, John Wiley & Sons, New York.

Windell, J.T., D.O. Norris, J.F. Kitchell & J.S. Norris. 1969. Digestive response of rainbow trout, *Salmo gairdneri,* to pellet diets. J. Fish. Res. Board Can. 26: 1801–1812.

Received 15.12.1984 *Accepted 24.9.1985*

Appendix

A criticism of fitting the straight line function to gastric evacuation data is the lack of a sound biological basis for the linear model. Advocates of the surface-area model have a priori grounds for suggesting that digestion is limited by the surface area of the food bolus (Tyler 1970). The surface-area model assumes that digestive fluids are not limiting. We propose a modification of the surface-area model which would result in a very nearly linear gastric evacuation profile.

Suppose that there is a single limiting component D of digestive fluid in the stomach at level D_R when the food enters the stomach. If the level drops below D_R then more fluid is secreted at a rate S until the level D_R is again achieved. The component D of the fluid is exhausted in the digestion process, the rate of which is proportional to D times $W^{2/3}$, the surface area of the food W. Then,

$$dD/dT = -a\ D\ W^{2/3} \quad \text{while } D \geqslant D_R,$$
$$= S - a\ D\ W^{2/3} \text{ otherwise},$$

and

$$dW/dT = -b\ D\ W^{2/3}.$$

With certain model parameters (Fig. 5), initially the exhaustion of D exceeds the replenishment and the level of D drops quickly. Later, as the surface area of the food decreases, the rate of exhaustion of D through the digestion process decreases, and D slowly returns toward its initial level. Figure 5 (line A) shows that simulation of the model produces nearly linear absolute gastric evacuation rates ($g\ h^{-1}$). Hunt (1960), Kitchell & Windell (1968), Windell et al. (1969), Bagge (1977), and Olson & Boggs (1986) found linear gastric evacuation rates ($\%\ h^{-1}$) that were independent of meal size. Smit (1967) and Norris et al. (1973) found S to increase with meal size. Making S proportional to meal size, and doubling that meal size gives line B (Fig. 5), which would be experimentally indistinguishable from line A.

We stress that this model is proposed only to offer a possible explanation for the linear model of gastric evacuation. There may be other scenarios that would also result in linear evacuation rates.

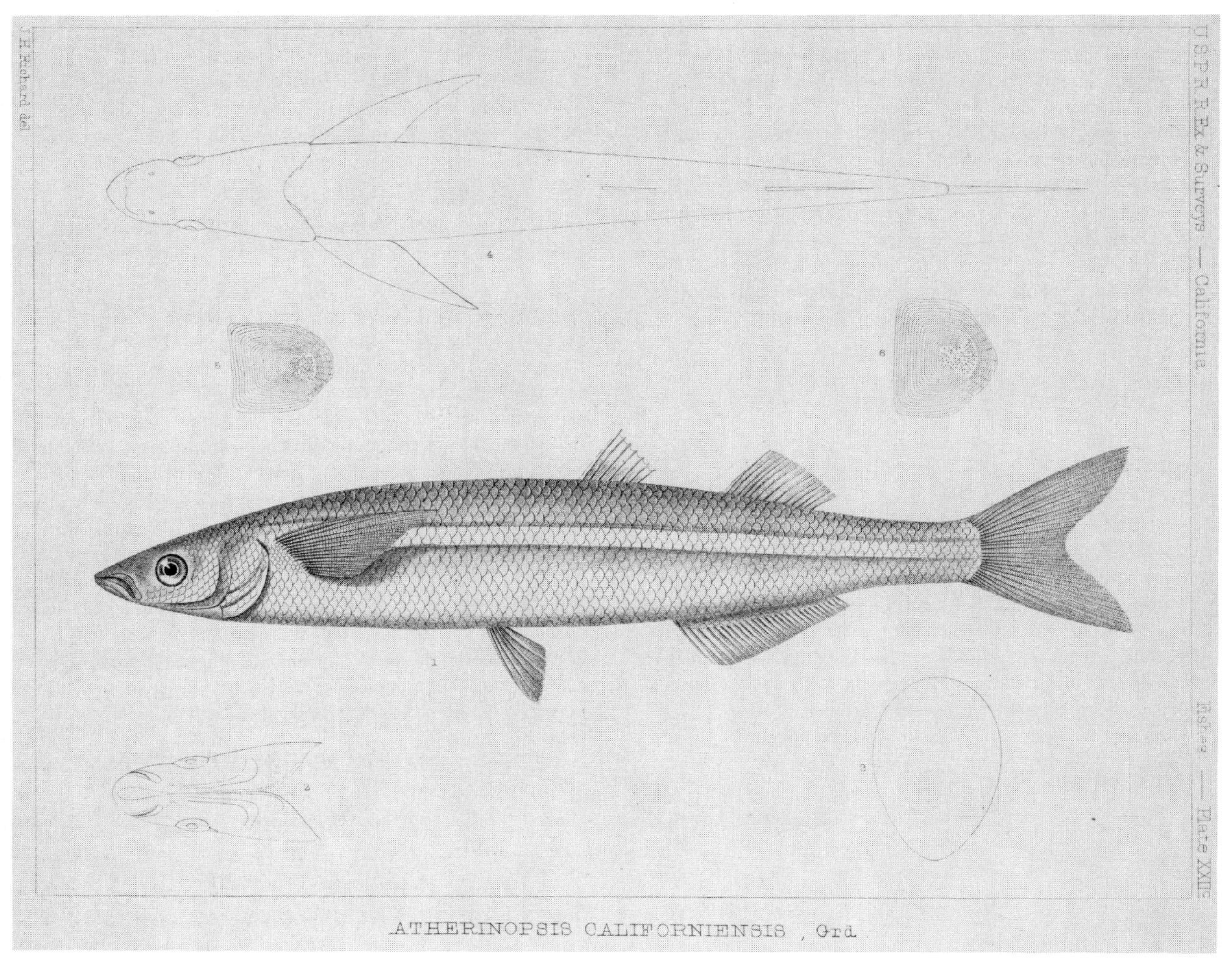

ATHERINOPSIS CALIFORNIENSIS, Grd.

Site-specific differences in the feeding ecology of the California sheephead, *Semicossyphus pulcher* (Labridae)

Robert K. Cowen
Scripps Institution of Oceanography, University of California, San Diego, Mail code A-008, La Jolla, CA 92093, U.S.A.

Keywords: Food availability, Feeding habits, Community structure, Predation, Predator-prey interactions, Switching, Kelp-forests, Urchins

Synopsis

The feeding ecology of four populations of *Semicossyphus pulcher* was examined with respect to such factors as site-specific prey availability, density of the sheephead population, and size of the sheephead. The diets of the sheephead were typically broad, though only a few prey categories dominated. There was considerable between-site differences in the dominant prey. The availability of potential prey (in terms of abundance) also differed between sites both in total abundance and abundance of specific prey types. The diets of the sheephead generally reflected the availability of prey where prey were abundant (i.e. San Nicolas Island) but not where prey were scarce (e.g. Cabo Thurloe). Where prey were scarce and the sheephead population biomass was large (e.g. Cabo Thurloe and Isla Guadalupe), the sheephead apparently switched to alternative, and presumably, lower priority prey (i.e., bryozoan encrusted algae). Some abundant, potential prey were avoided in areas where sheephead were typically small, suggesting an inability of the small fish to handle large prey. The interaction of the kind and abundance of prey, and the size and abundance of the sheephead is discussed in relation to the influence of sheephead predation on their prey populations and community structure.

Introduction

The influence of predators on their prey populations and on community structure has long interested ecologists. Much work has been done on the role of predation in structuring marine communities (see reviews by Menge & Sutherland 1976, Peterson 1979, Paine 1980), yet, little is known about the role of temperate fish predators in their respective communities (Choat 1982). An exception to this is recent work on the California sheephead, *Semicossyphus pulcher,* as a predator on sea urchins (Nelson & Vance 1979, Tegner & Dayton 1977, 1981, Cowen 1983). In all of these studies, sheephead were considered to be important predators on urchins, capable of regulating the urchins' behavior, population structure and density. But what is the effect of such predation on community structure and how could that effect be modified?

Between-site variability in the kinds and abundance of prey, as well as density of the predator's population and that of their competitors may all influence the diet and, ultimately, the role of the predator in one community relative to another. Previous studies on the feeding behavior of the sheephead suggest considerable variation in diet between different sheephead populations (Quast 1968, McCleneghan 1968, Winget 1968). Considering the potential influence sheephead may have on their prey populations and communities as reported above, and the observed variability of their diets, they are an appropriate predator to examine the question of how and to what extent a predator's influence may vary. Therefore, the present study (1) examines shifts in the diets of sheephead at four different sites, (2) relates these dietary differences to changes in available prey, and (3) discusses some of the possible implications on the biology of the sheephead and the sheephead's role as predators within these different communities.

Methods

The four widely separated study sites were: Dutch Harbor, San Nicolas Island (119°28′ W, 33°12′ N); northeast portion of Isla Guadalupe, Baja California (118°17′ W, 29°9′ N); Isla San Benitos, B.C. (115°34′ W, 28°18′ N); and Cabo Thurloe, B.C. (114°51′ W, 27°37′ N). All sites had a rocky substrate of moderate to high relief, (San Nicolas Island the highest and Cabo Thurloe the lowest). At all sites macroalgae grew on the rocky substrate. At Isla Guadalupe, the algal cover was dominated by browns (e.g. *Eisenia arborea* and *Dictyota* spp.) and greens (e.g. *Codium* spp.), whereas at the other three sites there was a significant coverage of fleshy and coralline reds. Algal coverage was most extensive at San Nicolas Island and least at Cabo Thurloe. All sites except Isla Guadalupe also had a *Macrocystis pyrifera* canopy. Several fish species which may have some diet overlap with sheephead (Quast 1968) were present at all sites (exceptions below) but were not as common as the sheephead; in particular: *Halichoeres semicinctus* (Labridae), *Damalichthys vacca, Rhacochilus toxotes* (rare at all sites), and *Embiotoca jacksoni* (Embiotocidae), and *Girella nigricans* (Kyphosidae). The most notable exception was the presence of a variety and abundance of rockfish, *Sebastes* spp. (Scorpaenidae), at San Nicolas Island and their virtual absence at all three of the other sites. Also, *H. semicinctus* was absent at San Nicolas Island. Lobsters, *Panulirus interruptus,* were not present at the San Nicolas Island site, but were common at the other three sites. None of the sites are close to major human population centers, and therefore are relatively free from extensive sportfishing for sheephead (though all areas are commercially fished for abalone, urchins, and lobsters).

At all sites sheephead were collected for diet analysis by spearing. Collections used in this study were made in May, 1982 for Islas San Benitos and Isla Guadalupe (82 and 92 fish, respectively), September, 1981 and May, 1982 for Cabo Thurloe (56 and 52 fish, respectively), and September, 1980, December-February, 1981, and May–July, 1981 for San Nicolas Island (128, 47, and 50 fish, respectively). An attempt was made to minimize bias towards a particular size class by collecting the first fish seen after each previous fish was collected. The tubular digestive tracts were removed and preserved in 4% formaldehyde and then transferred to 70% isopropanol. Fullness was estimated on a scale of 0 (empty) to 4 (100% full); only samples which were greater than 25% full were used in the analysis. The prey contents were removed and sorted into 27 different prey categories. Each prey type was enumerated and the proportion each contributed to the total volume of the gut contents was estimated by eye. The Index of Relative Importance (IRI) was then calculated to estimate the contribution of each prey type to the diet (Pinkas et al. 1971, Hyslop 1980). This index is calculated as:

$$IRI = (\%N + \%V) \times FO,$$

where %N is the numerical percentage a prey item contributes to the total diet, %V is volumetric percentage and FO is the frequency of occurence of the prey category. The adequacy of the sample sizes for diet analysis was determined by plotting cummulative prey species curves against randomly selected fish samples. To measure diet similarity between sites and between dates at San Nicolas Island and Cabo Thurloe, the Percent Similarity Index (PSI) was used. The PSI is calculated as:

$$PSI = \sum_{i=1}^{S} \min [(P_{i,1}), (P_{i,2})],$$

where S is the total number of species i in the two samples, and $P_{i,1}$ and $P_{i,2}$ are the proportions of species i in the two samples. To compare the evenness of prey utilization at each site, dietary breadth (B) was calculated as:

$$B = \frac{1}{\sum_{i=1}^{s} P_i^2},$$

where P_i is the proportion of item i in the diet. This value was then scaled as B/S such that the values will range between zero, the most uneven distribution, and unity, the most even.

Benthic samples were taken at each site to measure prey availability in terms of abundance. These samples were taken in May, 1982 for all sites. Ten $0.25\,m^2$ quadrats were haphazardly placed by a diver swimming 5 m off the bottom and blindly dropping the quadrat. Material within the quadrat was scraped and collected by an airlift suction device (Chess 1969). Because scraping was not to bare rock, the abundance of some organisms may have been underestimated (e.g. the calcareous tubed polychaete, *Spirobranchus spinosa,* was not adequately collected, though it was rather abundant at both Islas San Benitos and Isla Guadalupe). Also, due their cryptic behavior, sea urchins were not adequately sampled with this technique. Samples were preserved in 4% formaldehyde or frozen. The samples were later sorted into the same prey categories as above and volumes were determined by displacement.

Comparisons between available prey and observed diets were made to determine prey preference. For this, a forage ratio is calculated as P_{if}/P_{ib}, where P_i is the proportionate volume of prey i in the sheephead's diet (f) and the benthic sample (b) (Laur & Ebeling 1983). Laur & Ebeling considered a prey item to be selectively taken if its forage ratio was substantially greater than 1, and avoided if its forage ratio was less than 1.

Sheephead densities were estimated at each site from 5 × 50 m transects (Cowen 1983). The transects were counted at the same date that the initial collections were made, hence, density estimates are based on one time counts. In addition, total population biomass was calculated from sheephead densities and the size structure of the sample collected for diet analysis. Urchins were counted at each site in 1 × 5 m transects placed along crevices (Cowen 1983). Although the latter does not provide an estimate of density, it does provide a measure of the relative urchin abundance at each site.

Sheephead density, diet analysis and benthic sample data were combined to estimate the site-specific consumption rate of brachyuran crabs (see Paine 1969b). Since brachyuran crabs ranked either first or second in importance as prey at all sites, this estimate provides a measure of the predation pressure imposed by the different sheephead populations. The estimate was calculated in five steps: (1) From the diet analysis, the frequency of occurrence of crabs in the diet and the mean number of crabs per fish (with crabs in their diet) was calculated. (2) This value was combined with the density of the sheephead population to obtain an estimate of the number of crabs eaten per unit area per day ($hectare^{-1}\,day^{-1}$). (3) The crab standing crop was estimated from the benthic samples. (4) Dividing the number of brachyuran crabs eaten in one day by the standing crop yields an estimate of the percentage of the standing crop that was consumed per day by sheephead. (5) Finally, this last estimate was divided into 100 (percent) to calculate the total time required to consume the entire standing crop of crabs by sheephead foraging, assuming no other sources of mortality. The first step of this calculation assumes one main meal per day (Cowen 1983).

Results

Sample sizes were considered to be more than adequate according to the cummulative frequency analysis. The twenty most dominant prey (i.e. 90% of all prey categories) were accounted for within the first 30 samples at Islas San Benitos, Isla Guadalupe, and Cabo Thurloe, and 40 samples at San Nicolas Island. Subsequent samples added only a few rare species, most of which were probably taken incidentally with other prey (e.g. hydroids, anemones, sponges).

At both San Nicolas Island and Cabo Thurloe, sheephead diets were similar between different sampling periods (Table 1). The mean PSI value (95% CI) for the three possible season comparisons at San Nicolas Island was 75.4 (6.33), and at Cabo Thurloe the single comparison yielded a PSI value of 70.6. Based on this high similarity, within-

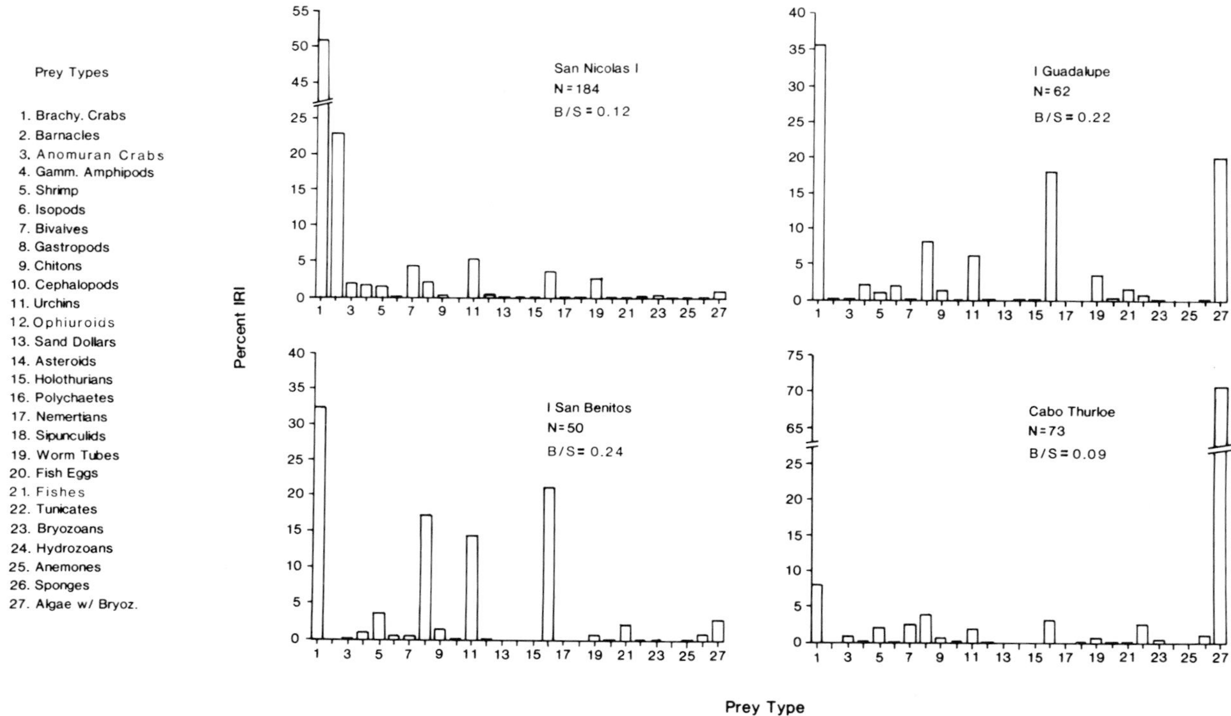

Fig. 1. Prey items of sheephead at the four study sites. Data are presented as percentage of total Index of Relative Importance (IRI) of 27 different prey categories.

Table 1. Within-site (i.e. seasonal) and between-site comparison of sheephead diets using the percent similarity index (PSI). N is the number of fish in each diet analysis sample in each pairwise comparison.

Comparison	N	N	PSI
Within-site			
CT			
9/81 vs. 5/82	37	36	0.71
SNI			
9/80 vs. 12/80–2/81	101	38	0.72
9/80 vs. 5–7/81	101	45	0.77
12/80–2/81 vs. 5–7/81	38	45	0.77
Between-site			
SNI vs. IG	184	62	0.54
SNI vs. ISB	184	50	0.49
SNI vs. CT	184	73	0.18
IG vs. CT	62	73	0.18
ISB vs. CT	50	73	0.22
IG vs. ISB	62	50	0.74

site samples of sheephead collected at different sampling periods at San Nicolas Island and Cabo Thurloe were clumped for the between-site comparisons.

Although sheephead ate a wide variety of prey items at all sites (S = 20 prey categories), the diets were dominated by one or a few prey items (B/S from 0.09–0.24, Fig. 1). Also, the dominant prey (based on percent of total IRI) differed considerably between sites. At San Nicolas Island, the dominant prey was small brachyuran crabs and barnacles (primarily *Balanus tintinnabulum,* 51% and 23% of total IRI, respectively). Urchins and bivalves ranked third and fourth, but were considerably less important than crabs and barnacles. Barnacles were either absent or insignificant in sheephead diets at the other three sites.

At Isla Guadalupe and Islas San Benitos, the diets were relatively more evenly distributed among prey categories (Fig. 1). Sheephead diets at Isla Guadalupe were dominated by brachyuran crabs, bryzoan (primarily *Membranipora membranacea*) encrusted algae, and polychaetes, 36, 20, and 18% of total IRI, respectively. Shelled gastropods and urchins were the fourth and fifth ranked items, respectively. At Islas San Benitos, four prey categories ranked high in importance. In decreasing order, these were: brachyuran crabs, polychaetes, shelled gastropods and urchins. Bivalves, which were relatively important at San Nicolas Island, were unimportant at both Isla Guadalupe and Islas San Benitos, where shelled gastropods were the dominant molluscs.

The dominant prey items in the sheephead's diet at Cabo Thurloe differed greatly from that at the other sites. Here, only one prey category dominated, namely bryzoan encrusted algae (almost exclusively *M. membranaceae* on *Macrocystis pyrifera*), which accounted for 71% of total IRI. Brachyuran crabs, which were the main prey items at all three other sites, only accounted for 8% of total IRI at Cabo Thurloe, and all other prey items were of minor importance.

Overall, the diet of the sheephead at Cabo Thurloe was the most divergent, followed in order by San Nicolas Island, Islas San Benitos and Isla Guadalupe (Table 1). The low similarity of the Cabo Thurloe sheephead's diet to the other sites is primarily due to the extensive utilization of bryozoan encrusted algae. The only other site where sheephead utilize this resource to any extent (i.e. Isla Guadalupe, Fig. 1) yielded the diet most similar to that of the Cabo Thurloe sheephead, though the similarity was still low. The sheephead diets at Cabo Thurloe and San Nicolas Island were the least similar (PSI = 0.19). To the other extreme, Isla Guadalupe and Islas San Benitos sheephead had the most similar diets (PSI = 0.74).

Mean total prey volumes in the benthic samples were significantly higher at San Nicolas Island compared to each of the other sites (t-test, $p<0.01$), while prey abundance at Isla Guadalupe, Islas San Benitos and Cabo Thurloe did not differ ($p>0.20$; Fig. 2). The sites differed in dominant potential prey items (Table 2). Brachyuran crabs dominated at both San Nicolas Island and Isla Guadalupe but were relatively rare at Islas San Benitos and Cabo Thurloe. At Islas San Benitos, gastropods were the

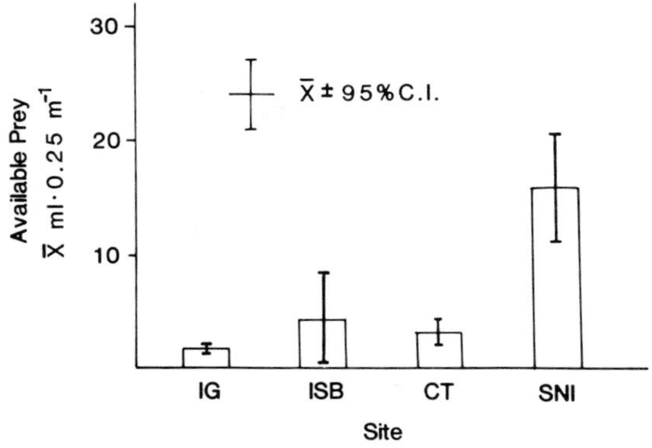

Fig. 2. Mean total volumes (ml) of available prey collected in $0.25\ m^2$ quadrats. N = 10 for each site, IG = Isla Guadalupe, ISB = Islas San Benitos, CT = Cabo Thurloe, SNI = San Nicolas Island.

most abundant potential prey, with polychaetes and small fish ranking second and third, respectively. At Cabo Thurloe, the dominant potential prey were anomuran crabs (i.e. hermit crabs in 1–2 mm gastropod shells). Though barnacles were only collected at San Nicolas Island, I also observed a few isolated barnacles on *Eisenia* at Isla Guadalupe. Lack of occurrence of a given prey item in the benthic samples probably indicates their extreme rarity rather than total absence at the site.

Comparison of the observed diets with site-specific availability of potential prey showed relatively low similarity at Cabo Thurloe (PSI = 0.37) and Islas San Benitos (PSI = 0.48) and somewhat higher similarity at both San Nicolas Island and Isla Guadalupe (PSI = 0.63 and 0.65, respectively). It is important to remember that this comparison does not include sea urchins or bryozoan encrusted algae, neither of which were adequately sampled with the benthic sampler (see below).

The forage ratios indicate that some degree of prey preference and avoidance occured at each site, though there were distinct site differences (Table 3). At all sites, chitons had the highest forage ratios, though in all cases chitons were rare as both potential and actual prey. Field observations suggest that most chitons are under rocks or urchins, or deep within crevices. It is possible that chitons become available to sheephead only when they are exposed by the removal of an urchin

Table 2. Mean displacement volumes (± 95% confidence limits) and percentage of total sample of potential prey items collected in 0.25 m² quadrats. N = 10 for each site. Prey type numbers correspond to those listed in Figure 1.

Prey type	Displacement volumes (ml)							
	SNI		IG		ISB		CT	
	V	%	V	%	V	%	V	%
1. Br. crab	5.9 (2.26)	37.0	0.6 (0.29)	26.0	0.3 (0.21)	7.0	0.1 (0.06)	3.0
2. Barnacle	1.2 (0.94)	8.0	0	0	0	0	0	0
3. An. crab	0.2 (0.28)	1.0	0	0	0	0	2.0 (1.02)	62.0
4. Amphipods	1.1 (0.50)	7.0	0.4 (0.41)	17.0	0.3 (0.28)	7.0	0.1 (0.06)	3.0
5. Shrimp	1.6 (0.86)	10.0	0.5 (0.34)	22.0	0.2 (0.25)	5.0	0.2 (0.13)	6.0
6. Isopods	0.1 (0.13)	1.0	0.1 (0.22)	4.0	0.1 (0.12)	2.0	0.2 (0.21)	6.0
7. Bivalves	1.0 (0.72)	6.0	0	0	0.02 (0.02)	0.5	0.1 (0.11)	3.0
8. Gastropods	1.7 (0.73)	11.0	0.1 (0.22)	4.0	2.3 (3.86)	55.0	0.4 (0.39)	12.0
9. Chitons	0.02 (0.03)	0.01	0	0	0.01 (0.01)	0.2	0.01 (0.02)	0.3
10. Cephalopod	0	0	0.01 (0.01)	0.4	0	0	0	0
12. Ophiuroids	1.6 (0.74)	10.0	0.01 (0.003)	0.4	0.01 (0.01)	0.2	0.01 (0.01)	0.3
14. Asteroids	0.1 (0.03)	1.0	0	0	0.1 (0.23)	2.0	0	0
16. Polychaetes	0.9 (0.71)	6.0	0.4 (0.05)	17.0	0.5 (0.14)	12.0	0.1 (0.04)	3.0
21. Fishes	0.3 (0.29)	2.0	0.2 (0.13)	9.0	0.4 (0.33)	10.0	0	0

(which also results from predation by the sheephead). At San Nicolas Island, other preferred prey included barnacles and anomuran crabs (both hermit crabs and mole crabs, the latter occur in the sand and were not sampled with the benthic sampler, therefore the forage ratio would likely be lower). At Isla Guadalupe, Islas San Benitos and Cabo Thurloe, there was selection for both brachyuran crabs and ophiuroids, but at San Nicolas Island crabs were only slightly preferred and ophiuroids were avoided (though the latter were relatively abundant, see Table 2). At both Islas San Benitos and Cabo Thurloe, sheephead also selectively foraged on bivalves. Of interest, anomuran crabs (i.e. hermit crabs) were distinctly avoided at Cabo Thurloe, though these crabs dominated the potential prey in the benthic samples (Table 2).

There was a substantial difference between sites in the relative abundance of urchins (Table 4). San Nicolas Island had the most urchins within crevices, while Cabo Thurloe had the least. At San Nicolas Island, the urchins were *Strongylocentrotus franciscanus* and *S. purpuratus,* but at the other three sites, the urchins were primarily *Centrostephanus coronatus* and/or *S. franciscanus*. Availability of these urchins to the sheephead is probably proportional to their abundance since, in areas where urchins are abundant (i.e. San Nicolas Island), the urchins are crowded within crevices and closer to the openings. In areas where the urchins

Table 3. Forage ratios (proportionate volume prey in fish per proportionate volume prey in benthic sample) for each site. Values marked '0/0' represent no prey consumed/none collected; values marked 'X/0' (where X is a letter a through c) represent a small quantity consumed/none collected. Actual values of 'X' are given in the footnote.

Prey	Forage ratio			
	SNI	IG	ISB	CT
1. Br. crab	1.42	1.93	5.50	6.23
2. Barnacles	2.46	a/0	0/0	a/0
3. An crabs	9.42	a/0	a/0	0.09
4. Amphipods	0.09	0.08	0.19	0.93
5. Shrimp	0.24	0.19	1.44	1.67
6. Isopods	0.71	1.45	0.84	0.26
7. Bivalve	0.60	b/0	4.73	9.02
8. Gastropods	0.15	0.97	0.20	1.63
9. Chitons	12.80	24.00	27.80	10.23
10. Cephalopods	0/0	0.75	b/0	a/0
12. Ophiuroids	0.24	6.00	5.56	6.14
14. Asteroids	0.21	c/0	0/0	0/0
16. Polychaetes	0.69	1.17	1.39	10.17
21. Fishes	0.53	1.13	0.99	b/0

IG – a = 0.0007; b = 0.002; c = 0.003
ISB – a = 0.007; b = 0.015
CT – a = 0.0003; b = 0.0015

are rare (i.e. Cabo Thurloe), they are usually found deep within the crevice and hence, unavailable to the sheephead (Cowen 1983).

Density and size of sheephead also differed considerably between sites (Table 5). San Nicolas Island and Islas San Benitos had the least dense sheephead populations, while Cabo Thurloe had the densest population. In contrast, sheephead at San Nicolas Island were much larger than those at

Table 4. Relative sea urchin abundances based on number of urchins in 5 × 1 m transects along crevices.

Site	N	$\bar{X}$	95% CI
SNI	10	72.6	14.2
IG	10	15.5	3.1
ISB	10	18.9	3.5
CT	14	5.4	1.6

the other sites. The sheephead at San Nicolas Island were almost 10 times as large (mean weight) as those at Isla Guadalupe. Total biomass differed by up to a factor of four (Table 5). Even though San Nicolas Island had a relatively low density of sheephead, their large size resulted in the largest biomass.

Variation in the biomass of the different sheephead populations may be a potentially important influence on site-specific prey availability. To test this idea, an estimate of the brachyuran crab standing crop consumed per day by sheephead was calculated (Table 6). The calculation suggests that there is an order of magnitude difference in consumption rates of crabs between San Nicolas Island and Cabo Thurloe. Even though fewer crabs are eaten per fish at Cabo Thurloe than San Nicolas Island (2.4 and 6.5, respectively), the larger number of fish and low crab standing crop results in a higher proportion of the crab population being consumed by sheephead every day. This is further exemplified by the great difference in time that would be required for the different sheephead populations to consume the entire crab population.

Table 5. Density (95% CI), mean weight (95% CI), and biomass of sheephead at each site.

Site	Number transects	Density (no. hectare^{-1})	Mean weight (g)	Total biomass (kg hectare^{-1})
SNI	30	204 (26.8)	1751.3 (187.2)	324
IG	10	355 (120.0)	206.5 (25.1)	73.3
ISB	10	210 (74.0)	519.3 (87.2)	109
CT	12	510 (96.0)	455.9 (82.2)	232.5

Table 6. Estimate of the percent of the brachyuran crab standing stock eaten by sheephead at each site.

	Site			
	SNI	IG	ISB	CT
Sheephead density (1)	204	355	210	510
Number of crabs eaten hectare^{-1} day^{-1} (2)	1163	852	336	561
Crab standing crop: number hectare (95% CI) (3)	2 015 840 (766 019)	192 640 (98 381)	103 200 (72 240)	34 400 (20 640)
Percent of crab standing crop consumed by sheephead day^{-1}	0.06	0.42	0.32	1.63
Days required to consume standing crop (4)	1667	238	307	61

(1) Data from Table 5. (2) Based on the mean number (95% CI) of crabs eaten per fish (with crabs in their diet) and the percent of fish with carbs (SNI – 6.5 (0.33) and 88%; IG – 3.2 (0.30) and 75%; ISB – 2.2 (0.28) and 76%; CT – 2.4 (0.23) and 46%). (3) Based on 0.25 m^2 benthic samples, Table 2. (4) Assuming consumption only by sheephead.

This estimate only considers mortality on the crabs by sheephead and not any other predators, therefore, the actual time required to consume the entire crab population if no new crabs recruited would be less.

Discussion

The results at all the sites suggest that sheephead are generalist predators that are flexible in their site specific choice of prey. Their diets reflect the availability of prey at each site within the bounds of some prey preference and negative selectivity. There are several possible explanations for the apparent preference or avoidance of certain prey observed in this study. First, some of the prey may be harder to find and thus less available than their relative abundance may suggest. For example, mobile gastropods may hide within the algal turf by day when sheephead are active, whereas filter-feeding barnacles must remain relatively exposed in order to have adequate access to water currents and, hence, food. Another possibility for prey preference is that capture of some prey may require a disproportionately large expenditure of energy (Estabrook & Dunham 1976, Pyke et al. 1977). At Cabo Thurloe, where the fish are relatively small, the very abundant hermit crabs may be too large and hard to handle. In contrast, even though brachyuran crabs are typically cryptic, they are easy to handle and the energetic return is probably high.

Since sheephead are flexible predators, but with some possible prey preferences, their diets will vary as prey abundances and species change. Prey switching, especially in areas with an overall low prey abundance, may result in radical changes in diet prey (e.g. crabs to bryozoan encrusted algae). Switching to alternate prey when a preferred prey item is sparse permits the predator to maintain a relatively stable population size. In this study there is no evidence for a numerical response (sensu Holling 1959) since the most dense population is where there is the least food (i.e. Cabo Thurloe). In fact, there is evidence that the density of these populations is independent of food resources, and is related primarily to recruitment (Cowen 1985). Instead, food availability (and possibly food quality) may be influencing the growth rates of the fish. Cowen (unpublished data) has found the sheephead to grow slower, and thereby remain smaller, at the sites with low prey availability (i.e., Isla Guadalupe, Cabo Thurloe and Islas San Benitos).

The influence of predators on their prey populations and on community structure has interested ecologists for many years. In one situation, a predator may impose an indirect control on community structure by selectively foraging on a prey species that is itself important to the structure of the community (Paine 1966, 1969a, Glasser 1979).

As an example, by foraging on a competitive dominant, the predator may release space for other organisms to utilize, thereby increasing community diversity (Connell 1961, Dayton 1971). In another situation, a predator may influence its community directly. In the extreme case, the predator population forages on a variety of prey to the extent of reducing the standing crop of all species, including, but not limited to the competitive dominants (e.g. Lubchenco 1978). Both mechanisms can ultimately influence community diversity.

To what extent does the impact of predation by sheephead differ from one site to the next? At San Nicolas Island, prey availability is very high. Even with the large sheephead biomass, there appears to be no shortage of their main prey, brachyuran crabs, as suggested by the very low consumption rate. It must be remembered that besides sheephead, San Nicolas Island also has a large population of rockfish, *Sebastes* spp., which also forage on the crabs. Therefore, the observed crab standing-crop at San Nicolas Island (and presumably that of other benthic prey) is supporting a substantially larger biomass of fish than the prey populations at the other sites. In this situation, we would not expect any direct effects on the community due to overexploitation by sheephead. However, in a previous study at San Nicolas Island, sheephead were found to have an important influence on the abundance and microhabitat distribution of urchins (Cowen 1983). A similar role for sheephead has been suggested in other studies at other sites: San Diego, California (Tegner & Dayton 1981) and Santa Catalina Island, California (Nelson & Vance 1979). In addition, Cowen (1983) found that in areas where sheephead densities were low, urchins were not restricted to crevices and they could be found foraging out in the open. In the extreme case, the urchins had overgrazed the local community and only a 'barren' area existed. Ebeling et al. (1985) found similar results at their site near Santa Barbara, California where sheephead densities where low and, following removal of drift algae by storms, the urchins foraged in the open, apparently free of predation by sheephead. Therefore, it appears that when sheephead are present in sufficient numbers, they may impose an important, indirect influence on the community by maintaining an important prey species (i.e. urchins) at a low and constant density. When sheephead are only present in low densities, then their role is minimal and, instead, populations of such prey as sea urchins are primarily controlled by other, less predictable events such as storms (Ebeling et al. 1985).

At the other three sites, prey abundance was low, suggesting possible overexploitation by the sheephead. At Cabo Thurloe, where prey was least abundant, the typically dominant prey, brachyuran crabs, was relatively uncommon and, consequently, little utilized. Regardless, the very high estimated consumption suggests that the crab population was being consumed at, or above its maximum sustainable rate as defined by Hines (1982), even though the sheephead there were mostly eating algae encrusted with bryozoans. In addition, other normally abundant prey (e.g. urchins) are also relatively rare. This evidence suggests that the sheephead at Cabo Thurloe may be imposing a strong, direct impact on their community by maintaining it in an overexploited state.

At both Isla Guadalupe and Islas San Benitos, the impact of sheephead predation appears to be intermediate between that at San Nicolas Island and Cabo Thurloe. Though the abundance of urchins is less at these two sites than at San Nicolas Island, the importance of urchins in the diet at both Isla Guadalupe and Islas San Benitos is equal to or greater than at San Nicolas Island. This may be due to different behavioral patterns of the dominant urchins at these sites. At San Nicolas Island, *S. franciscanus* and *S. purpuratus* do not forage outside crevices when algal drift is present. In contrast, *C. coronatus* regularly forage at night up to 1 m from their daytime resting holes (Nelson & Vance 1979). Failure of *C. coronatus* to return to a protected habitat prior to the beginning of the sheepheads active period may increase their susceptability to predation by sheephead relative to *Strongylocentrotus* spp. Also, though prey availability is low, the crab populations are not being consumed at an excessive rate.

Predation which leads to overexploitation would not necessarily result in local extinction of prey species within the community. When prey densities are sufficiently low, the predators may switch to less desirable, but more abundant prey (Murdoch 1969, Schoener 1971). This appears to be the case at both Isla Guadalupe and Cabo Thurloe where the sheephead have started to utilize bryozoan encrusted algae, a presumably low priority prey. In addition, the rocky reef is a highly heterogeneous surface which provides a variety of refugia for prey, thereby minimizing the chances of extinction (Paine 1969b, Mittelbach 1981).

In conclusion, sheephead exhibit considerable between-site variation in their diet and, ultimately, their influence as predators. Their influence appears to be greatly modified by the density of the sheephead population. Where sheephead are scarce, they have little apparent influence on the abundance of their prey populations. At intermediate densities, the sheephead may play an indirect role in the structuring of their local community by regulating urchin populations. In areas with very dense populations of sheephead, these fish may actually overexploit much of their benthic prey. There is insufficient data within this study to make conclusions about the community diversity within these different situations. But theory would predict (Connell 1978), and observations would suggest (Lubchenco 1978, Sousa 1979, Hixon & Brostoff 1983) that diversity is greatest in the community where predators are present in intermediate densities. Of course, a variety of factors may mediate the role of sheephead as predators, including the abundance of other predators, abundance and production of prey populations, and even storms. But the influence of these factors will be greatest where the sheephead are the least abundant.

Acknowledgements

This paper benefitted from careful reading and/or critical discussion with G. Cailliet, P. Dayton, J. Estes, A. Genin, N. Holland, R. Rosenblatt, C. Simenstad, and an anonymous reviewer. The Commander of the Pacific Missile Test Center, U.S. Navy, provided permission to work on San Nicolas Island and R. Dow facilitated operations on the island. I would like to thank J. Estes of the U.S. Fish and Wildlife Service for support to work on San Nicolas Island and in Mexico.

References cited

Chess, J.R. 1969. an airlift sampling device for benthic organisms. Research Memo 69-151-Ocean-M1, Westinghouse Res. Lab., Pittsburg.

Choat, J.H. 1982. Fish feeding and the structure of benthic communities in temperate waters. Ann. Rev. Ecol. Syst. 13: 423–449.

Connell, J.H. 1961. The effects of competition, predation by *Thais lapillus*, and other factors on natural populations of the barnacle *Balanus balanoides*. Ecol. Monogr. 31: 61–104.

Connell, J.H. 1978. Diversity in tropical rain forests and coral reefs. Science 199: 1302–1310.

Cowen, R.K. 1983. The effect of sheephead (*Semicossyphus pulcher*) predation on red sea urchin (*Strongylocentrotus franciscanus*) populations: an experimental analysis. Oecologia 58: 249–255.

Cowen, R.K. 1985. Large scale pattern of recruitment by the labrid, *Semicossyphus pulcher*: causes and implications. J. Mar. Res. (in press).

Dayton, P.K. 1971. Competition, disturbance, and community organization: the provision and subsequent utilization of space in a rocky intertidal community. Ecol. Monogr. 41: 341–389.

Ebeling, A.W., D.R. Laur & R.J. Rowley. 1985. Severe storm disturbances and reversal of community structure in a southern California kelp forest. Mar. Biol. 84: 287–294.

Estabrook, G.F. & A.E. Dunham. 1976. Optimal diet as a function of absolute abundance, relative abundance, and relative value of available prey. Amer. Nat. 110: 401–413.

Glasser, J.W. 1979. The role of predation in shaping and maintaining the structure of communities. Amer. Nat. 113: 631–641.

Hines, A.H. 1982. Coexistence in a kelp forest: size, population dynamics, and resource partitioning in a guild of spider crabs (Brachyura, Majidae). Ecol. Monogr. 52: 179–198.

Hixon, M.A. & W.N. Brostoff. 1983. Damselfish as keystone species in reverse – intermediate disturbance and diversity of reef algae. Science 220: 511–513.

Holling, C.S. 1959. The components of predation as revealed by a study of small-mammal predation of the European pine sawfly. Can. Entomol. 91: 293–320.

Hyslop, E.J. 1980. Stomach content analysis – a review of methods and their application. J. Fish Biol. 17: 411–429.

Laur, D.R. & A.W. Ebeling. 1983. Predator-prey relationships in surfperches. Env. Biol. Fish. 8: 217–229.

Lubchenco, J. 1978. Plant species diversity in a marine intertidal community: importance of herbivore food preference and algal competitive abilities. Amer. Nat. 112: 23–39.

McCleneghan, K. 1968. An analysis of stomach contents of three species of wrasses (family Labridae) from the waters off southern California. MSc. Thesis, University of Southern California, Los Angeles. 00 pp.

Menge, B.A. & J.P. Sutherland. 1976. Species diversity gradients: synthesis of the roles of predation, competition, and temporal heterogeneity. Amer. Nat. 110: 351–369.

Mittelbach, G.G. 1981. Foraging efficiency and body size: a study of optimal diet and habitat use by sungills. Ecology 62: 1370–1386.

Murdoch, W.W. 1969. Switching in general predators: experiments of predator specificity and stability of prey populations. Ecol. Monogr. 39: 335–354.

Nelson, B.V. & R.R. Vance. 1979. Diel foraging patterns of the sea urchin *Centrostephanus coronatus* as a predator avoidance strategy. Mar. Biol. 51: 251–258.

Paine, R.T. 1966. Food web complexity and species diversity. Amer. Nat. 100: 65–75.

Paine, R.T. 1969a. A note on trophic complexity and community stability. Amer. Nat. 103: 91–93.

Paine, R.T. 1969b. The *Pisaster-Tegula* interaction: prey patches, predator food preference, and intertidal community structure. Ecology 50: 950–961.

Paine, R.T. 1980. Food webs: linkage, interaction strength and community infrastructure. J. Anim. Ecol. 49: 667–685.

Peterson, C.H. 1979. Predation, competitive exclusion and diversity in the soft sediment benthic communities of estuaries lagoons. pp. 233–264. *In:* R.J. Livingston (ed.) Ecological Processes in Coastal and Marine Systems, Plenum, New York.

Pinkas, L., M.S. Oliphant & I.L.K. Iverson. 1971. Food habits of the albacore, bluefin tuna, and bonita in California waters. Calif. Dept. Fish and Game, Fish Bull. 152: 1–105.

Pyke, G.H., H.R. Pulliam & E.L. Charnov. 1977. Optimal foraging: a selective review of theory and tests. Q. Rev. Biol. 52: 137–154.

Quast, J.C. 1968. Observations on the food of the kelp-bed fishes. pp. 109–142. *In:* W.J. North & C.L. Hubbs (ed.) Utilization of Kelp-Bed Resources in Southern California, Calif. Dept. Fish Game, Fish Bull. 139, Sacramento.

Schoener, T.W. 1971. Theory of feeding strategies. Annu. Rev. Ecol. Syst. 2: 369–404.

Sousa, W.P. 1979. Experimental investigations of disturbance and ecological succession in a rocky intertidal algal community. Ecol. Monogr. 49: 227–254.

Tegner, M.J. & P.K. Dayton. 1977. Sea urchin recruitment patterns and implications of commercial fishing. Science 196: 324–326.

Tegner, M.J. & P.K. Dayton. 1981. Population structure, recruitment and mortality of two sea urchins (*Strongylocentrotus franciscanus* and *S. purpuratus*) in a kelp forest. Mar. Ecol. Prog. Ser. 5: 255–268.

Winget, R.R. 1968. Trophic relationships and metabolic energy budget of the California spiny lobster, *Panulirus interruptus* (Randall). MSc. Thesis, San Diego State University, San Diego. 232 pp.

Received 28.12.1984 *Accepted 11.7.1985*

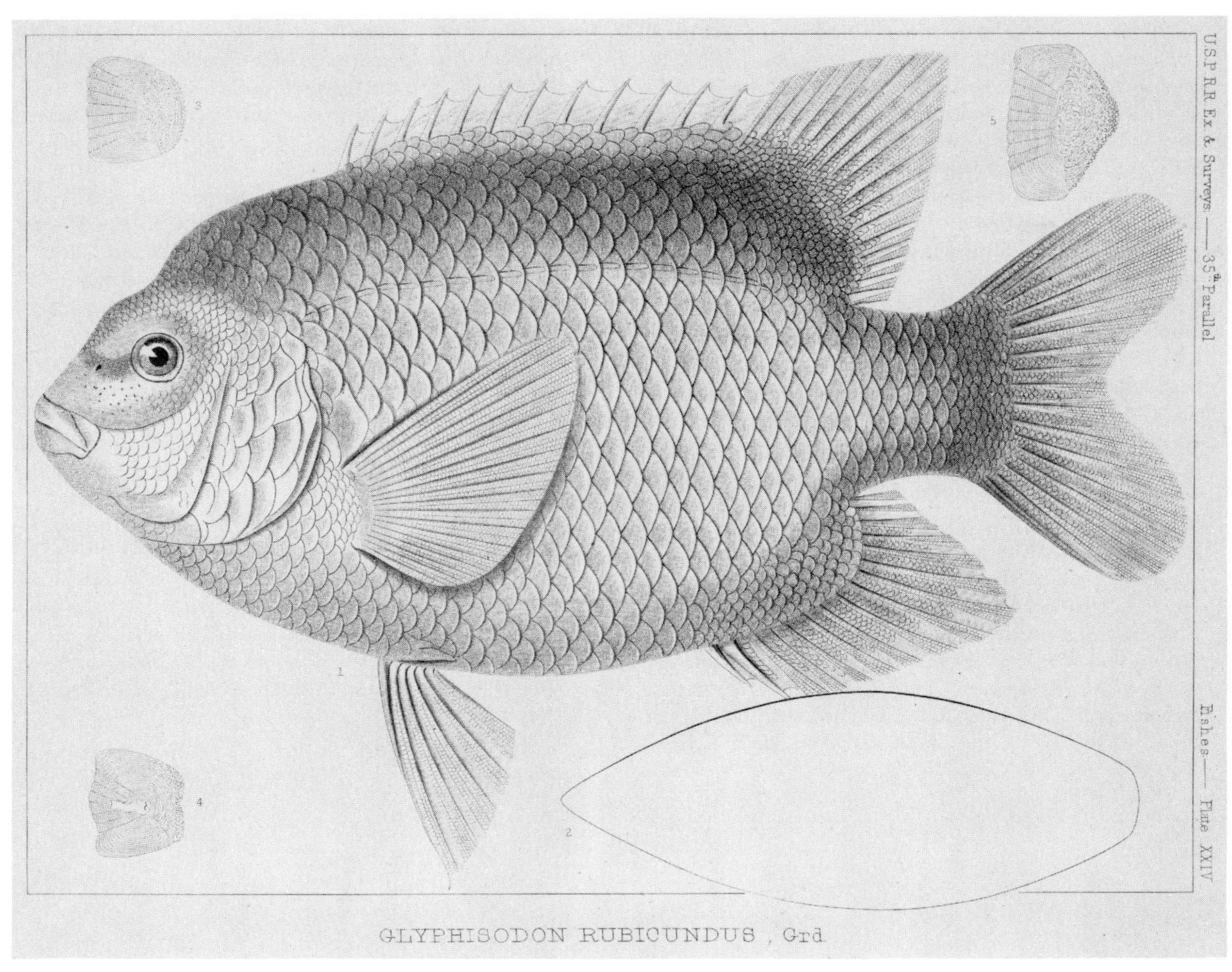

GLYPHISODON RUBICUNDUS, Grd.

Predator-prey interactions in Lake Michigan: model predictions and recent dynamics

James F. Kitchell[1] & Larry B. Crowder[2]
[1] *Center for Limnology, University of Wisconsin-Madison, Madison, WI 53706, U.S.A.*
[2] *Department of Zoology, North Carolina State University, Box 7617, Raleigh, NC 27695, U.S.A.*

Keywords: Alewife, Bioenergetics modeling, Forage fishes, Predation, Salmonids, Zooplankton

Synopsis

Several years ago, we used a bioenergetics model to evaluate the impact of increasing salmonid stocking on the highly variable alewife forage base in Lake Michigan. At that time, we forecast an alewife population decline and the following system-wide effects: increased abundances of large zooplankton, decreased salmonid growth rates, increased diet breadth of salmonids, niche shifts among competitors of the alewife, increased alewife growth rates and increased densities of fishes suppressed by alewife. Alewives have continued to decline steadily since 1981 and are now reduced to a density similar to early outbreak levels in the early 1960s. Recent reports on fish growth rates, zooplankton size and fish community structure support our projections regarding system-wide responses to the alewife decline.

Introduction

In the past century, all of the Laurentian Great Lakes have undergone major ecological changes directly associated with overexploitation by commercial fisheries and introduction of anadromous, marine fishes. As a result, many endemic species and fish stocks have been driven to local extinction; both the biological communities and major food webs have changed dramatically and permanently (Christie 1974, Smith 1968, Wells & McLain 1973).

The purposes of this paper are to: (1) review causes of major changes in the Great Lakes ecosystems during the past century, (2) briefly describe an analysis of predator-prey interactions used to develop forecasts of imminent changes in species interactions, and (3) report recent observations as a test of the forecast.

A primary cause of these changes was invasion by the sea lamprey *Petromyzon marinus*. Although there were endemic lamprey species present in the lakes, they achieved only small adult body sizes. The exotic sea lamprey becomes much larger and had a devastating effect on lake charr, *Salvelinus namaycush*, stocks which were stressed by growing commercial exploitation. Thus, the major piscivore in these ecosystems was nearly or entirely removed. Sea lampreys and the commercial fisheries subsequently contributed to the decline or local extinction of many of the endemic *Coregonus* species (Smith & Tibbles 1980).

In the absence of significant piscivore populations, the exotic smelt, *Osmerus mordax,* and/or alewife, *Alosa pseudoharengus*, flourished. Their effects as predators on and competitors with native planktivores remain unresolved (Crowder 1980) but apparently they were a major component in the decline of approximately half of the native zooplanktivorous fishes (Smith 1970). As a consequence of extremely high densities, massive alewife die-offs occurred fouling water intakes and beaches. By the mid 1960s, the lakes offered few fisheries of economic importance.

Development of a highly successful sea lamprey control program (Lawrie 1970) allowed re-establishment of piscivore populations. Native lake charr and a group of exotic salmonids including coho *Oncorhynchus kisutch*, and chinook salmon *O. tshawytscha*, brown trout *Salmo trutta* and rainbow trout *Salmo gairdneri* were stocked in an attempt to rehabilitate the lake charr stocks, to develop a biological control of the alewife nuisance and to establish potential new fisheries. The results were remarkable. A billion-dollar sport fishery has developed based on the salmon and trout that use alewife and smelt as their primary prey resources (Anon. 1984, Talhelm et al. 1979). These dynamic changes were generally true for those habitats that maintain oxygenated hypolimnia during the summer months in each of the five major lakes. Although alewife is uncommon in Lake Superior, smelt appear to have performed much the same role there. The timing and duration of these dynamics differed among lakes but can be generalized graphically as in Figure 1.

Many of the elements of these case histories are available in the proceedings of a series of symposia (Loftus & Regier 1972, Colby 1977, Smith 1980, Spangler et al. 1981). The specific components of this general history are best known for the dramatic changes that have occurred in Lake Michigan (Wells & McLain 1973). Thus, the alewife has changed from a novelty, to a nuisance and finally to an important prey resource. In a provocative paper, Smith (1970) described some of the major species interactions involving alewife and speculated about the consequences of increasing predation pressure due to salmonid stocking. He argued that reductions in alewife stocks would be an essential requisite for restoring native fish communities and that the reversals of declining native species stocks were possible but would involve substantial lags (on the order of a decade) once piscivory had suppressed alewife populations.

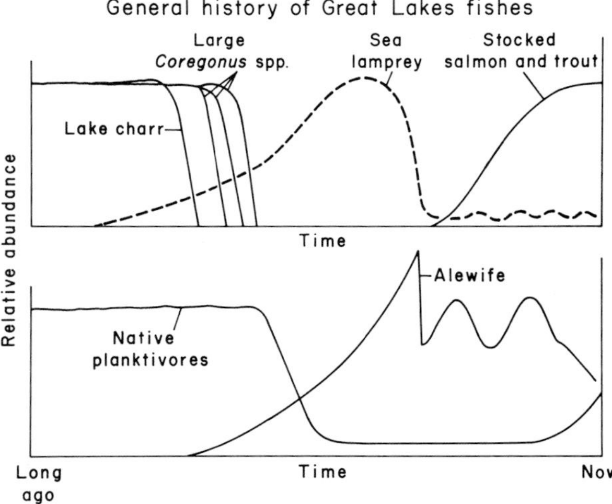

Fig. 1. A generalized, graphical representation of changes in abundance for major fish populations in the Laurentian Great Lakes. Time and abundance scales differ among the lakes.

The modeling approach

As represented in Figure 2, Lake Michigan's food web interactions are now dominated by exotic species. Forecasting the ecological consequences of increased fishery manipulations of this system offers some major challenges. Toward that end, we attempted to estimate the current and future levels of piscivory as one of the major ecological forces operating in this artificial system where predator populations are dominated by stocked salmonids and are thus uncoupled from the numerical dynamics of their prey resources.

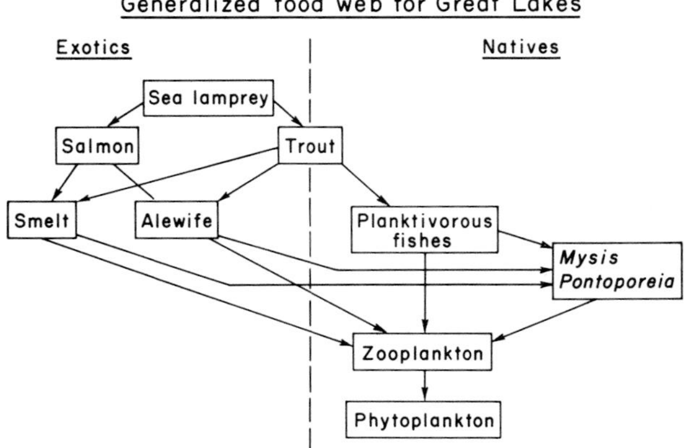

Fig. 2. A general food web for the pelagic system of the Laurentian Great Lakes.

Using an energetics-based modeling approach, Stewart et al. (1981) estimated the current and projected levels of mortality imposed on alewife and other forage species due to the increasing levels of salmonid stocking. The original analyses were conducted in 1978 (Stewart 1980) and subsequently published in 1981 as part of a forage fish symposium (Stewart et al. 1981). Full documentation of the modeling approach followed (Stewart et al. 1983). An assessment of the assumptions employed in this approach is presented in Kitchell (1983), and an independent evaluation of the adequacy of these models when used to quantify predator-prey interactions is presented by Rice & Cochran (1984).

Briefly restated, the energetics modeling approach is an empirical method designed to measure the amount of food consumed. A balanced energy budget is created for individuals by using the observed growth rate, thermal history, and diet to estimate predation rates. Estimates of mortality rates allow calculation of total population effects for each predator species and species of predators can be assembled as guilds. This approach contains no functional or mechanistic representation of predator-prey interactions; it simply expands a typical energy budget to allow estimates of total predation required to yield the observed growth and mortality rates. The resulting estimates are used as a basis for inference and extrapolation.

The major conclusions of our earlier modeling studies were that alewife stocks in Lake Michigan were experiencing substantial and increasing mortality due to predation by stocked salmonids. Given the highly variable recruitment success of alewife, we (Stewart et al. 1981) suggested that the alewife population might well exhibit a continuous decline and/or collapse. Because salmonids do not have their major effect on the alewife population until two to five years after stocking, we argued that the 'predatory inertia' exhibited in this system would prevent an effective management response because of the time lags in the effect of predators on prey populations.

Ecological forecasts

Stewart et al. (1981) developed a series of hypotheses and made forecasts of the ecological effects of the predicted alewife decline. The purpose of this paper is to review those predictions and to evaluate their adequacy in light of recent changes in Lake Michigan's trophic system.

The original forecasts of changes that would harbinger and accompany alewife decline were: (1) Increased abundances of large zooplankton; (2) reduced growth rates of salmon species, particularly of coho salmon whose short life history would be most responsive to declining prey resources; (3) increased diet breadth of salmon and trout accompanying the decline in preferred prey; (4) niche shifts among those fishes (e.g. ciscoes and yellow perch) competing with alewife; (5) increased growth rates of alewife; and (6) increased populations of those fishes suppressed during alewife expansion.

It is worth re-stating that these forecasts were developed in 1978, a year of above-average alewife abundance (Hatch et al. 1981) and published in 1981, another year of above-average alewife abundance (Eck & Brown 1985).

Our view of this system is based on predatory-prey interactions. It differs from an alternate, more conventional view based on population dynamics presented by Eck & Wells (1983) and that of Eck & Brown (1985). They assume that the observed fluctuations in alewife reflect variability around an equilibrium population size dating to the late 1960's and estimate that alewife stocks in Lake Michigan could sustain two to three times the level of predation imposed by current salmonid stocking policy. The central assumption of their approach was that alewife mortality was largely density-independent and that its sources were substitutable; if predator-induced mortality were increased, it would simply compensate for other agents of natural mortality and net alewife population dynamics would be little affected (Eck & Brown 1985). In strong contrast, our view was that increased predation could operate as a depensatory process (Walters et al. 1980), that predator-induced mortality was additive, and that the combination of one or two years of poor recruitment success or heavy over-winter mortality coupled with increasing levels of predation could operate as a positive feedback system which would ultimately yield a catastrophic decline in the alewife population. We predicted that a series of dramatic ecological changes in Lake Michigan would result.

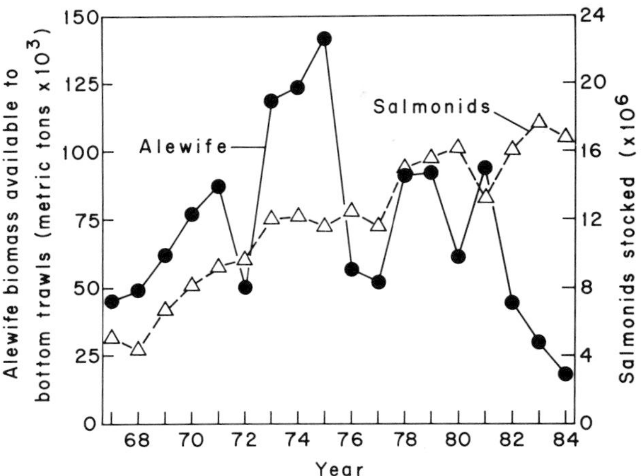

Fig. 3. Estimates of alewife biomass available to bottom trawls in Lake Michigan (U.S. Fish and Wildlife Service) compared with total annual stocking of salmonids for the period 1967–84. Sources are given in text.

Ecological results

Recent events in Lake Michigan confirm our forecasts. Alewife stocks have declined abruptly since 1981 (Wells & Hatch 1985) and are now at levels lower than those seen since the early 1960's when stocks began expanding. In 1984, alewife stocks were less than 20% of those observed after the 1966–67 die-off, which is the period of stasis presumed by Eck & Brown (1985).

Our earlier estimates of predation intensity were based on a 1976–77 compilation of salmonid stocking records plus extrapolations of future stocking based on agency planning and hatchery development programs. Since then both the plans and the results have changed. Based on a recently-published compilation (Anon. 1984) and telephone interviews with fishery management personnel from the states of Michigan, Wisconsin, Illinois, Indiana and the federal hatcheries, we updated Stewart's (1980) original compilation to include the recorded stocking rates for 1977–84. The result is presented in Figure 3 for comparison with the more recent estimates of alewife abundance reported by Wells & Hatch (1985) for Lake Michigan. Alewife biomass presented in Figure 3 is only an indicator of abundance as represented by availability to bottom trawls and is presented for comparative purposes. The annual production dynamics of alewife must be known for appropriate quantitative assessment of mortality due to salmonid predation. In addition, a comprehensive analysis of the 1977–84 period would require complete consideration of the stocking rates for different sizes, ages, species, times of the year and sites. That analysis is currently in progress.

Actual stocking rates generally exceeded those projected in 1976–77 by an average of about 10%. They ranged from 90% of anticipated levels for 1981 when 13.2×10^6 juvenile salmonids were stocked in Lake Michigan to 17.6×10^6 for 1983 which was 20% greater than expected. As in the previous compilation, agency personnel generally expect future production and stocking rates to be similar to those for 1984. In fact, both the effective stocking rates and the total predatory effects on the system will likely continue to rise even if empirical stocking rates do not change. The reasons for this allegation are several: (1) hatchery personnel are becoming more proficient at raising and releasing these exotic species which should result in greater survivorship, (2) because no stocking quota system is yet established, the surplus fish will probably be stocked in Lake Michigan or one of the other Great Lakes, (3) for various, complex reasons, the mix of species stocked has changed to the extent that those now favored (chinook salmon and rainbow trout) place a greater bioenergetic demand on alewife per fish stocked than those that have been proportionately reduced (coho salmon), (4) unknown but likely increasing recruitment of salmonids from natural reproduction in streams are not presently included in our analysis but would accelerate the expected changes.

In overview, all features of the management practices involved in stocking salmonid predators in Lake Michigan appear to be working as well or better than anticipated a decade ago, so that pressure on the forage base continues to increase. The major question remaining is: What is the capacity of this predator-prey system? We argue that the recently observed decline in Lake Michigan's alewife population is substantially if not entirely due to increased levels of predation by salmon and trout. The ability of the alewife population to compensate for this increased mortality will continue to decline and may be exceeded by the depensatory mortality imposed by a continuously growing populations of predators.

Given the general forecast of an alewife decline, how do recent conditions compare with those we predicted (Stewart et al. 1981)? We expect a cascade of effects stemming from piscivore-induced changes in the food web. We offer here a brief tabulation of observations made since the Stewart et al. (1981) paper was prepared (Table 1). In all cases, the ecological expectations of density-dependent species interactions continue to be borne out. The effects of increased piscivory are expressed through reduced competitive dominance by alewife and through increased populations of large zooplankton.

Unlike the open pelagic marine systems where large-scale manipulations of species interactions are often muted or diffuse (Ursin 1982), the key species interactions in Lake Michigan appear highly responsive to keystone-predator effects. The Great Lakes are large but relatively less diverse than marine systems. They offer a substantial opportunity for gaining greater understanding of both the vertical and horizontal interactions within a complex food web. There is also the advantage of comparisons of similar systems. With some slight differences, each of the five major Great Lakes is currently experiencing manipulations similar to those occurring in Lake Michigan.

In summary, an empirical, energetics-based model has been used to forecast the quantitative effect of a group of predators on their primary prey. In the context of Lake Michigan, where much of food web dynamics can be influenced by management practices, we previously forecast a suite of ecological consequences associated with increased predation pressure by stocked salmonids. These forecasts were made several years ago (Stewart et al. 1981) and are compared with responses observed in the recent past. In all cases, the anticipated effect was subsequently observed.

Table 1. Predictions based on Stewart et al. (1981) compared with independent, recent observations regarding trophic interactions in Lake Michigan.

Predictions	Recent observations
Increase of large zooplankton	Large copepods have generally increased (up to 10x) since 1980 (Evans et al. 1980). Cladocerans increased 35% in body size from 1979 to 1981 (Gitter 1982). Large *Daphnia* spp. dominated the 1983 summer zooplankton (Scavia et al. 1986).
Reduced growth rate of salmonids	Size at age and condition factors have declined continuously since the 1960's for all salmon and trout (Hagar 1984).
Increase diet breadth in salmonids	Alewife remain the primary prey but are less important and diets generally more diverse in each of the recent years. Late summer diets are most diverse with alewife a minor component (Hagar 1984).
Niche shifts among alewife competitors	Crowder (1986) documents recent changes among competitors. Morphological and behavioral changes by bloater chub *Coregonus hoyi* occurred in response to alewife competition.
Increase growth rates of alewife	Hagar (1984) documents an increase in alewife condition and growth.
Increased populations of other native fishes	Wells & Hatch (1985) and Crowder (1986) report strong recoveries of several species that declined during alewife expansion. In particular, yellow perch and deepwater ciscoes have increased several orders of magnitude since the late 70's.

Although we were successful in forecasting that conditions in Lake Michigan would soon exhibit major changes, we are not yet capable of predicting the magnitude or duration of those changes. The trophic system is now dominated by non-native predators and an uncoupled piscivore-planktivore interaction. The latter can be managed through stocking policy. However, the former is of concern because it is difficult to imagine what, if any, steady state may be established in this non-coevolved system. Our energetics models can be used to quantify interactions but they cannot predict dynamics. We suggest that evidence from the remaining native species may be a source of guidance and expectation. In particular, if the zooplankton assemblage begins to resemble its composition during the pre-alewife years, then there is at least some indication that the remaining native components of the food web are operating in ways represented by conditions of the past. Re-establishment of the previous, more diverse zooplankton base may also act to reduce recruitment variation in native fishes. We offer this as an hypothesis whose first test is now underway in Lake Michigan.

Acknowledgements

We thank Donald J. Stewart for encouraging us to build on his original analysis. We thank the supporting agencies and organizers of GUTSHOP '84 for providing for and encouraging our contribution. This work was supported by grants from the University of Wisconsin Sea Grant program.

References cited

Anonymous. 1984. Annual Report Great Lakes Fishery Commission, 1982. Ann Arbor. 153 pp.

Christie, W.J. 1974. Changes in fish species composition of the Great Lakes. J. Fish. Res. Board Can. 31: 827–854.

Colby, P.J. (ed.) 1977. Proceedings of the 1976 Percid International Symposium (PERCIS). J. Fish. Res. Board Can. 34: 1447–1999.

Crowder, L.B. 1980. Alewife, rainbow smelt and native fishes in Lake Michigan: competition or predation? Env. Biol. Fish. 5: 225–233.

Eck, G.W. & L. Wells. 1983. Biology, population structure, and estimated forage requirements of lake trout in Lake Michigan. U.S. Fish and Wildlife Serv. Techn. Rep. 111. 18 pp.

Eck, G.W. & E.H. Brown Jr. 1985. Lake Michigan's capacity to support lake trout (*Salvelinus namaycush*) and other salmonines: an estimate based on the status of prey populations in the 1970's. Can. J. Fish. Aquat. Sci. 42: 449–454.

Evans, M., B.E. Hawkins & D.W. Sell. 1980. Seasonal features of zooplankton assemblages in the nearshore of southeastern Lake Michigan. J. Great Lakes Res. 6: 275–289.

Gitter, M.J. 1982. Thermal distribution and community structure of Lake Michigan zooplankton with emphasis on interactions with young-of-year fishes. M.S. Thesis, University of Wisconsin, Madison. 129 pp.

Hagar, J.M. 1984. Diets of Lake Michigan salmonids: an assessment of the dynamics of predator-prey interaction. M.S. Thesis, University of Wisconsin, Madison. 97 pp.

Hatch, R.W., P.M. Haack & E.H. Brown, Jr. 1981. Estimation of alewife biomass in Lake Michigan, 1967–1978. Tran. Amer. Fish. Soc. 110: 575–584.

Kitchell, J.F. 1983. Energetics. pp. 312–338. *In*: P.W. Webb & D. Weihs (ed.) Fish Biomechanics, Praeger Publ., New York.

Lawrie, A.H. 1970. The sea lamprey in the Great Lakes. Tran. Amer. Fish. Soc. 99: 766–775.

Loftus, K.H. & H.A. Regier (ed.) 1972. Proceedings of the 1971 symposium on salmonid communities in oligotrophic lakes (SCOL). J. Fish. Res. Board Can. 29: 611–986.

Rice, J.A. & P.A. Cochran. 1984. Independent evaluation of a bioenergetics model for largemouth bass. Ecol. 65: 732–739.

Scavia, D., G.L. Fahnenstiel, M.S. Evans, D. Jude & J.T. Lehman. 1986. Influence of salmonine predation and weather on long-term water quality trends in Lake Michigan. Can. J. Fish. Aquat. Sci. 43 (in press).

Smith, B.R. (ed.) 1980. Proceedings of the 1979 Sea Lamprey International Symposium (SLIS). Can. J. Fish. Aquat. Sci. 37: 1585–2214.

Smith, B.R. & J.J. Tibbles. 1980. Sea lamprey (*Petromyzon marinus*) in lakes Huron, Michigan, and Superior: history of invasion and control, 1936–1978. Can. J. Fish. Aquat. Sci. 37: 1780–1801.

Smith, S.H. 1968. Species succession and fishery exploitation in the Great Lakes. J. Fish. Res. Board Can. 25: 667–693.

Smith, S.H. 1970. Species interactions of the alewife in the Great Lakes. Trans. Amer. Fish. Soc. 99: 754–765.

Spangler, G.R., A.H. Berst & J.F. Koonce. 1981. Perspective and policy recommendations on the relevance of the stock concept to fishery management. Can. J. Fish. Aquat. Sci. 38: 1908–1914.

Stewart, D.J. 1980. Salmonid predators and their forage base in Lake Michigan: a bioenergetics modeling synthesis. Ph.D. Thesis, University of Wisconsin, Madison. 225 pp.

Stewart, D.J., J.F. Kitchell & L.B. Crowder. 1981. Forage fishes and their salmonid predators in Lake Michigan. Trans. Amer. Fish. Soc. 110: 751–763.

Stewart. D.J., D. Weininger, D.V. Rothers & T.A. Edsall.

1983. An energetics model for lake trout, *Salvelinus namaycush*: application to the Lake Michigan population. Can. J. Fish. Aquat. Sci. 40: 681–698.

Talhelm, D.R., R.C. Bishop, K.W. Cox, N.W. Smith, D.N. Steinnes & A.L.W. Tuomi. 1979. Current estimates of Great Lakes fisheries values: 1979 status report. Great Lakes Fishery Commission. Ann Arbor. 79–1: 17 pp. (mimeo)

Ursin, E. 1982. Stability and variability in the marine ecosystem. Dana 2: 51–67.

Walters, C.J., G. Spangler, W.J. Christie, P.J. Manion & J.F. Kitchell. 1980. A synthesis of the knowns, unknowns, and policy recommendations from the Sea Lamprey International symposium. Can. J. Fish. Aquat. Sci. 37: 2202–2208.

Wells, L. & A.L. McLain. 1973. Lake Michigan: man's effect on native fish stocks and other biota. Great Lakes Fish. Comm. Tech. Rep. 20. 55 pp.

Wells, L. & R.W. Hatch. 1985. Status of bloater chubs, alewives, smelt, slimy sculpins and yellow perch in Lake Michigan, 1984. Mimeo report to the Great Lakes Fishery Commission, Lake Michigan Committee, March 19, 1985.

Received 25.2.1985 *Accepted 5.9.1985*

SALAR LEWISI, Grd.

The effect of the choice of evacuation model on the estimation of feeding rate

Ashley J. Mullen
Inter-American Tropical Tuna Commission, c/o Scripps Institution of Oceanography, La Jolla, CA 92037, U.S.A.

Keywords: Consumption, Experimental design, Field data, Ration

Synopsis

This paper considers: various models for describing evacuation from the stomachs of fishes; methods used to estimate feeding rates; and the type of field data required in each case.

Introduction

In an experiment of the type discussed here the stomach contents of fish are estimated at consecutive times. During the intervening period each fish is prevented from feeding so that the evacuation over the interval may be calculated.

The discussion within the literature concerning evacuation appears to have focussed upon two aspects of choosing the model: the physiological plausibility of the model; and some measure of the 'goodness of fit' such as the residual sum of squares. These criteria isolate the laboratory experiments from their rationale, which is usually to estimate the natural feeding rate, and neglect consideration of the required field samples.

Olson & Mullen (1986) had experimental data which were clearly linear and indicated that time for evacuation was independant of meal size. This was physiologically implausible, but a more complex model which had a physiological rationalization, gave a similar fit to the data. Since the more plausible model was nonetheless speculative, the simpler model was chosen for estimating the feeding rate.

If it is accepted that whatever model is chosen it will be only an approximate description of the evacuation process then there remains a danger of putting too much emphasis upon minimising the error involved in modelling the evacuation process within the laboratory experiment. Two different functions when fitted to evacuation data may give similar residual errors, but may create different requirements of the field data if they are used to estimate feeding rates. Three methods for estimating feeding rates are discussed (Elliott & Persson 1978, Olson & Mullen 1986, and Pennington 1985). In each case the method is introduced in a different way from the original, either to enhance comparability or to generalise the technique.

Elliott & Persson (1978)

If the proportion of food remaining in the stomach transformed by taking the logarithm can be approximated by a straight line when plotted against time then those data suggest the convenient exponential model. One does not need to consider meal size. If a meal, M, is consumed at time $t = 0$, the measure of food left in the stomach at $t = t_1$, W_{t1}, is:

$$W_{t1} = M^{e^{-rt1}}.$$

If

$$f(t_1) = W_{t1}/M$$

then

$$f(t_2)/f(t_1) = f(t_2 - t_1).$$

This seemingly arcane property (f is said to be a homomorphism) enables the subsequent contents of the stomach to be predicted from the contents at any particular time.

Consider a fish whose stomach contents are known twice, at t_1 and t_2. If it feeds at a constant rate, C, during the interval $[t_1, t_2]$ then

$$W_{t2} = W_{t1}\ f(t_2 - t_1) + C\int_{t1}^{t2} f(t)dt$$

and

$$C = \frac{W_{t2} - W_{t1}\ f(t_2 - t_1)}{\int_{t1}^{t2} f(t)dt}.$$

Substituting for f(t), denoting t_1 by 0, and t_2 by t; the total consumption over the interval, Ct, is given in the familiar form due to Elliott & Persson (1978),

$$Ct = \frac{(W_t - W_0\ e^{-rt})rt}{1 - e^{-rt}} \tag{1}$$

With a model such as (1), if samples are taken from the field at times $t_1, t_2, \ldots t_n$ the consumption rates may be calculated for each interval. Alternatively, if the consumption rate is constant, the interval may be treated as infinite, then, if $f(t) \rightarrow 0$ as $t \rightarrow \infty$, equation (1) gives

$$C = \frac{\bar{W}}{\int_0^\infty f(t)dt}, \tag{2}$$

where $\bar{W}$ is the (constant) mean weight of food in the stomachs of the sampled fish.

Olson & Mullen (1986)

Olson & Mullen described the following model for evacuation by yellowfin tuna:

$$f(t_1) = W_{t1}/M = 1 - rt_1 \quad \text{while} \quad 0 \leq t_1 \leq 1/r. \tag{3}$$

Note $f(t_1)$ is independent of M, the mean size, but the absolute rate of evacuation, $M \cdot f(t)$ is, of course, proportional to meal size. For this model

$$f(t_2)/f(t_1) = (1 - rt_1)/(1 - rt_2),$$

which does not equal

$$f(t_2 - t_1) = 1 - r(t_2 - t_1),$$

thus $f(t_2 - t_1)$ in equation (1) cannot be assertained without knowing the size of the original meal(s) which W_{t1} represents. This precludes using that equation to investigate the rates of consumption over a short period. Over a longer period, however, the food remaining within the stomach throughout the interval will form a lesser proportion of the quantity found at the end of that interval. Thus (2) can be used to obtain the mean daily feeding rate if either: feeding is assumed to be constant, i.e. meals are taken at random; or the samples are taken from the field at random intervals, and $\bar{W}$ is the mean of these samples. Substituting equation (3) for f(t) into equation (2) gives, simply, $\bar{C} = \bar{W} \cdot 2r$.

Pennington's (1985) method

This method may be used for evacuation data which may be described:

$$g(W_{t2}) = g(W_{t1}) - r(t_2 - t_1), \tag{4}$$

i.e., if there is a function, g, such that the transformed data lie upon a straight line. The exponential model is a special case of this model with $g(W_t) = \log(W_t)$, but the Olson & Mullen model cannot be formulated in this manner because the absolute rate of evacuation is dependent upon meal size. In any model of the form (4) the absolute rate of evacuation at any time is dependent only upon the stomach contents at that moment (the period for evacuation to, say 10% of present contents will, in general, vary).

From (4), differentiating with respect to t_2, using the chain rule and dropping the subscripts

$$\frac{dg}{dW}\frac{dW}{dt} = -r$$

so

$$dW/dt = -r(dg/dW)^{-1}.$$

If fish are feeding at rate C, let $(dg/dW)^{-1} = h$

then

$$dW/dt = C - rh.$$

Integrating from t_1 to t_2, and dividing by $(t_2 - t_1)$ gives

$$\frac{1}{(t_2 - t_1)} \int_{t1}^{t2} C\,dt =$$

$$\frac{(W_{t2} - W_{t1})}{(t_2 - t_1)} + \frac{r}{(t_2 - t_1)} \int_{t1}^{t2} h(W_t)dt.$$

Integrating a function over an interval and dividing by that interval gives the mean of the function over that interval therefore, if $\bar{C}$ is the mean feeding rate over the interval $[t_1, t_2]$ and $\overline{h(W)}$ is the mean of h(W), then

$$\bar{C} = r\,\overline{h(W)} + (W_{t2} - W_{t1})/(t_2 - t_1).$$

Thus, if a transformation can be found such that the transformed data suggest a straight line, with gradient $-r$, then the mean rate of consumption may be found over any period by the following rules: (a) find the formula for the derivative of that transformation with respect to the stomach content, $g'(W)$; (b) apply this formula to each field datum collected randomly throughout the interval; (c) take the reciprocal for each transformed datum $1/g'(W)$ and find the mean of this value for all the data; (d) multiply by r, the absolute value of the gradient; (e) Finally add a correction factor consisting of the estimated rate of change in the stomach contents over the period. If the period was extended to a complete day then one might feel justified in neglecting the correction term, and in the absence of trend then

$$W_{t1} = W_{t2} \text{ where } t_2 = t_1 = 24\,h.$$

Pennington gives a formulation for the variance of the rate of consumption when estimated by this method, but this is somewhat optimistic. The formula gives the variance for the transformation data obtained from the field, but it ignores the uncertainty of the transformation which was obtained from experimental data.

Discussion

Table 1 summarises the methods for estimating the feeding rates which have been described. The field data requirements may be inferred from the formula given in the table.

If the exponential model can be fitted satisfactorily to the data obtained from experiments then, if field samples may be obtained throughout the day, the diurnal cycle of feeding may be investigated simply by using the formula of Elliot & Persson (1978). If the field data are not sufficient to obtain multiple estimates of W throughout the day then, provided the samples have been taken at random intervals, or at least distributed approx-

Table 1. Method and evacuation model with the relevant equations. See text for details.

Method	Evacuation model	Equation for estimating diurnal feeding cycle	Equation for estimating daily consumption
E & P	$W_{t_2} = W_{t_1}\,e^{-r(t_2-t_1)}$	$C = \frac{W_{t_2} - W_{t_1}e^{-r(t_2-t_1)}}{1 - e^{-r(t_2-t_1)}}$	$C = r\bar{W}$ Special case of 0 & M and P
O & M	$f(t) = W_t/M$ independent of M	No general equation If possible then as E & P	$C = \frac{\bar{W}}{\int_0^\infty f(t)dt}$
P	$g(W_{t_2}) = g(W_{t_1}) - r(t_2 - t_1)$	$\bar{C} = r\,\overline{h(W)} + (W_{t_2} - W_{t_1})/(t_2 - t_1)$	$\bar{C} = r\,\overline{h(W)}$

imately uniformly through the day, one may obtain a mean feeding rate. Field samples are often clumped, one may be forced to calculate the mean feeding rate which would pertain if the fish took their meals at random.

Deviation from the exponential design model necessitates considering the effect of meal size upon the rate of evacuation. If the rate of evacuation is proportional to the residual *as a proportion of the original meal* then the method of Olson & Mullen (1986) may be used to obtain only a mean daily ration. This figure is estimated by dividing the mean measure of stomach content, from field data, by the total area under the evacuation curve, f(t), where f(t) is the proportion of meal remaining at time t.

It is unreasonable to assume that every species deviating from the exponential model will continue to evacuate at a rate proportional to initial meal size. It is more likely that the rate of evacuation would be proportional to some function of the food in the stomach at any given time; then Pennington's (1985) method may be applied. In its original form this entails finding to which power, q, the stomach contents need to be raised in order that when the data are plotted they form a straight line. The gradient, −r, of the line is noted. To calculate the rate of consumption during any interval the sampled weights w_i are transformed to $(r/q)\ W_i^{1-q}$, equivalent to

$$r\left(\frac{d(W^q)}{dW}\right)^{-1}.$$

The mean is found and then corrected for shorter intervals by adding the estimated rate of change of W over the interval.

To obtain the feeding rates throughout the day is possible but it requires a lot of field data. For each interval, not only is a mean necessary but also estimates for the stomach contents of the population at each end of that interval.

Few models, when fitted, go through the value of the original meal, the line usually appears to pass below the original meal size. If they consistently pass at a given percentage below the meal size then it seems reasonable not to constrain the fit, but to revise the feeding rate upwards by the same proportion, after it has been calculated.

One problem which has not been addressed so far is that of consecutive meals affecting each others' evacuation rates. If the rate of evacuation is dependent only upon the contents of the stomach at that time, then the rate will be independent of meal size and, the rate of evacuation of one meal will be independent of the presence of another meal. For example consider two meals, M_0 and M_1, ingested at $t = 0$ and $t = t_1$ and their remains RM_{0t} and RM_{1t} at $t > t_1$. Suppose an exponential decline in stomach contents, then at $t = t_2$

$$\begin{aligned} W_{t2} &= RM_{0t_2} + RM_{1t_2} \\ &= M_0 e^{-rt2} + M_1 e^{-r(t2-t1)}. \end{aligned}$$

The rates of change of RM_0 and RM_{1t} at time t_2 are simply

$$\frac{d(RM_{0t})}{dt} = -r(RM_{0t_2}) = -rM_0 e^{-rt2}$$

and

$$\frac{d(RM_{1t})}{dt} = -r(RM_{1t_2}) = -rM_1 e^{-r(t2-t1)}.$$

Thus each rate is independent of the other. Persson (1984) has shown however that for Perch, *Perca fluviatilis,* upon ingesting a meal before another was completely digested, the remnants of the first meal were evacuated faster than the previously validated exponential model predicted, while the second meal was retarded. Nevertheless the exponential function fitted the data for the combined meal so the Elliot & Persson (1978) formula may still be used.

Persson's (1984) work shows that the exponential model, which can be interpreted very simply physiologically, may lose that simple rationalisation upon close scrutiny. The underlying physiological model may be much more complicated than the exponential which is merely a crude approximation of the reality. Finding that deeper model may be interesting, but it is not necessary in order to estimate feeding rates. Physiology should not exclude an evacuation model from consideration for the purpose of estimating consumption.

Acknowledgements

Bob Olson stimulated my interest in this topic and introduced the relevant literature to me. In addition to the anonymous referees, constructive comments concerning the manuscript were made by Greg Cailliet, Charles Simenstad, Martin Hall, Patrick Tomlinson and Alex Wild.

References cited

Elliot, J.M. & L. Persson. 1978. The estimation of daily rates of food consumption for fish. J. Anim. Ecol. 47: 977–991.

Olson, R.J. & A.J. Mullen. 1986. Recent developments for making gastric evacuation and daily ration determinations. Env. Biol. Fish. 15: 00–00 (this volume).

Pennington, M. 1985. Estimating the average food consumption by fish in the field from stomach contents data. Dana 4: 81–86.

Persson, L. 1984. Food evacuation and models for multiple meals in fishes. Env. Biol. Fish. 10: 305–309.

Received 31.12.1984 *Accepted 10.9.1985*

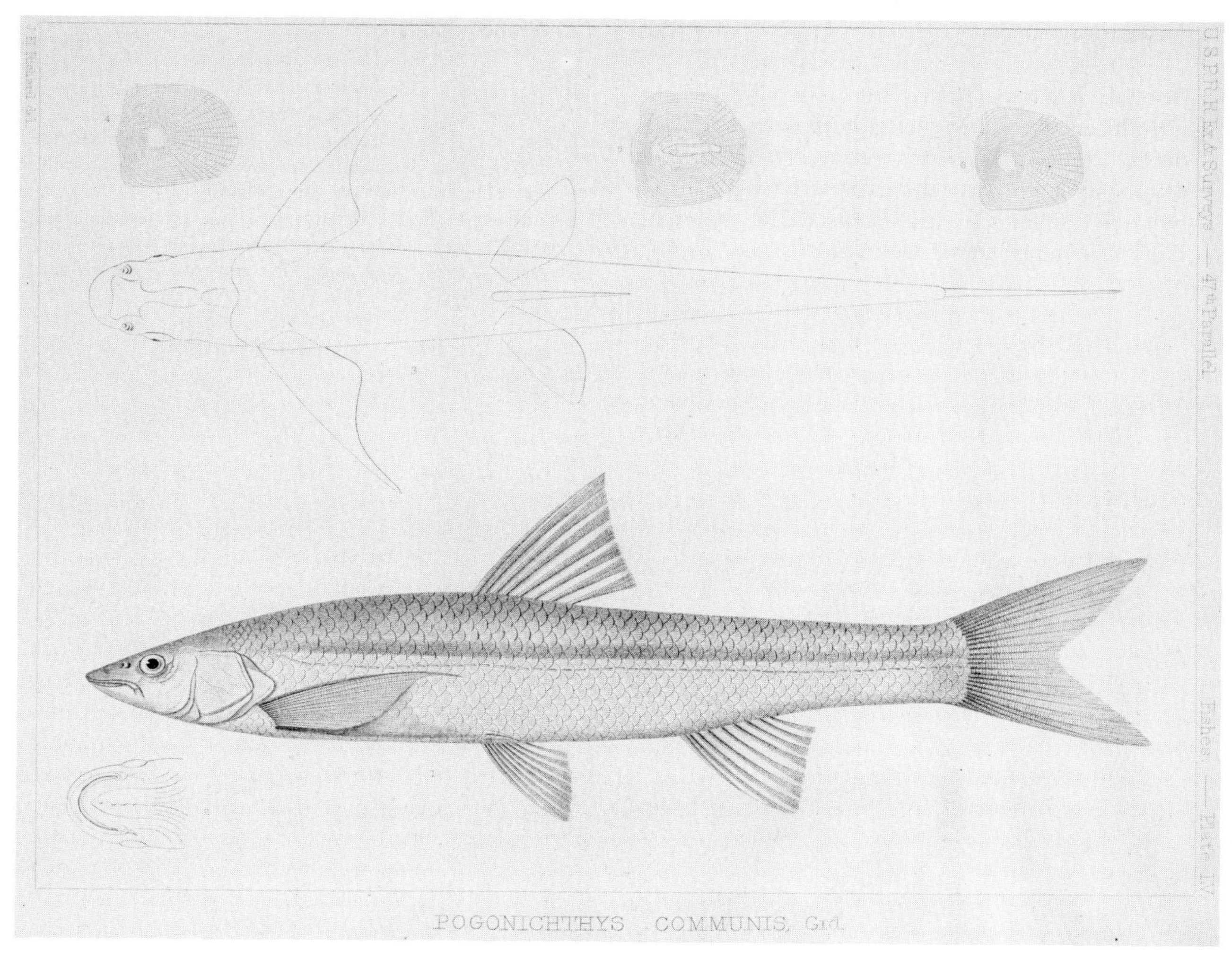

POGONICHTHYS COMMUNIS, Grd.

Use of modeling to investigate potential feeding strategies of parasitic lampreys

Philip A. Cochran[1] & James F. Kitchell
Center for Limnology, University of Wisconsin, Madison, WI 53706, U.S.A.
[1] *Present address: Division of Natural Sciences, St. Norbert College, De Pere, WI 54115, U.S.A.*

Keywords: Petromyzonidae, *Petromyzon, Ichthyomyzon,* Foraging behavior, Predator-prey, Optimal foraging, Energetics

Synopsis

A quantitative energetics model of feeding and growth by parasitic lampreys was used to assess overall rates of net energy intake under alternate feeding strategies and varying host availability. Our early attempts to predict optimal feeding behavior focused on the duration of attachment to the host, but model simulations have revealed that optima over this variable may be so flat that, as in one case, a deviation of nearly 30% from the optimal attachment time results in only a 3% decrease in the rate of net energy intake. This may contribute to the great variability that appears characteristic of feeding by parasitic lampreys. Further simulations have suggested that, because host blood is to some extent a renewable resource, lampreys should extend host survival when alternate hosts are scarce by removing blood at reduced rates. In other words, maximization of instantaneous rate of net energy intake is not equivalent to maximization of long term gain. Thus, in predicting optimal lamprey feeding behavior, it is necessary to consider simultaneously both attachment time and rate of host blood removal. Behavioral or evolutionary adjustment of these variables can have important consequences with respect to lamprey growth rates, the functional response of lampreys to changing host densities, mortality rates of host populations, and the expression of other lamprey behavioral traits.

Introduction

Energetics models have proven increasingly useful in the investigation of the ecology of fishes (Cochran & Rice 1981, Kitchell 1983, Rice & Cochran 1983, Stewart et al. 1983). We recently developed a quantitative energetics model that relates feeding and growth by parasitic lampreys to lamprey mass, host mass, and duration of attachment by lamprey to host (Cochran & Kitchell unpublished, Cochran 1984). This model incorporates the following observations and reasonable assumptions about the lamprey feeding process: (1) percentage of host blood volume removed per day is proportional to lamprey mass and inversely proportional to host mass; (2) host blood quality (in terms of energy concentration) is maintained for some time in the face of an ongoing lamprey attack before declining over a period that ends with host death; and (3) both the length of time that host blood quality is maintained and the length of time that a host survives an attack are related negatively to the percentage of host blood volume removed per day (Appendix). The lamprey energy budget of Kitchell & Breck (1980) is used to partition ingested energy among respiration, egestion, excretion, and growth. Model-generated predictions of growth by individual sea lampreys, *Petromyzon marinus,* with known feeding histories in the laboratory compared favorably with observed growth (Cochran & Kitchell unpublished).

Our original priority in developing the lamprey feeding model was to investigate the relationship of attachment time to the interval of time between successive feedings. Due to declining host blood energy concentration and increasing metabolic demand that results from increasing lamprey mass, the instantaneous rate of net energy intake declines during the course of a feeding bout (Fig. 1). Optimal foraging theory, as previously applied to the partial consumption of prey (Cook & Cockrell 1978, Sih 1980), predicts that in such a situation the time spent feeding on an individual prey item should increase with increases in energy expenditure during the non-feeding interval (Fig. 1). Variables that are correlated with energy expenditure during the non-feeding interval include any factors, such as prey availability, that affect its duration. For parasitic lampreys, greater host availability generally should result in shorter intervals between feedings and a concomitant decrease in the time spent feeding on individual hosts (Fig. 1).

A laboratory experiment to test for the predicted relationship between host density and lamprey attachment time was inconclusive (Cochran & Kitchell unpublished), at least partially due to extreme variability in lamprey feeding behavior. Moreover, some sea lampreys remained attached to single hosts for extended periods of time, apparently by feeding at low rates. For example, one lamprey followed a 62-day period without feeding with an attachment that lasted 70 days, during which time it more than doubled in weight but did not kill its host. We have also recorded extended attachments (>35 days) by captive chestnut lampreys, *Ichthyomyzon castaneus,* that could have killed their hosts in much shorter periods of time. It is apparent that lampreys exhibit behavioral plasticity in the rate at which host blood is removed.

Our observations on the variability of lamprey feeding prompted us to use the lamprey feeding model to investigate: (1) the predicted energetic consequences of deviations from optimal attachment times; and, (2) the potential energetic advantage of adjusting the rate of host blood removal in

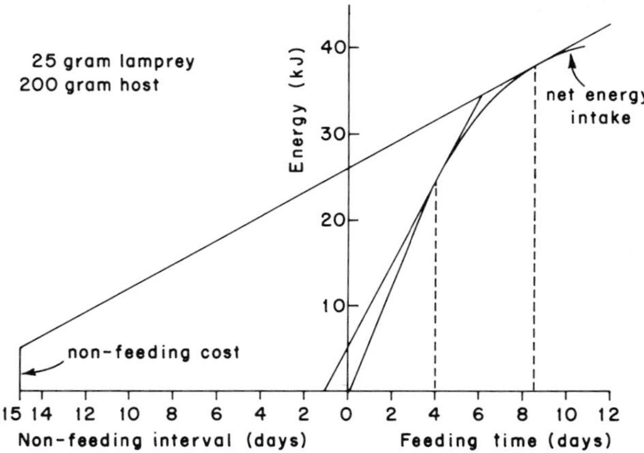

Fig. 1. Net energy intake by a 25 g lamprey feeding on a 200 g host as a function of time spent feeding on the host. The lamprey is assumed to remove 31.75% of the host's blood volume per day. The host dies after a feeding period of nearly 11 days. A tangent drawn from a non-feeding interval on the x-axis locates the corresponding optimum feeding time in terms of overall rate of net energy intake (slope of tangent). Non-feeding costs are estimated from metabolic rates of starved lampreys (Kitchell & Breck 1980). Optimum feeding time increases as non-feeding interval and/or costs increase. Adapted from Cochran 1984.

response to changes in host availability. Here we report the results of these investigations.

Methods

Two sets of model simulations of lamprey feeding were performed. Model parameters corresponded to *Petromyzon marinus* feeding on trout at 10° C. For all simulations we arbitrarily assumed an initial lamprey mass of 25 g and a host mass of 200 g (wet weight). For each simulated feeding bout, the rate of net energy intake over the feeding bout (hereafter referred to as 'the overall rate of net energy intake') was calculated as the intake of net energy (energy available for growth or reproduction after metabolic costs have been met) divided by total time (feeding time plus the non-feeding interval between successive attachments).

The first set of model simulations explored the energetic penalty, in terms of reduced overall rate of net energy intake, for deviating from the predicted optimal attachment time. Lampreys were assumed to remove blood from their hosts at a fixed rate determined by the ratio of lamprey to host mass (Appendix equation 1). In this case, therefore, a 25 g lamprey was assumed to remove 31.75% of a 200 g host's blood volume per day. Percentage of the maximum possible overall net rate of energy intake was calculated for lampreys that remained attached to their hosts for intervals of time that were greater or less than the optimal attachment time. We compared results for lampreys assumed to spend one day and fifteen days, respectively, between successive feedings.

In the second set of simulations, we made no assumption about the relationship between the ratio of lamprey to host mass and the percentage of host blood volume removed per day. Rather, both the percentage removal of host blood and attachment time were varied among simulations. Resulting rates of net energy intake were expressed as percentages of the maximum achievable rate and were plotted as isopleths on a graph of attachment time versus daily percentage removal of host blood. As in the first set of simulations, we compared results for lampreys assumed to spend one day and fifteen days between successive feedings, representing relatively high and low host availability respectively.

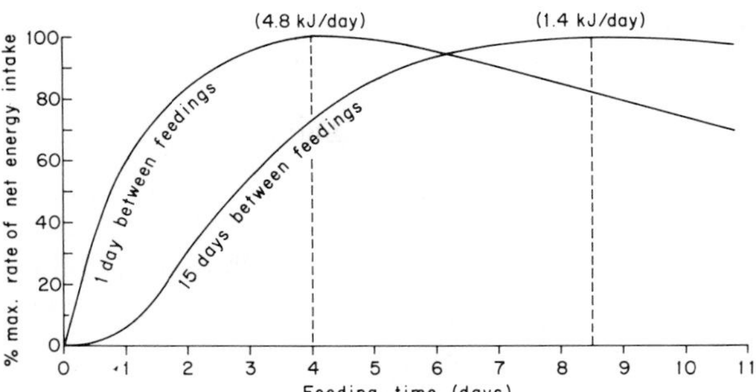

Fig. 2. Percentage of maximum possible rate of net energy intake achieved as a function of feeding time for a 25 g lamprey removing 31.75% of a 200 g host's blood volume per day. Maximum possible rates of net energy intake are indicated over corresponding attachment times for non-feeding intervals of 1 and 15 days.

Results and discussion

Deviations from the optimum attachment time

For a lamprey that removes 31.75% of its host's blood volume per day, the predicted optimal attachment time is 4 days if the interval between feedings is one day and 8.5 days if the interval between feedings is 15 days (Fig. 1). Rates of net energy intake achieved under these optimal strategies are 4.8 kJ per day and 1.4 kJ per day, respectively. Deviations from the optimal attachment time do not result in greatly reduced rates of net energy intake (Fig. 2). This is especially true for attachment times that exceed the optimum and when intervals between successive feedings are long. For example, when the interval between feedings is 15 days, a lamprey that remains attached until the host dies after 10.8 days, an increase of 27% over the optimal attachment time, achieves a rate of net energy intake only 3% below the maximum. Given that variation in such factors as the interval between feedings, the lamprey to host mass ratio, and the daily percentage removal of host blood is inevitable in the real world, differences of this magnitude are not likely to be statistically or biologically significant.

Based on this analysis, we can predict that lamprey attachment times will be highly variable. We also predict that, in general, distributions of attachment times will be skewed in the direction of longer feeding. Not only do attachment times longer than the optimum incur smaller energetic penalties than attachment times less than the optimum (Fig. 2), but the available energy remaining in the host to which a lamprey is currently attached is, all else equal, a more certain reward than that attainable by detaching and seeking out a new host (see discussion of risk aversion in Real 1980). Any tendency toward longer attachment times at a fixed rate of host blood removal will result in a greater probability of host mortality.

Simultaneous adjustment of attachment time and host blood removal

If it is assumed that a lamprey can adjust the rate at which it removes blood from a host as well as its attachment time in response to changes in the interval between attachments, then optimal feeding strategies are different than those discussed in the section above. With a one-day interval between successive feedings, a 25 g lamprey should remove blood from a 200 g host at a rate of 100% of the host blood volume per day for a period of one day (Fig. 3). With a 15-day interval between feedings, the same lamprey should remove blood at a rate of 45% of the host blood volume per day for 6 days (Fig. 4). Thus, it may be advantageous for a lamprey to sacrifice its instantaneous rate of energy intake in order that a greater overall rate of net energy intake be achieved. Longer intervals between attachments (lower host availability) should lead to both longer attachment times and lower percentage rates of host blood removal. Shorter intervals between attachments (greater host availability) should lead to shorter attachment times and higher percentage rates of host blood removal [a lamprey can increase its rate of energy intake only to some finite point, however; Farmer et al. (1975) recorded a maximum daily ration for *P. marinus* of 29.8% wet body weight].

A fortuitous consequence of reduced blood re-

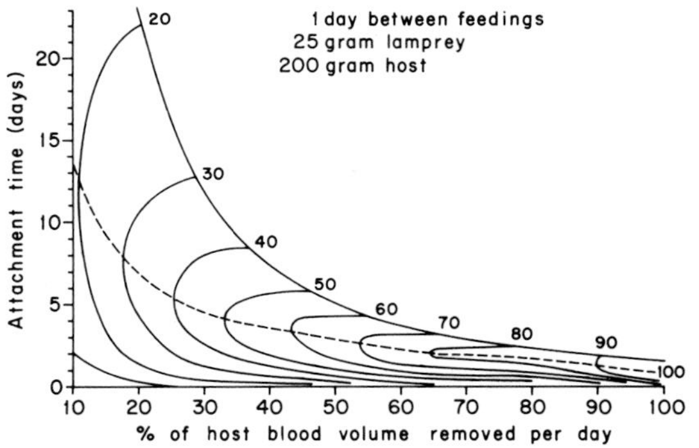

Fig. 3. Rate of net energy intake, expressed as a percentage of the maximum achievable rate (numbers along uppermost curve), as a function of rate of host blood removal and attachment time for a 25 g lamprey feeding on a 200 g host with a 1-day interval between feedings. Maximum rate of net energy intake is 10.00 kJ per day. Isopleths were mapped by calculating rate of net energy intake at intervals of 5% on the x-axis and 1 day on the y-axis. The dotted line connects optimal attachment times for fixed rates of host blood removal. The uppermost descending curve represents the time of host death for any fixed rate of host blood removal.

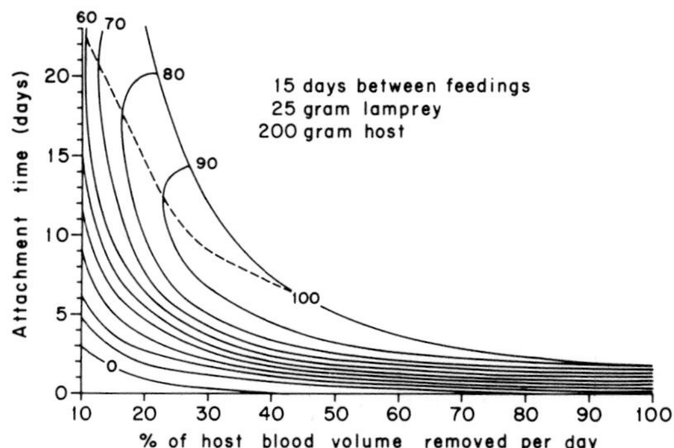

Fig. 4. Rate of net energy intake, expressed as a percentage of the maximum achievable (numbers along uppermost curve), as a function of rate of host blood removal and attachment time for a 25 g lamprey feeding on a 200 g host with a 15-day interval between feedings. Maximum rate of net energy intake is 1.44 kJ per day. See Figure 3 for other details.

moval rates and longer attachment times at lower host densities would be a decreased impact by a population of lampreys on its host population. Even though lampreys might still kill their hosts (e.g., Fig. 4), increased handling time per attack would result in a lower overall rate of host mortality. Moreover, reduced growth rates of lampreys would lessen the seasonal increase in their impact on the host population that results from increases in the lamprey/host mass ratio. Whether on an evolutionary scale, this mechanism has contributed to differences in adult body size of parasitic lamprey species in marine and freshwater habitats, differences that seem correlated with differences in host availability among habitats (Hubbs & Potter 1971), is still a matter for conjecture. It might be instructive to compare host blood removal rates for similarly-sized individuals of freshwater (e.g., *Ichthyomyzon castaneus*) and anadromous (e.g., *Petromyzon marinus*) species under identical laboratory conditions.

A reduction in the optimum rate of host blood removal at low host densities could shift the balance of selective pressures that affect other feeding behaviors. For example, attachment site on the host is highly variable both within and among species but seems to reflect a compromise between high rates of energy intake (ventral attachments) and avoidance of injury or premature detachment through abrasion on the bottom (dorsal attachments). Marine or freshwater lentic lampreys tend to attach ventrally, whereas lampreys from shallow freshwater lotic habitats tend to attach dorsally (Cochran 1986). If the optimum rate of host blood removal is lower under conditions of reduced host availability, this might result in a relaxation of selective pressure to feed ventrally. This in turn might be reflected in a greater proportion of dorsal attachments by lampreys in shallow lotic habitats or more variation in attachment sites chosen by lampreys in deepwater habitats.

The logistics of studying lampreys in the field or in the laboratory can be difficult to surmount. It is imperative that any experimental effort be directed at testing key hypotheses under appropriate conditions. We hope we have demonstrated the utility of the lamprey feeding model in identifying such hypotheses.

Acknowledgements

This research was partially funded by the National Oceanic and Atmospheric Administration's Office of Sea Grant, Department of Commerce, through an institutional grant to the University of Wisconsin, and by St. Norbert College.

References cited

Cochran, P.A. 1984. The foraging behavior of parasitic lampreys. Ph.D. Thesis, University of Wisconsin, Madison. 129 pp.

Cochran, P.A. 1986. Attachment sites of parasitic lampreys: comparisons among species. Env. Biol. Fish. (in press).

Cochran, P.A. & J.A. Rice. 1982. A comparison of bioenergetics and direct field estimates of cumulative seasonal food consumption by largemouth bass (*Micropterus salmoides*). pp. 88–96. *In*: G. Cailliet & C. Simenstad (ed.) Gutshop '81: Fish Food Habits Studies, Proceedings of the Third Pacific Technical Workshop, Washington Sea Grant, University of Washington, Seattle.

Cook, R.M. & B.J. Cockrell. 1978. Predator ingestion rate and its bearing on the theory of optimal diets. J. Anim. Ecol. 47: 529–549.

Farmer, G.J., F.W.H. Beamish & G.A. Robinson. 1975. Food consumption of the adult landlocked sea lamprey, *Petromyzon marinus,* L. Comp. Biochem. Physiol. 50A: 753–757.

Hubbs, C.L. & I.C. Potter. 1971. Distribution, phylogeny and taxonomy. pp. 1–65. *In*: M.W. Hardisty & I.C. Potter (ed.) The Biology of Lampreys, Volume I, Academic Press, London.

Kitchell, J.F. 1983. Energetics. pp. 312–338. *In*: P.W. Webb & D. Weihs (ed.) Fish Biomechanics, Praeger Publishers, New York.

Kitchell, J.F. & J.E. Breck. 1980. Bioenergetics model and foraging hypothesis for sea lamprey (*Petromyzon marinus*). Can. J. Fish. Aquat. Sci. 37: 2159–2168.

Real, L.A. 1980. On uncertainty and the law of diminshing returns in evolution and behavior. pp. 37–64. *In*: J.E.R. Staddon (ed.) Limits to Action: The Allocation of Individual Behavior, Academic Press, New York.

Rice, J.A. & P.A. Cochran. 1984. Independent evaluation of a bioenergetics model for largemouth bass. Ecology 65: 732–739.

Sih, A. 1980. Optimal foraging: partial consumption of prey. Amer. Nat. 116: 281–290.

Stewart, D.J., D. Weininger, D.V. Rottiers & T.A. Edsall. 1983. An energetics model for lake trout, *Salvelinus namaycush;* application to the Lake Michigan population. Can. J. Fish. Aquat. Sci. 40: 681–698.

Received 6.12.1984 *Accepted 18.6.1985*

Appendix

The lamprey feeding model is embodied in the following equations, which are justified in Cochran (1984) and Cochran & Kitchell (unpublished):

(1) $BV = 254\ (l)\ (f^{-1})$,

where BV is the percentage of the host's blood volume removed daily, l is lamprey wet mass, and f is host wet mass. Host blood volume is assumed to be 4.7% of host mass and to be maintained by osmosis as blood is removed by the lamprey. Host blood quality (energy concentration) is maintained at a constant level (3.29 kJ g^{-1} wet weight) for DQ days following the onset of an attack, where

(2) $\ln(DQ) = 4.92–1.08\ (\ln(BV))$.

After DQ days, host blood quality declines exponentially to a value of 0.53 kJ g^{-1} wet weight at the time of host mortality DD days after the onset of the lamprey attack; where

(3) $\ln(DD) = 8.03–1.63\ (\ln(BV))$.

The fate of energy ingested by the lamprey is described by the lamprey energetics model of Kitchell & Breck (1980).

Cochran (1984) and Cochran & Kitchell (unpublished) detail the derivation of the lamprey feeding model, compare its predictions to experimental results, and discuss its implications for selective attachment to large hosts.

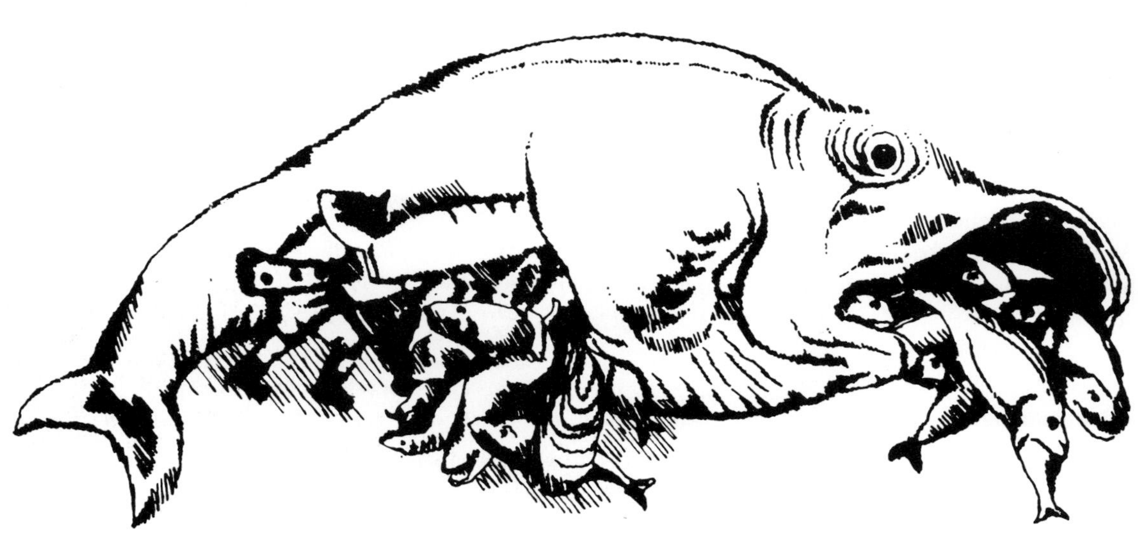

Incorporating fish food habits data into fish population assessment models

Patricia A. Livingston
Northwest and Alaska Fisheries Center, Resource Ecology and Fisheries Management Division, 7600 Sand Point Way, N.E., BIN C15700, Building 4, Seattle, WA 98115, U.S.A.

Keywords: Multispecies ecosystem models, Predation mortality, Predator-prey, Diet, Ration

Synopsis

Marine fisheries management is turning more to multispecies assessment models for management advice. Examples of such models include Laevastu's DYNUMES model, Andersen-Ursin's multispecies Beverton-Holt model, several multispecies VPA models, and singlespecies models. The main links between species in the models are submodels describing diet composition and food intake. These submodels are described in detail and the sampling requirements for estimating the various food habit parameters are discussed. Diet composition parameters are estimated from field collections of stomachs taken over large areas with sampling stratified by fish size, season, area, and time of day. Food intake parameters are usually estimated either from laboratory feeding experiments which determine maintenance and growth rations by temperature or from a combination of field stomach content weight data and laboratory gastric evacuation rate studies. Modelling population level feeding dynamics requires a re-assessment of how food habits data are collected and analyzed. The scale of the sampling program depends on the complexity of questions being asked about the system which in part determines which model will be used to respond to the questions. Sensitivity analysis can provide information about the sources of uncertainty in models due to model structure and model parameter estimates. Sensitivity analysis results can provide guidance as to which model to use and where to allocate research priorities. The ultimate choice of models and sampling programs depends on many variables including the complexity and size of the system, the availability of existing data, and the cost and ease of obtaining the most important data.

Introduction

Marine fisheries management has been dominated by the use of single-species assessment techniques and models. With recent improvements and modifications of modelling theory, multispecies models have become more attractive to scientists providing fishery management advice (Mercer 1982) although the actual use of the models in fishery management plans (FMP's) is not yet widespread. In addition, the Magnuson Fishery Conservation and Management Act of 1976, which dictates the course of marine fishery management in the United States, mandates that assessments should take into account ecological factors. This has placed further pressure on scientists to consider marine fisheries management in a multispecies assessment environment. The main data elements used to tie fish species together in a multispecies assessment are the trophic connections among fish species. Although food habits data have traditionally been used to develop predator-prey models, during recent years fish population assessment models which use fish feeding data have placed special demands on the format and quality of those data. Therefore food habits data ultimately used for management purposes should be collected and analyzed with the particular model requirements in mind.

It is the purpose of this paper to outline some of the common fishery assessment models in use and to provide details of the type of food habits data required for each model. In particular, Laevastu's DYNUMES model (Laevastu et al. 1982), Andersen & Ursin's (1977) multispecies Beverton-Holt model, the multispecies virtual population analysis (VPA) models (e.g. Helgason & Gislason 1979, Pope 1979, Majkowski 1981), and a reduced single-species model case (e.g. Majkowski & Waiwood 1981, Minet & Perodou 1978, Mehl & Westgard 1983, Dwyer et al. unpublished) will be discussed in terms of the food habits data requirements of each model and the feeding parameters which need to be estimated for each model. Special sampling and analysis requirements of food habits data for each model are discussed.

General model descriptions

DYNUMES model

The DYNUMES (Dynamic numerical ecosystem simulation) model developed by Taivo Laevastu at the NOAA-NMFS Northwest and Alaska Fisheries Center in Seattle, Washinton is a large-scale computer simulation model applied to the eastern Bering Sea marine ecosystem. Figure 1 depicts the general scheme of energy flow in the model. The main consideration in the model is the processes which define the growth and removal of fish and large invertebrates. Growth of fish is influenced by temperature, underlying age distribution, and available food. Fish mortality can occur through fishing, predation by marine birds or mammals, old age and disease, and predation by other fish. Plankton is available as food for fish but nutrient regeneration is not modelled.

Although age composition is implicit in the model it is used to externally calculate the initial input growth coefficient for a species. Thereafter, changes in temperature and starvation can affect the growth rate. Food composition and food requirements are the major processes which deter-

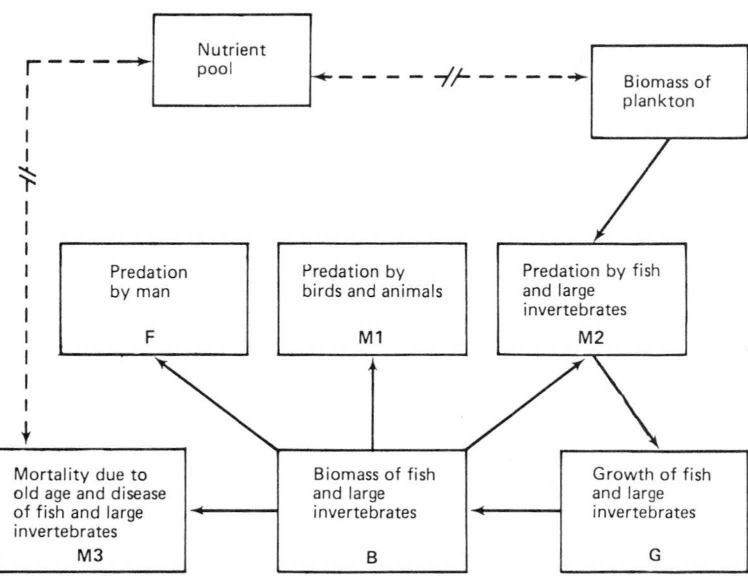

Fig. 1. Simplified view of biomass flow in Laevastu's DYNUMES model.

mine the degree of predation mortality of a species. Changes in food availability can cause starvation or can modify food composition to ultimately affect predation mortality. Table 1 outlines the diet composition submodels and sampling considerations. Food preference is expressed as the proportion by weight of preferred prey items in the ration of a given predator. These preferences are separated into two types for each predator: a deep water (>500 m) and shallow water (<500 m) food preference. If preferred items are not available in the necessary quantity at a given point some prey substitution occurs. Since the model has age structure defined implicitly the diet composition must also implicitly reflect age structure. This requires sampling all ages and weighting the resulting diet by the proportions of each predator age group in the model. The diet must be representative of a large area, the eastern Bering Sea, and must be an annual average diet. Thus, sampling must be spread out over the area and must be conducted over all seasons. Technically, only one year of sampling should be necessary but several years of stomach content data would be desirable to verify the model's prey switching due to availability at various grid points.

Food intake submodels are shown in Table 2. In DYNUMES food intake is calculated as the percentages of predator's body weight consumed per day for maintenance and for growth. Maintenance

Table 1. Description of diet composition submodels and general sampling considerations.

Model	Diet composition submodel	Parameter descriptions	Sampling considerations
DYNUMES (Laevastu)	$C_{ij} = R_i P'_{ij}$ $P'_{ij} = f(P_{ij} \cdot B_{jn,\,m})$	C_{ij} – consumption of prey j by predator i R_i – ration of predator i P_{ij} – proportion by weight of prey j in the diet of predator i $B_{jn,\,m}$ – biomass of prey j at a given coordinate (n, m)	P_{ij} must represent the composite diet of all predator age groups over a large region over a year. (Same P_{ij} is used for all years.) Separate diets for shallow (<500 m) and deep water.
Multispecies Beverton-Holt (Andersen-Ursin)	$C_{ij} = f(G_{ij}, R_i)$ $G_{ij} = p_{ij}\, g_{ij}$ $g_{ij} = e^{\frac{-(\ln w_i/w_j - \eta)^2}{2\sigma^2}}$	G_{ij} – suitability of prey j to predator i p_{ij} – vulnerability of prey j to predator i g_{ij} – prey size preference as a function of mean prey size preference η and σ the degree of size selectivity of predator i	g_{ij} should be estimated from intensive sampling of all predator size groups in a small area. Concurrent estimation of the relation between prey numbers and size in the environment. Model verification requires large coverage sampling. Pelagic and demersal diets are also defined.
VPA (Helgason-Gislason)	Same as above	Yearly average w_i and w_j values are used	g_{ij} can be estimated from large scale sampling of all predator sizes over a year and feeding parameters are tuned to fit observations.
VPA (Pope)	$C_{ij} = f(G_{ij}, R_i)$ $0 \leqslant G_{ij} \leqslant 1$	G_{ij} – suitability of prey j to predator i	G_{ij} is obtained from large scale sampling of all predator sizes over a year and is tuned to fit these observations. (same G_{ij} is used for all years)
VPA (Majkowski)	$C_{ij} = f(\gamma_{ij}, R_i)$	γ_{ij} – proportion by weight of prey j in the ration of predator i	γ_{ij} must represent diet of a predator over a large geographic area over a year. (Different γ_{ij} are estimated for each year in the analysis).
Reduced single-species cases	$C_{ij} = f(R_i, P_{ij})$	P_{ij} – proportion by weight of prey j in the diet of predator i	P_{ij} must represent the composite diet of all predator sizes over the whole region for a particular year.

Table 2. Description of food intake submodels and general sampling considerations.

Model	Food intake submodel	Parameter descriptions	Sampling considerations
DYNUMES (Laevastu)	$R_i = r_m B_i + r_g(\Delta B_i)$	R_i – ration of predator i B_i – biomass of predator i r_m – proportion of body weight consumed for maintenance (varies with latitude) r_g – food required for growth	Requires laboratory determination of maintenance requirements versus temperature by fish size and growth food requirements by size (reproductive growth confounded with somatic growth)
Multispecies Beverton-Holt (Andersen-Ursin), VPA (Helgason-Gislason, Majkowski), Reduced single species case (Waiwood-Majkowski)	$\frac{dR_i}{dt} = f_i h_i w_i^m$	R_i – ration of predator i f_i – feeding level (fraction of maximum food consumption) h_i – coefficient of gross anabolism m_i – power of weight which defines the food absorbing relationship	f is a function of search rate and could be estimated from laboratory experiments h is a species specific coefficient and can be estimated from laboratory feeding experiments m can be estimated from feeding experiments or length-at-age data (No temperature relationships, reproductive growth confounded with somatic growth)
Other reduced single-species cases (Mehl-Westgard)	$\sum C_i = 24\bar{S}_i R_i$	$\sum C_i$ – daily ration of predator i $\bar{S}_i$ – average hourly stomach content weight R_i – temperature dependent gastric evacuation rate	$\bar{S}$ is estimated from field stomach content data spaced over all periods of the day and which are representative of all size groups, seasons, and subareas. R is estimated from laboratory experiments which vary temperature and food type.

ration is also a function of temperature. Laboratory experiments which estimate maintenance and growth rations by temperature and fish size are required for each species in the model. At the present time the model includes food for reproductive growth with food for somatic growth but could explicitly define each if separate estimates are available.

There are several important facets of the model which must be emphasized to understand model behavior and data requirements. First, the model is biomass-based so that all equations and parameters must correspond to changes in biomass and not numbers. Second, the model has spatial resolution, thus migration of animals within a defined 24 × 24 grid of the Bering Sea can occur. Finally, the geographic extent of the modelled area is large, many species are included and the model can be run to simulate several years' events using monthly time steps.

Multispecies Beverton-Holt model

The multispecies Beverton-Holt model is a large-scale computer simulation model which is used to predict or describe events in the North Sea marine ecosystem. The complete version of the model deals with the population dynamics of phytoplankton to fish. It includes a description of nutrient cycling and maintains mass balance within the system. Age structure is explicitly defined and differential equations are used to describe the instantaneous rates of change in numbers of a fish species at age as functions of fishing and natural mortality. Natural mortality is subdivided into several sources in the model but the main source is predation. The degree of predation on a particular species depends on the ration of that species' predators, the predators' prey size preference, g_{ij}, and the vulnerability coefficient, p_j, of that species to predation which depends on prey behavior (i.e., schooling, burial in sand, immobile, etc.). A suit-

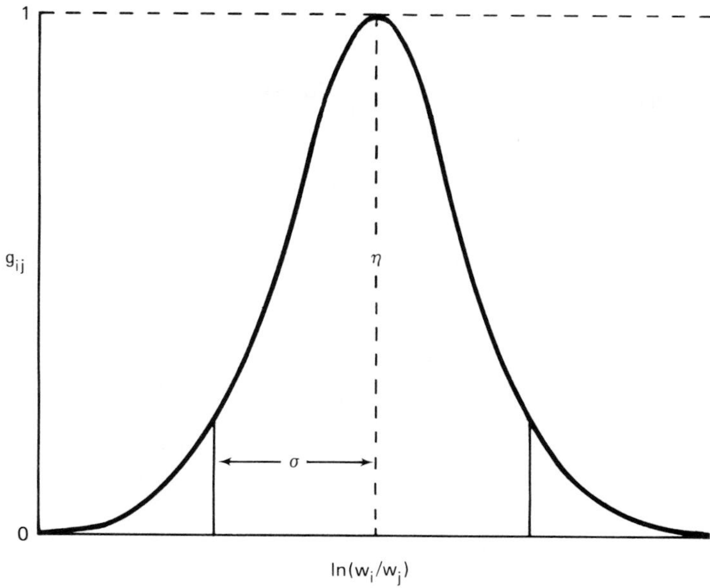

Fig. 2. Shape of Andersen-Ursin's prey size selection curve where η is the mean prey size preference.

ability index of a particular prey is formed from its vulnerability and the prey size preference index. In simple terms, prey size preference is a measure of a predator's selection of prey determined by the individual weight (size) of the prey (w_j) relative to the individual weight (size) of the predator (w_i). Figure 2 depicts the theoretical shape of the relationship, where the shape of the preference curve is determined by η, the mean prey size parameter, and σ, the preference parameter which describes the steepness of the curve. These size preference parameters should be estimated from intense short-term sampling of all predator size groups in a small area. The prey size distributions from the stomach contents do not reflect true preferences since prey do not occur in equal numbers in the environment. Concurrent sampling of prey in the area is also desirable to adjust estimates from the stomach contents by actual prey abundance. Large-scale sampling over the whole model area is still necessary to describe prey vulnerability, pelagic and demersal diets, and to verify model estimates.

Growth is described as the change in individual body weight as a function of the rates of energy uptake and breakdown. Rate of energy uptake, or ration, is calculated as a power function related to individual body weight using a modified bioenergetic approach described in detail by Ursin (1967). Three parameters need to be estimated for the ration equation. Feeding level, f, can be estimated from laboratory experiments such as in Sperber et al. (1977). The same feeding level for all species is used in the multispecies Beverton-Holt model. The coefficient of gross anabolism, h, is considered species specific in the model and laboratory feeding experiments should be performed for each species to estimate this coefficient. The exponent m can be estimated several ways according to Ursin (1967). Species specific laboratory feeding experiments or length-at-age data could be used to estimate this parameter but in the multispecies Beverton-Holt model it is constant for all species. In general, the model ignores temperature effects and the separation of food for reproduction and somatic growth. Seasonal growth is not a direct function of temperature but a result of variation in food supply originating from primary production.

In the model the North Sea is considered as one unit and stock migration or geographic separation of stocks is not explicit. Some separation of animals (and therefore food) is accomplished by defining the fraction of a species' population which is pelagic or demersal at a given time. Another important consideration of the model is that growth, ration, and prey preference are functions of individual weight. Finally, the model covers a large geographic area and simulates events occurring over several years for many species by integrating equations describing instantaneous rates of change.

Multispecies VPA models

Several multispecies virtual population analysis (VPA) models are in existence. A VPA is a set of equations describing the catches at age of a given species as a function of natural and fishing mortality at age and numbers at age at a given point in time. The most widely known versions are those of Pope (1979), Helgason & Gislason (1979), and Majkowski (1981). All of these models are essentially the same in their general structure: they are sets of simultaneous VPA's for fish species which are interconnected by predation. The main innovation of multispecies VPA models is that natural mortality at age of a given species is no longer externally estimated but is defined as a function of ration and prey consumption patterns of the other species in the model. The three models differ somewhat in definitions of ration and prey consumption as follows:

(1) Pope's (1979) model defines ration at age of a species as a fixed quantity of mass per year. A fixed fraction of the ration is composed of the other fish contained in the analysis. Prey consumption by a predator is determined by a prey suitability index, a coefficient having values from 0 to 1. The suitability index is similar in form to the one used in the multispecies Beverton-Holt model, but is not derived from a prey-size preference index. Instead it is derived by obtaining empirical food composition data obtained from large-scale stomach sampling of all predator sizes from a particular year and adjusting the suitability values in the model until the field observed food composition is obtained. Presumably only one year's sampling is necessary because a particular aged prey species is considered to have a constant suitability to a particular aged predator species.

(2) Helgason & Gislason's (1979) model defines ration as a power function of the average weight of fish in an age class. The consumption of prey that are not part of the analysis is considered as a constant amount as opposed to Pope's model where other food is a constant proportion of the ration. Prey selection is calculated in the same manner as in the multispecies Beverton-Holt model by the use of a suitability index derived by defining a predator's prey size preference and prey vulnerability. They claim that prey size preference can be estimated from large-scale stomach sampling of all predator sizes over a year. The feeding parameters can then be adjusted in the model until the model's estimates of prey consumed match the field observations.

(3) Majkowski (1981) calculates ration in a fashion similar to Helgason & Gislason as a power function of mean individual weight. He specifically recommends that this ration be based on mean individual weights and temperature dependent coefficients calculated for each year in the analysis. No allowance is made for prey other than the species being modelled. Each predator's average annual food composition is expressed as the fractions by weight of each age group of prey in the predator's ration. This average annual diet composition is estimated directly from field stomach sampling data over the whole area and for all ages of predators. In addition, diet composition is assumed to change every year and therefore it must be estimated for each year in the analysis.

Reduced single-species model

When only one predator is thought to have large predation impact on a prey population a reduced single-species model can be used to calculate the predation mortality induced by that predator alone. The calculation is usually made for a particular year and requires several pieces of information; an estimate of the annual food requirement or ration of the predator, the proportion by weight of the ration that is composed of the prey item of interest, and some kind of estimate of the amount of the total predator population that consumes that particular prey. Several examples of these calculations are shown by Mehl & Westgard (1983), Majkowski & Waiwood (1981), Waiwood & Majkowski (1984), Minet & Perodou (1978) and Dwyer et al. (unpublished). Various methods for estimating ration are used in these models. Ration is calculated by using bioenergetic considerations or using a combination of an evacuation rate model with field estimated stomach content weight data. Since gastric evacuation rates can vary depending on temperature and food type (Elliott & Persson 1978, Jobling 1980, 1981) these rates should be calculated for the range of temperatures and food items encountered for a given fish predator. The distribution of field sampling effort for the estimation of average stomach content weight should be distributed over all times of day and over all seasons, subareas, and fish sizes under consideration. Similarly, the predator diet and population estimates are usually computed directly from field data. Each of these estimates may be broken down into estimates for a particular area, season, and predator length group if sufficient information is available.

Summary of model characteristics

There are many similarities among the diet composition submodels both in form and in sampling requirements. All of the models use a fractional quantity to define either directly the amount of each prey type in a predator's diet as in DYNUMES, the VPA of Majkowski, and in the reduced single-species case or indirectly the amount of prey chosen by a predator through a prey suitability index as originally defined by Andersen & Ursin (1977). DYNUMES, the multispecies Beverton-Holt and Helgason-Gislason's VPA calculations are affected by the amount of available prey; these models are the most dynamic of all the models in this respect. The remaining models use the diet composition in a more fixed computational fashion.

The multispecies Beverton-Holt model's diet composition submodel requires the smallest scale field sampling program in terms of its geographic coverage and short time frame to estimate its parameters. It also demands however, that pelagic and demersal diets be estimated for a given predator which would require bottom and midwater sampling in the same area. The remaining models demand that sampling be carried out a minimum of one year and in some cases several years. All of the models require multi-year sampling for model verification. Additionally, most of the models require that sampling be stratified by predator size, season, and area.

The food intake submodels fall into three different types. DYNUMES uses laboratory estimates of growth and maintenance food requirement coefficients by fish size and temperature to calculate food consumption of a species with a given growth rate. Several models use Ursin's (1967) bioenergetic approach where net growth depends on the bioenergetic values estimated from laboratory experiments. Some of the submodel users have simplified the submodel by ignoring the effect of temperature on the values of the energy coefficients. The single-species model food intake submodel uses the combination of field stomach content data with temperature and food item dependent evacuation rate parameters.

Since Ursin's (1967) bioenergetic food intake submodel uses many species independent parameters it requires the least effort to estimate. The other two types of food intake submodels require several data items for each species. DYNUMES' food intake submodel requires experiments stratified by predator size and temperature. The single-species model's food intake submodel needs laboratory estimates stratified by temperature and food type and field sampling stratified by diel period, season, area, and predator size.

The focus of all the models discussed here is on population level feeding dynamics. Although the submodels used in each of the models are sometimes used for small-scale description of feeding behavior of a species in a restricted area, time scale, or size range their application to population level estimation is a data intensive effort. When several species of multiple ages need to be sampled in areas the size of the Bering or North Seas with time stratification by hour, season, and year the problem takes on overwhelming proportions. Very few field-oriented food habit studies have attempted to sample on such a large scale. Therefore most large-scale models still lack adequate food parameter estimates. In comparison, estimating parameters for the food intake submodels via laboratory experiments is not such a large-scale task. Laboratory experiments are time-intensive, though, and may take months or years to complete. Appropriate temperature ranges, test species, and size ranges of the test species need to be chosen beforehand to be useful to large-scale ecosystem models.

Model choice and research effort allocation

The discussion above presumes that a model has been chosen and a large-scale food habits data collection program has been approved for obtaining parameter estimates of the chosen models. Deciding upon a model depends mostly on the complexity of questions being asked about the system. For instance, if concern is expressed over the current impact in terms of predation mortality of a specific predator population on a particular prey population a reduced single-species model would suffice. However, as questions increase in complexity by including concerns about the dynamics of more species and other important processes such as migration, growth, and reproduction, the type of model chosen also increases in complexity. Whether the use of a more complex model provides a more accurate or valid measure is unclear. Lack of information regarding the mechanisms of processes which need to be modelled and lack of adequate data to parameterize models may actually increase uncertainty about model results.

Sensitivity analysis is a method used to validate model behavior and evaluate the importance of various data inputs in determining model outcomes (Miller 1974, Waide & Webster 1976, Wiens & Innis 1974). This type of analysis can provide information about error in model outputs and help identify model parameters which cause the most change in model outputs, thereby directing research effort toward obtaining more precise estimates of these parameters. Thus, sensitivity analysis provides information which could help in the decisions of model choice and research effort allocation for model parameter estimation.

Sensitivity analyses have been performed on the two most complex models discussed in this paper; the Andersen-Ursin multispecies Beverton-Holt model (Livingston 1983) and a simplified version of Laevastu's DYNUMES model (Livingston 1980, 1985). These sensitivity analyses were of the type called individual parameter perturbation (IPP) where input parameters are perturbed by an amount equal to their range of error. The analysis gives an indication of the amount of error in model outputs, assuming interaction effects among parameters are not significant (Rose 1981).

Table 3 shows some selected results from the sensitivity analysis on a simplified version of the multispecies Beverton-Holt model. Parameters which caused the most change in model outputs are shown along with the amount of the parameter perturbation and the resulting maximum change in model outputs. Changes of 10% in the model input parameters which define diet, food intake, and growth of fishes caused the largest changes in

Table 3. Input parameters in a simplified version of the Andersen-Ursin multispecies Beverton-Holt model which caused the largest changes in the model output parameter estimates of total ecosystem weight.

Input parameter description		Input parameter change (%)	Maximum output parameter change (%)
v	– fraction of consumed food which is assimilated	10	37
η	– mean prey size preference	10	20
h_3	– coefficient in anabolic term of growth equation for a 'cod-like' fish	10	26
m	– power in anabolic term of growth equation	10	98
n	– power in catabolic term of growth equation	10	– 26

model estimates of total weight relative to the perturbation. The largest change in model estimates of weight was produced by the change in m, the power term in the anabolic portion of the growth equation which is also used in food intake calculations.

Table 4 shows the input parameters which caused the largest changes in a much-simplified version of DYNUMES. Again, model outputs were most sensitive to changes in input parameters which defined model processes of growth, food intake, and diet. In particular, model outputs varied most when subregional variations in diet for each fish species were no longer allowed and a fish species had the same diet regardless of geographic location.

These results have several implications with regard to the decisions of research effort allocation and model choice. First, in terms of research effort allocation, both these model depend heavily on input estimates of parameters which describe diet, food intake, and resulting growth of fishes. This implies these parameters should be well estimated if either of these models are used. Thus, research effort should be directed toward obtaining more precise estimates of these parameters.

The results also highlight some problems with model structure which might influence the choice of a model. The results from testing the DYNUMES-like model indicate that a model which can describe subregional variations in diet may have an advantage over a model which has no such capability if the model is to be used for describing dynamics in large geographic areas where subregional diet variation is bound to occur. Similarly, a model which has explicit definition of size may be more favorable to use if prey choice is thought to be a function not only of geographic overlap of predator and prey but also of the relationship between predator size and prey size.

Thus, this sensitivity analysis provides information about the sources of uncertainty in these models; uncertainty due to model structure and uncertainty due to dependence on certain model parameter estimates. The analysis does not provide a definitive answer as to which specific model should be used or whether a large-scale stomach sampling program should be implemented to provide estimates to the chosen model. Ultimately, these choices will depend on a large set of factors such as the nature of the questions asked about a system, the complexity and size of the system, the availability of existing data, and the cost and ease of obtaining the most important data.

References cited

Andersen, K.P. & E. Ursin. 1977. A multi-species extension to the Beverton and Holt theory of fishing with accounts of phosphorous circulation and primary production. Medd. Dan. Fisk. Havunders N.S. 7: 319–435.

Elliott, J.M. & L. Persson. 1978. The estimation of daily rates of food consumption for fish. J. Anim. Ecol. 47: 977–991.

Helgason, T. & H. Gislason. 1979. VPA analysis with species interaction due to predation. Intern. Counc. Explor. Sea C.M. 1979/G: 52.

Jobling, M. 1980. Gastric evacuation in plaice, *Pleuronectes platessa* L.: effects of temperature and fish size. J. Fish Biol. 17: 547–551.

Jobling, M. 1981. Dietary digestibility and the influence of food components on gastric evacuation in plaice, *Pleuronectes platessa* L. J. Fish Biol. 19: 29–36.

Laevastu, T., F. Favorite & H.A. Larkins. 1982. Resource assessment and evaluation of the dynamics of the fisheries resources in the Northeastern Pacific with numerical eco-

Table 4. Input parameters in a simplified version of Laevastu's DYNUMES model which caused the largest changes in the model output parameter estimates of total ecosystem biomass.

Input parameter description		Input parameter change (%)	Maximum output parameter change (%)
G	– growth coefficient	10	– 33
K_g, K_m	– food requirement coefficients for growth and maintenance	20	88
$p_{i,j}$	– proportion of prey i in diet of predator j in each subregion	disregard subregional variation	127

system models pp. 70–81. *In:* M.C. Mercer (ed.) Multispecies Approaches to Fisheries Management Advice, Can. Spec. Publ. Fish. Aquat. Sci. 59. 169 pp.

Livingston, P.A. 1980. The Bulk Biomass Model: A stock assessment tool? M.S. Thesis, University of Washington, Seattle. 65 pp.

Livingston, P.A. 1983. Potential use of the Andersen-Ursin multispecies Beverton-Holt model for modeling North Pacific fish interactions. U.S. Dep. Commer., NOAA Tech. Memo. NMFS F/NWC–43. 31 pp.

Livingston, P.A. 1985. An ecosystem model evaluation: The importance of fish food habits data. Mar. Fish. Rev. 47: 9–12.

Majkowski, J. 1981. Application of a multispecies approach for assessing the population abundance and the age structure of fish stocks. Can. J. Fish. Aquat. Sci. 38: 424–431.

Majkowski, J. & K.G. Waiwood. 1981. A procedure for evaluating the food biomass consumed by a fish population. Can. J. Fish. Aquat. Sci. 38: 1199–1208.

Mehl, S. & T. Westgard. 1983. The diet and consumption of mackerel in the North Sea. Intern. Counc. Explor. Sea C.M. 1983/H: 34.

Mercer, M.C. (ed.) 1982. Multispecies approaches to fisheries management advice. Can. Spec. Publ. Fish. Aquat. Sci. 59. 169 pp.

Miller, D.R. 1974. Sensitivity analysis and validation of simulation models. J. Theor. Biol. 43: 345–360.

Minet, J.P. & J.B. Perodou. 1978. Predation of cod, *Gadus morhua,* on capelin, *Mallotus villosus,* off eastern Newfoundland and in the Gulf of St. Lawrence. ICNAF Res. Bull. 13: 11–20.

Pope, J.G. 1979. A modified cohort analysis in which constant natural mortality is replaced by estimates of predation levels. Intern. Counc. Explor. Sea C.M. 1979/H: 16.

Rose, K.A. 1981. A review and comparison of parameter sensitivity methods applicable to large simulation models. M.S. Thesis, University of Washington, Seattle. 50 pp.

Sperber, O., J. From & P. Sparre. 1977. A method to estimate growth rate of fishes, as a function of temperature and feeding level, applied to rainbow trout. Medd. Dan. Fisk. Havunders N.S. 7: 275–317.

Ursin, E. 1967. A mathematical model of some aspects of fish growth, respiration, and mortality. J. Fish. Res. Board. Can. 24: 2355–2453.

Waide, J.B. & J.R. Webster. 1976. Engineering systems analysis: Applicability to ecosystems. pp. 329–371. *In:* B.C. Patten (ed.) Systems Analysis and Simulation Ecology, Vol. 4. Academic Press, New York.

Waiwood, K. & J. Majkowski. 1984. Food consumption and diet composition of cod, *Gadus morhua,* inhabiting the southwestern Gulf of St. Lawrence. Env. Biol. Fish. 11: 63–78.

Wiens, J.A. & G.S. Innis. 1974. Estimation of energy flow in bird communities. A population bioenergetics model. Ecology 55: 730–746.

Received 15.12.1984 *Accepted 5.7.1985*

Fish predation and local prey diversity

Mark A. Hixon
Department of Zoology and College of Oceanography, Oregon State University, Corvallis, OR 97331, U.S.A.

Keywords: Species richness, Species evenness, Piscivory, Planktivory, Benthivory, Competition, Disturbance, Recruitment, Spatial refuges, Freshwater systems, Marine systems

Synopsis

Local species richness and evenness of a prey assemblage can change in a variety of patterns as the intensity of predation by fishes increases. Documenting this pattern requires measurements of prey diversity at a minimum of three widely spaced predation intensities. The particular pattern that occurs depends largely upon: (1) the processes structuring the prey assemblage in the absence of fish predation (recruitment limitation, physical disturbances, non-fish predators, mutualisms, competitive networks, or competitive hierarchies); (2) the pattern of prey population reduction by fish predation (equivalent, disproportionate on competitive dominants, or disproportionate on competitive subordinates); and (3) the processes governing local extinctions and immigrations in the prey assemblage (prey refuges, recruitment patterns, and local versus global species richness). A literature review of the impact of fish predation on other fishes, plankton, and benthos in a variety of freshwater and marine systems indicated that local effects on prey diversity are seldom documented adequately. Only seven experiments documented both prey diversity as a function of at least three predation intensities, as well as the probable mechanisms underlying the prey diversity response. Of these, two studies of freshwater benthos detected virtually no effects on prey diversity, apparently due to rapid prey recolonization or predator inefficiency. Two studies of freshwater zooplankton and three studies of coral-reef algae detected keystone-predator induced hump-shaped patterns characteristic of the intermediate-disturbance hypothesis. Two additional studies of coral-reef benthos detected negative effects on prey diversity over a range of at least three predation intensities, but the underlying mechanisms were unclear. Spatial prey refuges, particularly those provided by aquatic plants on soft bottoms and crevices on hard bottoms, have been shown to minimize the effects of fish predation on the diversity of benthic assemblages in both freshwater and marine systems. The concepts summarized in this paper suggest improvements for future studies of the impact of fishes on prey community structure.

Introduction

Species diversity is one of the major parameters characterizing ecological communities (Pielou 1975). The fact that predators can influence the local or within-habitat diversity of their prey has been known at least since the time of Darwin, and the general importance of predators in structuring prey communities is well documented (review by Connell 1975). However, only recently have the effects of predation by fishes on prey diversity been studied explicitly. (Here I include both carnivory and herbivory as forms of predation [Lubchenco 1979].) Early field manipulations in freshwater systems showed that fish predation can influence the local abundance of plants (Anderson 1950, Threinen & Helm 1954) and invertebrates (Ball & Hayne 1952, Hayne & Ball 1956), but effects on prey diversity per se were not reported. Not until the 1960's, with the surge of interest in community ecology, did researchers design studies to determine how fishes influence the local species richness and relative abundances of their prey.

The goals of this review are threefold. First, I will synthesize the general concepts on how predators may affect local prey diversity. This synthesis indicates the kinds of information needed for determining both the patterns by which prey diversity responds to increasing predation intensity, as well as the mechanisms underlying those patterns. Second, I will review field studies of the effects of fish predation on prey diversity in various freshwater and marine systems in an effort to document the extent of our knowledge on this subject. This survey includes papers appearing in major English-language journals and books published through 1984. Third, I will suggest approaches for future research. My ultimate intent is to provide a general conceptual framework for further studies.

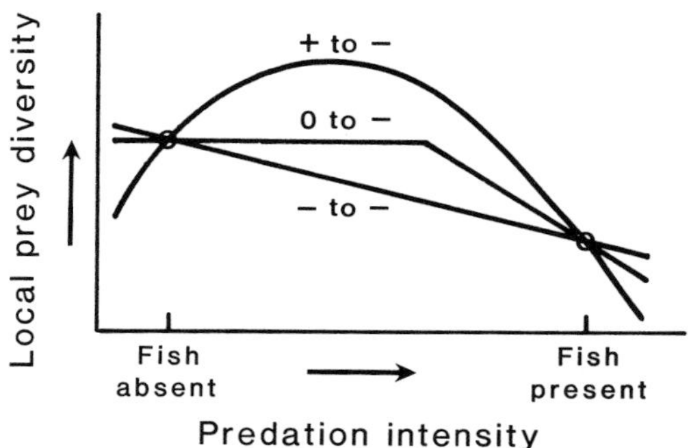

Fig. 1. Fish removal or exclosure experiments provide data on only two levels of predation intensity (fish present and fish absent), which are insufficient to determine the overall shape of the prey diversity-response curve. In this example, the same two points could lie on any one of three different curves: a hump-shaped response (+ to −), no response shifting to a negative response (0 to −), and a purely negative response (− to −).

Definitions

The pattern by which predation affects prey diversity is determined by examining the 'local species diversity' of the prey assemblage as a function of 'predation intensity', that is, by examining what I will call the prey 'diversity-response' curve (see Fig. 1 for examples). Before proceeding, these key variables must be defined. 'Species diversity' consists of two components (Pielou 1975): species number (richness) and equitability of relative abundances among species (evenness). Popular indices of diversity (such as the Shannon-Wiener H') combine these components into a single composite measure (reviews by Peet 1974, Pielou 1975). I will call such measures 'composite diversity' to distinguish them from measures of richness and evenness. Because richness and evenness can vary independently and because different composite measures differ in their sensitivity to changes in these components, it is necessary to consider richness and evenness separately if the mechanisms underlying prey diversity responses are to be elucidated.

'Local' species diversity is a more problematical and subjective concept. Although 'local' generally means 'within-habitat', the spatial scale of the local system can be either the entire habitat or a subsection of any size, determined entirely by the area or volume of the habitat that a researcher studies. Thus, for two examples, 'local' can refer to the entire plankton community in a small pond or to the benthic assemblage occupying a single settling plate on a large coral reef. This fact renders comparisons between different systems difficult, if not impossible. In any case, this paper deals with local or within-habitat (as opposed to global or between-habitat) species diversity.

Because the prey diversity response includes changes in evenness, the definition of predation intensity must include the effects of predators on the sizes of prey populations. Therefore, 'predation intensity' is defined here as a measure of the extent to which the combined abundances of the affected prey species are reduced by predation relative to their combined abundances in the absence of predation. The term 'affected' is important because some prey populations may not be reduced significantly by predators. Thus, if a prey population responds to increased predation pressure by compensatory increases in reproduction and/or recruitment, such that the population size is unaffected, then the effective predation intensity is zero as far as any change in prey diversity is concerned. Similarly, if spatial refuges or prey defenses prevent predators from affecting prey population sizes, then the effective predation intensity is also zero. Because of these factors, the most common measure of predation intensity – the population density of predators – may be inaccurate in some cases and is best replaced by direct measures of prey population reductions.

General concepts on predation and prey diversity

Our knowledge of the patterns and mechanisms by which predators affect the diversity of their prey has come from a large number of empirical studies (reviews by Harper 1969, Connell 1975, Murdoch & Oaten 1975, Lubchenco & Gaines 1981, Sousa 1984). Particular progress in this area has been made in studies of rocky intertidal systems (Paine 1966, Paine & Vadas 1969, Menge & Sutherland 1976, Lubchenco 1978, reviews by Paine 1977, Hughes 1980). Close behind the fieldworkers have been the theoreticians, who have modelled various scenarios concerning the effects of different kinds of predation on the diversity of different kinds of prey assemblages (MacArthur 1972, Smith 1972, Emlen 1973, Van Valen 1974, Holt 1977, Caswell 1978, Hastings 1978, Yodzis 1978, Crowley 1979, Huston 1979). These empirical and theoretical studies have produced a set of concepts that cannot be attributed to any single investigator; the following synthesis reflects and extends the work of many individuals.

As with any conceptual generalization, many complicating factors of real systems are excluded from this review. Two such complexities are perhaps most relevant to the present analysis. The first is the role of ontogenetic shifts in species interactions, such as cases where one species may prey upon the young of another species, yet be competitively dominated by the surviving adults of that same species (review by Werner & Gilliam 1984). Thus, the boundary between predation and competition is not always as obvious as the general concepts would lead us to believe. The second complexity is the temporal scale of the prey diversity response. Such responses do not occur instantly. If predation intensity changes suddenly, as in the case of all predator manipulation experiments, a certain (often unknown) period of time will pass before prey diversity stabilizes. Although few field studies have determined whether a documented prey diversity response is transient or equilibrial, long-term studies should be regarded as more appropriate than short-term studies in evaluating the concepts discussed below. This is because these concepts are generally based on the assumption that prey diversity reaches some equilibrium value at each successive level of predation intensity.

To determine how fish may affect the local diversity of their prey, the structure of the prey assemblage in the absence of fish predation must first be known. With fish absent, the prey assemblage at any given time will be structured predominantly by one of five categories of factors: (1) pre-settlement factors (recruitment limitation, e.g., Keough 1983) or post-settlement events (physical disturbances, e.g., Sousa 1979) which limit population abundances to levels where substantial interactions among the prey species are precluded; (2) predation (or parasitism) by animals other than fishes; (3) mutualisms among the prey species; (4) nontransitive competitive networks (Buss & Jackson 1979); or (5) linear competitive hierarchies. Note that, while one of these factors should predominate at any one time, any given prey assemblage is probably structured by a cyclical or noncyclical sequence of several factors over the course of ecological time. Thus, considering these factors separately is a further (albeit necessary) conceptual simplification. In any case, once the determinants of the richness and evenness of the prey assemblage in the absence of fish predation are known, the effects and underlying mechanisms of the addition of fishes can be inferred by comparison.

Little information is available on the effects of adding fish predators to prey systems structured by recruitment limitation, physical disturbances, or non-fish predation, and virtually none is available for systems structured by mutualisms or competitive networks. The following discussion hypothesizes the prey diversity responses that would occur in prey assemblages structured by each of these categories of factors, as summarized in Table 1.

In the first category (Table 1A), limited recruitment and/or physical disturbances prevent the prey populations from reaching densities where they interact, so the local population size of any one prey species is unaffected by that of any other prey species. Therefore, the addition of fish to the system may or may not increase prey evenness, depending upon which prey populations are reduced. At high predation intensities, prey richness will decline as some species become locally extinct. Due to the lack of interactions between the prey populations, the extinction of one prey species cannot allow the immigration of any new prey species, so predation should not increase prey richness in such systems (Lubchenco & Gaines 1981).

In the second category (Table 1B), the prey assemblage in the absence of fish predation is structured by non-fish predators. Here, the effects of adding fish depend to a large extent on whether or not the fish consume the non-fish predators as well as the original prey. If they do not (e.g., if the non-fish predators are larger than the fish), then the fish may intensify the effects of the non-fish predators, be they positive or negative, only if the diets of the fish and non-fish predators are similar. Otherwise, the combined impact of all predators must be known to determine the prey diversity response and its underlying mechanisms. If the fish do consume the non-fish predators, then the effects on prey diversity are still unpredictable without extensive knowledge of the system. For example, if the non-fish predators enhance prey diversity in the absence of fish, then disproportionate consumption of these predators by fish may indirectly cause a decrease in overall prey diversity. On the other hand, if non-fish predators reduce prey diversity, then consumption of these predators by fish may enhance overall prey diversity indirectly by freeing the lowest trophic levels from predation by intermediate levels. This latter outcome, involving three trophic levels, has been suggested as an alternative to the 'size-efficiency hypothesis' discussed below for freshwater zooplankton.

In the third category, mutualisms (Table 1C), the population size of each prey species is positively affected by those of the others. Therefore, the addition of predators to the system will probably not affect prey evenness; as one mutualist is reduced in abundance, the others will decline proportionately. However, if predation intensity reaches a level where a prey species becomes locally extinct, so may one or more mutualist species, causing a decrease in prey richness. Because mutualisms are often coevolved (Thompson 1982), any immigration of new species into the system facilitated by predation removing the mutualists would possibly change the character of the system to one of the other categories. Note that our knowledge of the role of mutualisms in most communities is virtually nonexistent (Addicott 1984).

In the fourth category, competitive networks (Table 1D), a kind of competitive balance is maintained among prey species. For example, species A outcompetes species B, species B outcompetes species C, but species C outcompetes species A

Table 1. Changes in the local prey diversity of systems not structured by competitive hierarchies following increases in fish predation intensity.

Process structuring prey assemblage in the absence of fishes	Response[1] of prey diversity to increased fish predation intensity		
	Richness	Evenness	Composite
A. Recruitment limitation or physical disturbance	0 to −	? to ?	? to ?
B. Non-fish predation	? to ?	? to ?	? to ?
C. Mutualism	0 to −	0 to 0	0 to −
D. Competitive network	0 to −	? to −	? to −

[1] 0: none, −: decrease, ?: unpredictable without additional information (see text). The format 'a to b' indicates response 'a' at low predation intensities shifting to response 'b' at high intensities (see Fig. 1 for examples).

(Buss & Jackson 1979). Unless the intensity of predation is exactly equivalent on all prey species (Lubchenco & Gaines 1981), the addition of predators to the system can only break this competitive balance, eventually reducing both prey richness and evenness at high predation intensities. If predation facilitated the immigration of new species into the system, then the character of the system would likely change to that of a competitive hierarchy.

In summary, adding fish to prey systems structured by recruitment limitation, physical disturbances, non-fish predators, mutualisms, or nontransitive competitive networks may or may not affect prey evenness, but should usually cause a decrease in prey richness at high predation intensities (Table 1).

The effects of adding predators to prey assemblages structured predominantly by linear competitive hierarchies has received the most attention in the literature. In the absence of predation, the superior competitors among the prey should come to numerically dominate such a system (Paine 1984). Competitively subordinate prey species may persist in the system due to some low level of physical disturbance, with fugitive strategies or priority effects (sensu Paine 1977) perhaps allowing established subordinates to hold space against dominants. In any case, the prey system in the absence of predators should be dominated by the superior competitors, with subordinate competitors being rare or even locally excluded. If predators are added to the system and proceed to effectively reduce the population size of one or more prey species, then the effect on prey diversity depends largely upon two factors: first, which prey species are reduced with respect to the competitive hierarchy of the prey (Lubchenco 1978); and second, the pattern by which predation and other factors effect the local extinction of species in the prey assemblage and/or the immigration of new prey species.

First, consider the pattern of prey reduction by predation. Predators may: (1) reduce all prey by the same proportion (called 'equivalent predation' by Van Valen 1974); (2) disproportionately reduce one or more competitively dominant prey species; or (3) disproportionately reduce one or more competitively subordinate species. Obviously, an equivalent predator is a trophic generalist. However, the other two kinds of predators may be either generalists or specialists. All that matters in terms of the diversity responses of the prey assemblage is that either competitively dominant or subordinate species are differentially reduced in abundance relative to the other prey species.

Now, consider the pattern of local extinction and immigration of prey species. If predation reduces the total abundance of prey under any of the above patterns, then the possibility exists that species will become locally extinct and/or new species will invade the system. Four general patterns are possible (Table 2): (I) no extinctions or immigrations; (II) extinctions with no immigrations; (III) no extinctions with immigrations of new species; and (IV) both extinctions of established species and immigrations of new species. Local extinctions are likely when prey refuges are locally scarce and recruitment of members of the original prey assemblage is limited. Immigrations of new species are likely when the local species pool is smaller than the regional species pool and barriers to dispersal and establishment are few. Thus, scenarios (I) and (II), lacking immigration, are most likely to occur in isolated systems, such as small ponds where the local system is the entire pond, or in larger systems where the availability of immigrants of new species is limited.

Scenario (IV) in Table 2 is the most complex situation, but as far as responses in prey diversity are concerned, all that is important is the net change in the number of prey species after each successive increase in predation intensity. Thus, if the number of extinctions of original prey species is exactly balanced by immigrations of new species, then the net outcome is identical to that of scenario (I). Similarly, if the numbers of extinctions and immigrations are not equal, then the net outcome can be identical to either scenario (II) or (III). Of greater interest is an outcome where immigrations exceed extinctions at moderate predation intensities, and vice versa at high intensities. In such a case, as predation intensity increases from zero, prey richness initially will be unaffected, then in-

Table 2. Local extinction and immigration scenarios in prey assemblages subjected to predation, each listed with contributing causative factors.

Immigration of new prey species?	Extinction of established prey species? No:		Yes:	
No:	I:	many refuges; local = regional species pool	II:	few refuges; local = regional species pool
Yes:	III:	many refuges; local < regional species pool	IV:	few refuges; local < regional species pool

crease, and ultimately decrease. I will consider this outcome in detail.

Considering all combinations of the three patterns of prey population reduction and the four patterns of prey extinction and immigration presented above, 12 general patterns of prey diversity responses to increasing predation intensity result (Table 3). The apparent complexity of Table 3 disappears if each column is examined separately. Entries in the column labelled 'richness' are set by the definitions of the extinction/immigration scenarios (see Table 2). Presumably, prey richness is not affected at low predation intensities, indicated by '0' entries. Each entry in the column labelled 'composite' combines the adjacent listings in the richness and evenness columns. In cases where a positive response ('+') in one column is combined with a negative response ('−') in the other (e.g., Case A.II), the composite outcome is unpredictable ('?'). Most entries in the column labelled 'evenness' have not been considered in previous publications, so an explanation for each follows:

Case A.I: By definition, equivalent predation decreases all prey populations by the same proportion. If richness does not change, then evenness will be unaffected (Van Valen 1974).

Case A.II: If predation intensity reaches a level where rare species in the system go extinct, then evenness will increase with each extinction because the remaining species will exhibit more equitable relative abundances.

Case A.III: If new species invade the system, the

Table 3. Predator induced changes in local prey diversity of systems structured by competitive hierachies.

Pattern of prey reduction by predation	Extinction/immigration scenario (Table 2)	Response[1] of prey diversity to increased predation intensity		
		Richness	Evenness	Composite
A. Equivalent	I	0 to 0	0 to 0	0 to 0
	II	0 to −	0 to +	0 to ?
	III	0 to +	0 to ?	0 to ?
	IV	0 to + to −	0 to ? to +	0 to ? to ?
B. Disproportionate on competitively dominant species	I	0 to 0	+ to −	+ to −
	II	0 to −	+ to −	+ to −
	III	0 to +	+ to ?	+ to ?
	IV	0 to + to −	+ to ? to −	+ to ? to −
C. Disproportionate on competitively subordinate species	I	0 to 0	− to −	− to −
	II	0 to −	− to −	− to −
	III	0 to +	− to ?	− to ?
	IV	0 to + to −	− to ? to −	− to ? to −

[1] 0: none, +: increase, −: decrease, ?: unpredictable (see text). The format 'a to b' indicates response 'a' at low predation intensities shifting to response 'b' at high intensities. The format 'a to b to c' indicates responses at low ('a'), moderate ('b'), and high ('c') predation intensities. Thus, '+ to −' and '+ to ? to −' both indicate hump-shaped responses (see Fig. 1).

effect on evenness depends upon the number and relative abundances of the immigrants relative to the original assemblage. The pattern of immigration cannot be predicted a priori.

Case A.IV: This situation is a combination of Case A.I at low predation intensities, Case A.III at moderate intensities, and Case A.II at high intensities.

Case B.I: Disproportionate predation on the competitively and numerically dominant species increases evenness at low intensities. This outcome occurs whether or not the subordinate prey increase in abundance due to competitive release because the abundances of the dominants are reduced toward those of the subordinates, thus increasing the equitability of relative abundances. This outcome is the basis of Paine's (1969) 'keystone species' concept. However, at greater predation intensities the competitive dominants will be reduced to the point of becoming relatively rare, and evenness will subsequently decline.

Case B.II: This situation will be identical to the previous case, except that as each dominant species becomes extinct, evenness may increase slightly (as in Case A.II) before continuing to decline as predation intensity progressively increases and reduces the next most dominant species.

Case B.III: As the abundances of the dominant species are reduced toward those of the subordinates at low predation intensities, evenness will increase (as in Case B.I). However, once new species immigrate, the effect on evenness becomes unpredictable (as in Case A.III).

Case B.IV: This situation is a combination of Case B.I at low predation intensities, Case B.III at moderate intensities, and Case B.II at high intensities.

Case C.I: Disproportionate predation on the relatively rare, competitively subordinate species can only decrease evenness because the rare species in the system become even rarer.

Case C.II: This situation is similar to the previous case in the same way Case B.II resembles Case B.I (see above).

Case C.III: As the relatively rare subordinate species become even rarer at low predation intensities, evenness will decrease (as in Case C.I). However, once new species immigrate, the effect on evenness becomes unpredictable (as in Case A.III).

Case C.IV: This situation is a combination of Case C.I at low predation intensities, Case C.III at moderate intensities, and Case C.II at high intensities.

An overview of Table 3 provides two major conclusions. First, only half of the 12 prey diversity-response patterns listed produce outcomes that are uniequivocal for all three diversity measures: case A.I, where equivalent predation has no effect on prey diversity (Van Valen 1974); cases B.I, B.II, and B.IV, where disproportionate predation on competitively dominant species causes hump-shaped (+ to −) responses in prey evenness and composite diversity, characteristic of the 'intermediate-disturbance hypothesis' (Connell 1978); and cases C.I and C.II, where disproportionate predation on competitively subordinate species decreases prey evenness and composite diversity (Lubchenco 1978). No case unequivocally produces a positive response over a full range of predation intensities, although such responses are possible (e.g. case B.III).

Second, the shape of the curve describing prey evenness or composite diversity as a function of predation intensity can, at least theoretically, assume any shape in situations where new species immigrate to the affected system (scenarios III and IV in Tables 2 and 3). Therefore, if new species become established in the system, the shape of the prey diversity-response curve cannot be predicted a priori.

Of what use are these ideas and patterns in understanding the effects of fish predation on prey diversity? Although derived from non-fish systems, these concepts indicate that, first, predicting the shape of the prey diversity-response curve is often impossible, and second, understanding the mechanisms causing observed patterns requires considerable knowledge of both the prey assemblage and the fishes. Of particular practical importance is the fact that at least three widely spaced levels of predation intensively must be investigated to adequately determine the shape of the prey diversity-response curve. As illustrated in Figure 1, the common experimental design in which fish are simply removed from a system or locally excluded by cages, providing only two levels of predation intensity, is inadequate. As will become clear below, this presence/absence problem appears repeatedly in the literatue.

I will now review the literature on the effects of fish predation on prey diversity in an effort to extract any general patterns, and to determine which studies have adequately determined both the shape of the prey diversity-response curve and the underlying mechanisms. I will not list the particular design flaws of each individual study. For such critiques, the reader is referred to Allan (1983, 1984) for freshwater systems and Choat (1982) for marine systems. The review is divided into two major sections: freshwater systems and marine systems (including estuaries). Within each section, I will consider three general prey categories (other fishes, plankton, and benthos).

Freshwater systems

General reviews of predator-prey interactions involving freshwater fishes are provided by Macan (1977), Clepper (1979), and Healey (1984). Most published studies have not considered the effects of fish predation on prey diversity.

Piscine prey

Despite a number of experimental studies showing that predatory fishes can affect the standing stocks of prey fishes (Foerster & Ricker 1941, Garman & Nielsen 1982), few studies have investigated the effects of native predatory fishes on the local diversity of native prey fishes. Perhaps the most relevant study is that of Clady & Nielsen (1978), who sampled the fish fauna of Oneida Lake, New York, over a 13-year period. They found that the species richness and composite diversity of prey fishes other than yellow perch was positively correlated with the abundance of young-of-the-year yellow perch, *Perca flavescens*. The walleye, *Stizostedion vitreum,* which was the dominant piscivore in this system, apparently preferred the young yellow perch as prey, so predation intensity on other fishes was reduced when yellow perch were abundant (Forney 1974). Clady & Nielsen thus suggested that their pattern was interpretable as a negative relationship between the evenness of prey species other than yellow perch and the intensity of walleye predation on those species, with apparent changes in prey richness due to some species becoming so rare that they were not sampled. They proposed that the mechanisms causing this pattern were those described for a negative response in prey diversity in a system where few competitive interactions occur among the prey (Table 1A). However, as the authors noted, no data were gathered on either the food habits of walleye during this period or the interactions between the prey fishes, so the proposed pattern and mechanisms were entirely speculative.

The most compelling evidence of predatory fishes affecting the local diversity of prey fishes comes from studies of introduced piscivores (see Courtenay & Stauffer 1984, Moyle 1985 for reviews). Because introduced predators and native prey usually have not coevolved, the prey often lack refuges from these new sources of mortality. In such circumstances, the impact of novel predators on prey assemblages can be extreme. For example, the introduction of brown trout, *Salmo trutta,* to the continental United States has resulted in the local decline of a variety of native prey fishes in a variety of systems (review by Taylor et al. 1984). Similarly, Christie (1974) among others reviewed the general decline and extirpation of native fishes following the introduction of the sea lamprey, *Petromyzon marinus,* to the Great Lakes (see also Smith 1980 and included papers). Crowder (1980) has hypothesized that exotic planktivorous fishes have contributed to this decline by consuming the pelagic eggs and larvae of native fishes. In one of the most dramatic documented impacts of an exotic predatory fish, the cichlid *Cichla ocellaris* caused the local extinction of six native fish species in man-made Gatun Lake, Panama (Zaret & Paine 1973). While such examples are definite cases of predators decreasing prey richness, the mechanisms causing these patterns have not been documented.

Once clear pattern that has emerged from studies of predator-prey interactions among freshwater fishes is the importance of spatial refuges in moderating the effects of piscivores. In the Gatun Lake study, a turbid river flowing into the lake provided a refuge for native species; apparently *Cichla* was an ineffective predator in murky water (Zaret 1979). Other observations (Jackson 1961), recently bolstered by experimental studies (Fraser & Cerri 1982, Savino & Stein 1982, Werner et al. 1983, Mittelbach 1984), have demonstrated that aquatic vegetation provides an effective refuge for prey fishes. However, I could find no studies documenting prey fish diversity as a function of predation intensity over a range of refuge availabilities.

Planktonic prey

Interactions between planktivorous fishes and freshwater zooplankton have been studied more intensively than any other predator-prey relationship involving fishes (see Nilsson 1978, Kerfoot 1980, Zaret 1980 for reviews). A well-documented pattern is that zooplankton species composition shifts toward smaller forms when planktivorous fishes are abundant (Hrbaček et al. 1961, Hrbaček 1962, Brooks & Dodson 1965, Reif & Tappa 1966, Galbraith 1967, Hall et al. 1970, Wells 1970, Hutchinson 1971, Nilsson 1972, Sprules 1972, Stenson 1972, Warshaw 1972, Nilsson & Pejler 1973, Andersson et al. 1978, Stenson et al. 1978, Lynch 1979, Stenson 1982, Hamrin 1983, Spencer & King 1984, but see Drenner et al. 1982). Recently, this pattern has also been documented in phytoplankton systems (Drenner et al. 1984). However, very few of these studies reported explicit changes in measures of the diversity of the plankton community at different intensities of fish predation.

Of the few studies that did report responses in plankton diversity, most investigated only two levels of fish predation (fish present and fish absent). This fact makes it impossible to know the overall shape of the prey diversity-response curve (Fig. 1). In any case, most studies comparing zooplankton communities in lakes with and without planktivorous fishes, or before and after planktivorous fishes were introduced, reported a decrease in the species richness of the zooplankters investigated (Brooks & Dodson 1965, Galbraith 1967, Pope et al. 1973, Carter & Kwik 1977, Northcote et al. 1978, Stenson 1978, Von Ende 1979, Doroshev in Li & Moyle 1981, Hurlbert & Mulla 1981). In contrast, planktivory by fishes enhanced all components of zooplankton diversity in an extensive experimental study of small pond systems (Hall et al. 1970).

Several correlative studies provided numerous between- or within-system comparisons of zooplankton diversity, possibly caused by a range of (i.e. more than two) predation intensities. For example, Nilsson & Pejler (1973) made comparisons between twenty-eight Swedish lakes. Zaret (1980) interpreted their data as demonstrating that zooplankton species richness increased montonically with increasing fish predation intensity. Within Swedish lakes, Hamrin (1983) correlated seasonal changes in zooplankton diversity with seasonal changes in the abundance of an obligate planktivore, the vendace or ciscoe, *Coregonus albula*. Zooplankton species richness decreased montonically as vendace populations increased. Of course, such studies share the problem of all 'natural experiments': lack of controls (Connell 1975).

The general conclusion that emerges from such studies is that predation has a definite impact on local plankton diversity, but the nature of that impact is usually unknown (or at least unreported). However, the behavioral components of fish planktivory have been well-documented: particulate-feeding fishes select large and/or highly visible zooplankters lacking morphological defenses (review by O'Brien 1979). (This generalization may not apply to filter-feeding fishes [Drenner et al. 1982, see also Janssen 1980].) To understand the mechanisms by which such size-selective predation affects zooplankton diversity, the processes structuring the plankton community in the absence of fishes must be known. Brooks & Dodson (1965) suggested that large zooplankters dominate smaller species in competition for food. Thus, as fishes reduce the abundance of large species, the smaller species would increase in abundance due to competitive release (review by Hall et al. 1976). The resulting effects on zooplankton diversity would be the same as those listed in Table 3B, often assuming a hump-shaped response over a full range of predation intensities.

This idea, known as the 'size-efficiency hypothesis', has been challenged by the alternative hypothesis that fishes may reduce not only large zooplankters but also the invertebrate predators of small zooplankters. Therefore, small species may increase in abundance due to release from predation rather than competition (Dodson 1974). Moreover, both Neill (1975) and Lynch (1978, 1979) determined experimentally that the moderate-sized cladoceran *Ceriodaphnia* competitively dominates larger zooplankters in some systems. These two studies are the only I found that not only documented the processes structuring the prey assemblage, but also investigated more than two intensities of fish predation.

Neill's (1975) laboratory experiments were conducted in replicate two-liter microcosms originally stocked with twelve species of crustacean zooplankton, four of which were at detectable abundances at the beginning of the predation tests (Table 4). The microcosms were subjected to predation by zero, one, or two mosquitofish, which preyed disproportionately on the competitively dominant zooplankters. Explicit diversity indices were not reported. However, the equilibrium zooplankton richnesses after 13 weeks were four species with no fish, seven species with one fish, and six species with two fish (Neill's Table 3). Thus, the response of prey richness (and probably composite diversity) to increasing predation intensity was hump-shaped. Although these systems were unrealistic, given the artificial composition of the initial prey assemblages, this experiment documented a hump-shaped response similar to case B.IV (Table 3).

Lynch's (1979) field experiments were conducted in approximately cubic-meter enclosures containing natural pond-water zooplankton assemblages subjected to predation by bluegill sunfish (Table 4). There were basically three intensities of predation: no fish, one or two fish, and four or five fish. Fish preyed disproportionately on the competitively dominant zooplankters, as well as on the planktonic predator *Chaoborus*. Again, explicit diversity indices were not reported. Nonetheless, by the end of the experiment, the enclosures with no fish contained a total of nine zooplankton species, those with moderate fish predation contained twelve species, and those with the greatest fish predation contained eleven species (Lynch's Fig. 3 and 4). Similar to Neill's study, these results suggest that the response of prey richness (and perhaps composite diversity) to increasing predation intensity was hump-shaped. However, the mechanisms causing this response were probably a combination of those described for both case B.IV (Table 3), since Lynch's enclosures appeared to be open to immigration, as well as a release from invertebrate predation (Table 1B), since the fish removed *Chaoborus*.

Benthic prey

Most studies of the effects of fish predation on benthic plant and animal communities in freshwater reported a negative impact on the abundance of benthos, but no explicit data on benthic species diversity (Anderson 1950, Ball & Hayne 1952, Threinen & Helm 1954, Hayne & Ball 1956, Straškraba 1965, Kajak et al. 1972, Andersson et al. 1978, Stenson et al. 1978, Crowder & Cooper 1982, Power & Matthews 1983, Hemphill & Cooper 1984, Post & Cucin 1984, Power et al. 1985, reviews by Prejs 1984 for herbivorous fishes and Taylor et al. 1984 for introduced fishes). An apparently common pattern is that the abundances of large invertebrates are reduced more extensively than those of smaller species (Hall et al. 1970, Crowder & Cooper 1982, Hemphill & Cooper 1984, Morin 1984a, Post & Cucin 1984), similar to the effect of size-selective predation by planktivores (see above).

Most experimental studies that did report diversity information investigated only two levels of fish predation (fish present and fish absent), so the overall shape of the prey diversity-response curve was unknown (Fig. 1). No differences in the species richness and minor differences in the evenness of benthic invertebrates between the two treatments were reported in most experiments, which included fish density manipulations in ponds (Hall et al. 1970) and streams (Zelinka 1974, Allan 1982), as

Table 4. Field and microcosm experiments investigating the effects of at least three levels of fish predation intensity upon local prey diversity.

Source & System	Predator[1]	Prey	Predation intensity[2]		Prey diversity response[3]			Apparent mechanism
			Manipulation	Duration	Richness	Evenness	Composite	
FIELD EXPERIMENTS:								
Lynch 1978, 1979 freshwater pond (Minnesota)	sunfish[a]	zoo-plankton	enclosures: L: 0 fish, M: 1–2 fish, H: 4–5 fish	40 days	+ to −	nr	nr	Table 3, Case B.IV; & Table 1B
Flecker 1984 freshwater stream (W. Virginia)	sculpins[b]	aquatic insects	enclosures: L: 0 fish, M: 3–6 fish, H: 12 fish & open	2 weeks	0 to 0	nr	nr	rapid prey recolonization
Gilinsky 1984 freshwater pond (N. Carolina)	sunfish[a]	benthic inverts.	enclosures: L: 0 fish, M: 20 fish, H: 60 fish	1 year	0 to 0	nr	nr	predator inefficiency
Lassuy 1980 marine coral reef (Yap Island)	damsel-fish[c], other fishes	benthic algae	L: caged, M: exposed to other fish, H: weeded by damselfish	2 months	nr	− to −	− to −	predator weeding behavior
Lassuy 1980 marine coral reef (Guam)	damsel-fish[d], other fishes	benthic algae	L: caged, M: defended by d-fish, H: exposed to other fish	2 months	nr	+ to −	+ to −	Table 3, Case B.IV
Hixon & Brostoff 1981, 1982, 1983 marine coral reef (Hawaii)	damsel-fish[e], parrot-fishes, surgeon-fishes	benthic algae	L: caged, M: defended by d-fish, H: exposed to p-fishes & s-fishes[4]	1 year	+ to −	+ to −	+ to −	Table 3, Case B.IV
Sammarco 1983 marine coral reef (Australia)	damsel-fish[c], other fishes	benthic algae	L: caged, M: defended by d-fish H: exposed to other fish	11 months	+ to −	+ to −	+ to −	Table 3, Case B.IV
MICROCOSM EXPERIMENTS:								
Neill 1975 freshwater (Texas)	mosquito-fish[f]	zoo-plankton	bowls: L: 0 fish, M: 1 fish, H: 2 fish	13 weeks	+ to −	nr	nr	Table 3, Case B.IV
Brock 1979 marine (Hawaii)	parrotfish[g]	reef benthos	tanks: L: 0–2 fish, M: 3–5 fish, H: 6–8 fish	36 days	− to −	nr	− to −	Table 3, Case A.II?

[1] Predator species: a: *Lepomis macrochirus*, b: *Cottus bairdi* and *C. girardi*, c: *Hemiglyphidodon plagiometopon*, d: *Eupomacentrus* (now *Stegastes*) *lividus*, e: *Stegastes fasciolatus*, f: *Gambusia affinis*, g: *Scarus taeniurus*.

[2] Relative predation intensities: L: low, M: moderate, H: high. See text for enclosure sizes.

[3] Prey diversity response: see Tables 1 & 3 for explanation and Fig. 1 for examples (nr: not reported).

[4] Predation intensity measured directly.

well as caging experiments in ponds (Thorp & Bergey 1981) and streams (Reice 1983, Flecker & Allan 1984).

Only two studies investigated three levels of fish predation intensity, thus allowing a determination of the shape of the prey diversity-response curve. Flecker (1984) placed 97 × 38 × 45 cm enclosures containing 0, 3, 6, or 12 sculpins (including open controls) in a West Virginia stream (Table 4). After two weeks, the numbers of aquatic insect taxa were statistically indistinguishable among the replicated treatments. Species evenness and composite diversity values were not reported, although the populations of some taxa (especially Chironomidae) were reduced more than others. Flecker suggested that prey diversity was largely unaffected because of rapid recolonization of the enclosures by drifting insects.

In a similar experiment, Gilinsky (1984) placed 2 × 3 × 1.5 m enclosures containing 0, 20, or 60 bluegill sunfish in a North Carolina pond (Table 4). Additionally, each fish treatment compared enclosures stocked with artificial (rope) 'macrophytes' (high heterogeneity) with bare enclosures (low heterogeneity). After one year, the numbers of benthic invertebrate species were virtually identical among the replicated fish treatments within each heterogeneity treatment. That is, all the high-heterogeneity enclosures yielded about the same species richness regardless of fish density, as did all the low-heterogeneity enclosures. Species evenness and composite diversity values were not reported. Gilinsky (1984) suggested that prey diversity was largely unaffected because fish alone were not sufficiently efficient predators. Overall, the high-heterogeneity enclosures supported more benthic species than the low-heterogeneity enclosures, indicating that the 'macrophytes' enhanced diversity by supplying additional microhabitats rather than prey refuges per se.

Reviewing experimental studies in stream systems, Allan (1983) concluded that benthic communities are seldom structured by predation. He suggested that fish may not interact strongly with the benthos due to either the abundance of spatial refuges available to bottom-living organisms or the evolution of anti-predatory adaptations by the prey. The importance of aquatic vegetation as a spatial refuge preventing the local extinction of the benthic prey of freshwater fishes has been suggested in a variety of field observations (Hemphill & Cooper 1984), laboratory experiments (Ware 1973, Brusven & Rose 1981, Cook & Streams 1984), and field manipulations (Macan 1966, 1977, Hall et al. 1970, Cooper & Crowder 1979, Crowder & Cooper 1982, but see Flecker & Allan 1984, Gilinsky 1984).

I found only three experiments, all in lake systems, which documented changes in benthic species richness, all indicating a decrease in the number of species investigated in the presence of fish (Macan 1966, 1977, Henrikson & Oscarson 1978, Morin 1984a, b). By far, the most detailed of these studies was that of Morin (1984a, b), who compared the diversity of dragonfly larvae metamorphosing from pens accessible and closed to predation mostly by bluegill sunfish, *Lepomis macrochirus*. Four species metamorphosed from closed pens, three abundantly, while a very low number of only two species metamorphosed from the pens accessible to fishes. In the absence of fish predation, the species composition of dragonfly larvae was governed by priority effects, suggesting that early breeding species competitively inhibited the establishment of late breeding species. However, fish predation was so intense that it lowered the abundance of all species, indicating that no species were able to increase in abundance due to competitive release. Thus, while this system may be structured by a competitive hierarchy of sorts in the absence of predation, the lack of data on intermediate predation intensities precludes a determination of either the entire prey diversity-response pattern (Fig. 1) or the underlying mechanisms (Table 3).

Marine systems

Relatively few studies have been published on the effects of marine fishes on prey diversity. Virtually all studies investigated benthic prey assemblages.

Piscine prey

I found no studies reporting an impact of predatory marine fishes on the diversity of other fishes. Although predatory fishes apparently influence the behavior of other reef fishes (Hobson 1972, Ebeling & Bray 1976, Ebeling & Laur 1985, Schmitt & Holbrook 1985), attempts to investigate experimentally the impact of such piscivores upon reef-fish communities have not yielded clear results (Bohnsack 1982, J. Stimson, personal communication). The logistic constraints of such experiments are obvious. Nonetheless, because some reef fishes have been shown to compete interspecifically (Hixon 1980, Larson 1980), and such fishes are prey of both teleost and elasmobranch predators (Bray & Hixon 1978), local predation effects on prey diversity by the mechanisms in Table 3 seem possible.

Planktonic prey

I also found no studies reporting an effect of planktivorous marine fishes on the diversity of their prey. Although local effects of particulate-feeding fishes on the abundance of marine zooplankton assemblages follow the general pattern of size-selective predation documented in freshwater systems (Hobson & Chess 1976, 1978, Bray 1981), no diversity responses have yet been demonstrated.

Benthic prey

Unlike freshwater studies, which have been restricted largely to temperate systems, analyses of interactions between marine fishes and benthic communities have been conducted in a wide variety of geographical settings. Because fundamental differences between geographic regions are evident, I will review polar, temperate, and tropical studies separately.

Polar systems

I found only one study of the effect of polar fishes on the diversity of their prey. Comparing intertidal pools in Antarctica, Duarte & Moreno (1981) documented a positive correlation between the population density of the amphipod-specialist nototheniid, *Harpagifer bispinis,* and the species richness, evenness, and composite diversity of amphipods. They hypothesized that the mechanisms creating this pattern were those of a case B.III prey diversity response (Table 3).

Temperate systems

The effect of marine fishes on benthic prey in temperate waters appears to depend on the substrate of the habitat. Virnstein (1978) and Peterson (1979) reviewed experimental caging studies in coastal soft-bottom habitats. The general effect of excluding fishes and other large predators from unvegetated areas appears to be an increase in infaunal species richness, with no tendency for any competitively dominant species excluding others in the absence of predation (Naqvi 1968, Reise 1977, Virnstein 1977). This pattern suggests that the infaunal communities of estuaries may be structured by two of the processes described above for Table 1A and B: physical disturbance and/or non-fish predation. In the latter case, large predators like fishes may consume mostly predatory infauna, so that when large predators are excluded, the predatory infauna increase in abundance and reduce the diversity of prey infauna (Ambrose 1984, Summerson & Peterson 1984).

Infaunal diversity generally is greater in vegetated than exposed areas of estuaries (Orth 1977, Reise 1977, Virnstein 1977, Summerson & Peterson 1984). This pattern may be due to vegetation providing refuges from predation by fishes and other large predators, increased microhabitat complexity providing opportunities for additional prey species, and/or increased substrate density, with increases in diversity being a simple artifact of increases in the overall density of organisms (Summerson & Peterson 1984). In any case, caging experiments in vegetated areas generally fail to detect an effect on prey diversity (e.g., Young et al. 1976, Orth 1977, Reise 1977, Virnstein 1978, Young & Young 1978, Summerson & Peterson 1984). These and other results from salt-marsh and seagrass systems (review by Cooper & Crowder 1979, more recently Nelson 1979, Stoner 1979, Heck & Thoman 1981, Kneib & Stiven 1982, Minello & Zimmerman 1983) indicate that spatial prey refuges provided by vegetation prevent fishes or any other large predators from significantly affecting species diversity in soft-bottom benthic communities.

Regardless of the true mechanisms operating in soft-bottom predator-prey interactions, most studies in estuaries do not allow a clear demonstration of the impact of fishes. First, the cages in such studies usually exclude birds and large invertebrate predators as well as fishes. Quammen (1984) separated the impact of fishes from other large predators experimentally and detected no effects of fishes on invertebrate diversity in two different California estuaries. Second, even with adequate controls, cages present the familiar fish-present versus fish-absent problem: investigating only two levels of predation intensity cannot determine the shape of the prey diversity-response curve (Fig. 1). It appears that any general effects of fishes on the diversity of soft-bottom infauna are yet to be documented.

Choat (1982) reviewed studies of the effects of fish predation on temperate hard-bottom communities. He concluded that 'the potential for temperate fishes to have general and unequivocally recognizable effects on the distribution and abundance of their benthic prey is not apparently realized' (p. 441). For example, although seasonal changes in the abundance of amphipods and polychaetes on rock reefs off New Zealand varied inversely with those of the sparid *Chrysophrys auratus*, caging experiments detected little evidence for a fish predation effect upon either the magnitude or seasonality of invertebrate abundance (Choat & Kingett 1982). In the same general study area, Ayling (1981) determined experimentally that intense grazing by the monacanthid *Parika scaber* may decrease the abundance of sponges and ascidians, but found that grazing by urchins was the most influential factor affecting benthic communities. Off Australia, Keough (1984) found that monocanthids inhibited the successful establishment of tunicates on bivalve shells. Although the tunicates were the potential competitive dominants among the epifauna, they were usually recruitment limited, so that the fish simply eliminated already rare species by the mechanisms described above for Table 1A.

The local diversity of temperate algae also appears to be seldom influenced by fishes. For example, caging experiments by Foster (1975) and Kennelly (1983) revealed no consistent relationship between fish grazing intensity and algal diversity in Californian and Australian kelp beds, respectively (see also Wheeler 1980).

Nonetheless, significant fish predation effects have been documented in other temperate systems, especially on relatively simple substrates. For example, Russ (1980) found that settling plates protected from fish grazing off Australia became covered by ascidians, the apparent competitive dominants. At the same time, plates exposed to fishes exhibited a greater composite diversity of epifauna, suggesting a prey diversity response similar to case B.III (Table 3). The opposite response was documented by Bernstein & Jung (1979), who found that predation by the labrid *Oxyjulis californica* decreased the diversity of epifauna on blades of kelp off California. This fish differentially reduced the abundance of organisms such as barnacles, which were inferior competitors for space on kelp blades relative to bryozoans. Thus, this system may illustrate a prey diversity response similar to case C.II (Table 3). However, the overall shape of the diversity-response curves in both these studies are unknown, since only two levels of fish predation (fish present and fish absent) were investigated (Fig. 1).

In summary, available evidence suggests that temperate marine fishes may have negligible effects on the diversity of benthic prey inhabiting both soft and hard bottoms. For soft-bottom systems, this pattern may be due to the prey communities being structured mostly by physical disturbances or invertebrate predation. For hard-bottom systems, this pattern may be a result of the complexity of natural substrates (both primary and secondary space) providing ample prey refuges, since studies of relatively homogeneous surfaces have demonstrated significant effects of fish predation. Clearly, more data are needed to substantiate or refute these tentative generalizations.

Tropical systems

Fishes appear to have more pronounced effects on the benthic diversity of coral-reef systems, with the greatest impact made by herbivorous fishes on macroalgal assemblages (reviews by Ogden & Lobel 1978, Borowitzka 1981, Lubchenco & Gaines 1981, Gaines & Lubchenco 1982, Hixon 1983). Unlike temperate regions, tropical reefs support a great diversity of herbivorous fishes, especially parrotfishes (Scaridae) and surgeonfishes (Acanthuridae). (Gaines & Lubchenco [1982] have reviewed the hypotheses proposed to account for this striking zoogeographic pattern.) Field experiments have shown that such fishes are intense grazers, greatly affecting the local abundance (Stephenson & Searles 1960, Randall 1961, Hatcher 1981, Hatcher & Larkum 1983) and between-habitat distribution of macroalgae (John & Pople 1973, Hay 1981, Hay et al. 1983). Within-habitat caging experiments have demonstrated that algal diversity is greater on substrates protected from parrotfish and surgeonfish grazing (John & Pople 1973, Day 1977, Lassuy 1980, Hixon & Brostoff 1982, 1983), with grazer-resistant crustose forms dominating exposed substrates (Vine 1974, Wanders 1977, Hixon & Brostoff 1981, 1985).

Three similar experiments – in Guam (Lassuy 1980), Hawaii (Hixon & Brostoff 1981, 1982, 1983), and Australia (Sammarco 1983) – investigated three levels of fish grazing intensity, permitting determinations of the overall shape of the prey diversity-response curve (Table 4). Each study examined the algal assemblages which developed on substrates exposed to each of three treatments: (1) protected within cages (low intensity); (2) inside the defended territories of individual herbivorous damselfishes (moderate intensity); and (3) exposed outside territories to herbivorous parrotfishes, surgeonfishes, and others (high intensity). All these fishes have similar diets, yet parrotfishes and surgeonfishes are more destructive grazers and typically occur at greater densities than damselfishes. Only Hixon & Brostoff (1982, 1983) quantified grazing intensity by counting the 'standing crop' of fish bite marks on experimental surfaces. Destructive grazing was more than an order of magnitude greater on surfaces exposed outside territories (about 270 bites per 50 cm^2) than on those protected by damselfish aggression inside territories (about 15 bites per 50 cm^2), and no bite marks were found on surfaces within the cages. Moreover, urchins and other large invertebrate grazers were rare in the Hawaiian system, so the cages manipulated only fishes. Cage controls indicated no secondary effects on algal diversity.

The result of all three experiments was that all components of algal diversity were greatest inside damselfish territories. Thus, the general prey diversity response appeared to be hump-shaped, despite the fact that the experiments involved different reef systems and species. In the Hawaiian study, the caged surfaces became dominated by a few coarsely branched erect algal species, while intensely grazed exposed surfaces became covered by a few crustose and prostrate forms. Similar patterns occurred in the studies off Guam and Australia. There appears to be a general evolutionary trade-off among macroalgae between competitive ability, characteristic of many erect species, and resistance to grazing, characteristic of mostly crustose and prostrate species (Hay 1981, Lubchenco & Gaines 1981, Littler et al. 1983). To the extent that this generalization is true, the results of Lassuy (1980), Hixon & Brostoff (1982, 1983) and Sammarco (1983) illustrate the mechanisms of a case B.IV hump-shaped prey diversity response (Table 3).

However, territorial damselfishes do not always cause hump-shaped patterns. In some systems, predation intensity can be greater inside territories than outside. For example, damselfish off Yap Island actively 'weeded' certain algal species from their algal mats (Lassuy 1980, see also Irvine 1982). Using the same experimental design as the Guam, Hawaii, and Australia studies discussed above, Lassuy (1980) found that algal diversity decreased as predation intensity (grazing plus weeding) apparently increased from cages to outside territories to inside territories. Thus, the overall prey diversity response appeared to be negative (Table 4). Similarly, Montgomery (1980) found in the Gulf of California that intensive grazing by the damselfish *Microspathodon dorsalis* maintained a near monoculture of a single alga within its territory. Finally, note that the overall intensity of grazing occurring within damselfish territories is a function of two factors: the grazing by the damselfish itself and the extent to which the damselfish is successful in excluding other grazers. In situations where parrotfishes and surgeonfishes form large schools, they can successfully invade and perhaps overgraze territories (Barlow 1974, Robertson et al. 1976), although some damselfish can effectively exclude grazing sea urchins from their territories (Williams 1981). Thus, despite the fact that damselfish territories generally exhibit a greater biomass of algae than surrounding areas (Vine 1974, Brawley & Adey 1977, Lassuy 1980, Montgomery 1980, Hixon & Brostoff 1981, 1982, Sammarco 1983), algal diversity within territories may or may not be greater than in surrounding areas.

The few studies demonstrating fish predation effects on corals and other tropical invertebrate assemblages (Neudecker 1979, Wellington 1982, Fitz et al. 1983, Wolf et al. 1983) usually have not reported explicit impacts on prey diversity (review by Hixon 1983). Day (1977) compared the epifauna which grew on caged and uncaged settling plates placed in a subtidal cave on the Great Barrier Reef. He found that species richness was greater on the uncaged plates, which were exposed to grazing by pomacanthid and balistid fishes. Day concluded that the mechanisms causing the pattern were basically those described for case B.III (Table 3), but no data on the mechanisms were provided. In Hawaii, Brock (1979) compared the benthic assemblages which developed after 36 days in 117 × 117 × 40 cm outdoor microcosms containing a range of 0 to 8 parrotfish (Table 4). He found that the species richness and composite diversity of benthic flora and fauna on exposed surfaces declined with increasing parrotfish density. Because Brock investigated a wide range of fish densities, his data demonstrate a definite negative response in prey diversity, although the mechanisms causing this pattern were not documented. However, given that parrotfish are highly destructive grazers (Hixon & Brostoff 1982, 1983), it seems likely that predation was equivalent on the bare tank surfaces, suggesting a pattern similar to case A.II (Table 3). Finally, recent experimental studies of Panamanian rocky intertidal systems demonstrated that fishes and invertebrate predators together caused a hump-shaped response in benthic prey diversity (Menge et al. 1985), but the role of fishes alone was not investigated over a range of predation intensities.

In summary, herbivorous fishes strongly affect the local diversity of tropical algae, with several experimental studies documenting hump-shaped prey diversity responses. Relatively few studies of the diversity responses of corals and other tropical invertebrates have been reported, with available evidence indicating definite impacts by fishes. However, the overall patterns of these impacts and their underlying mechanisms are largely unknown. Given the major effects of fishes on reef benthos documented in some systems, spatial prey refuges could be expected to enhance benthic diversity in areas where fishes are abundant. In fact, experimental studies have shown that benthic diversity is greater in crevices than on exposed surfaces when grazing intensity is high, but not when grazing intensity is low (Brock 1979, Hixon & Brostoff 1985).

Discussion

The preceding review provides four major conclusions. First, most studies investigating the impact of fishes on prey systems did not report explicit diversity measures. Second, most studies that did provide diversity data either relied on comparisons of different systems or experimentally investigated only two levels of predation intensity (fish present and fish absent). Third, most studies found no or few fish effects on prey diversity, often because of an abundance of spatial refuges in the habitat. Refuges, particularly those provided by aquatic plants on soft bottoms and crevices in hard bottoms, have been found to be important in preventing the local extinction of prey species in a variety of freshwater and marine systems where fishes are the dominant predators.

Fourth, I found only nine experiments (summarized in Table 4) which investigated at least three levels of fish predation intensity, three levels being the minimum necessary to determine the overall shape of the prey diversity-response curve (Fig. 1). Of these, two studies of freshwater benthic systems detected virtually no fish effects on prey diversity (Flecker 1984, Gilinsky 1984). Two studies of coral-reef benthos (Brock 1979, Lassuy 1980 [off Yap]) detected negative effects of fish on prey diversity, but the exact mechanisms causing this pattern were not documented. Finally, two studies of temperate freshwater zooplankton (Neill 1975, Lynch 1978, 1979) and three studies of coral-reef benthic algae (Lassuy 1980 [off Guam], Hixon & Brostoff 1981, 1982, 1983, Sammarco 1983) detected hump-shaped responses characteristic of the intermediate-disturbance hypothesis (sensu Connell 1978). These last five studies also demonstrated keystone-species effects (sensu Paine 1969) caused by the predators disproportionately reducing the abundance of the competitively dominant prey species (Table 3, case B.IV).

Of the nine experiments, two took place in microcosms (Neill 1975, Brock 1979), so the results were of unknown applicability to natural systems. Four of the studies were of short duration (two months or less; Lynch 1978, 1979, Brock 1979, Lassuy 1980, Flecker 1984), so whether the results represented stabilized prey diversity responses was unknown. Finally, only Hixon & Brostoff (1982, 1983) included direct measurements of fish grazing intensity.

I will now discuss the ramifications of these conclusions in suggesting approaches for future research. A striking pattern from surveying the literature was that studies of benthic systems reported species diversity indices much more often than studies of planktonic systems. This pattern occurred despite the fact that plankton studies usually provided data on the relative abundances of species. Composite diversity indices certainly have their drawbacks (reviews by Peet 1974, Pielou 1975), but explicit richness and evenness measures are required if the precise impacts of fishes on local prey diversity are to be documented.

One of the greatest logistic challenges for future studies will be to determine adequately the overall shape of the prey diversity-response curve. This will require the analysis of a full range of fish predation intensities, the endpoints being, at one extreme, the absence of fish, and at the other extreme, the highest predation intensity encountered at natural fish densities. (Note that, if fish predation intensity is limited by factors other than the abundance of prey, then the highest naturally occurring predation intensities may not be greater than the 'low' levels listed in Tables 1 and 3. The possible result of such a situation could be the documentation of only the ascending part of an otherwise hump-shaped prey diversity response.) In any event, comparisons of natural systems with different fish population densities can provide only correlations. Simple fish-present versus fish-absent data (as pointed-out repeatedly above) can determine only whether or not there is a fish predation effect, not the general nature of that effect. Experimental subdivisions of natural lakes (cf. Henrikson & Oscarson 1978) or replicate experimental ponds (cf. Hall et al. 1970) containing different fish densities may be the most effective manipulations in small freshwater systems. Large replicate enclosures with different fish densities may be the best approach in marine and large freshwater systems. Investigating one fish species at a time is advisable. Under any circumstances, the smaller the enclosures used and the less extensive the experimental controls employed, the greater the probability of unrealistic artifacts. In any case, the mechanisms by which fish affect the local diversity of their prey cannot be elucidated unless the overall pattern of that effect is first documented.

Given the documented importance of spatial refuges in a variety of predator-prey systems involving fishes, especially benthic systems, the role of refuges in altering prey diversity responses should be a fruitful area of future research. In general, adding refuges should act to lower the probabilities of local prey extinctions, shifting the extinction/immigration patterns in Table 2 from scenarios II and IV to scenarios I and III. Computer simulations suggest that, as the proportion of the habitat providing spatial refuges increases, the shape of the prey diversity-response curve changes in predictable ways (Hixon unpublished). Experimental studies will require factorial designs in which the effects of a range of predation intensities are analyzed over a range of refuge availabilities (Brock 1979,

Gilinsky 1984, Hixon & Brostoff 1985).

Finally, focusing not only on the fishes as the predators (determining patterns of selectivity and prey population reduction), but also on the prey assemblage itself will be essential for elucidating the mechanisms by which prey diversity responds to varying predation intensity. In particular, the processes structuring prey assemblages in the absence of fish predation must be known, as well as the factors contributing to local patterns of extinction and immigration. Such studies will undoubtedly require collaboration between fish ecologists and experts on the prey species.

The bottom line of this review is that, considering the effects of predation by fishes on local prey diversity, we know a fair amount about several systems and very little or nothing about most. A substantial body of concepts and theory stands largely untested for fish predator-prey systems. Not simply more data are needed, but more studies designed to test explicit mechanistic hypotheses.

Acknowledgements

I am grateful for constructive comments from G.M. Cailliet, F.L. Carpenter, J.H. Choat, T. Farrell, S.D. Gaines, J. Lubchenco, B.A. Menge, P.J. Morin, B.K. Orr, C.A. Simenstad, and T. Turner, who reviewed the entire manuscript, and D.J. Hall, H.W. Li, P.B. Moyle, C.H. Peterson, and M.E. Power, who reviewed specific sections of an earlier draft. Special thanks to B. Goldowitz, H. Li, B. Orr, and L. Persson for providing many references. I sincerely apologize to those authors whose relevant work I either overlooked or misinterpreted. Ultimate thanks to G.M. Cailliet and C.A. Simenstad for organizing an excellent symposium. This paper is dedicated to the future memory of Sarah Connor.

References cited

Addicott, J.F. 1984. Mutualistic interactions in population and community processes. pp. 437–455. *In:* P.W. Price, C.N. Slobodchikoff & W.S. Gaud (ed.) A New Ecology: Novel Approaches to Interactive Systems, Wiley-Interscience, New York.

Allan, J.D. 1982. The effects of reduction in trout density on the invertebrate community of a mountain stream. Ecology 63: 1444–1455.

Allan, J.D. 1983. Predator-prey relationships in streams. pp. 191–229. *In:* J.R. Barnes & G.W. Minshall (ed.) Stream Ecology: Application and Testing of General Ecological Theory, Plenum Press, New York.

Allan, J.D. 1984. Hypothesis testing in ecological studies of aquatic insects. pp. 484–507. *In:* J.H. Resh & D.M. Rosenberg (ed.) The Ecology of Aquatic Insects, Praeger Press, New York.

Ambrose, W.G. 1984. Role of predatory infauna in structuring marine soft-bottom communities. Mar. Ecol. Prog. Ser. 17: 109–115.

Anderson, J.M. 1950. Some aquatic vegetation changes following fish removal. J. Wildl. Mgt. 14: 206–209.

Andersson, G., H. Berggren, G. Cronberg & C. Gelin. 1978. Effects of planktivorous and benthivorous fish on organisms and water chemistry in eutrophic lakes. Hydrobiologia 59: 9–15.

Ayling, A.M. 1981. The role of biological disturbance in temperate subtidal encrusting communities. Ecology 62: 830–847.

Ball, R.C. & D.W. Hayne. 1952. Effects of the removal of the fish population on the fish-food organisms of a lake. Ecology 33: 41–48.

Barlow, G.W. 1974. Extraspecific imposition of social grouping among surgeonfishes (Pisces: Acanthuridae). J. Zool. (Lond.) 174: 330–340.

Bernstein, B.B. & N. Jung. 1979. Selective pressures and coevolution in a kelp canopy community in southern California. Ecol. Monogr. 49: 335–355.

Bohnsack, J.A. 1982. Effects of piscivorous predator removal on coral reef fish community structure. pp. 258–267. *In:* G.M. Cailliet & C.A. Simenstad (ed.) Fish Food Habit Studies, University of Washington Sea Grant Press, Seattle.

Borowitzka, M.A. 1981. Algae and grazing in coral reef ecosystems. Endeavour 5: 99–106.

Brawley, S.H. & W.H. Adey. 1977. Territorial behavior of threespot damselfish (*Eupomacentrus planifrons*) increases reef algal biomass and productivity. Env. Biol. Fish. 2: 45–51.

Bray, R.N. 1981. Influence of water currents and zooplankton densities on daily foraging movements of blacksmith, *Chromis punctipinnis*, a planktivorous reef fish. U.S. Fish. Bull. 78: 829–841.

Bray, R.N. & M.A. Hixon. 1978. Night-shocker: predatory behavior of the Pacific electric ray (*Torpedo californica*). Science 200: 333–334.

Brock, R.E. 1979. An experimental study on the effects of grazing by parrotfishes and role of refuges in benthic community structure. Mar. Biol. 51: 381–388.

Brooks, J.L. & S.I. Dodson. 1965. Predation, body size, and compositon of plankton. Science 150: 28–35.

Brusven, M.A. & S.T. Rose. 1981. Influence of substrate composition and suspended sediment on insect predation by the torrent sculpin, *Cottus rhotheus*. Can. J. Fish. Aquat. Sci. 38: 1444–1448.

Buss, L.W. & J.B.C. Jackson. 1979. Competitive networks: nontransitive competitive relationships in cryptic coral reef environments. Amer. Nat. 113: 223–234.

Carter, J.C.H. & J.K. Kwik. 1977. Instar distribution, vertical distribution, and interspecific competition among four species of *Chaoborus*. J. Fish. Res. Board Can. 34: 113–118.

Caswell, H. 1978. Predator-mediated coexistence: a nonequilibrium model. Amer. Nat. 112: 127–154.

Choat, J.H. 1982. Fish feeding and the structure of benthic communities in temperate waters. Ann. Rev. Ecol. Syst. 13: 423–449.

Choat, J.H. & P.D. Kingett. 1982. The influence of fish predation on the abundance cycles of an algal turf invertebrate fauna. Oecologia 54: 88–95.

Christie, W.J. 1974. Changes in the fish species composition of the Great Lakes. J. Fish. Res. Board Can. 31: 827–854.

Clady, M.D. & L.A. Nielsen. 1978. Diversity of a community of small fishes as related to abundance of the dominant percid fishes. pp. 109–113. *In:* R.L. Kendall (ed.) Selected Coolwater Fishes of North America, Amer. Fish. Soc. Spec. Publ. 11.

Clepper, H. (ed.) 1979. Predator-prey systems in fisheries management. Sport Fishing Institute, Washington, D.C. 504 pp.

Connell, J.H. 1975. Some mechanisms producing structure in natural communities: a model and evidence from field experiments. pp. 460–490. *In:* M.L. Cody & J.M. Diamond (ed.)Ecology and Evolution of Communities, Belknap-Harvard University Press, Cambridge.

Connell, J.H. 1978. Diversity in tropical rain forests and coral reefs. Science 199: 1302–1310.

Cook, W.L. & F.A. Streams. 1984. Fish predation on *Notonecta* (Hemiptera): relationship between prey risk and habitat utilization. Oecologia 64: 177–183.

Cooper, W.E. & L.B. Crowder. 1979. Patterns of predation in simple and complex environments. pp. 257–267. *In:* H. Clepper (ed.) Predator-Prey Systems in Fisheries Management. Sport Fishing Institute, Washington, D.C.

Courtenay, W.R. & J.R. Stauffer (ed.) 1984. Distribution, biology, and management of exotic fishes. Johns Hopkins University Press, Baltimore. 430 pp.

Crowder, L.B. 1980. Alewife, rainbow smelt and native fishes in Lake Michigan: competition or predation? Env. Bio. Fish. 5: 225–233.

Crowder, L.B. & W.E. Cooper. 1982. Habitat structural complexity and the interaction between bluegills and their prey. Ecology 63: 1802–1813.

Crowley, P.H. 1979. Predator-mediated coexistence: an equilibrium interpretation. J. Theor. Biol. 80: 129–144.

Day, R.W. 1977. Two contrasting effects of predation on species richness in coral reef habitats. Mar. Biol. 44: 1–5.

Dodson, S.I. 1974. Zooplankton competition and predation: an experimental test of the size-efficiency hypothesis. Ecology 55: 605–613.

Drenner, R.W., F. DeNoyelles & D. Kettle. 1982. Selective impact of filter-feeding gizzard shad on zooplankton community structure. Limnol. Oceanogr. 27: 965–968.

Drenner, R.W., J.R. Mummert, F. DeNoyelles & D. Kettle. 1984. Selective particle ingestion by a filter-feeding fish and its impact on phytoplankton community structure. Limnol. Oceanogr. 29: 941–948.

Duarte, W.E. & C.A. Moreno. 1981. The specialized diet of *Harpagifer bispinis*: its effect on the diversity of Antarctic intertidal amphipods. Hydrobiologia 80: 241–250.

Ebeling, A.W. & R.N. Bray. 1976. Day versus night activity of reef fishes in a kelp forest off Santa Barbara, California. U.S. Fish. Bull. 74: 703–717.

Ebeling, A.W. & D.R. Laur. 1985. The influence of plant cover on surfperch abundance at an offshore temperate reef. Env. Biol. Fish. 12: 169–179.

Emlen, J.M. 1973. Ecology: an evolutionary approach. Addison-Wesley, Reading. 493 pp.

Fitz, H.C., M.L. Reaka, E. Bermingham & N.G. Wolf. 1983. Coral recruitment at moderate depths: the influence of grazing. pp. 89–96. *In:* M.L. Reaka (ed.) The Ecology of Deep and Shallow Coral Reefs, NOAA Symp. Ser. Undersea Res. 1(1).

Flecker, A.S. 1984. The effects of predation and detritus on the structure of a stream insect community: a field test. Oecologia 64: 300–305.

Flecker, A.S. & J.D. Allan. 1984. The importance of predation, substrate and spatial refugia in determining lotic insect distributions. Oecologia 64: 306–313.

Foerster, R.E. & W.E. Ricker. 1941. The effect of reduction of predaceous fish on survival of young sockeye salmon at Cultus Lake. J. Fish. Res. Board Can. 5: 315–336.

Forney, J.L. 1974. Interactions between yellow perch abundance, walleye predation, and survival of alternate prey in Oneida Lake, New York. Trans. Amer. Fish. Soc. 103: 15–24.

Foster, M.S. 1975. Regulation of algal community development in a *Macrocystis pyrifera* forest. Mar. Biol. 32: 331–342.

Fraser, D.F. & R.D. Cerri. 1982. Experimental evaluation of predator-prey relationships in a patchy environment: consequences for habitat use patterns in minnows. Ecology 63: 307–313.

Gaines, S.D. & J. Lubchenco. 1982. A unified approach to marine plant-herbivore interactions. II. Biogeography. Ann. Rev. Ecol. Syst. 13: 111–138.

Galbraith, M.G. 1967. Size-selective predation on *Daphnia* by rainbow trout and yellow perch. Trans. Amer. Fish. Soc. 96: 1–10.

Garman, G.C. & L.A. Nielsen. 1982. Piscivority by stocked brown trout (*Salmo trutta*) and its impact on the nongame fish community of Bottom Creek, Virginia. Can. J. Fish. Aquat. Sci. 39: 862–869.

Gilinsky, E. 1984. The role of fish predation and spatial heterogeneity in determining benthic community structure. Ecology 65: 455–468.

Hall, D.J., W.E. Cooper & E.E. Werner. 1970. An experimental approach to the production dynamics and structure of freshwater animal communities. Limnol. Oceanogr. 15: 839–928.

Hall, D.J., S.T. Threlkeld, C.W. Burns & P.H. Crowley. 1976. The size-efficiency hypothesis and the size structure of zooplankton communities. Ann. Rev. Ecol. Syst. 7: 177–208.

Hamrin, S.F. 1983. The food preference of vendace (*Coregonus albula*) in south Swedish forest lakes including the predation effect on zooplankton populations. Hydrobiologia 101: 121–128.

Harper, J.L. 1969. The role of predation in vegetational diversity. Brookhaven Symp. Biol. 22: 48–62.

Hastings, A. 1978. Spatial heterogeneity and the stability of predator-prey systems: predator-mediated coexistence. Theor. Pop. Biol. 14: 380–395.

Hatcher, B.G. 1981. The interaction between grazing organisms and the epilithic algal community of a coral reef: a quantitative assessment. Proc. 4th Int. Coral Reef Symp. 2: 515–524.

Hatcher, B.G. & A.W.D. Larkum. 1983. An experimental analysis of factors controlling the standing crop of the epilithic algal community on a coral reef. J. Exp. Mar. Biol. Ecol. 69: 61–84.

Hay, M.E. 1981. Herbivory, algal distribution, and the maintenance of between-habitat diversity on a tropical fringing reef. Amer. Nat. 118: 520–540.

Hay, M.E., T. Colburn & D. Downing. 1983. Spatial and temporal patterns in herbivory on a Caribbean fringing reef: the effects on plant distribution. Oecologia 58: 299–308.

Hayne, D.W. & R.C. Ball. 1956. Benthic productivity as influenced by fish predation. Limnol. Oceanogr. 1: 162–175.

Healey, M. 1984. Fish predation on aquatic insects. pp. 255–288. *In:* V.H. Resh & D.M. Rosenberg (ed.) The Ecology of Aquatic Insects, Praeger Press, New York.

Heck, K.L. & T.A. Thoman. 1981. Experiments on predator-prey interactions in vegetated aquatic habitats. J. Exp. Mar. Biol. Ecol. 53: 125–134.

Hemphill, N. & S.D. Cooper. 1984. Differences in the community structure of stream pools containing or lacking trout. Verh. Internat. Verein. Limnol. 22: 1858–1861.

Henrikson, L. & H.G. Oscarson. 1978. Fish predation limiting abundance and distribution of *Glaenocorisa p. propinqua*. Oikos 31: 102–105.

Hixon, M.A. 1980. Competitive interactions between California reef fishes of the genus *Embiotoca*. Ecology 61: 918–931.

Hixon, M.A. 1983. Fish grazing and community structure of reef corals and algae: a synthesis of recent studies. pp. 79–87. *In:* M.L. Reaka (ed.) The Ecology of Deep and Shallow Coral Reefs, NOAA Symp. Ser. Undersea Res. 1(1).

Hixon, M.A. & W.N. Brostoff. 1981. Fish grazing and community structure of Hawaiian reef algae. Proc. 4th Int. Coral Reef Symp. 2: 507–514.

Hixon, M.A. & W.N. Brostoff. 1982. Differential fish grazing and benthic community structure on Hawaiian reefs. pp. 249–257. *In:* G.M. Cailliet & C.A. Simenstad (ed.) Fish Food Habit Studies, University of Washington Sea Grant Press, Seattle.

Hixon, M.A. & W.N. Brostoff. 1983. Damselfish as keystone species in reverse: intermediate disturbance and diversity of reef algae. Science 220: 511–513.

Hixon, M.A. & W.N. Brostoff. 1985. Substrate characteristics, fish grazing, and epibenthic reef assemblages off Hawaii. Bull. Mar. Sci. 37: 200–213.

Hobson, E.S. 1972. Activity of Hawaiian reef fishes during the evening and morning transitions between daylight and darkness. U.S. Fish. Bull. 70: 715–740.

Hobson, E.S. & J.R. Chess. 1976. Trophic interactions among fishes and zooplankters near shore at Catalina Island, California. U.S. Fish. Bull. 74: 567–598.

Hobson, E.S. & J.R. Chess. 1978. Trophic relationships among fishes and plankton in the lagoon at Enewetak Atoll, Marshall Islands. U.S. Fish. Bull. 76: 133–153.

Holt, R.D. 1977. Predation, apparent competition, and the structure of prey communities. Theor. Pop. Biol. 12: 197–229.

Hrbaček, J. 1962. Species composition and the amount of the zooplankton in relation to the fish stock. Rozpr. ČSAV, Ser. Mat. Nat. Sci. 72: 1–117.

Hrbaček, J., M. Dvořaková, V. Kořinek & L. Prochazková. 1961. Demonstration of the effect of fish stock on the species composition of zooplankton and the intensity of metabolism of the whole plankton association. Verh. Internat. Verein. Limnol. 14:192–195.

Hughes, R.N. 1980. Predation and community structure. pp. 699–728. *In:* J.H. Price, D.E.G. Irvine & W.F. Farnham (ed.) The Shore Environment, Vol. 2: Ecosystems, Academic Press, London.

Hurlbert, S.H. & M.S. Mulla. 1981. Impacts of mosquitofish (*Gambusia affinis*) predation on plankton communities. Hydrobiologia 83: 125–151.

Huston, M. 1979. A general hypothesis of species diversity. Amer. Nat. 113: 81–101.

Hutchinson, B.P. 1971. The effect of fish predation on the zooplankton of ten Adirondack lakes, with particular references to the alewife, *Alosa pseudoharengus*. Trans. Amer. Fish. Soc. 100: 325–335.

Irvine, G.V. 1982. The importance of behavior in plant-herbivore interactions. pp. 240–248. *In:* G.M. Cailliet & C.A. Simenstad (ed.) Fish Food Habit Studies, University of Washington Sea Grant Press, Seattle.

Jackson, P.B.N. 1961. The impact of predation, especially by the tiger-fish (*Hydrocyon vittatus* Cast.) on African freshwater fishes. Proc. Zool. Soc. London 136: 603–622.

Janssen, J. 1980. Alewives (*Alosa pseudoharengus*) and ciscoes (*Coregonus artedii*) as selective and non-selective planktivores. pp. 580–586. *In:* W.C. Kerfoot (ed.) Evolution and Ecology of Zooplankton Communities, University Press of New England, Hanover.

John, D.M. & W. Pople. 1973. The fish grazing of rocky shore algae in the Gulf of Guinea. J. Exp. Mar. Biol. Ecol. 11: 81–90.

Kajak, Z., K. Dusoge, A. Hillbricht-Ilkowska, E. Pieczynski, A. Prejs, I. Spodniewska & T. Weglenska. 1972. Influence of the artificially increased fish stock on the lake biocenosis. Verh. internat. Verein. theor. angew. Limnol. 18: 228–235.

Kennelly, S.J. 1983. An experimental approach to the study of factors affecting algal colonization in a sublittoral kelp forest.

J. Exp. Mar. Biol. Ecol. 68: 257–276.

Keough, M.J. 1983. Patterns of recruitment of sessile invertebrates in two subtidal habitats. J. Exp. Mar. Biol. Ecol. 66: 213–245.

Keough, M.J. 1984. Dynamics of the epifauna of the bivalve *Pinna bicolor*: interactions among recruitment, predation, and competition. Ecology 65: 677–688.

Kerfoot, W.C. (ed.) 1980. Evolution and ecology of zooplankton communities. University Press of New England, Hanover. 793 pp.

Kneib, R.T. & A.E. Stiven. 1982. Benthic invertebrate responses to size and density manipulations of the common mummichog, *Fundulus heteroclitus,* in an intertidal salt marsh. Ecology 63: 1518–1532.

Larson, R.J. 1980. Competition, habitat selection and the bathymetric segregation of two rockfish (*Sebastes*) species. Ecol. Monogr. 50: 221–239.

Lassuy, D.R. 1980. Effects of 'farming' behavior by *Eupomacentrus lividus* and *Hemiglyphidodon plagiometopon* on algal community structure. Bull. Mar. Sci. 30: 304–312.

Li, H.W. & P.B. Moyle. 1981. Ecological analysis of species introductions into aquatic systems. Trans. Amer. Fish. Soc. 110: 772–782.

Littler, M.M., D.S. Littler & P.R. Taylor. 1983. Evolutionary strategies in a tropical barrier reef system: functional-form groups of marine macroalgae. J. Phycol. 19: 229–237.

Lubchenco, J. 1978. Plant species diversity in a marine intertidal community: importance of herbivore food preference and algal competitive abilities. Amer. Nat. 112: 23–39.

Lubchenco, J. 1979. Consumer terms and concepts. Amer. Nat. 113: 315–317.

Lubchenco, J. & S.D. Gaines. 1981. A unified approach to marine plant-herbivore interactions. I. Populations and communities. Ann. Rev. Ecol. Syst. 12: 405–437.

Lynch, M. 1978. Complex interactions between natural coexploiters – *Daphnia* and *Ceriodaphnia.* Ecology 59: 552–564.

Lynch, M. 1979. Predation, competition, and zooplankton community structure: an experimental study. Limnol. Oceanogr. 24: 253–272.

Macan, T.T. 1966. Predation by *Salmo trutta* in a moorland fish pond. Verh. internat. Verein. theor. angew. Limnol. 16: 1081–1087.

Macan, T.T. 1977. The influence of predation on the composition of fresh-water animal communities. Biol. Rev. 52: 45–70.

MacArthur, R.H. 1972. Geographical ecology. Harper and Row, New York. 269 pp.

Menge, B.A., J. Lubchenco & L.R. Ashkenas. 1985. Diversity, heterogeneity and consumer pressure in a tropical rocky intertidal community. Oecologia 65: 394–405.

Menge, B.A. & J.P. Sutherland. 1976. Species diversity gradients: synthesis of the roles of predation, competition and temporal heterogeneity. Amer. Nat. 110: 351–369.

Minello, T.J. & R.J. Zimmerman. 1983. Fish predation on juvenile brown shrimp, *Penaeus aztecus* Ives: the effect of simulated *Spartina* structure on predation rates. J. Exp. Mar. Biol. Ecol. 72: 211–231.

Mittelbach, G.G. 1984. Predation and resource partitioning in two sunfishes (Centrarchidae). Ecology 65: 499–513.

Montgomery, W.L. 1980. The impact of non-selective grazing by the giant blue damselfish, *Microspathodon dorsalis,* on algal communities in the Gulf of California, Mexico. Bull. Mar. Sci. 30: 290–303.

Morin, P.J. 1984a. The impact of fish exclusion on the abundance and species composition of larval odonates: results of short-term experiments in a North Carolina farm pond. Ecology 65: 53–60.

Morin, P.J. 1984b. Odonate guild composition: experiments with colonization history and fish predation. Ecology 65: 1866–1873.

Moyle, P.B. 1985. Fish introductions into North America: patterns and ecological impact. *In:* H. Mooney (ed.) Biological Invasions in North America. (in press).

Murdoch, W.W. & A. Oaten. 1975. Predation and population stability. Adv. Ecol. Res. 9: 2–132.

Naqvi, S.M.Z. 1968. Effects of predation on infaunal invertebrates of Alligator Harbor, Florida. Gulf Res. Rep. 2: 313–321.

Neill, W.E. 1975. Experimental studies of microcrustacean competition, community composition and efficiency of resource utilization. Ecology 56: 809–826.

Nelson, W.G. 1979. Experimental studies of selective predation on amphipods: consequences for amphipod distribution and abundance. J. Exp. Mar. Biol. Ecol. 38: 225–245.

Neudecker, S. 1979. Effects of grazing and browsing fishes on the zonation of corals in Guam. Ecology 60: 666–672.

Nilsson, N.A. 1972. Effects of introductions of salmonids into barren lakes. J. Fish. Res. Board Can. 29: 693–697.

Nilsson, N.A. 1978. The role of size-biased predation in competition and interactive segregation in fish. pp. 303–325. *In:* S.D. Gerking (ed.) Ecology of Freshwater Fish Production, Wiley and Sons, New York.

Nilsson, N.A. & B. Pejler. 1973. On the relation between fish fauna and zooplankton composition in north Swedish lakes. Rep. Inst. Freshwater Res. Drottningholm 53: 51–77.

Northcote, T.G., C.J. Walter & J.M.B. Hume. 1978. Initial impacts of experimental fish introductions on macrozooplankton communities of small oligotrophic lakes. Verh. int. Verein. theor. angew. Limnol. 20: 2003–2012.

O'Brien, W.J. 1979. The predator-prey interaction of planktivorous fish and zooplankton. Amer. Sci. 67: 572–581.

Ogden, J.C. & P.S. Lobel. 1978. The role of herbivorous fishes and urchins in coral reef communities. Env. Biol. Fish. 3: 49–63.

Orth, R.J. 1977. The importance of sediment stability in seagrass communities. pp. 281–300. *In:* B.C. Coull (ed.) Ecology of Marine Benthos, University of South Carolina Press, Columbia.

Paine, R.T. 1966. Food web complexity and species diversity. Amer. Nat. 100: 65–75.

Paine, R.T. 1969. A note on trophic complexity and community stability. Amer. Nat. 103: 91–93.

Paine, R.T. 1977. Controlled manipulations in the marine intertidal zone, and their contributions to ecological theory. Acad. Nat. Sci. Spec. Publ. 12: 245–270.

Paine, R.T. 1984. Ecological determinism in the competition for space. Ecology 65: 1339–1348.

Paine, R.T. & R.L. Vadas. 1969. The effects of grazing by sea urchins, *Strongylocentrotus* spp., on benthic algal populations. Limnol. Oceanogr. 14: 710–719.

Peet, R.K. 1974. The measurement of species diversity. Ann. Rev. Ecol. Syst. 5: 285–307.

Peterson, C.H. 1979. Predation, competitive exclusion, and diversity in the soft- sediment benthic communities of estuaries and lagoons. pp. 233–264. *In:* R.J. Livingston (ed.) Ecological Processes in Coastal and Marine Systems, Plenum Press, New York.

Pielou, E.C. 1975. Ecological diversity, Wiley, New York. 165 pp.

Pope, G.F., J.C.H. Carter & G. Power. 1973. The influence of fish on the distribution of *Chaoborus* spp. (Diptera) and the density of larvae in the Matamek River System, Quebec. Trans. Amer. Fish. Soc. 102: 707–714.

Post, J.R. & D. Cucin. 1984. Changes in the benthic community of a small precambrian lake following the introduction of yellow perch, *Perca flavescens*. Can. J. Fish. Aquat. Sci. 41: 1496–1501.

Power, M.E. & W.J. Matthews. 1983. Algae-grazing minnows (*Campostoma anomalum*), piscivorous bass (*Micropterus* spp.), and the distribution of attached algae in a small praire-margin stream. Oecologia 60: 328–332.

Power, M.E., W.J. Matthews & A.J. Stewart. 1985. Grazing minnows, piscivorous bass and stream algae: dynamics of a strong interaction. Ecology 66: 1448–1456.

Prejs, A. 1984. Herbivory by temperate freshwater fishes and its consequences. Env. Biol. Fish. 10: 281–296.

Quammen, M.L. 1984. Predation by shorebird, fish, and crabs on invertebrates in intertidal mudflats: an experimental test. Ecology 65: 529–537.

Randall, J.E. 1961. Overgrazing of algae by herbivorous marine fishes. Ecology 42: 812.

Reice, S.R. 1983. Predation and substratum: factors in lotic community structure. pp. 325–345. *In:* T.D. Fontaine & S.M. Bartell (ed.) Dynamics of Lotic Ecosystems, Ann Arbor Science, Ann Arbor.

Reif, C.B. & D.W. Tappa. 1966. Selective predation: smelt and cladocerans in Harveys Lake. Limnol. Oceanogr. 11: 437–438.

Reise, K. 1977. Predator exclusion experiments in an intertidal mud flat. Helgo. wiss. Meeresunters. 30: 263–271.

Robertson, D.R., H.P.A. Sweatman, E.A. Fletcher & M.G. Cleland. 1976. Schooling as a mechanism for circumventing the territoriality of competitors. Ecology 57: 1208–1220.

Russ, G.R. 1980. Effects of predation by fishes, competition, and structural complexity of the substratum on the establishment of a marine epifaunal community. J. Exp. Mar. Biol. Ecol. 42: 55–69.

Sammarco, P.W. 1983. Effects of fish grazing and damselfish territoriality on coral reef algae. I. Algal community structure. Mar. Ecol. Prog. Ser. 13: 1–14.

Savino, J.F. & R.A. Stein. 1982. Predator-prey interactions between largemouth bass and bluegills as influenced by simulated, submersed vegetation. Trans. Amer. Fish. Soc. 111: 255–266.

Schmitt, R.J. & S.J. Holbrook. 1985. Patch selection by juvenile black surfperch (Embiotocidae) under variable risk: interactive influence of food quality and structural complexity. J. Exp. Mar. Biol. Ecol. 85: 269–285.

Smith, B.R. 1980. Introduction to the proceedings of the 1979 Sea Lamprey International Symposium (SLIS). Can. J. Fish. Aquat. Sci. 37: 1585–1587.

Smith, F.E. 1972. Spatial heterogeneity, stability, and diversity in ecosystems. Trans. Connecticut Acad. Arts Sci. 44: 307–335.

Sousa, W.P. 1979. Experimental investigations of disturbance and ecological succession in a rocky intertidal community. Ecol. Monogr. 49: 227–254.

Sousa, W.P. 1984. The role of disturbance in natural communities. Ann. Rev. Ecol. Syst. 15: 353–391.

Spencer, C.N. & D.L. King. 1984. Role of fish in regulation of plant and animal communities in eutrophic ponds. Can. J. Fish. Aquat. Sci. 41: 1851–1855.

Sprules, W.G. 1972. Effects of size-selective predation and food competition on high altitude zooplankton communities. Ecology 53: 375–386.

Stenson, J.A.E. 1972. Fish predation effects on the species composition of the zooplankton community in eight small forest lakes. Rep. Inst. Freshwater Res. Drottningholm 52: 132–148.

Stenson, J.A.E. 1978. Differential predation by fish on two species of *Chaoborus* (Diptera, Chaoboridae). Oikos 31: 98–101.

Stenson, J.A.E. 1982. Fish impact on rotifer community structure. Hydrobiologia 87: 57–64.

Stenson, J.A.E., T. Bohlin, L. Henrikson, B.I. Nilsson, H.G. Nyman, H.G. Oscarson & P. Larsson. 1978. Effects of fish removal from a small lake. Verh. internat. Verein. Limnol. 20: 794–801.

Stephenson, W. & R.B. Searles. 1960. Experimental studies on the ecology of intertidal environments of Heron Island. I. Exclusion of fish from beach rock. Aust. J. Mar. Freshw. Res. 2: 241–267.

Stoner, A.W. 1979. Species specific predation on amphipod crustacea by pinfish (*Lagodon rhomboides*): mediation by macrophyte standing crop. Mar. Biol. 55: 201–207.

Straškraba, M. 1965. The effect of fish on the number of invertebrates in ponds and streams. Mitt. internat. Verein. Limnol. 13: 106–127.

Summerson, H.C. & C.H. Peterson. 1984. Role of predation in organizing benthic communities of a temperate-zone seagrass bed. Mar. Ecol. Prog. Ser. 15: 63–77.

Taylor, J.N., W.R. Courtenay & J.A. McCann. 1984. Known impacts of exotic fishes in the continental United States. pp. 322–373. *In:* W.R. Courtenay & J.R. Stauffer (ed.) Distribution, Biology, and Management of Exotic Fishes, Johns

Hopkins University Press, Baltimore.
Thompson, J.N. 1982. Interaction and coevolution. Wiley-Interscience, New York. 179 pp.
Thorp, J.H. & E.A. Bergey. 1981. Field experiments on responses of a freshwater, benthic macroinvertebrate community to vertebrate predators. Ecology 62: 365–375.
Threinen, C.W. & W.T. Helm. 1954. Experiments and observations designed to show carp destruction of aquatic vegetation. J. Wildl. Mgt. 18: 247–250.
Van Valen, L. 1974. Predation and species diversity. J. Theor. Biol. 44: 19–21.
Vine, P.J. 1974. Effects of algal grazing and aggressive behaviour of the fishes *Pomacentrus lividus* and *Acanthurus sohal* on coral-reef ecology. Mar. Biol. 24: 131–136.
Virnstein, R.W. 1977. The importance of predation by crabs and fishes on benthic infauna in Chesapeake Bay. Ecology 58: 1199–1217.
Virnstein, R.W. 1978. Predator caging experiments in soft sediments: caution advised. pp. 261–273. *In:* M.L. Wiley (ed.) Estuarine Interactions, Academic Press, New York.
Von Ende, C.N. 1979. Fish predation, interspecific predation, and the distribution of two *Chaoborus* species. Ecology 60: 119–128.
Wanders, J.B.W. 1977.The role of benthic algae in the shallow reef of Curacao (Netherlands Antilles) III: the significance of grazing. Aquat. Bot. 3: 357–390.
Ware, D.M. 1973. Risk of epibenthic prey to predation by rainbow trout (*Salmo gairdneri*). J. Fish. Res. Board Can. 30: 787–797.
Warshaw, S.J. 1972. Effects of alewives (*Alosa pseudoharengus*) on the zooplankton of Lake Wononskopomuc, Connecticut. Limnol. Oceanogr. 17: 816–825.
Wellington, G.M. 1982. Depth zonation of corals in the Gulf of Panama: control and facilitation by resident reef fishes. Ecol. Monogr. 52: 223–241.
Wells, L. 1970. Effects of alewife predation on zooplankton populations in Lake Michigan. Limnol. Oceanogr. 15: 556–565.
Werner, E.E. & J.F. Gilliam. 1984. The ontogenetic niche and species interactions in size-structured populations. Ann. Rev. Ecol. Syst. 15: 393–425.
Werner, E.E., J.F. Gilliam, D.J. Hall & G.G. Mittelbach. 1983. An experimental test of the effects of predation risk on habitat use in fish. Ecology 64: 1540–1548.
Wheeler, A. 1980. Fish-algal relationships in temperate waters. pp. 677–698. *In:* J.H. Price, D.E.G. Irvine & W.F. Farnham (ed.) The Shore Environment, Vol. 2: Ecosystems, Academic Press, London.
Williams, A.H. 1981. An analysis of competitive interactions in a patchy back-reef environment. Ecology 62: 1107–1120.
Wolf, N.G., E.B. Bermingham & M.L. Reaka. 1983. Relationships between fishes and mobile benthic invertebrates on coral reefs. pp. 69–78. *In:* M.L. Reaka (ed.) The Ecology of Deep and Shallow Coral Reefs, NOAA Symp. Ser. Undersea Res. 1(1).
Yodzis, P. 1978. Competition for space and the structure of ecological communities. Lec. Notes Biomath. 25: 1–191.
Young, D.K., M.A. Buzas & M.W. Young. 1976. Species densities of macrobenthos associated with seagrasses: a field experimental study of predation. J. Mar. Res. 34: 577–592.
Young, D.K. & M.W. Young. 1978. Regulation of species diversity of seagrass associated macrobenthos: evidence from field experiments in the Indian River Estuary, Florida. J. Mar. Res. 36: 569–593.
Zaret, T.M. 1979. Predation in freshwater fish communities. pp. 135–143. *In:* H. Clepper (ed.) Predator-Prey Systems in Fisheries Management, Sport Fishing Institute, Washington, D.C.
Zaret, T.M. 1980. Predation and freshwater communities. Yale University Press, New Haven. 187 pp.
Zaret, T.M. & R.T. Paine. 1973. Species introduction in a tropical lake. Science 182: 449–455.
Zelinka, M. 1974. Die Eintagsfliegen (Ephemeroptera) in Forellenbachen der Beskiden. III. Der Einfluss des verschiedenen Fischbestandes. Věst. Cs. spol. Zool. 38: 76–80.

Received 4.2.1985 *Accepted 10.7.1985*

PIMELODUS OLIVACEUS, Grd.

Physiology of feeding and starvation tolerance in overwintering freshwater fishes

Kathleen M. Sullivan
Department of Biology, University of Miami, P.O. Box 249118, Coral Gables, FL 33124, U.S.A.

Keywords: Energetics, Yellow perch, Largemouth bass, Respiration, Excretion, Lipid Stores, Low temperature, *Micropterus salmoides, Perca flavescens*

Synopsis

Overwintering conditions in north temperate lakes characterized by low temperatures, low oxygen tensions, and, for some species, low food supplies are thought to have effects on fish species composition and recruitment. Examination of respiration and excretion rates of various overwintering species as well as measurements of the amount of lipid stored in visceral and muscle tissue during the winter months suggest feshwater fishes can be divided into groups based on their tolerance to starvation. Cold-water species, such as yellow perch, *Perca flavescens*, which have low preferred body temperatures throughout the year (13° C or lower), are very sensitive to starvation during winter months. This species feeds actively throughout much of the winter and stores lipids primarily in the viscera and gonads prior to early spring spawning. Warm water fishes, such as largemouth bass, *Micropterus salmoides*, which have a high preferred body temperature during the summer (23–26° C) and appear not to feed at low winter temperatures, have an intermediate tolerance to starvation during winter months. Largemouth bass store lipids both in muscle and visceral tissue, and winter starvation tolerance is dependent on the amount of stored lipids. Fishes that are physiologically intolerant of long periods of food deprivation and actively feeding during the winter (e.g. yellow perch) are thus distinguished behaviorally and physiologically from fishes that are more tolerant of prolonged starvation (e.g. largemouth bass) and tend to be less active during the winter months.

Introduction

Freshwater fishes overwintering in north temperate lakes face not only extreme low temperatures (2–3° C), but also lower oxygen tensions and reduced food availability for periods of up to six months of ice cover (Greenbank 1945, Cooper & Washburn 1949). Any one of these conditions alone presents a formidable challenge to aquatic ectotherms, but, the combination represents a most rigorous environment with which fish must contend.

Throughout Wisconsin and Michigan, there is an overlap of fish fauna. Warm-water species that are at the northern limit of a predominantly southern distribution (e.g. largemouth bass, sunfish, and bowfin) overlap with cold water species that are at the southern end of a predominantly northern distribution (e.g. northern pike, yellow perch and muskellunge). It appears that the ability to overwinter segregates the two groups and contributes to distribution differences, though overlaps depend, to a large degree, on the type of lake setting in which the fish are found (Mooreman 1957). Differences in both the physical and ecological features of a lake can influence overwintering conditions and survival of different species (Magnuson & Karlen 1970, Rahel 1982, Tonn & Magnuson 1982).

Largemouth bass, *Micropterus salmoides,* from the southern and northern end of the species range show significant differences in the thermal tolerance of individuals. Laboratory experiments have demonstrated that the different thermal preferences among largemouth bass populations are correlated with specific enzyme phenotypes determined electrophoretically (Philipp et al. 1981). Fish with different thermal preferences may have different overwintering success rates.

Spawning in freshwater fishes can occur at almost any time of the year (Scott & Crossman 1973, Craig 1977). Regardless of time of spawning, representatives of both cold and warm water species hatch early in the year giving young-of-the-year [the new age class (0+)] the maximum time to feed and grow before the onset of winter.

Frozen lakes frequently show low oxygen tensions when water is sealed off from atmospheric contact and microbial activity is high. Size and shape of a lake as well as severity of winter determine the likelihood of anoxic conditions developing during winter months (Greenbank 1945). Cyprinids such as carp and goldfish are known to tolerate total anoxia in ice-covered lakes (Blažka 1958, Johnson & Bernard 1983) though it is unknown how food supply and lipid stores influence this ability.

Overwintering survival of fishes is dependent on both winter water temperatures and on the size of individual fish both within and between year classes (Hunt 1969, Oliver et al. 1979). In the case of young-of-the-year smallmouth bass, *Micropterus dolomieui*, overwintering survival is decreased with increased winter water temperatures presumably because of the higher utilization rates of energy stores at these higher temperatures (Oliver et al. 1979). In contrast, an 11-year study of the overwintering survival of wild juvenile brook charr, *Salvelinus fontinalis*, indicated an increase in overwintering survival with an increase in mean winter water temperatures. Hunt (1969) suggested that the higher mean winter water temperatures afforded the juvenile charr more opportunities to feed during the winter months. Both investigations of overwintering survival indicate a strong size-dependent component such that larger fish have a greater chance for survival.

Survival during winter months at low temperatures may depend on either the ability to continue feeding or the ability to slow metabolic demands on energy stores. Superimposed on this low temperature tolerance or dormancy is the addition problem of hypoxia. It appears that freshwater fish may employ one of two possible overwintering strategies. The first strategy, employed by the smallmouth bass, a warm water species, is to store sufficient energy by actively feeding in summer months to meet metabolic needs during the winter with little or no feeding. Thus, elevated winter temperatures can cause the fish to rapidly utilize energy stores when there is no food and result in higher winter mortalities. The second strategy, employed by the brook charr, a cold water species, is to feed actively throughout the winter, possibly exploiting reduced cover of potential prey.

The goals of this investigation are to evaluate in situ routine respiration rates and tissue lipid stores of both a warm water species, largemouth bass *Micropterus salmoides*, and a cold water species, yellow perch *Perca flavescens*, in late fall and throughout the winter. In particular, the role of lipid stores in survival of young-of-the-year and small size classes is examined.

Methods and materials

Collection of fish

Young-of-the-year largemouth bass were collected from the Michigan State Department of Natural Resources hatchery pond in Saline, Michigan. Fish of the same year class ranging from 2.0 to 6.0 g (58 to 94 mm total length) were seined from a single pond in October, November and May. Fish were transported and held in water maintained at pond temperatures. Yellow perch were collected in minnow traps at the Trout Lake Biological Station of the University of Wisconsin in Vilas County, Wisconsin. Largemouth bass ages were known; yellow perch ages were not, but sizes ranged from 9.0 to 35.0 g (82 to 165 mm total length).

Acclimation of fish

During the spring and summer months, largemouth bass collected from the hatchery ponds were held in large floating pens on Third Sister Lake near the University of Michigan. Hatchery largemouth bass were easily transported to research sites in both Wisconsin and Michigan with no mortalities. All three lakes (Saline Hatchery pond, Lake Mendota and Third Sister Lake) were eutrophic, slightly alkaline lakes. Yellow perch were moved from Trout Lake, an oligotrophic lake, to Lake Mendota, an eutrophic slightly alkaline lake, at nearly identical temperatures with no mortalities. Thus, moving and transplanting fish was not thought to have an adverse effect on their physiology.

Both the largemouth bass and yellow perch were held at the Center for Limnology on Lake Mendota at the University of Wisconsin's Madison campus. Fish were maintained under the same temperature, oxygen and light conditions that occurred under the ice in the littoral zone of Lake Mendota. Fish were held in large holding tanks floated in the boat slip circulated with either filtered or unfiltered lakewater pumped from under the ice. Tanks were dark to simulate under-ice light conditions. Ambient temperatures and oxygen concentrations were easily maintained by pumping.

Respiration and excretion measurements

Oxygen consumption measurements were made for both yellow perch and largemouth bass during winter months and for largemouth bass throughout the year. Respiration measurements were made on individual fish held in tubular chambers 24 h before the experiment (Sullivan & Smith 1982). Respiration experiments ran for 12 to 24 h during the winter months, and 3 to 5 h during the summer months at ambient temperatures. Oxygen tensions were determined by Winkler titrations with azide modification, and by Yellow Springs Instruments (Yellow Springs, Ohio) oxygen electrodes calibrated daily with nitrogen and air-purged water. Chamber control experiments were run without fish to determine the respiration of microbes in the water. Oxygen consumption rates were expressed in milligrams of oxygen consumed per kilogram fish per hour.

Fish were subjected to acute hypoxia by stripping oxygen from water entering respiration chambers with a nitrogen-gas column (Petrosky & Magnuson 1973). The oxygen tension was lowered over an eight hour period to 2.0 mg $O_2 l^{-1}$. Respiration rates were then measured at constant oxygen content of 2.0 mg $O_2 l^{-1}$ in flow-through chambers for 12 h. Hypoxia tolerance was arbitrarily judged as the ability to maintain equilibrium during the entire experiment (12 h). Tests of acute hypoxia tolerance were measured on four individual fish of each species in December and in late February.

Water samples were removed periodically in routine oxygen consumption experiments for ammonia and urea assays that were measured by the phenol-hypochlorate method and urease conversion of urea to ammonia (Strickland & Parsons 1972). Nitrogen excretion was calculated in mg nitrogen excreted per kg fish per hour.

Tissue lipid determinations

Fish were sacrificed by freezing on dry ice. Specimens were measured and weighed then stored in the freezer until percent lipid determinations were made. Lipid determinations were made on 0.3–2.0 grams of white muscle and visceral tissue homogenized with 4 parts distilled water. Aliquots were then removed for isopropanol/hexane (2:3) extraction of lipids (Kates 1972). After evaporation of the solvent under nitrogen, lipids were charred, and quantitative determinations were made spectrophotometrically (Marsh & Weinstein 1966). Tripalmitin was used as a standard for this assay. Lipid content is expressed as mg per g wet weight of tissue.

Results

Respiration and excretion rates.

Average oxygen consumption rates of starved and fed fish of both species along with the water temperature of the lake water showed a gradual decrease from November to February (Fig. 1). For largemouth bass held without food for a period of 29 weeks during the winter until mid-May, respiration rates remained less than 150 mg $O_2 kg^{-1} h^{-1}$. For yellow perch held only 7 weeks without food then offered cladocerans and frozen brine shrimp, respiration rates were lowest in February at 63 mg $O_2 kg^{-1} h^{-1}$ (± 16; $n = 4$), then sharply increased through March and April.

Nitrogen excretion rates for both species along with O:N ratios for respiration experiments from December, January and February (Table 1), showed that nitrogen excretion was six to ten times lower in non-feeding largemouth bass than it was feeding yellow perch. The low nitrogen excretion rates with the decreasing total body lipid stores during the winter indicate that largemouth bass utilize body lipid stores as their primary energy source during winter months when they are not feeding. Largemouth bass O:N ratios for the winter months were between 5 and 9. Yellow perch, in contrast, showed relatively high nitrogen excretion rates throughout the winter months, with O:N ratios between 1 and 2.

Hypoxia tolerance

For largemouth bass, mean respiration rates dropped from 96 ± 26 mg $O_2 kg^{-1} h^{-1}$ to 72 ± 15 mg $O_2 kg^{-1} h^{-1}$ when oxygen tensions were lowered from 12 mg $O_2 l^{-1}$ to 2 mg $O_2 l^{-1}$ at 4° C in December (Table 2). This decrease was not significant statis-

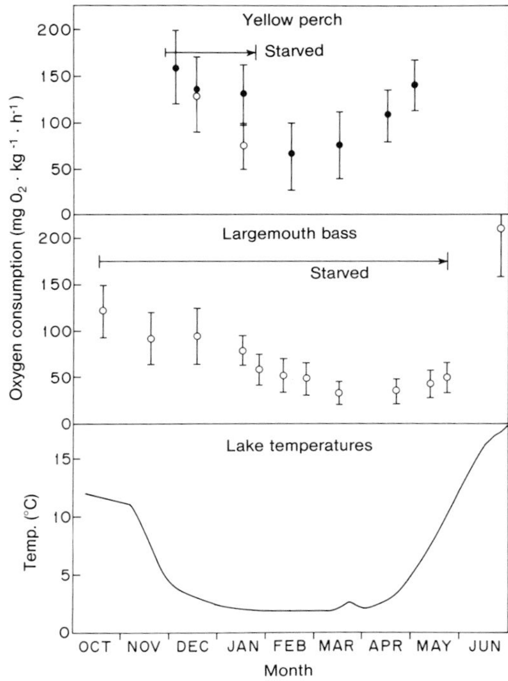

Fig. 1. Oxygen consumption rates of yellow perch, *Perca flavescens*, and largemouth bass, *Micropterus salmoides*, during winter months while fish are held in under-ice conditions. Six individual fish (October through December), then three individual fish (January through June) were used for each measurement, mean values with one standard deviation are plotted. Lake temperature taken during the winter months is given in the bottom panel (○ = starved; ● = post-absorptive rates of fed fish).

Table 1. Post-absorptive nitrogen excretion rates of yellow perch and largemouth bass during winter months. Mean, standard deviation and number of fish are given. Oxygen to nitrogen (O:N) ratios are calculated from oxygen consumption rates measured at the same time.

	Yellow perch		Largemouth bass	
	Nitrogen excretion $mg\ kg^{-1} h^{-1}$	O:N ratio	Nitrogen excretion $mg\ kg^{-1} h^{-1}$	O:N ratio
December	76.0 ± 41	1.8	13.0 ± 4.0	7.4
3.5° C	n = 4		n = 4	
January	67 ± 42	1.9	9.7 ± 2.3	5.6
2.0° C	n = 3		n = 6	
February	63 ± 39	1.4	5.6 ± 1.6	9.1
2.0° C	n = 4		n = 4	

tically (student t-test, $\alpha = 0.05$). Two fish lost equilibrium, and were on their sides after 8 h at 2.0 mg $O_2 l^{-1}$, but all four fish tested survived the 12 h exposure to this lowered oxygen concentration. In February when the oxygen concentration was lowered from 5.8 to 2.0 mg $O_2 l^{-1}$ in 2.0° C water, one of the four largemouth bass tested died after 6 h in the respiration chamber, thus only three fish were included in the respiration measurements. All three surviving largemouth bass showed loss of equilibrium after 12 h, but again there was no significant difference in mean oxygen consumption rates at the two oxygen tensions (student t-test, $\alpha = 0.05$).

For yellow perch, mean respiration rates dropped from 137 ± 17 mg $O_2 kg^{-1} h^{-1}$ to 112 ± 36 mg $O_2 kg^{-1} h^{-1}$ when the oxygen tensions were lowered from 12 mg $O_2 l^{-1}$ to 2 mg $O_2 l^{-1}$ at 4° C in December (Table 2). In February, when the oxygen tension was lowered from 5.8 mg $O_2 l^{-1}$ to 2 mg $O_2 l^{-1}$ in 2° C water, yellow perch mean oxygen consumption rates increased slightly. Yellow perch used in the hypoxia tolerance tests maintain equilibrium throughout the experiments.

There was a significant difference in respiration rate between the two species at the lowered oxygen tension (student t-test, $\alpha = 0.05$). Yellow perch appeared unaffected by the lowered oxygen tensions with no loss of equilibrium in either the December or February experiments. Both species mean respiration rates decreased slightly with a lowering of the oxygen tension to 2.0 mg $O_2 l^{-1}$ in December. However, in February, the mean respiration rate of yellow perch increased slightly with a decrease in the oxygen tension while the mean respiration rate of the largemouth bass decreased slightly.

Starvation tolerance

Largemouth bass were held throughout the winter without food, and mortalities of the 200+ young-of-the-year bass were less than 10% during the first 12 weeks of starvation. At the end of 24 weeks, 68% of the young-of-the-year bass had died apparently from starvation. Fish were extremely emaciated. These bass would not feed in the lake water tanks at temperatures lower than 6° C. A variety of food (chopped liver, cladocerans and tubifex worms) were offered to isolated largemouth bass throughout the winter with no success.

Largemouth bass are much more resistant to starvation during the fall and winter months (Fig. 2). During the winter months when water temperatures are below 5° C, young-of-the-year bass can survive up to 29 weeks before 50% of the fish expire. During the summer months when the fish are actively feeding, 50% of 1-year bass will starve to death in less than 10 days at temperatures above 25° C. The exact shape of the mortality curve is not clear.

Starved yellow perch held in filtered lake water without food as the lake was freezing showed very high mortalities; 70% of the 57 perch died in the first seven weeks. Dead fish appeared emaciated

Table 2. Respiration rate of yellow perch and largemouth bass at under-ice conditions in December and February under ambient and hypoxia (2.0 mg $O_2 l^{-1}$) oxygen tensions.

	Lake temperature (° C)	Ambient oxygen tension mg l^{-1}	Respiration[a] rate at ambient oxygen tension (mg $O_2 kg^{-1} h^{-1}$)	Respiration[a] rate at 2.0 mg $O_2 l^{-1}$
December				
Perch	4° C	12.8	137 ± 30	112 ± 36
Bass	4° C	12.8	96 ± 26	72 ± 15
February				
Perch	2° C	5.8	65 ± 28	97 ± 31
Bass	2° C	5.8	51 ± 11	33 ± 19

[a] Mean value of four individual fish (three for bass in February) ± standard deviation.

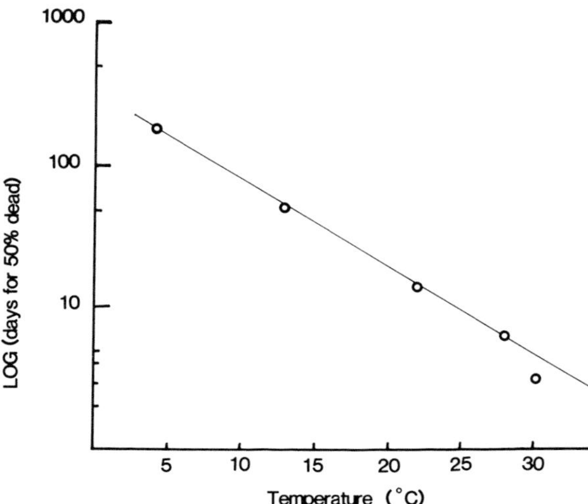

Fig. 2. Starvation mortality (log days to 50%) versus temperature (° C) for young-of-the-year largemouth bass. Fish were held at different times of the year: 5° C = December through March; 13° C = October; 21° C = June; and 26–30° C = July and August.

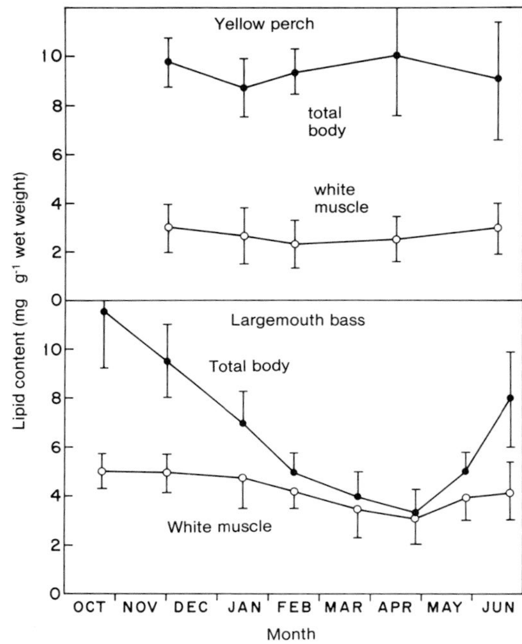

Fig. 3. Total body and white muscle lipid content of yellow perch and largemouth bass held in ambient lake conditions from October to June. Mean values with standard deviations are plotted; means represent lipid determinations made on three to five individual fish.

and had 2.3 mg lipid per gram wet weight visceral tissue (± 1.3, $n = 6$) compared to 8.3 mg lipid (± 0.8, $n = 3$) in field-caught fish collected by hook and line during that same time period.

Respiration rates for starved yellow perch (open symbols, Fig. 1) were 60% of the post-absorptive routine oxygen consumption rates for fed fish (closed symbols) after five weeks of starvation. Starved yellow perch were very susceptible to fin bacterial infections, which seemed to contribute to high mortalities. Largemouth bass remained infection-free under identical treatment.

Lipid storage and utilization

A clearer picture of the energy storage and utilization patterns of these two species emerged by examining changes in body lipid stores during the winter months (Fig. 3). Young-of-the-year largemouth bass had very high total body lipid content in the fall (11.4 ± 2.1 mg lipid per gram wet weight). The lipid content of largemouth bass dropped sharply throughout the winter until late April when total body lipid content was 3.3 ± 1.4 mg lipid per gram wet weight. In contrast, yellow perch held in the boat slip throughout the winter and starved for the first seven weeks of holding showed total body lipid contents of 9–10 mg lipid per gram wet weight throughout the winter; very little of this lipid was stored in the muscle mass (less than 3 mg g^{-1}) and lipid levels did not change with the season.

Seasonal changes in energy metabolism

Routine oxygen consumption rates decrease for both fed and starved largemouth bass from summer temperatures of 21° C to fall temperatures of 13° C. During the summer months at temperatures above 20° C, fish starved 5 to 7 days show a significantly lower rate of oxygen consumption compared to fed fish (student's t-test, $\alpha = 0.05$). In the fall, respiration measurements made at 13° C for starved and fed largemouth bass were not significantly different. Standard metabolic rate of largemouth bass was predicted from the model of Rice et al. (1983) which was based on data presented in Beamish (1970). Standard metabolic rate calculated here by substituting water temperature and using a body weight of 6 g in the equation:

$$Q_{O2} = 0.348\ W^{0.645}\ e^{0.0313T},$$

where Q_{O2} = standard metabolic rate (in mg O_2 h^{-1}), W = body weight (in g) and T = temperature (in ° Celcius) was generally higher than experimental values (Fig. 4).

Comparison of respiration and excretion rates of yellow perch to those rates predicted from a bioenergetics model applied to yellow perch by Kitchell et al. (1977) showed that experimentally determined respiration rates were higher than predicted [mean of 97 ± 31 mg $O_2\,kg^{-1}h^{-1}$ compared to 42 mg $O_2\,kg^{-1}h^{-1}$ predicted by Kitchell et al. (1977) at 2° C] and that excretion rates of perch were much lower for overwintering fish than those predicted by the bioenergetics model (63 to 76 mg $N\,kg^{-1}h^{-1}$ estimated experimentally compared to 833 mg $N\,kg^{-1}h^{-1}$ estimated by the bioenergetics model).

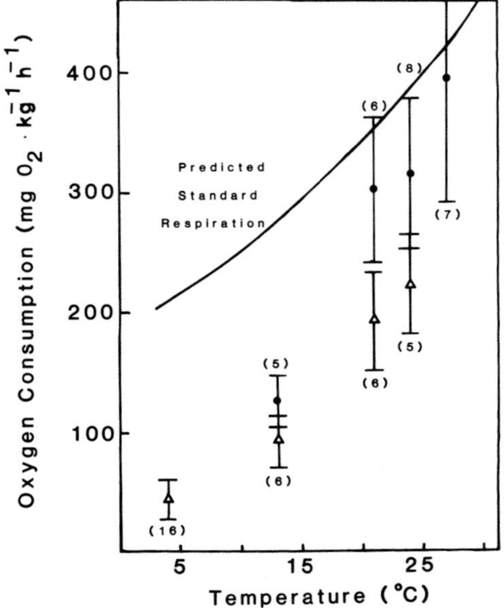

Fig. 4. In situ oxygen consumption of young-of-the-year largemouth bass plotted against environmental temperature. △ = starved; ● = fed within 36 h of respiration measurements; solid line = predicted rate for 5 g fish from Rice et al. (1983); sample sizes in parentheses.

Discussion

Freshwater fishes in north temperate lakes have long been known to have different abilities to survive low temperature, low oxygen and restricted food availability (Mooreman 1957, Moore 1942, Klinger et al. 1982, Crawshaw 1984). As indicated herein, largemouth bass and yellow perch are no exception. Overwintering mortalities depend on the severity of the winter, exceptionally mild winters resulting in the higher mortality of fish depending on stored energy such as the smallmouth bass (Oliver et al. 1979) and largemouth bass (Mooreman 1957). Extremely cold winters would increase the mortalities of fish that feed actively during the winter, such as brook charr (Hunter 1969) and possibly yellow perch (Magnuson & Karlen 1970).

The ability to survive up to 6 months of ice cover is not strictly related to a species' tolerance to low oxygen; often the non-uniform conditions within winterkill lakes allow less tolerant species to survive (Magnuson & Karlen 1970, Mills 1972, Petrosky & Magnuson 1973).

Klinger et al. (1982) looked at three species of freshwater fish that appear to survive anoxic conditions in winter lakes: the central mudminnow, *Umbra limi*, the fathead minnow, *Pimephales promelas*, and the brook stickleback, *Culaea inconstans*. The overwintering success of these species was enhanced by their small size, low metabolic rate, tolerance to low oxygen conditions and reduced locomotor activity. In addition, these fish utilized high oxygen microzones, and the central mudminnow was able to utilize oxygen directly from gas bubbles under the ice.

A factor not discussed by Klinger et al. (1982) that appears critical to overwintering survival of fish is total body lipid content at the start of the winter or ability to feed throughout the winter. Visceral lipid stores are related to both starvation tolerance, and perhaps hypoxia tolerance during the winter. Bass were clearly less tolerant to acute hypoxia in February than they were in December when total body lipid content was higher. The depletion of total body lipid content later in the winter may indicate limited energy stores in the form of other substrates such as glycogen. The decrease in glycogen stores severely limits acute hypoxia tolerance (Hochachka 1980). Magnuson & Karlen (1970) observing northern pike, *Esox lucius*, yellow perch and bluegill, *Lepomis macrochirus*, behavior beneath the ice of a shallow lake throughout the winter, found that yellow perch are the most active of the species observed, and that northern pike are the least active. Bluegill remained farther beneath the ice than the other two species. Northern pike and perch frequently sounded into the anoxic layers of the lake. Magnuson & Karlen (1970) concluded that differences in fish behavior significantly prolonged winter survival. The combination of little locomotory activity with a position directly beneath the ice apparently favored the longer survival of northern pike over bluegill over perch.

The role of lipids in acute and chronic hypoxia tolerance of fishes is not well established. At low temperatures lipid stores appear to be involved in anaerobic pathways (Ekberg 1962, Johnston & Bernard 1983), and thus may enhance tolerance to hypoxia. Largemouth bass were starvation tolerant and appeared to utilize lipid stores throughout the winter when they were not feeding and were inactive. Largemouth bass were very intolerant to acute hypoxia stress despite their low oxygen consumption rates. Petrosky & Magnuson (1973) suggested that the loss of equilibrium in a winterkill situation would be fatal for fish because if a fish becomes negatively buoyant (as the bluegill did when exposed to low oxygen tensions in their study) it would sink into the anoxic bottom waters. Thus, some winterkill resulting from even a temporary loss of equilibrium may occur even before critical low oxygen levels are reached. Yellow perch, in contrast, were very intolerant to food deprivation during the winter. Oxygen consumption rates were variable. Total body lipid content remained relatively constant in the yellow perch as they continued to feed throughout the winter on zooplankton and cladocerans in the unfiltered lake water. In addition, yellow perch were able to maintain equilibrium during periods of acute hypoxia. Overwintering success of young-of-the-year largemouth bass appears dependent on the amount of lipid stored by early winter, and the rate at which it is utilized. Overwintering success in small yellow perch is dependent on the availability of food.

Both yellow perch and largemouth bass suffered large mortality rates when total body lipid stores were depleted. Yellow perch deprived of food rapidly depleted their lipid stores, and died. Largemouth bass, which had lower metabolic rate and lower locomotory activity survived several months on stored lipid energy.

For both yellow perch and largemouth bass, prediction of metabolic rates during the winter months cannot be determined by the decrease in temperature alone. Possible changes in the energy metabolism of both species during the winter months appear to have altered the physiological energetics which are well documented for both species during the summer months (or higher experimental temperatures) (Beamish 1970, Rice et al. 1983, Kitchell et al. 1977). The largemouth bass seem to have the strategy of supressing metabolic rates during the winter months. The general strategy of storing lipids during warm weather and remaining dormant during winter months applies to both smallmouth and largemouth bass. Latitudinal and regional variations in the severity of winter conditions seems to dictate the amount of total body lipid stored.

Crawshaw (1984) examined the importance of low ambient temperature in winter dormancy of largemouth bass and brown bullhead, *Ictalurus nebulosus*, and reported a decrease in locomotory activity in both species with a decrease in temperature to 3° C. The likelihood of arousing the fish from this dormancy was inversely proportional to the ambient temperature. Crawshaw (1984) also noted that if given the opportunity, largemouth bass would move to temperatures above 25° C within one day in spite of consequent acid-base and metabolic imbalances. In a study of the overwintering mortality of smallmouth bass, Oliver et al. (1979) noted that in late fall and winter when temperatures were below 10° C, most fish were dormant, and did not feed. Young-of-the-year smallmouth bass ranging in size from 55 to 107 mm suffered size-selected mortality, that is, the smaller individuals were least likely to survive the winter. Oliver et al. (1979) discussed the hypothesis that smallmouth bass grow more during warm summers, and thus, a greater number of the year class can grow through a critical size threshold to permit survival through the winter. Smallmouth bass have a more northern distribution compared to largemouth bass, and experience longer winters. The fish used in Oliver et al. (1979) investigation were held in central Ontario where temperatures were below 10° C from mid-November until mid-May. The smallmouth bass had total body lipid stores of over 20% in October, almost twice the amount observed in the young-of-the-year largemouth bass collected in Michigan. Total body lipid stores in smallmouth bass at the end of the winter (mid-May) was still over 15% of the wet body weight.

In contrast, largemouth bass, ranging from 58 to 94 mm total length had an average of 12 mg lipid per gram wet weight (12%) in October, but only 3.5 mg lipid per gram wet weight at the end of the winter ice cover in April. It is possible that the milder winter during which the experiments were run contributed to a rapid depletion of lipid stores due to the length of time water temperatures were 4° C or higher between November and April. The winter of 1982–1983 was exceptionally mild in central Wisconsin, with ice covering Lake Mendota on January 13, 1983 and thawing out by April 12, 1983. The lake thawed in the center for 7 days in March (March 17–24) then re-froze. Thus, the lake was frozen for 72 days, which is lower than the mean number of days of ice cover for Lake Mendota based on records kept for the past 150 years, 102 ± 13 days ($\bar{x} \pm$ S.D.). This mild winter with warmer temperatures may have accounted for higher largemouth bass mortalities in the in situ experiments.

Fishes can be characterized by their lipid stores as starvation tolerant or starvation sensitive. An example of starvation sensitive fish is the plaice, *Pleuronectes platessa*, which can have up to 7% lipid by wet weight in the muscle, but depend primarily on visceral fat stores for short term food deprivation (Johnson 1981, Watanabe 1982). Plaice can be starved only for short periods of time (days) before contractile proteins in the white muscle are utilized as an energy source, and can experience a 60–80% decrease in the fractional volume of myofibrils (muscle proteins) when starved 35 days at 10° C (Johnston 1981). In contrast, salmon can store larger quantities of both muscle and visceral lipids in preparation for spawning migrations. Both salmon and trout can store large quantities of lipid, and are capable of tolerating starvation for up to six months (Jezierska et al. 1982). The most starvation resistant fish is the eel *Anguilla rostrata*, adults have been known to tolerate up to four years of food deprivation (Moon 1983). *Anguilla rostrata* deposits very little visceral lipid, and are thought to utilize primarily free amino acids in the muscle as an energy store during starvation.

Starvation tolerance is enhanced by the low temperatures and the inactivity or dormant state observed in many species during winter months. Overwintering survival of freshwater fishes is determined by a combination of temperature, oxygen availability and energy stores (either stored in the form of body lipids or availability of prey during the winter). Strategies of remaining dormant throughout the winter (largemouth bass) or actively feeding (yellow perch) will vary in success with yearly variations in winter severity.

Acknowledgements

This research was supported by a Guyer Postdoctoral Fellowship from the Department of Zoology and the Center for Limnology at University of Wisconsin, Madison, Wisconsin. The assistance and comments of M. Magnuson and M. Meyer are gratefully acknowledged.

References cited

Beamish, F. W. H. 1970. Oxygen consumption of the largemouth bass, *Micropterus salmoides*, in relation to swimming speed and temperature. Can. J. Zool. 48: 1221–1228.

Blažka, P. 1958. The anaerobic metabolism of fish. Physiol. Zool. 31: 117–128.

Cooper, G. P. & G. N. Washburn. 1949. Relation of dissolved oxygen to winter mortality of fish in Michigan lakes. Trans. Amer. Fish. Soc. 76: 23–33.

Craig, J. F. 1977. The body composition of adult perch, *Perca flaviatilis*, in Windermere, with reference to seasonal changes and reproduction. J. An. Ecol. 46: 617–632.

Crawshaw, L. I. 1984. Low temperature dormancy in fish. Amer. J. Physiol. 246: 479–486.

Ekberg, D. R. 1962. Anaerobic and aerobic metabolism in gills of the crucian carp adapted to high and low temperatures. Comp. Bioch. Physiol. 5: 123–128.

Greenbank, J. 1945. Limnological conditions in ice-covered lakes, especially as related to winterkill of fish. Ecol. Monogr. 15: 343–392.

Hochachka, P. W. 1980. Living without oxygen. Harvard University Press, Cambridge. 181 pp.

Hunt, R. L. 1969. Overwinter survival of wild fingerling brook trout in Lawrence Creek, Wisconsin. J. Fish. Res. Board Can 26: 1473–1483.

Jezierska, B., J. R. Hazel & S. D. Gerking. 1982. Lipid mobilization during starvation in the rainbow trout, *Salmo gairdneri*, with attention to fatty acids. J. Fish Biol. 21: 681–692.

Johnston, I. A. 1981. Quantitative analysis of muscle breakdown during starvation in the marine flatfish, *Pleuronectes platessa*. Cell Tissue Res. 214: 369–386.

Johnston, I. A. & L. M. Bernard. 1983. Utilization of the ethanol pathway by crucian carp during anoxia. J. Exp. Biol. 104: 73–78.

Kates, M. 1972. Techniques in lipidology. pp. 267–601. *In*: T.S. Work & E. Work (ed.) Laboratory Techniques in Biochemistry and Molecular Biology, Vol. 3, Part 2, Academic Press, New York.

Kitchell, J. F., D. J. Stewart & D. Weininger. 1977. Applications of a bioenergetics model to yellow perch (*Perca flavescens*) and walleye (*Stizostedion vitreum vitreum*) J. Fish. Res. Board Can. 34: 1922–1935.

Klinger, S. A., J. J. Magnuson & G. W. Gallepp. 1982. Survival mechanisms of the central mudminnow (*Umbra limi*), fathead minnow (*Pimephales promelas*) and brook stickleback (*Culaea inconstans*) for low oxygen in winter. Env. Biol. Fish. 7: 113–120.

Magnuson, J. J. & D. J. Karlen. 1970. Visual observation of fish beneath the ice in a winterkill lake. J. Fish Res. Board Can. 27: 1059–1068.

Marsh, J. B. & D. B. Weinstein. 1966. Simple charring method for determination of lipids. J. Lipid Res. 7: 574–576.

Mills, K. H. 1972. Distribution of fishes under the ice in relation to dissolved oxygen, temperature and free carbon dioxide in Mystery Lake, Wisconsin. M.Sc. Thesis, University of Wisconsin, Madison. 56 pp.

Moon, T. W. 1983. Changes in metabolic reserves and enzymes in food-deprived immature American eels, *Anguilla rostrata* (LeSueur). Can. J. Zool. 61: 802–811.

Moore, W. G. 1942. Field studies of the oxygen requirements of certain freshwater fishes. Ecology 23: 319–329.

Mooreman, R. B. 1957. Some factors related to success of fish populations in Iowa farm ponds. Trans. Amer. Fish. Soc. 86: 361–370.

Oliver, J. D., G. F. Holeton & K. E. Chua. 1979. Overwintering mortality of fingerling smallmouth bass in relation to size, relative energy stores, and environmental temperature. Trans. Amer. Fish. Soc. 108: 130–136.

Petrosky, B. R. & J. J. Magnuson. 1973. Behavioral responses of northern pike, yellow perch, and bluegill to oxygen concentration under simulated winterkill conditions. Copeia 1973: 124–133.

Phillips, D. P., W. F. Childers & G. S. Whitt. 1981. Management implications for different genetic stocks of largemouth bass (*Micropterus salmoides*) in the United States. Can. J. Fish. Aquat. Sci. 38: 1715–1723.

Rahel, F. J. 1982. Fish assemblages in Wisconsin bog lakes. Ph.D. Thesis, University of Wisconsin, Madison. 157 pp.

Rice, J. A., J. A. Breck, S. M. Bartell & J. F. Kitchell. 1983. Evaluating the constraints of temperature, activity and consumption on growth of largemouth bass. Env. Biol. Fish. 9: 263–275.

Scott, W. B. & E. J. Crossman. 1973. Freshwater fishes of Canada. Fish. Res. Board Can. Bull. 184. 966 pp.

Strickland, J. D. & T. R. Parsons. 1972. A practical handbook of seawater analysis. Fish. Res. Board Can. Bull. 167. 310 pp.

Sullivan, K. M. & K. L. Smith, Jr. 1982. Energetics of sablefish, *Anoplopoma fimbria* under laboratory conditions. Can. J. Fish. Aquat. Sci. 39: 1012–1020.

Tonn, W. M. & J. J. Magnuson. 1982. Patterns in species composition and richness of fish assemblages in northern Wisconsin lakes. Ecology 63: 1149–1166.

Watanabe, T. 1982. Lipid nutrition in fish. Comp. Biochem. Physiol. 73B: 3–15.

Received 15.12.1984 *Accepted 2.10.1985*

Diurnal feeding and prey size selection in Atlantic salmon, *Salmo salar*, alevins*

Howard I. Browman[1] & Brian M. Marcotte
McGill University, Institute of Oceanography, 3620 University Street, Montreal, Quebec H3A 2B2, Canada
[1] *Present address: Department of Systematics and Ecology, The University of Kansas, Lawrence, KS 66045, U.S.A.*

Keywords: Early life history, Fish feeding, Endogenous rhythms, Neuroethology, Feeding success, Sleep

Synopsis

Results of an experiment on diurnal feeding behaviour and prey size selection in Atlantic salmon, *Salmo salar,* alevins are reported. Peaks in total behavioural activity, ingestion rate, feeding success (# of ingestions per unit effort) and feeding error (total number of prey missed or rejected after an attack) occurred in the early morning (0600–0900 h), at mid-day (1100–1400 h) and in the evening (1900–2000 h). Variability in feeding success decreased as its mean value increased and variability in feeding error decreased as error decreased. This pattern in the behavioural indices which reflect the alevin's ability to execute spatially-oriented activities (i.e. prey size selection, success, error) is put forward as a possible manifestation of underlying rhythms in the animal's nervous system. In the early morning alevins fed selectively on the largest copepods (>0.8 mm). This restricted range of selectivity was associated with low light levels and high activity. Later in the day, a broader range of the larger size classes were selected (>0.7 mm). Intensity of selection was inversely associated with total activity. This evidence of diurnal pattern in alevin feeding behaviour shows that (1) fish exhibit behavioural rhythms early in their life history, and (2) prey size selection, in terms of both intensity and breadth of size class selected for, changes over the day.

Introduction

Rhythms in fish physiological functions, locomotor activity, schooling and feeding behaviour have been widely reported in the literature (e.g. Schwassmann 1971, Thorpe 1978). In general, fish are most active at dawn and dusk, though the timing of rhythms varies among fish species and habitat and is at least partially dependent upon a host of environmental variables (Manteifel et al. 1978, Schwassmann 1979).

Despite a growing body of information, endogenous and/or exogenous rhythms in fish feeding behaviour have been largely overlooked in studies not directly concerned with them. As a result, rhythms have not generally been controlled for in studies of fish feeding nor have their effects on patterns of size selection been closely examined.

Rhythms in larval fish behaviour, though not unknown (e.g. Oliphan 1957, Blaxter 1965, Godin 1981), have received far less attention, partly because the small size and fragility of larval fish limits their suitability for feeding experiments. Prey size selection in larval fish is also known (e.g. Beyer et

* A contribution from the Huntsman Marine Laboratory and the Institute of Oceanography, McGill University. This paper is dedicated to the memory of the late Mark Loomer, M.D., whose accidental death at the age of 25 was a tragedy for all who knew him.

al. 1980, Christensen et al. 1980, Checkley 1982, Mikheev 1984) but has not generally been examined in conjunction with, or in relation to, rhythms in feeding behaviour.

We report the results of an experimental study on Atlantic salmon alevins (sensu Balon 1984) with goals of (1) quantitatively describing diurnal patterns (rhythms) in activity, feeding behaviour and prey size selection; (2) establishing a connection among these patterns; and (3) discussing the significance of these results, and rhythms in general, to the study of fish feeding.

Materials and methods

Unfed alevins of Atlantic salmon, *Salmo salar* L., were obtained from the North American Salmon Research Center in St. Andrews, New Brunswick and the Florenceville Federal Fish Hatchery in Florenceville, New Brunswick in May 1983. Experimental fish, of approximately the same age and size (25 mm), were maintained under natural lighting conditions in identical holding tanks (77 L × 32 D × 37 cm H: 91 liters) at the Huntsman Marine Laboratory, St. Andrews, New Brunswick. All experiments were conducted under natural light. Aquaria were supplied with running freshwater from the hypolimnion of a nearby lake. The water supply was filtered through 64 μm mesh to prevent introduction of unwanted material and was maintained at approximately 12° C (Lemm & Hendrix 1981). A natural assemblage of live zooplankton was added to holding tanks twice daily at a concentration of approximately 200 items per liter. Alevins had been feeding for 4 days prior to the experiment. Feeding of fish was discontinued 8 h prior to experimentation by removing them to a holding tank containing only filtered water.

Experimental and holding aquaria were surrounded with aqua-green cloth to simulate light attenuation in fresh water (Lythgoe 1979, Gibson 1980), prevent external movement from alarming the fish, and dampen the effects of rapidly changing light intensity. Pale green, 1 cm^2 graph paper was placed on the outer floor of each aquarium to aid in the estimation of horizontal distances. Air stones placed in the center of each aquarium created vertical water currents (which were laminar along the aquarium floor) upon which prey items were transported. The bubble curtain created by the airstone had no observable effect on alevin feeding behaviour. Current speed in the areas where fish fed was maintained at approximately 6–10 $cm \cdot sec^{-1}$ (see Rimmer & Power 1978) by regulating the air flow. Estimates of current speed were obtained by measuring the distance, per unit time, travelled by a particle in the feeding current. Atlantic salmon alevins naturally orient into current (Kalleberg 1958, Rimmer & Power 1978, Wankowski & Thorpe 1979) so that this artificial situation provided a reasonable simulation of natural feeding conditions.

Alevins are approximately 25 mm long at this interval of development, have relatively large and well-developed eyes, are visual predators (Protasov 1970, Wankowski 1981) and feed in an easily recognized sequence (e.g. Mills 1971, Mighell 1981, Wankowski 1981). The following sequential behaviours were enumerated using finger key counters:

Eye fixation = biocular locking of an alevin's eyes on a target.

Movement towards target = motion of the entire body, or head region only, towards the item visually targeted.

Bite = opening-snapping-closing of the jaws on the item.

Miss = unsuccessful bite.

Spit = forceful rejection of an item from the buccal cavity.

Also enumerated were: *Social behaviours* = interactions with other alevins.

Ambiguous behaviours = non-feeding, non-social activity (e.g. yawn).

Alevins, oriented into the current, lay on the substrate using pectoral fins for support and stabilization. When a target carried by the current passed by, the fish either exhibited the feeding behaviour sequence or ignored it. Fish were observed through rectangular viewports from behind a cardboard baffle. A shade was attached to the baffle and hung over the observer to minimize contrast effects and detection by the fish.

To examine diurnal changes in feeding behaviour alevins were observed on an hourly basis between 0600 h and 2100 h, the hours during which enough light was available for reliable observation. Light intensity (lux) at the water surface was recorded hourly using a Gossen Lunasix 3 light meter.

At 0500 h, 30 naive alevins were removed from a holding tank and were placed in an observation tank. The following procedure was repeated hourly for the duration of the experiment. Ten alevins, selected at random from the 30, were observed for a period of 3 min each. After 10 observation periods, 10 alevins were removed from the observation tank and sacrificed by immersion in concentrated formalin (preliminary trials revealed no regurgitation). Ten unfed and naive alevins were added to the observation tank to replace those removed. A minimum of 15 min was allowed for acclimation (preliminary observations had revealed that the behaviour of added fish became indistinguishable from those already present after 10 to 15 min).

The fish were fed on zooplankton obtained each morning at the outflow of Wheaton Lake, New Brunswick, a small, eutrophic, pristine glacial lake. A plankton net (125 μm mesh) was placed in the lake outflow for 15 min. The contents were transferred into clean buckets, transported to the laboratory, and filtered through a 1 mm mesh to remove large organisms and extraneous material. To introduce a standard amount of food into the experimental aquarium, the filtrate was mixed thoroughly and equal volumes were poured into beakers using the four-beaker sub-sampling method (Van Guelpen et al. 1982). Known volumes from these beakers were poured into the tanks at the outset of the experiment and, to replenish prey numbers depleted by feeding, at 60 min intervals over its course. Preliminary trials established that this method was accurate to within ±15% (Table 3).

The guts of sacrificed fish were immediately injected with 10% formalin to arrest digestion. Whole fish were then preserved in 5% phosphate-buffered formalin (Markle 1984). Fork length, body width immediately behind the pectoral fins, and mouth width were recorded to the nearest millimeter. Guts were removed under a dissecting microscope. The numerical composition of gross taxonomic groups (cladocerans with and without eggs, copepods, water mites, miscellaneous) present in the foregut was recorded. Incomplete prey items were counted only if the anterior section was present. All well-preserved individuals were measured for length (copepods: prosome length; cladocerans: length without spine) using an ocular micrometer.

At the end of each hour, 11 water samples were removed from the tanks by rapidly immersing a beaker into the feeding current. Tank zooplankton samples were preserved in 5% buffered formalin and subsequently examined for taxonomic composition. The numerical composition of gross taxonomic groups (as above) was recorded by examination of the entire sample under a dissecting microscope. The first 50 individuals (or the total number present if less than 50) of each group were measured for length (as above).

Data analysis

Behaviour data. – The behavioural data matrix was composed of 10 observations for each of the behaviours listed above, recorded hourly over the 15 h of the experiment. As prey density varied somewhat between hours (Table 3) the behavioural variables were normalized by dividing the number of observations of each behaviour by the total number of prey items available in a given time period. Because the effect of density on these feeding parameters is linear at low values (Browman & Marcotte, unpublished), normalization should not have introduced bias.

The following composite variables were calculated from normalized data:

Total activity: = # eye fixations + # moves + # bites + # misses + # spits + # social + # ambiguous,

Miss-spit (feeding error): # of misses + # of spits,

Bite/fix: # of bites/# of fixes,

Ingestion: # of bites − (# of misses + # of spits),

Success 1: # of ingestions/# of bites,

Success 2: # of ingestions/# of fixes.

All variables were tested for normality using a one sample Kolmogorov-Smirnov test (UNIVARIATE procedure of the Statistical Analysis System, SAS User's Guide, 1982 edition). Because there was significant deviation from normality, the nonparametric Kruskal-Wallis statistic (Daniel 1978; SAS – NPAR1WAY, option WILCOXON) was employed to test the null hypothesis that time of day had no effect on a given composite variable. If the null hypothesis was rejected ($P<0.05$), nonparametric Tukey-Type multiple comparisons (Zar 1984) were calculated to establish the location(s) of any discernable difference(s).

Gut content data

Total number of items in the gut. – The number of individuals of each taxonomic group present in the gut were normalized by dividing each by the numbers of individuals of the corresponding group present in the tank. The Kruskal-Wallis statistic was calculated (as above) to test the null hypothesis that there were no diurnal differences in normalized numbers of each taxonomic group present in the gut. If the null hypothesis was rejected ($P<0.05$), multiple comparisons were calculated (as above).

Feeding selectivity on gross taxa (total numbers ingested, not normalized) was evaluated using the Standardized Forage Ratio (SFR) index (see below). The null hypothesis that there were no diurnal differences in these indices was tested as above.

Prey size selection. – Our goal here was to determine whether alevins selected among prey sizes nonrandomly and, if this was the case, to evaluate the intensity of size selection and determine which size classes were preferred. Towards this end, the following three-step approach was used.

Step 1. The Kolmogorov-Smirnov two sample test (Hollander & Wolfe 1983) was calculated to evaluate the size frequency distributions of gut and tank items. Its null hypothesis is identity of two relative cumulative frequency distributions, one expected (tank) and the other observed (gut). As no between fish differences in the size distributions of items in the gut were observed ($P>0.40$), tests for differences between size frequency distributions of gut and tank items were carried out on lumped data using the NPAR TESTS (option K-S) procedure of Statistical Package for the Social Sciences (SPSSX User's Guide, 1983 edition). In the same way, hourly changes in tank size distributions were tested (all combinations of comparison), and found lacking ($P>0.4$). Diurnal differences in gut size distributions (all combinations of comparison) were then evaluated.

Step 2. Once selection was established, its intensity was evaluated using Bartell's (1982) Index of Selection Intensity

$$S = \frac{L_I - L_T}{L_R},$$

where L_I = mean length of ingested prey, L_T = mean length of prey in tank, L_R = range of prey lengths over gut and tank samples, and S = selection intensity.

S ranges from −1 to +1; values near zero indicate no size selection; high positive values indicate intense selection for large prey; high negative values intense selection for small prey. Lengths beyond ±2 standard deviations of mean values were omitted as they affect the index nonlinearly (Bartell 1982).

The null hypothesis that there were no diurnal differences in intensity of size selection was tested using the nonparametric one sample Wilcoxon signed ranks test (Hollander & Wolfe 1983). A descernable difference ($P<0.05$) for any given test leads to the conclusion that the particular S-value tested (i.e. the intensity of selection at that hour) is different from all of the other S-values (i.e. the intensity of selection at all other hours). The tests were carried out using Statistics Plus (option WILCOXON SIGNED RANKS TEST), a general statistics package for the Apple II computer (Madigan & Lawrence 1982).

In addition to testing for diurnal differences in S, the degree of association (correlation) between time of day and intensity of size selection was evaluated. The nonparametric Spearman rank correlation coefficient (RHO) was calculated because (1) it could not be assumed that diurnal effects on selection intensity were linear, (2) the data were not bivariate normally distributed (tested as above) and (3) as a result of (1) and (2) the standard product-moment correlation coefficient (r) was not applicable (Daniel 1978). RHO varies between −1 and +1, with a value of +1 indicating perfect direct relationship between variables, and a value of −1 indicating perfect inverse relationship. The test was carried out using Statistics Plus (option Spearman RHO).

Step 3. To add further resolution, the specific size classes of prey preferred (0.2 mm size intervals) were determined using the Standardized Forage Ratio (SFR) index (Chesson 1978, 1983). The index yields measurements of preference which are independent of prey or size class abundance (Lechowitz 1982, Pearre 1982), allowing meaningful comparisons of index results, both within samples and between treatments. SFR was calculated as follows:

$$SFR_i = \frac{r_i/p_i}{\sum_{i=1}^{n} r_i/p_i},$$

where r_i = proportion of prey size class (or taxon) i in the gut, p_i = proportion of prey size class (or taxon) i in the tank, and n = total number of size classes evaluated.

SFR ranges between 0 and 1, with values above 1/n indicating preference, values below 1/n avoidance. If there were no individuals of a given size class in the gut, SFR was set to 0. If there were no individuals of a given size interval in the tank zooplankton sample, but this size class was present in the gut, a value of 0.1% was arbitrarily assigned the tank interval.

Results

Behaviour

Total activity (Fig. 1) was highest in the early morning, followed by less pronounced mid and late day rises. Variance (in the form of standard deviation) was positively related to the mean. This pattern

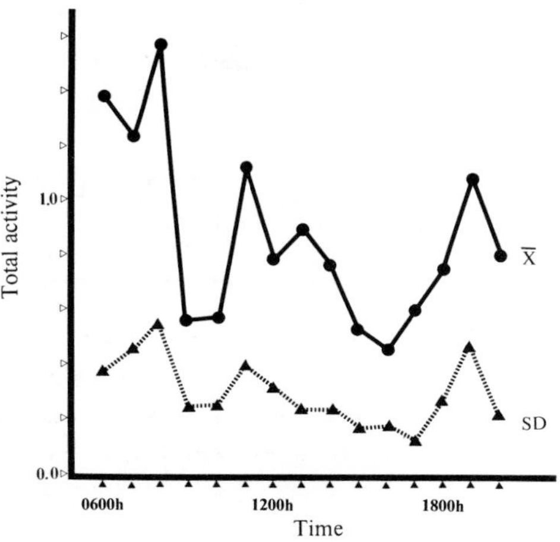

Fig. 1. The relationship between mean (for 10 fish) Total activity (# eye fixations + # moves + # bites + # misses + # spits + # social + # ambiguous) per 3 min observation period (normalized for tank zooplankton abundance) and time of day. SD = standard deviation (broken line) around the mean (solid line). The following hour pairs were statistically discernable, by multiple comparisons test, at the level of probability (P) indicated: 0600 and 0800 h vs 0900, 1000, 1500, 1600, 1700 h ($P<0.01$); 0700 h vs 0900, 1000, 1500, 1600 h ($P<0.06$); 0900 h vs 1100 h ($P<0.08$); 1100 h vs 1500, 1600 h ($P<0.05$); 1600 h vs 1900 h ($P<0.06$).

was similar for all of the singular behaviours of which this composite variable, and those which follow, were composed. Miss-spit (i.e. feeding error) followed the same general trend, though the late-day trough and peak were not statistically discernable (Fig. 2). Bite/fix (Fig. 3) exhibited an extended early morning peak followed by a sharp depression, then a gradual rise. The late day peak was not statistically discernable. Variance in Bite/fix was inversely related to the mean.

For Ingestion (Fig. 4) and Success 1 (Fig. 5) only early morning peaks were statistically discernable. The 1000 h peak in Success 1 was associated with an anomolously low number of Bites in that time period. The variance of Success 1 was positively related to its mean. Variance in Ingestion exhibited a 3 peak rhythm while its mean decreased monotonically. The early morning peak-trough of Success 2 was statistically discernable from values later in the day (Fig. 6). Mean values over the remainder of the day followed a pattern inverse of the trends in the other behavioural variables. Variance in Suc-

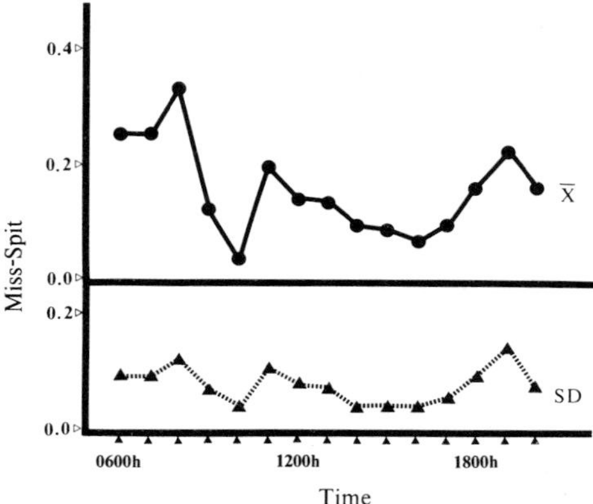

Fig. 2. The relationship between mean (for 10 fish) number of miss-spits (# misses + # spits, i.e. feeding error) per 3 minute observation period (normalized for tank zooplankton abundance) and time of day. SD = standard deviation (broken line) around the mean (solid line). The following hour pairs were statistically discernable, by multiple comparisons test, at the level of probability (P) indicated: 0600 and 0700 h vs 1000, 1500, 1600 h (P<0.05); 0800 h vs 0900, 1000, 1400, 1500, 1600, 1700 h (P<0.05); 1000 h vs 1100, 1800, 1900, 2000 h (P<0.06).

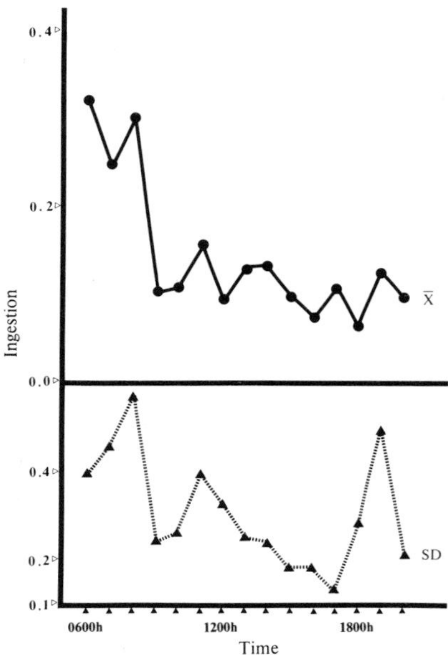

Fig. 4. The relationship between mean (for 10 fish) number of Ingestions (# Bites − # Miss − # Spits) per 3 minute observation period (normalized for tank zooplankton abundance) and time of day. SD = standard deviation (broken line) around the mean (solid line). The following hour pairs were statistically discernable, by multiple comparisons test, at the level of probability (P) indicated: 0600 h vs 1000, 1200, 1500, 1600, 1800, 2000 h (P<0.05); 0700 h vs 1600, 1800 h (P<0.05); 0800 h vs 1600, 1800 h (P<0.05).

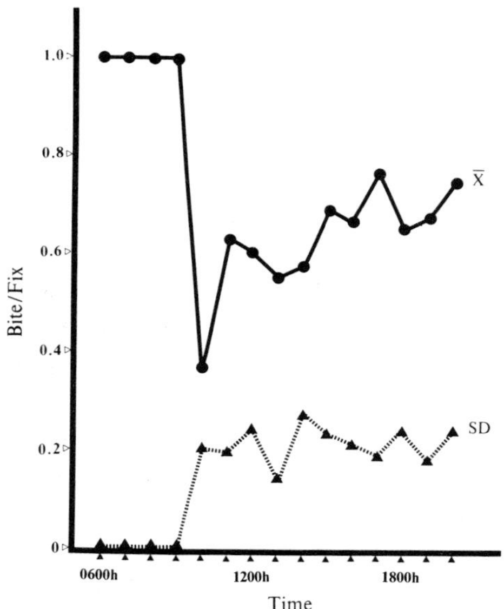

Fig. 3. The relationship between mean (for 10 fish) number of bite/fixes (# bites/# fixes) per 3 minute observation period (normalized for tank zooplankton abundance). The following hour pairs were statistically discernable, by multiple comparisons test, at the level of probability (P) indicated: 0600 h vs 1000, 1100, 12N0.08); 0700 h vs 000, 1100, 1200, 1300, 1400 h (P<0.05); 0700 h vs 1600, 1800, 1900 h (P<0.08).

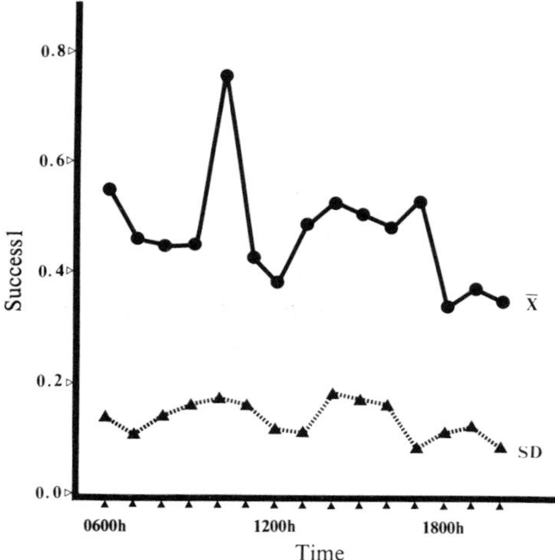

Fig. 5. The relationship between mean (for 10 fish) feeding success (# ingestions/# bites, i.e. Success 1) per 3 min observation period (normalized for tank zooplankton abundance) and time of day. SD = standard deviation (broken line) around the mean (solid line). The following hour pairs were statistically discernable, by multiple comparisons test, at the level of probability (P) indicated: 1000 h vs 1200, 1800, 1900, 2000 h (P<0.05).

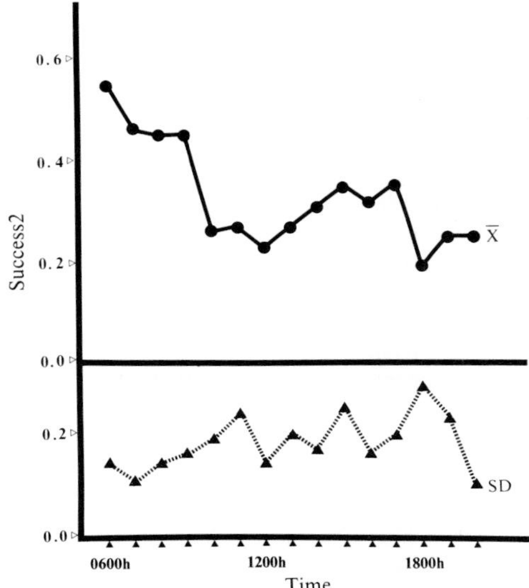

Fig. 6. The relationship between mean (for 10 fish) feeding success (# ingestions/# fixes, i.e. Success 2) per 3 minute observation period (normalized for tank zooplankton abundance) and time of day. SD = standard deviation (broken line) around the mean (solid line). The following hour pairs were statistically discernable, by multiple comparisons test, at the level of probability (P) indicated: 0600 h vs 1000, 1100, 1200, 1800, 1900, 2000 h ($P<0.05$); 0600 h vs 1300 h ($P<0.06$); 0700 h vs 1200, 1800 H ($P<0.06$); 0800 h vs 1800 h ($P<0.05$); 0900 h vs 1800 h ($P<0.05$).

cess 2 was inversely related to its mean.

In general, the most pronounced peaks occurred in the early morning (0600–0800 h), followed by a morning trough (0900–1000 h), a mid-day rise (1100–1400 h), another trough (1500–1800 h) and a late day rise (1900–2000 h). This trend in the mean values of each composite variable was consistent, although the entire trend was not statistically discernable in all cases.

Gut contents

Almost all component taxa of the tank zooplankton were found in alevin guts. Some were present in low numbers and could not be evaluated statistically. These were: water mites (2 species), the cladocerans *Holopedium sp.* and *Alona affinis*, chironomid larvae, other fly larvae and oligochaetes. Thus, statistical analysis was limited to: (1) all calanoid and cyclopoid copepods, with no more than three species of either appearing at any time; and (2) cladocerans, primarily *Bosmina sp.*, *Diaphanosoma sp.* and *Daphnia sp.* This composition of prey is similar to diets reported for Atlantic salmon alevins in the wild (White 1936, Allen 1941, Pinskii 1967, Arnemo et al. 1980, Williams 1981).

Total number of items in the gut

In general, there were no discernable diurnal differences in the total number of items present in the gut (Table 1); the number of copepods in the gut was discernably different for 0600 versus 0800 h ($P<0.05$). High variance in the numbers may have been responsible for this lack of resolution (e.g. a single fish could contain anywhere from 0 to 60 items).

There was strong selection for copepods, and avoidance of cladocerans, over all hours. Diurnal changes in selection were not statistically discernable, though variability in selection, as reflected in the standard deviation of the indices, was greater in the early morning (Table 1).

Prey size selection

Step 1 (presence of size selection): Sample sizes were sufficient to evaluate size selection for 8 of the 15 h of observations (Table 2). Sizes of items in the gut were generally greater than those in the tank. With the exception of 0600 h, all size distributions of items in the guts were statistically discernable from the size distributions of items in the tank, indicating that alevins were feeding selectively (Feller & Kaczynski 1975); sizes of items in the gut were greater than those in the tank. The distribution of sizes selected, compared among fish over different hours, was significantly different for 0600 versus 1700 h (K-S $Z = -0.490$, $P<0.05$), 0600 versus 1800 h (K-S $Z = +0.595$, $P<0.02$), and 1000 versus 1800 h (K-S $Z = -0.420$, $P<0.05$) indicating that, for these hour pairs, there were differences in size selection.

Step 2 (intensity of size selection): Selection intensity (Bartell's S) was significantly lower in the early morning and at mid-day (Fig. 7), with significantly higher selection intensities occurring in the

Table 1. Hourly summaries of (1) total number of items in alevin guts and (2) the Standardized Forage Ratio index (SFR) for copepods and cladocerans. SFR values >0.05 indicate preference, <0.05 avoidance. Values are ± standard deviations.

Hour	Total # in gut	Normalized total # in gut	SFR copepods	SFR cladocerans
0600	1.6 ± 1.2	0.06 ± 0.04	0.80 ± 0.35	0.20 ± 0.35
0700	2.2 ± 2.0	0.10 ± 0.09	0.94 ± 0.13	0.06 ± 0.13
0800	2.7 ± 1.9	0.16 ± 0.11	0.98 ± 0.04	0.02 ± 0.04
0900	2.6 ± 2.4	0.09 ± 0.08	0.83 ± 0.35	0.17 ± 0.35
1000	4.3 ± 3.7	0.14 ± 0.12	0.65 ± 0.42	0.35 ± 0.42
1100	2.9 ± 3.4	0.09 ± 0.11	0.94 ± 0.06	0.06 ± 0.06
1200	3.5 ± 2.7	0.12 ± 0.09	0.92 ± 0.14	0.08 ± 0.14
1300	2.3 ± 2.4	0.07 ± 0.07	0.97 ± 0.04	0.03 ± 0.04
1400	3.1 ± 3.2	0.10 ± 0.11	0.99 ± 0.04	0.01 ± 0.04
1500	1.8 ± 1.8	0.04 ± 0.04	0.99 ± 0.02	0.01 ± 0.02
1600	7.0 ± 6.0	0.11 ± 0.10	0.98 ± 0.03	0.02 ± 0.03
1700	3.3 ± 3.9	0.07 ± 0.08	0.98 ± 0.03	0.02 ± 0.03
1800	2.5 ± 3.1	0.06 ± 0.08	0.82 ± 0.40	0.18 ± 0.40
1900	2.0 ± 2.7	0.06 ± 0.08	0.94 ± 0.07	0.06 ± 0.07
2000	2.0 ± 2.4	0.05 ± 0.06	0.99 ± 0.02	0.01 ± 0.02

later day. There was an overall correlation between S and (1) time of day (RHO = 0.61, P<0.02); and (2) total activity (RHO = −0.48, P<0.08). Depressed selection intensity in the early morning (0600 h) coincided with low light levels; illumination at all other hours was above levels at which an effect on selection would be expected (Table 3).

Step 3 (preferred size classes): The largest size classes in the gut were composed almost entirely of copepods, while large cladocerans were consumed in small numbers and were almost always carrying eggs (Table 4, Fig. 8 and see Mellors 1975). In the gut, small cladocerans were found in association with larger copepods, indicating that at least some were ingested incidently (O'Brien 1979). The largest size classes (>0.9 mm) were preferred over all others in the early morning. In the afternoon and late day the preferred size range was extended downward to items greater than 0.7 mm and less than 1.0 mm. It should be noted that index values for the largest size intervals are suspect because of their limited representation in gut and tank samples.

Table 2. Hourly summary of Kolmogorov-Smirnov (Z) results testing for differences between the size frequency distributions of items in the tank versus items in alevin guts. P<0.05 indicates size selection. M-W = Mann-Whitney test. n = sample size.

Hour	n Gut	n Tank	n Total	K-S Z	Maximum difference		P<
0600	9	24	33	1.21	+0.472		0.108
1000	16	31	47	2.12	+0.651		0.0001
1100	13	31	44	1.92	+0.633		0.001
1200	17	15	32	1.20	−0.424		0.115
1200	17	15	32	–	–	M-W	0.034
1700	46	61	107	4.15	+0.809		0.0001
1800	28	47	75	3.53	+0.843		0.0001
1900	14	39	53	2.26	+0.703		0.0001
2000	16	34	50	2.29	+0.695		0.0001

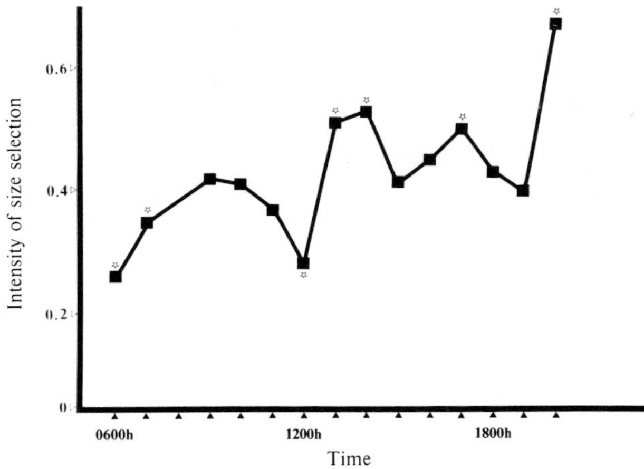

Fig. 7. The relationship between intensity of size selection (Bartell's S) and time of day. Stars (*) associated with points indicate that the S value for that hour is statistically different (P<0.05) from all other hours.

Discussion

Until recently, studies of feeding in larval fish had been limited to qualitative reports of feeding behaviour and/or gut contents. Contemporary investigations have begun to examine more closely the relationships between the feeding and biology of young fish and environmental variables with their inherent periodicities (Blaxter 1965, Rosenthal 1969, Houde & Shekter 1980, Hunter 1980, Godin 1981, Williams 1981, Balon 1986). The quantitative evidence of diurnal pattern in alevin feeding behaviour reported here shows that: (1) fish exhibit behavioural rhythms early in their life history; (2) their daily feeding behaviour is complex and variable; and (3) prey size selection, in terms of both intensity and breadth of size class selected for, begins soon after first exogenous feeding and changes over the day.

The observed pattern in total alevin activity was similar to that reported in the literature for Atlantic salmon. In a qualitative study, Pinskii (1961, 1962) observed two periods of intensified feeding activity, one in the early morning and another in the late day; the former was 1–1.5 times greater than the latter. As Atlantic salmon alevins do not feed at night (Hoar 1942, Pinskii 1961, 1967, personal observation), the early morning peak was not surprising. Under the conditions of our experiment, al-

Table 3. Hourly summaries of (1) Bartell's index of selection intensity (S), (2) the probability (P) that any given value of S is different from all others, (3) light intensity and (4) tank zooplankton density.

Hour	S	P<	Light intensity (lux)	Tank zooplankton density (number per liter)
0600	0.26	0.001	1.75	27
0700	0.35	0.024	66	23
0800*	–	–	175	27
0900	0.42	0.65	260	30
1000	0.41	0.59	700	31
1100	0.37	0.052	1200	31
1200	0.28	0.002	1500	30
1300	0.51	0.018	700	33
1400	0.53	0.011	1200	30
1500	0.41	0.59	1600	47
1600	0.45	0.47	480	41
1700	0.50	0.67	300	46
1800	0.43	0.60	500	39
1900	0.40	0.60	140	34
2000	0.67	0.001	66	38
			mean =	33.8
			S.D. =	7.0

* Insufficient data for calculation of S.

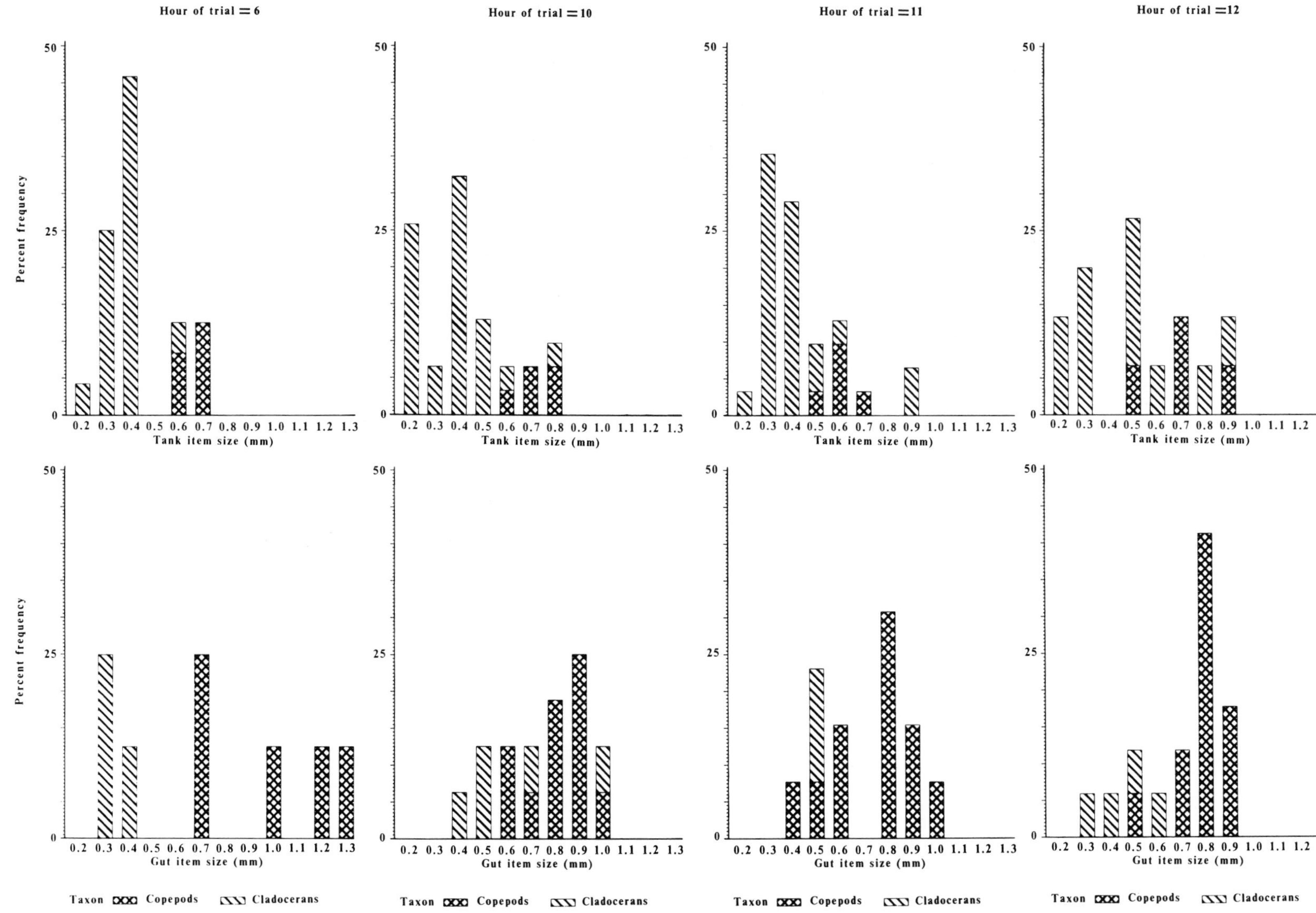
Hour of trial = 6
Hour of trial =10
Hour of trial =11
Hour of trial =12
Percent frequency
0
25
50
0.2 0.3 0.4 0.5 0.6 0.7 0.8 0.9 1.0 1.1 1.2 1.3
Tank item size (mm)
Gut item size (mm)
Taxon Copepods Cladocerans

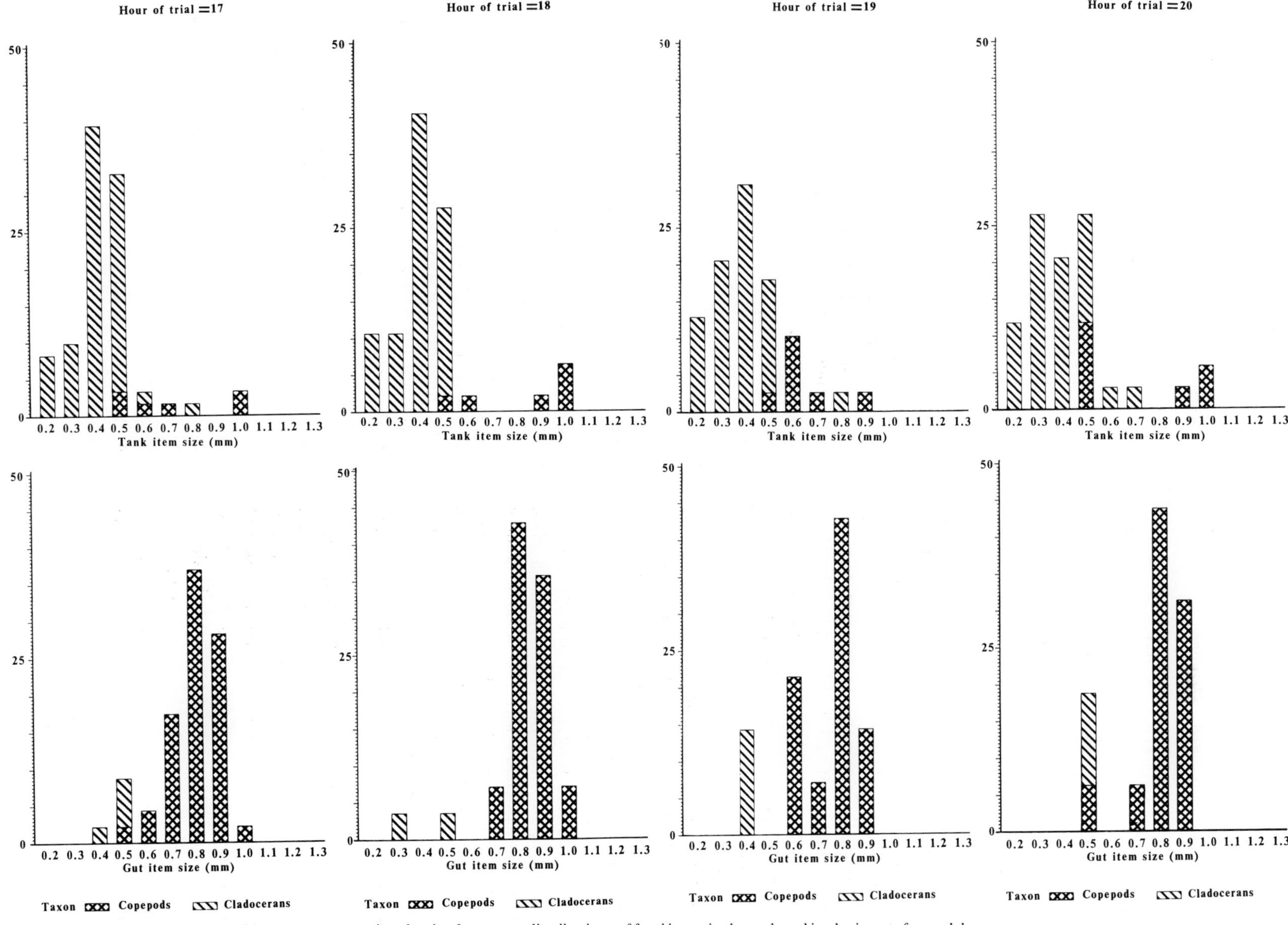

Fig. 8. Percent frequency histograms comparing the size frequency distributions of food items in the tank and in alevin guts for each hour analyzed. Data for each hour is lumped from the gut contents of 10 fish.

Table 4. Hourly summaries of the Standardized Forage Ratio index (SFR) for each 0.2 mm size interval of prey items available. Values >1/n indicate preference, <1/n avoidance.

Hour	0600	1000	1100	1200	1700	1800	1900	2000
1/n	0.17	0.20	0.20	0.25	0.20	0.20	0.25	0.20
SFR (.2–.3 mm)	0.005	0	0	0.04	0	0.004	0	0
SFR (.4–.5 mm)	0.003	0.006	0.009	0.14	0.003	0.001	0.02	0.02
SFR (.6–.7 mm)	0.005	0.03	0.01	0.19	0.09	0.08	0.16	0.04
SFR (.8–.9 mm)	0	0.07	0.08	0.63	0.89	0.89	0.82	0.94
SFR (1.0–1.1 mm)	0.33	0.90	0.90	–	0.02	0.03	–	0
SFR (1.2–1.3 mm)	0.66	–	–	–	–	–	–	–

evins had been without food for the 8 nocturnal hours immediately prior to the experiment, so that the pronounced early morning peak in activity was at least partially in response to an empty gut.

The pattern exhibited by Miss-spit, which was interpreted as an indicator of error in the feeding sequence, followed that of 'total activity'. As activity declined, error in feeding (and its associated variance) declined. The composite variables expressing feeding success (Success 1 and Success 2) also reflected this pattern; both declined with increasing activity. Bite/fix, which was interpreted as the likelihood that the fish would attack an item upon which they had visually fixed, was high in the first few hours of feeding; alevins attacked virtually everything upon which they visually fixed. High Bite/fix was associated with the peak in feeding error, the drop in both indicators of feeding success, the increased variability in selection for gross taxa and the reduction in the intensity of size selection. Also, as the likelihood of a bite-following-a-fix increased, the variance in this likelihood decreased.

In a study of the feeding habits of walleye larvae, Mathias & Li (1982) observed a three-peak diel pattern in feeding activity as well as a shift in the type of prey selected. Williams (1981), observed diel changes in the diets of Atlantic salmon alevins. A comparison of Figures 1 and 7 revealed a similar phenomenon; salmon alevins were less selective during their most active periods, indicating that diurnal rhythm in activity also affects prey selection.

It may be hypothesized that the accurate/efficient performance of spatial tasks (e.g. estimation of prey distances, angles, sizes, shapes, motility patterns) by alevins is rhythmic. That (1) error in the feeding sequence (Miss-spit) carried lower variance when its mean value was low; (2) variance in feeding success (Success 2) declined when its mean value increased; and (3) intensity of size selection was inversely associated with 'total activity' may indirectly support this hypothesis. Recent studies on the neuroethology of feeding behaviour (Ewert et al. 1983, Ewert 1985, Marcotte & Browman 1986), the physiological ecology of sleep (Karmanova 1982) and the brain mechanisms of behaviour (Laming 1981) have begun to suggest the underlying causes of these rhythms and their implications for behavioural ecology. Because these processes exist in fish, in at least a phylogenetically precursive form (Allison & Twyver 1970, Broughton 1972, Tauber 1974, Karmanova et al. 1981, Shapiro et al. 1981, Karmanova 1982, Davis & Northcutt 1983), it would seem appropriate to introduce some of them here.

Karmanova et al. (1981) reported on the 24 h rhythm in sleep states of the catfish, *Ictalurus nebulosus*. Both day and night rhythms were characterized by a pattern of wakeful states followed by periods of sleep-like states (Fig. 9D). The 24 h pattern in catfish sleep-like states and heart rate reported by Karmanova et al. (1981) was offset by 30 min (because of differences in the time zone locations of the studies) and compared to the activity pattern reported here for Atlantic salmon alevins (Fig. 9). Low alevin activity occurred during periods corresponding to sleep-like states (and

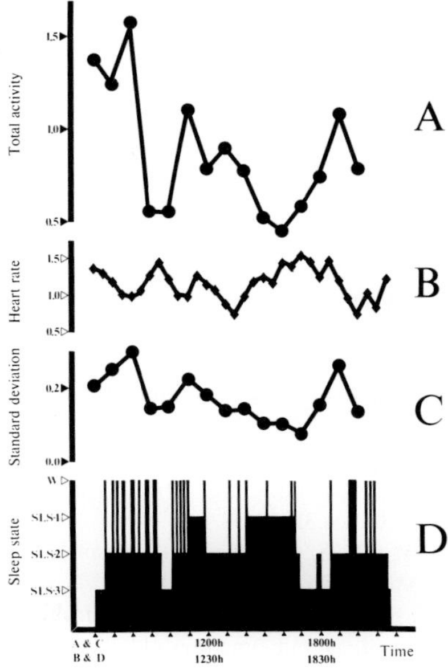

Fig. 9. (A) – The relationship between mean (for 10 fish) Total activity (# eye fixations + # moves + # Bites + # misses + # spits + # social + # ambiguous) per 3 minute observation period (normalized for tank zooplankton abundance) and time of day. SD = standard deviation (broken line) around the mean (solid line). The following hour pairs were statistically discernable, by multiple comparisons test, at the level of probability (P) indicated: 0600 and 0800 h vs 0900, 1000, 1500, 1600, 1700 h (P<0.01); 0700 h vs 0900, 1000, 1500, 1600 h (P<0.06); 0900 h vs 1100 h (P<0.08); 1100 h vs 1500, 1600 h (P<0.05); 1600 h vs 1900 h (P<0.06). (B) – Diurnal periodicity in the heart rate of the catfish. Ordinate = time interval (seconds) between two subsequent heartbeats (redrawn from Karmanova et al. 1981). (C) – Variance in total activity (Fig. 9A) as represented by its standard deviation. (D) – diurnal periodicity in the wakefulness-primary sleep cycle of the catfish. SLS = sleep-like-state; SLS-1 = characterized by immobility with plastic muscle tone; SLS-2 = characterized by immobility with rigid muscle tone; SLS-3 = characterized by immobility with muscle relaxation; W = wakefulness (redrawn from Karmanova et al. 1981).

long intervals between heart beats); high activity during periods of wakefulness (shorter intervals between heart beats). Peaks in the variance of alevin 'total activity' (Fig. 9C) corresponded to intervals of rapid alternation between wakefulness and sleep in the catfish (Fig. 9D). Interestingly, there is evidence which suggests that activity rhythms in piscivores are the opposite of these of their planktivorous prey (Spencer 1939, Thorpe 1978, Federova & Drozzhina 1982); piscivores, in general, are active immediately before dawn and after sunset, the periods during which their planktivorous prey may be in sleep-like states.

In an examination of EEG's from the optic tectum of a toad, Laming (1983) found that lower thresholds for neuronal discharge were associated with behaviourally active or aroused states and that these lowered thresholds enabled the animal to better detect and respond to environmental changes. Andrew (1983) and Andrew & Brennan (1983) reported right/left differences in the manner in which visual information was categorized in the brains of young chicks. Gur et al. (1982) have shown that verbal and spatial tasks in humans lead to increased blood flow to different cerebral hemispheres. Klein & Armitage (1979) have reported 90 to 100 minute oscillations in the performance of verbal and spatial tasks in humans. These oscillations were associated with lateral switching of the respective cerebral hemisphere which dominated observed behaviour (Gevins et al. 1983). This, in turn, has been linked to the REM/Non-REM rhythm in brain activity during sleep and its continuation during wakefulness, i.e. the Basic Rest Activity Cycle (BRAC) (Kleitman 1963, 1969, Jouvet 1973, Van Valen 1973, Broughton 1975, Klein & Armitage 1979). At certain points in the BRAC, performance of spatial tasks is enhanced, while at others spatial task performance is less effective. Processes such as these may well underly behavioural patterns such as those reported here for total activity, feeding error and feeding success.

Processes such as those discussed here are fundamental to an animals' performance of cognitive (sensu Marcotte 1983) and spatial tasks, such as feeding, and should be more closely examined so that they can be incorporated into studies of behavioural ecology. Rhythms in perceptual and cognitive skills should be carefully considered in the scheduling of experiments and/or observations on fish feeding. Current models/ideas of fish feeding (e.g. Ware 1971, Milinski 1977, Confer et al. 1978, O'Brien 1979, Giguere et al. 1982, Hairston et al. 1982, Dill 1983, Ringler 1983, Dunbrack & Dill 1984, Wright & O'Brien 1984), though they have begun to consider the animals' perceptual and cognitive abilities/limitations, may well benefit by proceeding these several steps further.

Acknowledgements

We thank Brian Glebe, North American Salmon Research Center, and Redge Hitcock, Florenceville Federal Fish Hatchery, for supplying alevins. Fred Purton and John Foster provided invaluable logistic aid and many useful suggestions. Thanks are due Mike Walsh for his knowledgeable advice on statistical packages. Our manuscript benefitted greatly from the constructive criticism of Wayne Hunte, Rob Peters, Charles A. Simenstad and two anonymous reviewers. This study was supported by scholarships from McGill University Factulty of Graduate studies and Research to HIB, by computer funds from the McGill University Computing Center, and by funds from Gaea Consultants to BMM.

References cited

Allen, K.R. 1941. Studies on the biology of the early stages of the salmon (*Salmo salar*). 2. Feeding habits. J. Anim. Ecol. 10: 47–76.

Allison, T. & Van Twyver. 1970. The evolution of sleep. Nat. Hist. 79: 56–65.

Andrew, R.J. 1983. Lateralization of emotional and cognitive function in higher vertebrates, with special reference to the domestic chick. pp. 477–509. *In:* J.P. Ewert, R.R. Capranica & D.J. Ingle (ed.) Advances in Vertebrate Neuroethology, Plenum Press, New York.

Andrew, R.J. & A. Brennan. 1983. The lateralization of fear behaviour in the male domestic chick: a developmental study. Anim. Behav. 31: 1166–1176.

Arnemo, R., C. Puke & N.G Steffner. 1980. Feeding during the first weeks of young salmon in a pond. Arch. Hydrobiol. 89: 265–273.

Balon, E.K. 1984. Reflections on some decisive events in the early life of fishes. Trans. Amer. Fish. Soc. 113: 178–185.

Balon, E.K. 1986. Types of feeding in the ontogeny of fishes and the life-history model. Env. Biol. Fish. 16: 11–24 (this volume).

Bartell, S.M. 1982. Influence of prey abundance on size-selective predation by bluegills. Trans. Amer. Fish. Soc. 111: 453–461.

Beyer, J.E. & P.M. Christensen. 1980. Food consumption by larval fish. Inter. Comm. Explor. Seas, Internal Commun. 1980/L: 28.

Blaxter, J.H.S. 1965. The feeding of herring larvae and their ecology in relation to feeding. Rep. Calif. Coop. Oceanic Fisheries Invest. 10: 78–88.

Broughton, R.J. 1972. Phylogenetic evolution of sleep systems. pp. 2–7. *In:* M.H. Chase (ed.) The Sleeping Brain, Brain Research Institute, Los Angeles.

Broughton, R.J. 1975. Biorhythmic variations in consciousness and psychological functions. Can. Psychol. Rev. 16: 217–239.

Checkley, D.M. Jr. 1982. Selective feeding by Atlantic herring (*Clupea harengus*) larvae on zooplankton in natural assemblages. Mar. Ecol. Prog. Ser. 9: 245–253.

Chesson, J. 1978. Measuring preference in selective predation. Ecology 59: 211–215.

Chesson, J. 1983. The estimation and analysis of preference and its relationship to foraging models. Ecology 64: 1297–1304.

Christensen, E.B., V. Christensen & J.E. Beyer. 1980. Size selective predation by larval fish. Inter. Comm. Explor. Seas. Internal Commun. 1980/L: 27.

Confer, J.L., G.L. Howick, M.H. Corzette, S.L. Kramer, S.L. Fitzgibbon & R. Landesberg. 1978. Visual predation by planktivores. Oikos 31: 27–38.

Daniel, W.W. 1978. Applied nonparametric statistics. Houghton Mifflin Co., Boston. 510 pp.

Davis, R.E. & R.G. Northcutt. 1983. Fish neurobiology, Vol. 2, Higher brain areas and functions. University of Michigan Press, Ann Arbor. 375 pp.

Dill, L.M. 1983. Adaptive flexibility in the foraging behavior of fishes. Can J. Fish. Aquat. Sci. 40: 398–408.

Dunbrack, R.L. & L.M. Dill. 1984. Three-dimensional prey reaction field of the juvenile coho salmon (*Oncorhynchus kisutch*). Can. J. Fish. Aquat. Sci. 41: 1176–1182.

Ewert, J.P. 1985. Concepts in vertebrate neuroethology. Anim. Behav. 33: 1–29.

Ewert, J.P., R.R. Capranica & D.J. Ingle. 1983. Advances in Vertebrate Neuroethology. Plenum Press, New York. 1238 pp.

Federova, G.V. & K.S. Drozzhina. 1982. Daily feeding rhythm of pike-perch, *Stizostedion lucioperca,* and perch, *Perca fluviatilis,* from lake Ladoga. J. Ichthyol. 22: 52–60.

Feller, R.J. & V.W. Kaczynski. 1975. Size selective predation by juvenile chum salmon (*Oncorhynchus keta*) on epibenthic prey in Puget Sound. J. Fish. Res. Board. Can. 32: 1419–1429.

Gevins, A.S., R.E. Schaffer, J.C. Doyle, B.A. Cutillo, R.S. Tannehill & S.L. Bressler. 1983. Shadows of thought: shifting lateralization of human brain electrical patterns during brief visuomotor task. Science 220: 97–99.

Gibson, J.J. 1980. The ecological approach to visual perception. Houghton Mifflin Co., Boston. 332 pp.

Giguere, L.A., A. Delage, L.M. Dill & J. Gerritsen. 1982. Predicting encounter rates for zooplankton: a model assuming a cylindrical encounter field. Can J. Fish. Aquat. Sci. 39: 237–242.

Godin, J-G.J. 1981. Daily patterns of feeding behavior, daily rations, and diets of juvenile pink salmon (*Oncorhynchus gorbuscha*) in two marine bays of British Columbia. Can. J. Fish. Aquat. Sci. 38: 10–15.

Gur, R.C. R.E. Rosen, B.E. Obrist, J.P. Hungerbuhler, D. Younkin, A.D. Rosen, B.E. Skolnick & M. Reivich. 1982. Sex and handedness differences in cerebral blood flow during rest and cognitive activity. Science 217: 659–661.

Hairston, N.G. Jr., T. LiKao & S.S. Easter Jr. 1982. Fish vision and the detection of planktonic prey. Science 218: 1240–1242.

Hoar, W.S. 1942. Diurnal variations in feeding activity of young salmon and trout. J. Fish. Res. Board Can. 6: 90–101.

Hollander, M. & D.A. Wolfe. 1983. Nonparametric statistical methods. John Wiley & Sons, New York. 503 pp.

Houde, E.D. & R.C. Schekter. 1980. Feeding by marine fish larvae: developmental and functional responses. Env. Biol. Fish. 5: 315–334.

Hunter, J.R. 1980. The feeding behavior and ecology of marine fish larvae. pp. 287–330. *In:* J.E. Bardach, J.J. Magnuson, R.C. May & J.M. Reinhart (ed.) Fish Behavior and Its Use in the Capture and Culture of Fishes, ICLARM Conf. Proc. 5, Manila.

Jouvet, M. 1973. Essai sur le reve. Archives Italien de Biologie 111: 564–576.

Kalleberg, H. 1958. Observations in a stream tank of territoriality and competition in juvenile salmon and trout (*Salmo salar* L. and *S. trutta* L.). Rep. Inst. Freshw. Res. Drottningholm 39: 55–98.

Karmanova, I.G. 1982. Evolution of sleep. Stages of the formation of the 'wakefulness-sleep' cycle in vertebrates. Karger, New York. 164 pp.

Karmanova, I.H., A.I. Belich & S.G. Lazarev. 1981. An electrophysiological study of wakefulness and sleep-like states in fish and amphibians. pp. 181–200. *In:* P.R. Laming (ed.) Brain Mechanisms of Behavior in Lower Vertebrates, Cambridge University Press, Cambridge.

Klein, R. & R. Armitage. 1979. Rhythms in human performance: one and a half hour oscillations in cognitive style. Science 204: 1326–1328.

Kleitman, N. 1963. Sleep and wakefulness. University of Chicago Press, Chicago. 552 pp.

Kleitman, N. 1969. Basic rest-activity cycle in relation to sleep and wakefulness. pp. 33–38. *In:* A. Kales (ed.) Sleep: Physiology and Pathology, J.B. Lippincott Co., Philadelphia. 360 pp.

Laming, P.R. 1981. Brain mechanisms of Behavior in lower vertebrates. Cambridge University Press, Cambridge. 318 pp.

Laming, P.R. 1983. Relationships between the responses of visual units, EEGs and slow potential shifts in the optic tectum of the toad. pp. 595–602. *In:* J.P. Ewert, R.R. Capranica & D.J. Ingle (ed.) Advances in Vertebrate Neuroethology, Plenum Press, New York.

Lechowitz, M.J. 1982. The sampling characteristics of electivity indices. Oecologia 52: 22–30.

Lemm, C.A. & M.A. Hendrix. 1981. Growth and survival of Atlantic salmon (*Salmo salar*) fed various starter diets. Prog. Fish-Cult. 43: 195–199.

Lythgoe, J.N. 1979. The ecology of vision. Oxford University Press, Oxford. 244 pp.

Madigan, S. & V. Lawrence. 1982. Stats plus. A general statistics package for the Apple II. Human Systems Dynamics. 143 pp.

Manteifel, B.P., I.I. Girsa & D.S. Pavlov. 1978. On rhythms in fish behaviour. pp. 215–224. *In:* J.E. Thorpe (ed.) Rhythmic Activity of Fishes, Academic Press, New York.

Marcotte, B.M. 1983. The imperatives of copepod diversity: perception, cognition, competition and predation. pp: 49–72. *In:* F. Schram (ed.) Crustacean Phylogeny, Balkema Press, Rotterdam.

Marcotte, B.M. & H.I. Browman. 1986. Foraging behaviour in fishes: perspectives on variance. Env. Biol. Fish. (in press).

Markle, D.F. 1984. Phosphate buffered formalin for long term preservation of formalin fixed ichthyoplankton. Copeia 1984: 525–528.

Mathias, J.A. & S. Li. 1982. Feeding habits of walleye larvae and juveniles: comparative laboratory and field studies. Trans. Amer. Fish. Soc. 111: 722–735.

Mellors, W.K. 1975. Selective predation of ephippial daphnia and the resistance of ephippial eggs to digestion. Ecology 56: 974–980.

Mighell, J.L. 1981. Culture of Atlantic salmon, *Salmo salar,* in Puget Sound. Mar. Fisher. Rev. 43: 1–8.

Mikheev, V.N. 1984. Prey size and food selectivity in young fishes. J. Ichthyol. 24: 66–76.

Milinski, M. 1977. Experiments on the selection of predators against spatial oddity of their prey. Zool. Tierpsychol. 43: 311–325.

Mills, D. 1971. Salmon and trout: a resource, its ecology, conservation and management. Oliver & Boyd, Edinburgh. 351 pp.

O'Brien, W.J. 1979. The predator-prey interaction of planktivorous fish and zooplankton. Amer. Scient. 67: 572–581.

Oliphan, V.I. 1957. On the 24-hour feeding rhythms of the larvae of the Baikal grayling and on 24-hour rhythms in fish larvae in general. Fish. Res. Board Can. Transl. Ser. 834. 9 pp. (translated from Russian).

Pearre, S. Jr. 1982. Estimating prey preference by predators: uses of various indices, and a proposal of another based on x^2. Can J. Fish. Aquat. Sci. 39: 914–923.

Pinskii, F. Ya. 1961. The daily rhythm of feeding by Atlantic salmon fingerlings (*Salmo salar* L.). Fish. Res. Board Can. Trans. Ser. 400. 12 pp. (translated from Russian).

Pinskii, F. Ya. 1962. The daily feeding rhythms of salmon young (*Salmo salar* L.). Bull. Moskov. Obshchestva Ispytatelei Prirody, Otdel. Biol. 67: 152. (In Russian).

Pinskii, F. Ya . 1967. Daily feeding rhythm and diets of young of the salmon (*Salmo salar* L.) when raised in ponds. Fish. Res. Board Can. Transl. Ser. 1143. 11 pp. (translated from Russian).

Protasov, V.R. 1970. Vision and near orientation in fish. Israel Program for Scient. Transl., Jerusalem. 175 pp.

Rimmer, D.M. & G. Power. 1978. Feeding response of Atlantic salmon (*Salmo salar*) alevins in flowing and still water. J. Fish. Res. Board Can. 35: 329–332.

Ringler, N.H. 1983. Variation in foraging tactics of fishes. pp. 159–171. *In:* D.L.G. Noakes et al. (ed.) Predators and Prey in Fishes, Dev. Eur. Biol. Fish. 2, Dr W. Junk Publishers, The Hague.

Rosenthal, H. 1969. Investigations regarding the prey catching behavior in larvae of the herring (*Clupea harengus*). Mar. Biol. 3: 208–221.

Schwassmann, H.O. 1971. Biological rhythms. pp. 371–428. *In:* W.S. Hoar & D.J. Randall (ed.) Fish Physiology, Vol. 6, Academic Press, New York.

Schwassmann, H.O. 1979. Biological rhythms: their adaptive significance. pp. 613–630. *In:* M.A. Ali (ed.) Environmental Physiology of Fishes, Plenum Press, New York.

Shapiro, C.M., C.J. Woolf & D. Borsook. 1981. Sleep ontogeny in fish. pp. 171–180. *In:* P.R. Laming (ed.) Brain Mechanisms of Behavior in Lower Vertebrates, Cambridge University Press, Cambridge.

Spencer, W.P. 1939. Diurnal activity in fresh-water fishes. Ohio J. Sci. 39: 119–132.

Statistical Analysis System. 1982. SAS User's Guide: Basics. SAS Institute Inc., Cary. 923 pp.

Statistical Package for the Social Sciences. 1983. SPSSX User's Guide. McGraw-Hill Book Co. 806 pp.

Tauber, E.S. 1974. Phylogeny of sleep. pp. 133–172. *In:* E.D. Weitzman (ed.) Advances in Sleep Research I. Spectrum Publ., New York.

Thorpe, J.E. 1978. (ed.). Rhythmic activity in fishes. Acadmic Press, London. 312 pp.

Van Guelpen, L., D.F. Markle, & D.J. Duggan. 1982. An evaluation of accuracy, precision and speed of several zooplankton subsampling techniques. J. Cons. Perm. Int. Explor. Mer 40: 226–236.

Van Valen, L. 1973. A note on dreams. J. Biol. Psychol. 15: 19.

Wankowski, J.W.J. 1981. Behavioral aspects of predation by juvenile Atlantic salmon (*Salmo salar* L.) on particulate, drifting prey. Anim Behav. 28: 557–571.

Wankowski, J.W.J. & J.E. Thorpe. 1979. Spatial distribution and feeding in Atlantic salmon, *Salmo salar* L. juveniles. J. Fish Biol. 14: 239–247.

Ware, D.M. 1971. Predation by rainbow trout (*Salmo gairdneri*): the effect of experience. J. Fish. Res. Board Can. 29: 1847–1852.

White, H.C. 1936. The food of salmon fry in eastern Canada. J. Biol. Board Can. 2: 499–506.

Williams, D.D. 1981. The first diets of postemergent brook trout (*Salvelinus fontinalis*) and Atlantic salmon (*Salmo salar*) alevins in a Quebec river. Can. J. Fish. Aquat. Sci. 38: 765–771.

Wright, D.I. & W.J. O'Brien. 1984. The development and field test of a tactical model of the planktivorous feeding of white crappie (*Pomoxis annularis*). Ecolog. Monogr. 54: 65–98.

Received 15.12.1984 *Accepted 8.9. 1985*

Piscivory in a coral reef fish community

James D. Parrish, James E. Norris, Michael W. Callahan, Janice K. Callahan, Eric J. Magarifuji & Robert E. Schroeder
Hawaii Cooperative Fishery Research Unit, 2538 The Mall, University of Hawaii, Honolulu, HI 96822, U.S.A.

Keywords: Predation, Resident, Northwestern Hawaiian Islands, Trophic, Atoll, Gut analysis, Diets, Demersal, Carnivory

Synopsis

The community of demersal, resident reef fishes of the Northwestern Hawaiian Islands was studied to determine the various roles of fishes within the piscivorous trophic subweb. Relative abundance estimates for 160 fish species were made on the basis of extensive visual censuses and four chemical collections. Diet information was collected by gut analysis of 52 piscivorous species from 16 families. Twenty species were termed major piscivores and 32 minor piscivores. Twelve to 31% of the biomass of the entire fish community at the four chemical stations was piscivorous. The Apogonidae, Labridae, Mullidae, Scorpaenidae, Pomacentridae, Holocentridae and Gobiidae were the most important prey families, providing over 70% of all prey. The families producing largest piscivorous impacts were the Muraenidae, Synodontidae, Congridae, Labridae, Scorpaenidae, Holocentridae, Priacanthidae and Cirrhitidae. Several families contained a number of intermediate level species whose combination of relatively high abundance, moderate degree of piscivory, and substantial vulnerability to predation gave them a wide trophic scope and an important place in the total trophic activity. The quantitative approach used here to analyze community trophics by estimating predation by and upon whole interacting populations offers promise for improved understanding of these complex systems.

Introduction

Many fishes resident on coral reefs routinely consume fish. However, the exact nature of the roles that these piscivorous predators play and the extent of their influence on the trophic ecology of coral reef fish communities are poorly known. It is important to understand the full set of piscivorous interactions within the group. Both the mortality imposed upon various prey and the growth and reproductive potential gained by their predators may be important determinants of the community structure. Comprehensive trophic studies are lacking for entire fish assemblages from natural reef communities with structures unaffected by human disturbances such as fishing.

Several major works in recent years have cataloged fish diets and/or assigned trophic levels to fishes in various locations in the tropics (e.g. Hiatt & Strasburg 1960, Harmelin-Vivien 1981, Hobson 1974, Randall 1967, Talbot 1965, Williams & Hatcher 1983). However, these studies did not analyze the piscivorous portion of the fish community as an interacting assemblage, nor did they attempt to quantify the effects of the community of piscivorous predators on the community of potential prey.

The objectives of the present study were to (1)

quantitatively describe the populations and piscivorous diets of resident fishes in a remote, pristine, oceanic atoll and island region, (2) estimate the standing stock of piscivores in sample reef communities as a fraction of the total numbers and biomass of resident fish, and (3) quantify the importance of the various piscivores and prey to the entire fish community. The fish communities studied occupied inshore, shallow-water, coral reef tracts in the largely uninhabited Northwestern Hawaiian Islands (NWHI), from Midway to French Frigate Shoals. Most observations, measurements, and collections were made at the latter two atolls. Working at these locations and collecting data widely in a variety of common habitat subtypes, we attempted to characterize the trophics of NWHI fishes generally to a first approximation. This area of the Hawaiian archipelago, and these atolls in particular, were chosen for their relatively undisturbed conditions, their reasonable accessibility, and the importance of their marine fauna for wildlife preservation or resource exploitation. The first extensive studies of the ecology and resources of this region have recently been reported (Grigg & Tanoue 1984).

Materials and methods

Diet analysis

The entire gut contents of samples of all common, resident fish species were analyzed quantitatively, including all their fish and invertebrate prey. The invertebrate and general algal results are the subject of other reports (e.g. Parrish et al. 1984, Parrish et al. 1985). No prey taxa other than fish were included in the present analyses.

For each fish prey category in each predator species, the fraction of the total gut contents was calculated in terms of the following diet composition variables: numbers of prey individuals (% numbers), volume of prey (% volume), weight of prey (% weight), and the fraction of specimens containing a particular prey type (% occurrence). Prey volume was determined directly by liquid displacement of whole prey or their parts. Weight values were estimates of the weight of the original prey before ingestion and were derived from the length of the carcass in the gut using length-weight regressions. Percent numbers and % occurrence seem to be relatively reliable and accurate, although uncertainties in the interpretation of trace remains, such as scales and fins, may introduce bias. Determination of volume and weight composition involves considerable and variable negative bias due to the varying degrees of digestion of food remnants. A large fraction of prey remains could not be identified to species. Therefore, to maintain adequate sample sizes, analysis of prey fish data was not carried below the family level.

Indices of relative importance (IRI) were calculated for each piscivorous species, using 'fish' versus 'all other' as general prey categories. Following Pinkas et al. (1971), the index for each prey category was:

$$\mathrm{IRI_v} = (\%\ \text{numbers} + \%\ \text{volume}) \times (\%\ \text{occurrence})\ \text{or} \quad (1)$$

$$\mathrm{IRI_w} = (\%\ \text{numbers} + \%\ \text{weight}) \times (\%\ \text{occurrence}). \quad (2)$$

The $\mathrm{IRI_v}$ values for 'fish' and 'all other' were summed for each predator species, and fish consumption was reported as a percentage of that sum (% $\mathrm{IRI_v}$).

Estimation of the community

The definition of the entire demersal, resident reef fish community in this study was based on visual census of a variety of habitat subtypes and on chemical collections of fish communities on isolated patch reefs. It included only resident species and excluded large, wide ranging species (e.g. carangids and carcharhinids) and pelagic and surface dwelling species (e.g. scombrids, clupeids and belonids).

A total of 199 visual censuses ranging in area from 15 m^2 to 500 m^2 were conducted in the study areas. Censuses were taken using a modification of the method of Brock (1954), where sections of the reef were divided into quadrats using transect lines. Fishes were counted in adjacent quadrats as two divers swam along the sides of the line. This method is particularly effective for detecting conspicuous, diurnally active fishes but typically underestimates cryptic species (Brock 1982).

To census cryptic species more accurately and to provide a more complete insight into the composition of whole, relatively discrete fish communities, chemical collections were made using rotenone at four separate, isolated patch reefs ranging from 80 m^2 to 120 m^2 in area. Reefs were selected on the basis of their similarity in size and shape and apparent diversity and density of fish. They were also chosen as representative of other reefs in the lagoons. Collections were made by surrounding a reef with a barrier net of 12.7 mm square mesh and then applying rotenone powder premixed in seawater. Essentially all fish on the patch reef were killed, and all specimens were collected by divers within the barrier net. All specimens were identified to species, weighed, and body length (standard, fork and total) measured.

Estimates of density for each species were calculated from the visual censuses and chemical collections. The estimates used for diurnally active, conspicuous fishes such as acanthurids, mullids, chaetodontids, labrids, pomacentrids, and scarids were obtained from visual censuses. Estimates for cryptic, inconspicuous fishes such as eels, apogonids, gobiids, holocentrids, and scorpaenids were obtained from the chemical collections. What is herein called the 'relative abundance' was then computed for each group – i.e. the density of the group as a percent of the combined density of the entire resident community.

Abundance of piscivores in the community

Fish communities are traditionally divided into trophic categories by assigning each species in the community sample entirely to a single category, such as 'carnivore' or 'piscivore'. In the present study the typical diet composition of almost all species was known, and only a few were 100% piscivorous. Thus, only that portion of a species' abundance that represented its piscivory was assigned to the piscivorous category. Because weight estimates of individual fish were not available from visual censuses, the abundance of piscivores was taken from the four chemical collections. The total numbers and biomass of all specimens collected of each piscivorous species were weighted by the % IRI_v (Eqn. 1) as a measure of the degree of piscivory of the species. These weighted values were summed to obtain a total piscivorous number and a total piscivorous biomass for each reef. Each of these estimates was then divided by the total fish abundance for that reef to obtain the percent piscivorous abundance.

Community trophic parameters

The predation pressure exerted by piscivores is affected by the relative abundance of the predators. Theoretically it would be desirable to obtain sample sizes of predators reflecting the natural relative abundance of piscivores in the community. However, reef fish cannot be collected non-selectively, and processing of such samples would not be feasible. In fact, the sample sizes of piscivores used for gut analysis in this study bear no particular relation to their natural relative abundance. Hence, in analyzing community trophic relationships, the results of gut analyses were adjusted by using the estimated community composition (relative abundances) of predators.

The percent of the total piscivorous feeding (PTPF) done by the ith predator species was calculated as

$$\mathrm{PTPF} = \frac{\sum_J (C_{Ji} A_i F_i)}{\sum_i \sum_J (C_{Ji} A_i F_i)} \times 100. \tag{3}$$

The percent of the total prey (PTP) provided by the $_j$th prey family was calculated as

$$\mathrm{PTP} = \frac{\sum_i (S_{ij} A_i F_i)}{\sum_j \sum_i (S_{ij} A_i F_i)} \times 100, \tag{4}$$

where: C_{Ji} was the fraction of numbers, volume, weight or occurrence which predator species i consumed of the total amount of prey species j consumed by all predators, S_{iJ} was the fraction of numbers, volume, weight or occurrence which prey species $_j$ provided in the diet of predator species i,

A_i was the relative abundance (as a decimal fraction) of predator species i, and F_i was the fraction of all specimens of predator species i that contained fish.

PTPF and PTP were calculated separately for each diet composition variable (numbers, volume, weight, occurrence). The set of j values of PTP is analogous to the diet of a super predator – i.e. the entire predator community. Thus, for each prey family, the separate results from the four diet composition variables were also combined by calculating IRI_v and IRI_w.

Results

In the course of this study, 160 resident, demersal fish species were observed in the reef habitat in more than negligible numbers. This group was considered to represent the NWHI shallow-water, hard-substrate fish community (Table 1, column 1; Table 4, column 1).

Diet data were collected from 126 fish species containing identifiable food items. Fifty-two species contained some fish parts (Table 1). The degree of piscivory varied widely (range of % IRI_v: 0.01% to 100%). It seemed useful to separate species for which piscivory seemed a major and important trophic habit from species that only occasion-

Table 1. Abundance of piscivores in the community, the importance of fish in their diets, and their portions of total community piscivory. Major piscivores (% IRI_v 50%) are shown in bold face.

FAMILY Piscivore	1 Relative abundance[a] (%)	2 Sample size[b]	3 Fish as a portion of all prey (%) by number	4 by volume	5 by occurrence	6 by IRI_v	7 % of total piscivorous feeding[c]
MURAENIDAE							
Gymnothorax eurostus[d]	1.5	58	22.9	46.7	22.4	15.3	1.7
G. flavimarginatus[d]	0.2	6	87.5	99.7	100.0	98.9	7.8
G. meleagris[d]	0.1	1	100.0	100.0	100.0	100.0	–[e]
G. moluccensis[d]	0.8	12	7.7	<0.1	8.3	0.4	–[e]
G. steindachneri[d]	0.3	11	45.5	83.1	45.5	75.0	0.6
G. undulatus[d]	0.4	28	89.6	98.3	96.4	99.2	26.1
Muraena pardalis[d]	<0.1	2	33.3	93.9	50.0	63.6	<0.1
CONGRIDAE							
Conger cinereus[d]	0.6	14	65.0	96.8	71.4	90.5	10.9
C. oligoporus[d]	0.2	3	20.0	95.7	33.3	31.4	<0.1
OPHICHTHIDAE							
Ichthyapus vulturis[d]	1.7	4	75.0	75.0	75.0	90.0	–[e]
SYNODONTIDAE							
Saurida gracilis[d]	0.1	56	94.4	99.8	100.0	99.9	8.7
Synodus englemani	<0.1	6	100.0	100.0	100.0	100.0	<0.1
S. ulae[d]	0.2	107	87.4	99.3	94.4	99.3	20.4
ANTENNARIIDAE							
Antennarius drombus[d]	0.2	11	84.6	99.6	90.9	98.3	1.2
OPHIDIIDAE							
Brotula multibarbata[d]	0.8	15	11.5	6.1	20.0	2.1	0.1
HOLOCENTRIDAE							
Sargocentron tiere	<0.1	2	11.1	0.0	50.0	2.9	<0.1[f]
S. spiniferum	–	9	7.5	44.8	22.2	7.5	–[f]
S. diadema[d]	7.4	29	0.6	0.3	3.4	<0.1	–[e]
Myripristis berndti[d]	0.8	30	5.9	59.4	20.0	9.4	4.4
M. amaenus[d]	0.1	11	12.5	33.5	18.2	6.2	<0.1
Neoniphon sammara[d]	4.1	63	1.6	4.0	3.2	0.1	–[e]

Table 1. (Continued).

FAMILY Piscivore	1 Relative abundance[a] (%)	2 Sample size[b]	3 Fish as a portion of all prey (%) by number	4 by volume	5 by occurrence	6 by IRI_v	7 % of total piscivorous feeding[c]
AULOSTOMIDAE							
Aulostomus chinensis	<0.1	40	31.1	90.1	50.0	52.3	0.2
FISTULARIIDAE							
Fistularia commersoni	<0.1	13	37.3	84.0	69.2	66.5	<0.1
SCORPAENIDAE							
Pterois sphex	–	7	5.0	0.0	14.3	0.4	–[g]
Scorpaenodes kellogi[d]	0.1	4	75.0	9.1	75.0	68.5	–[e]
S. littoralis[d]	0.6	26	32.5	53.8	34.6	28.0	1.9
Scorpaenopsis diabolus[d]	0.1	31	82.2	97.9	93.5	97.8	2.9
S. brevifrons	–	1	100.0	100.0	100.0	100.0	–[g]
S. cacopsis[d]	0.1	9	54.5	41.8	55.6	67.2	–[e]
Scorpaena ballieui[d]	0.4	53	5.2	16.9	5.7	0.8	<0.1
Dendrochirus barberi[d]	0.8	27	11.4	13.9	11.1	2.0	–[e]
PRIACANTHIDAE							
Priacanthus alalaua	–	19	11.1	52.9	31.6	14.9	4.2[f]
P. meeki[d]	0.3	91	1.3	10.1	7.7	0.9	–[f]
MULLIDAE							
Mulloides vanicolensis	1.2	27	0.7	0.8	3.7	<0.1	–[e]
Pseudupeneus cyclostomus	<0.1	22	48.1	89.1	63.6	71.4	0.1
P. porphyreus	<0.1	28	1.1	3.8	3.6	0.1	–[e]
P. multifasciatus[d]	0.2	63	0.9	0.8	6.3	0.1	<0.1
P. bifasciatus	<0.1	7	4.5	66.0	14.3	7.4	<0.1
CIRRHITIDAE							
Paracirrhites forsteri	0.2	16	50.0	73.7	62.5	66.9	1.2
Cirrhitus pinnulatus	0.1	64	9.1	21.0	18.8	4.3	0.8
Cirrhitops fasciatus	0.3	17	5.6	24.5	11.8	2.7	0.7
LABRIDAE							
Bodianus bilunulatus[d]	0.2	50	0.5	2.0	16.0	0.2	–[e]
Epibulus insidiator[d]	<0.1	21	8.6	25.7	23.8	5.0	–[e]
Cheilinus unifasciatus[d]	0.1	98	35.1	33.5	45.9	29.7	1.7
Cymolutes leclusei	<0.1	7	4.3	68.1	28.6	25.6	–[e]
Thalassoma duperrey	7.3	42	0.7	1.3	9.5	0.1	–[e]
T. fuscum	–	4	66.7	97.1	100.0	94.8	–[g]
T. purpureum	<0.1	26	21.5	22.0	53.8	24.8	0.5
T. ballieui	1.0	77	3.7	19.1	26.0	5.4	3.4
BOTHIDAE							
Bothus sp.	0.3	11	9.4	56.9	27.3	14.2	0.3
BALISTIDAE							
Melichthys niger	<0.1	33	0.7	<0.1	12.1	0.1	<0.1
M. vidua	<0.1	15	1.8	2.7	13.3	0.4	–[g]

[a] Per cent of all fish in the community.
[b] Number of piscivore specimens that had identifiable gut contents.
[c] Per cent of all community piscivory imposed.
[d] Occurred in chemical samples of patch reefs for estimation of piscivorous abundance.
[e] Inadequate prey identification for analysis involving prey families.
[f] Pooled for community trophic analysis.
[g] No estimate of relative abundance available; not used in community trophic analysis.

ally ate fish. The predators divided naturally into a group of major piscivores, with IRI_v >50%, and a group of minor piscivores (Table 1, column 6). More detailed results of the extensive analysis of the diets and feeding ecology of these piscivorous species have been previously reported (Norris 1985).

The chemical samples from the four patch reefs contained 13 major and 16 minor piscivores (Table 1). The piscivorous portion of these communities accounted for 19.4% (11.7%–30.6%) of the total biomass and 6.0% (2.2%–8.7%) of the total numbers of fish present (Table 2). Although there was considerable variability among the four stations, these results suggest that the overall level of piscivorous activity in the community is quite high.

Calculation of PTP for all prey families indicated that over 73% of the prey consisted of seven families, irrespective of the measure of diet used. The distribution of predation over the seven families was much the same whether based on % occurrence or % numbers (Fig. 1). The order of the top seven prey families was the same by both measures, and the actual percentage values were not greatly different for any of these prey families. No other fully resident prey family accounted for more than 3% of the total prey by % occurrence or % numbers; most families accounted for much less.

The distribution based on volume was somewhat similar to that for weight, but both were noticeably different from the distributions by % occurrence and % numbers. Each of these four measures can potentially provide a different and significant insight into community trophic relationships. Because of methodological biases, interpretations of volume and weight results are somewhat difficult. For example, the consumption of pomacentrids was much lower as measured by % volume than as measured by % occurrence, % numbers or % weight. This seems to be primarily because most damselfish prey specimens left an insufficient bulk of remains in predator guts to permit a full representation of their importance by direct volume measurement. However, lengths of remnants permitted reasonable weight estimation by regression. Since analyses based on numbers of prey individuals are less prone to bias, most of the following results are presented using this prey measure.

Table 2. Abundance of the piscivorous portion of reef fish communities in the Northwestern Hawaiian Islands.

Location	% of total numbers	% of total biomass
French Frigate Shoals	8.1	30.6
French Frigate Shoals	2.2	11.7
French Frigate Shoals	5.0	22.8
Midway	8.7	12.5
Mean	6.0	19.4
S.D.	3.0	9.0

IRI values based on all predators combined showed trends much like those demonstrated by % occurrence and % numbers of prey. The same seven prey families were dominant, with % IRI values ranging from 39% to 2.3%. No other family value was greater than 1.3% for volume or weight, and most were much less. The rank order of sizes of % IRI_v values was essentially the same as that of % occurrence and % numbers except that the % IRI_v values for the Pomacentridae and Holocentridae were effectively equal. Except for the Pomacentridae, the general magnitudes of % IRI_v and % IRI_w values for a family were fairly close, and there was much similarity in the rank order of sizes. For the three families that ranked differently, the % IRI_w values were very nearly equal.

Fishes in most of the major prey families were eaten by a considerable variety of predators (Table 3). For example, labrids were eaten by 18 species from 9 families and scorpaenids by 12 species from 8 families. However, for most prey groups, a few predator species dominated the community consumption of the prey. For some families a dominant predator (Table 3, columns 2–4) had highly piscivorous habits (Table 1, column 6) and was moderately abundant (Table 1, column 1), e.g. *Conger cinereus* preying on apogonids. In most cases, in fact, the dominant predators were highly piscivorous – e.g. Muraenidae, Congridae and Synodontidae. However, the abundance of any predator had a major effect on the level of predation. In some cases a predator with minor piscivorous habits was a large factor in the consumption of a prey group.

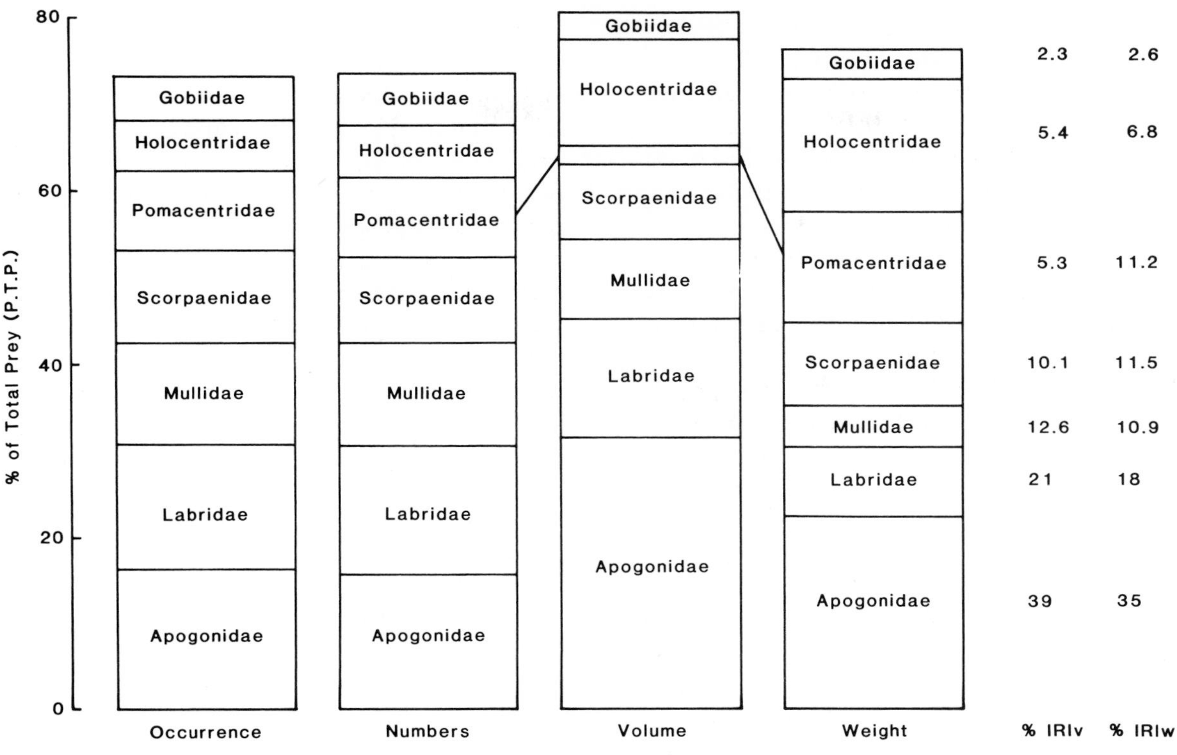

Fig. 1. Consumption of the seven major fish prey families of the community by all predators combined.

For example, gut analysis of *Thalassoma ballieui* gave evidence of only moderate piscivory (3% numbers). But because of the high abundance of this wrasse, it exerted fairly significant predation on apogonids and pomacentrids, and particularly on labrids and mullids. At least two other wrasses had considerably more piscivorous habits. *Cheilinus unifasciatus* ate fish as 35% (numbers) of its prey, but was only one-sixth as abundant as *T. ballieui*. Thus, it had a relatively minor influence on the overall community predation of only three of the major prey families. Since a large fraction of its prey was not identifiable to the family level, its impact as a predator on some prey families may have been underrepresented.

Seven families of major predators (Table 3) accounted for the great majority of all predation on the seven major prey families (range, 51–96%; over 90% for most families). They were responsible for most of the piscivory in the entire community (combined PTPF about 91%). These predator families, together with the Mullidae, Holocentridae, Priacanthidae, Bothidae, Aulostomidae and Fistulariidae accounted for over 99% of all piscivory in the analyzed community (Table 4, column 3; Table 1, column 7).

A full spectrum of trophic roles was found in the fish community (Table 4). The Scaridae, Chaetodontidae and Acanthuridae were not piscivorous, were relatively minor prey for piscivores, and were only moderately abundant. Thus, they contributed little (PTP range 1.2–1.6%) to the food base that supported the piscivorous community. The Gobiidae, Pomacentridae and Apogonidae were also not piscivorous. However, they were important in fish diets and quite abundant. Thus, they were among the largest contributors to the trophic support of the piscivorous community (PTP range 6–16%).

The other four abundant prey families also had rather high values of PTP (6–15%) and provided major support to the rest of the piscivorous community. They were also somewhat piscivorous, with PTPF values of 0.12, 5, 4.5 and 6%). The trophic subwebs involving many species of these families – and to some extent the Priacanthidae and Cirrhitidae – were rather complex. Some ate small amounts of highly piscivorous species that were otherwise at the highest trophic levels. For example, Mullidae ate Synodontidae, Bothidae, and Cirrhitidae; Cirrhitidae and Holocentridae ate Muraenidae; Scorpaenidae ate Antennariidae. Scorpaenidae were one of the major families that ate Scorpaenidae (Table 3). Cirrhitidae were also major consumers of Cirrhitidae, and cannibalism was found at the species level for *Paracirrhites forsteri.* Scorpaenidae ate Mullidae, but Mullidae also ate Scorpaenidae. In fact this study demonstrated that Scorpaenidae – often viewed as obvious and well-adapted piscivores – were at least as important to other resident fish species in the local community in their role as prey. Congridae appeared in a similar dual role as important prey and predators. These results were less solid because the sample size for Congridae as prey was small; they occurred in only a few guts of abundant predators such as *Myripristis berndti.*

The Priacanthidae and all families below them (trophically higher) in Table 4 were not particularly abundant with the exception of the Muraenidae.

Table 3. Distribution of predation on the seven major prey families. Values are percents of all consumption of fish in the entire community.

Prey family	1 % of Total prey by all predators	2 by predator family	3 Major predator families (No. species)	4 Dominant predator species	5 No. other predator family (species)
APOGONIDAE	16	4	Muraenidae (2)	*Gymnothorax undulatus*	1 (5)
		8	Congridae (2)	*Conger cinereus*	
		1.5	Synodontidae (2)	*Saurida gracilis*	
		2	Labridae (1)	*Thalassoma ballieui*	
LABRIDAE	15	4	Muraenidae (3)	*Gymnothorax eurostus*	3 (11)
		3	Congridae (1)	*Conger cinereus*	
		1	Synodontidae (2)	*Synodus ulae*	
		0.5	Scorpaenidae (2)	*Scorpaenopsis diabolus/Scorpaena ballieui*	
		1.2	Cirrhitidae (3)	*Paracirrhites forsteri*	
		4	Labridae (4)	*Thalassoma ballieui*	
MULLIDAE	12	7	Muraenidae (4)	*Gymnothorax steindachneri/ Gymnothorax undulatus*	4 (8)
		1.5	Synodontidae (2)	*Synodus ulae*	
		3	Labridae (2)	*Thalassoma ballieui*	
SCOPAENIDAE	10	0.4	Synodontidae (2)	*Saurida gracilis/Synodus ulae*	4 (7)
		3	Antennariidae (1)	*Antennarius drombus*	
		2	Scorpaenidae (4)	*Scorpaenodes littoralis*	
		0.6	Cirrhitidae (1)	*Paracirrhites forsteri*	
POMACENTRIDAE	9	3	Muraenidae (3)		6 (11)
		3	Congridae (1)	*Conger cinereus*	
		1.4	Synodontidae (3)	*Saurida gracilis*	
		1.3	Labridae (2)	*Cheilinus unifasciatus/Thalassoma ballieui*	
HOLOCENTRIDAE	6	2.4	Muraenidae (2)		3 (6)
		1.1	Synodontidae (2)	*Saurida gracilis*	
		2	Antennariidae (1)	*Antennarius drombus*	
GOBIIDAE	6	1.4	Muraenidae (1)		
		1.6	Synodontidae (2)	*Synodus ulae*	4 (7)

Table 4. Abundance, consumption of, and predation by the major fish families of the community.

Fish family	1 Relative abundance	2 % of total prey	3 % of total piscivorous feeding	4 % occurrence of fish in diet
Scaridae	2	1.2		
Chaetodontidae	2	1.4		
Acanthuridae	2	1.6		
Gobiidae	19	6		
Pomacentridae	16	9		
Apogonidae	16	16		
Mullidae	2	12	0.12	9
Scorpaenidae	2.4	10	5	28
Holocentridae	12	6	4.5	20
Labridae	12	15	6	29
Congridae	2	3	11	58
Priacanthidae	0.7	1.2	4	16
Cirrhitidae	0.6	2	3	27
Bothidae	0.3	0.15	0.3	13
Muraenidae	4	0.3	36	44
Synodontidae	0.3	0.8	29	96
Antennariidae	0.2	0.07	1.2	91
Aulostomidae	0.1	0.5	0.2	49
Fistulariidae	<0.1		<0.1	69

They were increasingly piscivorous, and some had sizable values of PTPF – e.g. the Priacanthidae, Muraenidae and Synodontidae. None of the higher level predators contributed significantly as prey to the rest of the piscivorous community.

Discussion

This study required integrating data from disparate sources, each involving some sampling problems and inherent biases. The chemical collections probably gave the best estimate obtainable for cryptic fish species, but some individuals may have escaped collection within the reef structure. Use of tested census protocols and many widespread censuses with replication helped control the considerable bias inherent in visual census work. Advanced states of digestion reduced the sample sizes of identifiable prey items; 71% of piscivorous species contained some prey that could be identified to family. For a few abundant species with poorly resolved fish diets (e.g. *Thalassoma duperrey, Ichthyapus vulturis*), this bias caused their trophic position and importance in the community to remain in some doubt.

It was not feasible to directly measure the rate of prey consumption in terms of size of prey or frequency of feeding. The limited range of sizes of predator species occurring in the community and the general agreement of trends in prey importance by % numbers versus % volume suggest that our results are not merely artifacts of predator-prey size relationships. A range of sizes of adult fish of most predator species was analyzed, and results should reflect the diets of fish averaged over much of adult life.

The set of PTPF values indicates how the piscivory practiced on the prey as a whole is distributed among the various predators, i.e. the share of the total prey taken by each. The value of PTPF depends upon both the diet characteristics of the predator and the predator's abundance. Thus, the relative values of PTPF for the various predators need not be similar to the relative values of their diet composition variables. For this community, this fact can be simply demonstrated by comparing the rank order of the values in Table 1, column 7 with those in columns 3, 4, 5 or 6 or by similarly comparing columns 3 and 4 of Table 4. Some correlation is introduced by the computation of PTPF (Eqn. 3). Nevertheless, it is obvious that the concordance of PTPF and diet composition variables is far from complete.

For example, muraenids produced a considerably higher PTPF than their % occurrence would suggest (and considerably higher than the more piscivorous congrids) because they were fairly abundant (Table 4). On the other hand, the aulostomids, fistulariids and antennariids were among the most piscivorous predators (e.g. highest % occurrence values), and yet none were abundant enough to produce a major predatory force on the prey community. These results suggest that, although comparisons of piscivores based on diet data as the sole criterion may be useful, they must be interpreted with caution when considering the effects of interactions of complete populations.

Estimates of the piscivorous portion of fish community standing stock have been attempted in a few studies (Table 5), although results have usually been reported as carnivory or some other, larger category. Harmelin-Vivien (1981) reported the total percent carnivore individuals and, in various reef habitats, the piscivorous fraction of this total. These values were used to estimate piscivorous '% of total numbers' for each of the habitats on her reef tract, and then a simple mean of the habitat values was calculated. The technique of collecting fish with explosives may produce serious underestimates of some large, rather numerous piscivores that lack swim bladders (e.g. eels). Williams & Hatcher (1983) recognized different degrees of piscivory by separating their piscivores into 'con-

Table 5. Estimates of the piscivorous portion of the standing stock in various fish communities.

Source	Total species, all trophic types	% of all species piscivorous	% of total numbers piscivorous	% of total biomass piscivorous
Harmelin-Vivien (1981)[1,2,3]	142	13.4	2.8[4] (1.0–5.2)	–
Goldman & Talbot (1976)[2,5]	–	–	–	54
Talbot (1965)[2,6]	106	13.2	–	11[7] (3, 9, 24)
Williams & Hatcher (1983)[2,8]	323	8.0	1.0[9]	5.7[9] (2.0, 6.1, 11.2)
Stimson et al. (1983), calculated from Brock et al. (1979)[1,10]				
1966 Collection	–	–	–	16
1977 Collection	–	–	–	13
Calculated from Brock et al. (1979), 1966 collection[1,10,11]	76	18.4	4.0	8.3
Present study[1,11,12]	126	41.3	6.0	19.4

[1] Chemical collections.
[2] Explosive collections.
[3] Long transect on large reef at Tulear, Madagascar.
[4] Simple mean and range of values for serveral habitat types.
[5] One Tree Island, Great Barrier Reef.
[6] Three transects at one reef, Tutia Reef, Tanzania.
[7] Overall values for 3 transects pooled and individual values.
[8] Three locations in transect across full Great Barrier Reef.
[9] Overall values for 3 locations (30 stations) pooled, and individual values.
[10] One patch reef, Kaneohe Bay, Oahu, Hawaii.
[11] Using degrees of piscivory based on the present study.
[12] Four patch reefs in Northwestern Hawaiian Islands.

firmed' and 'facultative' groups, but these categories do not permit direct comparison with our results.

In all these previous reports each species was placed entirely in one category according to what type of prey was believed to dominate its diet, based on diet studies or other information in the literature. Thus, some partially piscivorous species must have been left out of the piscivorous category and some partially piscivorous species counted in that category at full abundance. The net bias (direction or magnitude) of these procedures cannot be estimated. Thus, there is some difficulty with interpretation of these results, in comparison with results from our procedure. Our sample size is not large (chemical stations at four patch reefs at two atolls), and it may not be fully representative of the diversity of NWHI habitats. However, these are the only data in existence for the NWHI that include the important cryptic component. Most published studies on piscivorous standing stock seem to involve rather few quantitative collections, often scattered over considerable areas and diverse habitats. Despite the difficulties associated with all these data sets, the comparison of our results with others in Table 5 seems useful as a first approximation.

The full range of results of all these studies is considerable: 1–4% piscivores by number and 5.7–54% by biomass (Table 5). The biomass values from the present study fall within the full range of results from other studies and are somewhat high compared with results for the high Hawaiian island (Kaneohe Bay). Both Hawaiian areas lack almost entirely the shallow-water snappers and groupers that are prominent and rather piscivorous on the Great Barrier Reef, in East Africa, and throughout most of the tropics. The NWHI results for the piscivorous '% of total numbers' in the community are somewhat high compared with all other values reported. (The calculated '% of all species piscivorous' in the NWHI [41.3%] is not directly comparable with the values from other studies since all species that are piscivorous at any level in the NWHI are included in it. This value is much lower [15.9%] if only major piscivores [defined in Table 1] are included.) Comparison of the meager data for biomass and numerical fractions suggests that piscivores may be smaller on the average in Hawaiian communities than in the other areas, and that piscivores may be larger in the NWHI than in Kaneohe Bay. Both these possibilities are reasonable in terms of the faunal composition in the various areas and casual observations of fish sizes during underwater work at the Hawaiian sites.

There is still no complete or unequivocal evidence that piscivory is an ubiquitous or important force in structuring reef fish communities, although the possibility has been much discussed (e.g. Nolan 1975, Sale 1980, Lassig 1982, Stimson 1983, Parrish 1985). One kind of circumstantial evidence – perhaps an ecologically necessary but not sufficient condition – is that the community contain a substantial number of fairly piscivorous fishes. The diet data, abundance, and piscivorous biomass estimates from this study suggest that this condition is met in the NWHI. Field experimental work now in progress to selectively remove piscivores should shed further light on the question.

Some general structural features of the NWHI piscivorous trophic subweb seem similar to other known trophic systems. Several predators consume a common prey or prey group. A single predator commonly uses a number of prey types. Thus, some predators may produce large effects widely through the community, e.g., those with broad diets and high PTPF such as the Muraenidae and Synodontidae. The trophic subweb contains some loops (the simplest of which are simply 'A eats B eats A'), and trophic interactions among some groups are complex.

There are probably few if any trophic studies of similar systems with which to compare our quantitative estimates of interactions. It seems likely that in other situations, as here, a few trophic linkages are quantitatively quite important to the system as a whole and others are much less so. For other systems for which such data are lacking (and require long and arduous work to obtain), the general magnitudes of linkages from this study might be used to estimate those for the unmeasured systems. This should help identify the important relationships and may permit better understanding of how an unstudied system functions and better prediction of how it might respond to disturbance (natural or human). Where population abundance information can be obtained for an unstudied system (much more quickly and cheaply than diet information), those results could be used with diet results from the present study to provide a first estimate for simulating the unstudied system.

Most of the knowledge of fish trophics on coral reefs at present is (at best) at the level: the diet of Predator A consists of X% Prey B, Y% Prey C and Z% Prey D. Lacking other information, this knowledge may give some impression of which fishes are important in some trophic roles. The present study demonstrates that if we account for the abundance of the predators, we can derive more complete and sometimes rather different impressions of the magnitude of the trophic processes. For example, some species that have well established reputations as important piscivores are in fact individually highly piscivorous, but their small populations produce nearly negligible effects on the community. Some non-piscivorous species that seem very vulnerable and are reasonably common (e.g., some of the herbivores and the omnivorous Chaetodontidae) are not eaten in significant quantity by abundant predators. These non-piscivorous fishes consume considerable food resources; the route by which these resources pass on through the trophic system remains in doubt. A number of fishes have been shown to have considerable trophic scope, i.e. the total predation upon them by the local fish community and upon the community by them are both sizable. These results suggest that the study of fish communities as a whole, quantifying and integrating species diets and species populations, is an important key to understanding trophic systems.

Acknowledgements

This research received partial financial support as Project NI/R-4, sponsored by the University of Hawaii Sea Grant College Program under Institutional Grant No. NA79AA-D-00085 and NA81AA-D-00070 from NOAA Office of Sea Grant, Department of Commerce. This is Sea Grant publication UNIHI-SEA GRANT-JC-85-11. The work was supported in part by the Marine Affairs Coordinator, subsequently the Ocean Resources Office, State of Hawaii Department of Planning and Economic Development. This is Ocean Resource Office Contribution No. 28.

Laboratory, equipment, logistics and administrative support were provided through the Hawaii Cooperative Fishery Research Unit by its permanent sponsors, the Hawaii Division of Aquatic Resources, the U.S. Fish and Wildlife Service, and the University of Hawaii. The U.S. Fish and Wildlife Service and the U.S. Navy (through its contractor, Base Services Inc.), generously provided housing, subsistence, transportation and field logistic support at French Frigate Shoals and Midway, respectively. Susan Nakamura prepared the figure. This research was made possible in a major way by the willing and competent help in the field and laboratory of a large group of research assistants (see Parrish et al. 1984).

References cited

Brock, R.E. 1982. A critique of the visual census method for assessing coral reef fish populations. Bull. Mar. Sci. 32: 269–276.

Brock, R.E., C. Lewis & R.C. Wass. 1979. Stability and structure of a fish community on a coral patch reef in Hawaii. Mar. Biol. (Berl.) 54: 281–292.

Brock, V.E. 1954. A preliminary report on a method of estimating reef fish populations. J. Wildlife Mgmt. 18: 297–308.

Goldman, B. & F.H. Talbot. 1976. Aspects of the ecology of coral reef fishes. pp. 125–154. *In:* O.A. Jones & R. Endean (ed.) Biology and Geology of Coral Reefs, Volume 3, Biology 2, Academic Press, New York.

Grigg, R.W. & K.Y. Tanoue. 1984. Proceedings of the second symposium on resource investigations in the Northwestern Hawaiian Islands. Vol. 1 & 2. UNIHI-SEAGRANT-MR-84-01. 491 pp.

Harmelin-Vivien, M.L. 1981. Trophic relationships of reef fishes in Tulear (Madagascar). Oceanologica Acta 3: 365–374.

Hiatt, R.W. & D.W. Strasburg. 1960. Ecological relationships of the fish fauna on coral reefs of the Marshall Islands. Ecol. Monogr. 30: 65–127.

Hobson, E.S. 1974. Feeding relationships of teleostean fishes on coral reefs in Kona, Hawaii. U.S. Fish. Bull. 72: 915–1031.

Lassig, B.R. 1982. The minor role of large transient fishes in structuring small-scale coral patch reef fish assemblages.

Doctoral Dissertation, Macquarie University, North Ryde. 223 pp.

Nolan, R.S. 1975. The ecology of patch reef fishes. Doctoral Dissertation, University of California, San Diego. 230 pp.

Norris, J.E. 1985. Trophic relationships of piscivorous coral reef fishes from the Northwestern Hawaiian Islands. M.S. Thesis, University of Hawaii, Honolulu. 71 pp.

Parrish, J.D. 1985. Effects of fishing on predation and the structure of the fish resource community. Proposal to University of Hawaii Sea Grant College Program, Honolulu. 17 pp.

Parrish, J.D., M.W. Callahan, J.M. Kurz, J.E. Norris, A.E. Sudekum, G.Y. Akita, J.W. Banta, A. Chun, S.K. Coffee, M.A. DeCrosta, S.D. Feldkamp, A.E. Henry, C.J. Lau, W.G. Lyle, E.J. Magarifuji, T.J. Mirenda, S.L. Sanderson, R.E. Schroeder, M.H. Seigaku, A.C. Solonsky, C.T. Sorden, L.R. Taylor, Jr. & A.K. Tomita. 1984. Trophic relationships of nearshore fishes in the Northwestern Hawaiian Islands. pp. 221–225. *In:* R.W. Grigg & K.Y. Tanoue (ed.) Proc. 2nd Symp. Resource Investigations Northwestern Hawaiian Islands, Vol 1, UNIHI-SEAGRANT-MR-84-01, University of Hawaii, Honolulu.

Parrish, J.D., M.W. Callahan & J.E. Norris. 1985. Fish trophic relationships that structure reef communities. Proc. 5th International Coral Reef Symposium (in press).

Pinkas, L., M.S. Oliphant & I. Iverson. 1971. Food habits of albacore, bluefin tuna and bonito in California waters. California Dept. Fish and Game Fishery Bulletin 152: 5–12.

Randall, J.E. 1967. Food habits of reef fishes of the West Indies. Studies in Tropical Oceanography (Miami) 5: 665–847.

Sale, P.F. 1980. The ecology of fishes on coral reefs. Oceanogr. Mar. Biol. Ann. Rev. 18: 367–421.

Stimson, J., S. Blum & R.E. Brock. 1983. An experimental study of the influence of muraenid eels on reef fish sizes and abundance. University of Hawaii Sea Grant Quarterly 4: 4. 6 pp.

Talbot, F.H. 1965. A description of the coral structure of Tutia Reef (Tanganyika Territory, East Africa), and its fish fauna. J. Zoology 145: 431–470.

Williams, D.M. & A.I. Hatcher. 1983. Structure of fish communities on outer slopes of inshore, mid-shelf and outer shelf reefs of the Great Barrier Reef. Mar. Ecol. Prog. Ser. 10: 239–250.

Received 6.12.1985 *Accepted 22.7.1985*

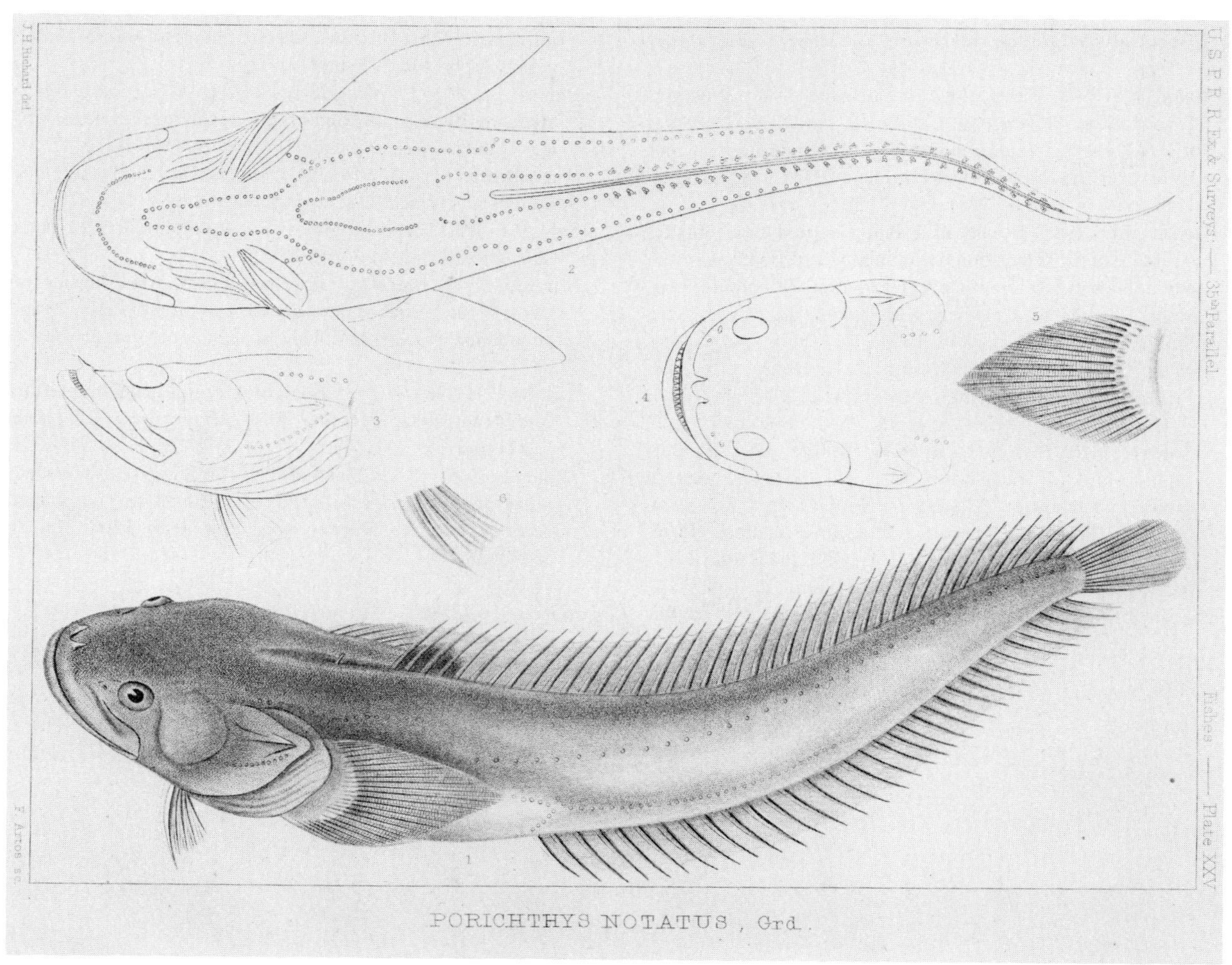

PORICHTHYS NOTATUS, Grd.

An analysis of the feeding rate of white crappie

Barbara I. Evans & W. John O'Brien
Department of Systematics and Ecology, University of Kansas, Lawrence, KS 66045, U.S.A.

Keywords: Centrarchidae, Foraging, Handling time, Planktivorous fish, *Pomoxis annularis*, Search, Zooplankton

Synopsis

The diet composition and feeding rate of animals is an important determinant of their fitness. We have studied the feeding behavior of white crappie, *Pomoxis annularis*, using videotapes of crappie feeding in the laboratory on various types and densities of zooplankton under different temperature regimes. We have determined the time taken by white crappie to complete the cycle of searching for prey, pursuing and attacking located prey, and reinitiating search. The time spent in search increases with decreasing temperature, but decreases when white crappie are feeding on large prey. Run and pursuit swimming speeds increase from 2 cm sec^{-1} at 4° C to 6 cm sec^{-1} at 20° C and then declines. Run and pursuit swimming speeds also increase when white crappie are feeding on large prey. These rates have been developed into a simple time budget of white crappie feeding. This budget accurately predicts white crappie feeding rates when each individual search is almost always successful. Some of the ramifications of this work on current theories predicting planktivorous fish feeding rates are discussed.

Introduction

Feeding is often considered one of the most important activities animals undertake (Curio 1976). With some species, territoriality or competition for other resources creates a situation where food is not in short supply and does not limit fitness; however, this does not appear to be the case with planktivorous fish. As with other predators which feed on prey organisms much smaller than themselves, planktivorous fish must consume a great many prey to survive; thus, feeding must be a major component in determining their fitness.

Two key aspects of feeding are diet composition and feeding rate. What an animal feeds on is important in determining its energy and nutritional intake; however, feeding rate is also important because the energy intake must be greater than the energy expended. Holling (1959) presented a framework in which the feeding of predators could be analyzed to study diet composition. He divided the act of predation into a series of discrete, sequential behaviors: prey location, pursuit, attack and capture.

Holling (1966) also pointed out that a refinement of this approach could be used to understand and predict the feeding rate of a predator if the time spent in each of the stages of predation could be estimated. The components of a planktivorous fish feeding rate are:

$$FR^{-1} = T_L + T_P + T_A.$$

FR is the time to eat a given prey species; T_L is the time to locate that prey; T_P is the time to pursue the located prey and T_A is the time to attack and retain that prey. These last two components of the predation cycle are combined into T_A because they cannot be observationally separated. Maximum feeding rate is limited by how rapidly the fish can execute the steps of the predation cycle and by the volume of water actually searched. From such an analysis one can assess which factors regulating feeding rate are environmentally determined and which factors may be under the control of the predator. Those factors determining feeding rate under the control of the predator can potentially be maximized so as to allow the predator to forage optimally (Pyke 1984).

In this paper, we describe the application of the predation cycle approach to the feeding of white crappie, *Pomoxis annularis*. We measured the time spent in the various intervals of the cycle, recorded the actual volume from which they locate and pursue prey, and developed a time budget for feeding on an individual prey. White crappie were chosen for this study because when under 15 cm in size (TL) they are entirely pelagic in distribution and feed exclusively on zooplankton (O'Brien et al. 1984). Thus, the crappie feed in a simple environment which is easy to study and should be relatively easy to model.

Methods

The feeding cycle of white crappie was determined by observing a total of 4 fish (9–15 cm), feeding singly or in pairs in a 90 cm square aquarium filled to a depth of 30 cm for periods ranging from 30 min to 2 h. The sides and bottom of the aquarium were insulated with white styrofoam, which also visually isolated the fish. The behavior of the fish was videotaped using a JVC 100 color television camera mounted 1.5 m above the bottom of the tank, giving a field of view (on the television screen) of 90% of the bottom of the aquarium. As the fish fed, two observers described their activity onto the audio portion of the tape. During 5–10% of their pursuits, the fish moved vertically more than 15 cm; these pursuits were noted and disregarded during subsequent analysis to avoid error due to parallax change.

At no time did we observe clumps of prey nor did the fish appear to feed on a clump of prey. That is, the fish never remained in one locale for multiple attacks but rather always moved forward or rotated before attacking another prey.

Prior to an experiment, the fish were starved for at least 24 hours. Experiments consisted of adding to the aquarium a known number of either small lake zooplankton or large zooplankton, *Daphnia magna* or *Daphnia pulex*. Small lake zooplankton were natural zooplankton assemblages collected from one of several local ponds or reservoirs; the density in the aquarium was kept at around 5 zooplankton per liter. The size distribution of the lake zooplankton prey in these experiments ranged from 0.5 to 1.5 mm and was composed mainly of *Daphnia galeata, Daphnia ambigua* and *Diaptomus pallidus*. When *Daphnia magna* were used, the densities in the aquarium were either 0.1 or 1.0 prey l^{-1} and organisms ranged in size from 3 to 4 mm. In the experiments where the density was 0.1 prey l^{-1}, a new prey was introduced into the aquarium by siphon each time the fish ate a daphnid, thereby keeping the prey density nearly constant. When *Daphnia pulex* were used, the density was 7.0 prey l^{-1}, and the *D. pulex* ranged in size from 2.0 to 3.0 mm.

The light in the aquarium came from two rheostat controlled 150 w incandescent floodlights and was diffused by a thick white cloth arranged tent-like above the aquarium through which the lens of the TV camera projected. The light intensity was maintained between 100 and 300 lux, which was at or above the maximum reaction distance threshold of 100 lux observed for white crappie by Wright & O'Brien (1984). The temperature in the aquarium was held constant by a large aquarium heater and/or Blue M cold finger, both regulated by a Technilab thermostat. Fish were observed at temperatures ranging from 3° to 24° C, and were acclimated to a new temperature for at least one day per 1° C temperature change. All observations were made during the early afternoon.

Fish behavior was scored from the videotapes. Pursuit angles and distances, run lengths, and turn angles were marked on the screen, using a grease pencil, then distances were converted to centimeters by using the ratio of the actual length of the fish to the television screen length of the fish. Pursuit, run, and pause times were measured from the videotapes using a Cronus digital stopwatch, and run and pursuit speeds were expressed as cm sec^{-1}. A minimum of 20 data points were recorded for each parameter reported. Graphed data show means of at least 20 observations and a bar representing a 95% confidence interval. Thus on all figures where the 95% confidence intervals overlap neither mean, the means are significantly different at the 95% confidence level.

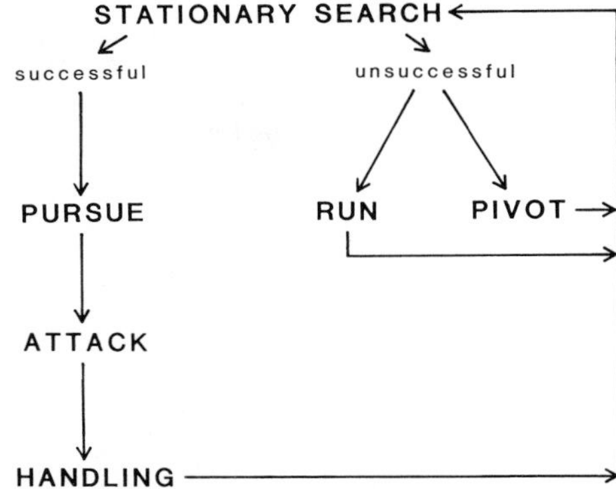

Fig. 1. Predation cycle of white crappie. Depicts our current understanding of behaviors followed by white crappie when feeding on various zooplankton prey. See text for details.

Results

To understand the dynamics of planktivorous fish feeding, it is necessary to determine how long the fish spends searching, the size and shape of the water volume searched, and the fish's run and pursuit speed. Earlier studies have suggested that sunfish are cruise searchers, stopping only when prey are sighted to aim directly at the prey and then pursue it. Our observations show that white crappie commonly stop even when prey are not located; thus they stop to search and likely do not search while swimming (O'Brien et al. 1985).

White crappie predation cycle. – White crappie and presumably other planktivorous sunfish have a simple, predictable predation cycle (Fig. 1). If search is successful, the crappie pursues the prey, attacks, and searches again. If search is unsuccessful, the crappie repositions itself and, once stationary, searches again. Repositioning by swimming a short distance is termed a 'run' while rotating the body on the spot is termed a 'pivot'.

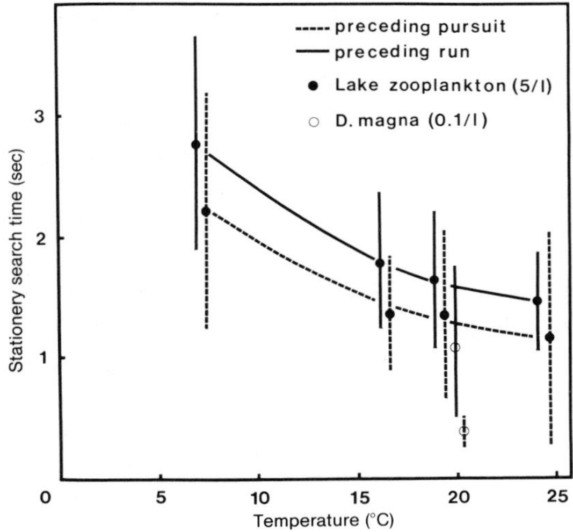

Fig. 2. Time spent in stationary search after a run both preceding a pursuit and a run. The vertical bars indicate the 95% confidence interval around the individual means. The solid and dashed lines represent linear regressions of $\log_e$ transformed data. R^2 for prerun is 0.9712, $p<0.05$. R^2 for prepursuit is 0.942, $p<0.05$.

Search time. – Time spent in stationary search varies with temperature and size of available prey. Times for both successful and unsuccessful search following a run increase with declining temperature (Fig. 2). While times for successful search are slightly shorter than times of unsuccessful search at all temperatures, an analysis of covariance gave no significant difference between these times ($F = 0.08$, d.f. = 1, 252, $p>0.75$). With these two cases combined, a t-test of the hypothesis that $B = 0$ as temperature varied yielded a significant trend and rejection of the hypothesis ($t = 4.31$, d.f. = 253, $p<0.001$). Search also occurs after an attack and ingestion of prey (O'Brien et al. 1985) (Fig. 3). Once again times required for successful search are slightly shorter than times for unsuccessful search at all temperatures (Fig. 3). An analysis of covariance gave no significant difference between

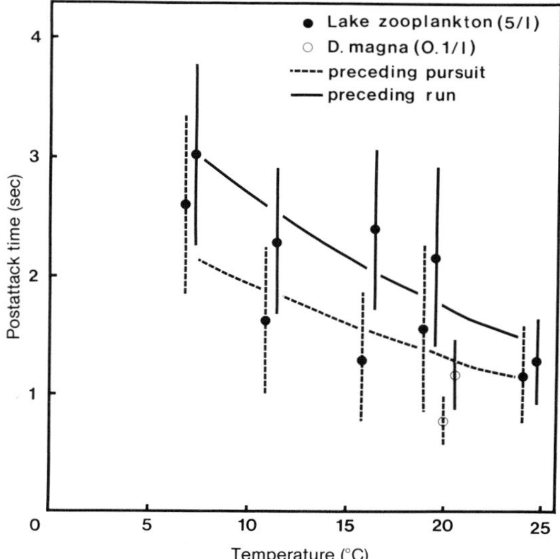

Fig. 3. Time spent in post attack search after a pursuit both preceding a pursuit and a run. The vertical bars indicate the 95% confidence interval around the means. The solid and dashed lines represent linear regressions of $\log_e$ transformed data. R^2 for prerun is 0.747, $p<0.1$. R^2 for prepursuit is 0.794, $p<0.05$.

these times (F = 0.00, d.f. = 1, 193, p >0.75). With these two cases combined, a t-test of the hypothesis that B = 0 as temperature varied yielded a significant trend and rejection of the hypothesis (t = 6.00, d.f. = 195, p<0.001). Furthermore, search time spent after a run (Fig. 2) is not significantly different from search time spent after an attack (Fig. 3) (F = 1.07, d.f. = 1, 448, p>0.5).

The size of prey available also affects search times. Search time is shorter when large prey are common, as shown in Figures 2 and 3. Search times followed by an attack are significantly different when white crappie are feeding on *D. magna* versus lake daphnids (Figs. 2, 3) (successful search following a run t = 2.31, d.f. = 36, p<0.05 and successful search following an attack t = 2.47, d.f. = 39, p<0.02). Search times that were unsuccessful and thus followed by a run are significantly different when white crappie are feeding on *D. magna* versus lake daphnids when search followed an attack (Fig. 3) (t = 2.47, d.f. = 39, p<0.02), but are not significantly different when search followed a run (Fig. 2). Time spent searching for large prey when no prey are located is slightly but not statistically significantly longer than time spent searching before an attack (Fig. 2). Similarly, successful and unsuccessful search times for large prey following attack (Fig. 3) are not reliably different.

Search volume. – The size and shape of the water volume searched was found to vary as a function of visual density of prey (Fig. 4). At high light intensities (200 lux) and high absolute prey density ($70\,l^{-1}$), and thus high visual density of prey, the volume within which pursuit occurred can be described by a shape 45° to the right and left of forward-directed. At low light (2.8 lux) and low absolute prey density ($5\,l^{-1}$), and thus low visual density, however, the volume within which pursuit was observed is described by a shape much like a semicircle. Crappie selected prey more than 5 to 10 cm above or below their horizontal plane in fewer than 10 percent of the observed pursuits.

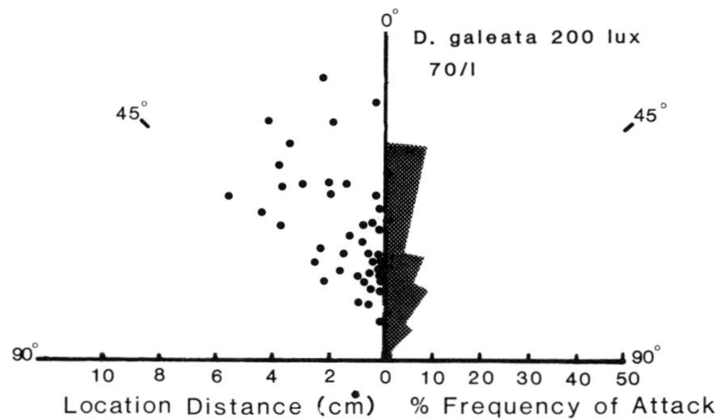

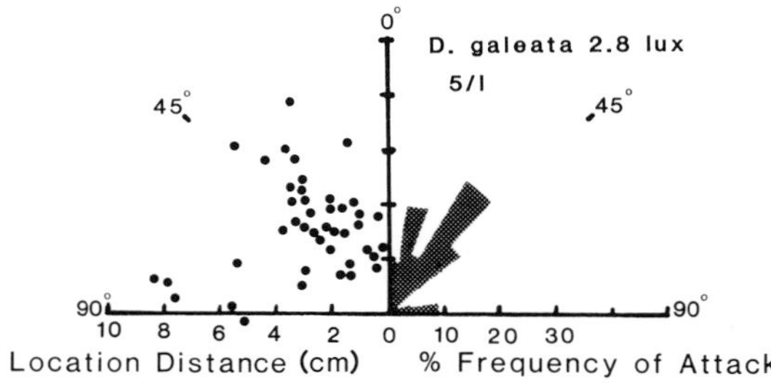

Fig. 4. Horizontal search areas at high light intensity and prey density (upper panel) and low light intensity and prey density (lower panel). The dots to the left of the center line indicate individual angles and distances to which prey were pursued. The shaded area to the right of the center line indicates the percentage of pursuits in 10 degree segments.

Run speed. – For stationary searchers such as white crappie, the time spent repositioning must also be considered in a time budget of feeding. Again, temperature and prey size are important determinants of repositioning time. Run speed was found to increase with increasing temperature until 20° C, and then to decrease slightly (Fig. 5). Linear regression of the means yielded a significant temperature effect (F = 10.8, d.f. = 1, 5, $p<0.025$). Run speed was found to be slower when crappie were feeding on lake zooplankton comprised primarily of copepods than on lake *Daphnia* and was much faster when *D. magna* were the prey. All three means were significantly different from one another: copepods versus lake *Daphnia*, t = 5.21, d.f. = 50, $p<0.001$); lake *Daphnia* versus D. magna t = 8.77, d.f. = 42, $p<0.001$.

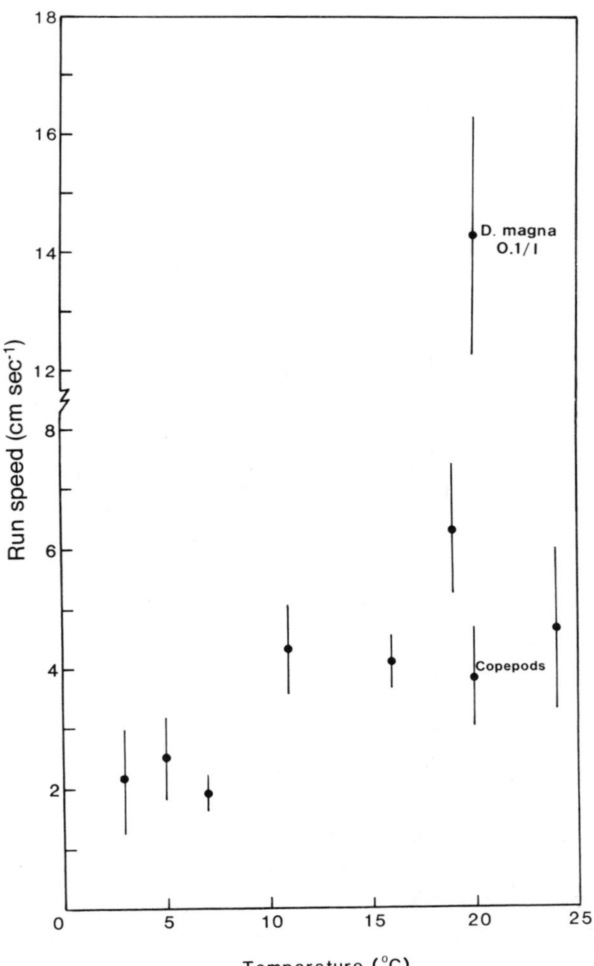

Fig. 5. Run speeds as a function of temperature and prey type. The vertical bars indicate the 95% confidence interval.

Pursuit speed. – The speed at which crappie pursued its prey was found to increase with increasing temperature until about 20° C, and then to decline slightly, similar to run speed (Fig. 6). Linear regression of means gave a significant regression (F-test = 27.07, d.f. = 1, 5, $p<0.005$). Pursuit speed was slowest when fish were feeding on lake zooplankton comprised mainly of copepods and fastest when feeding on *D. magna*: copepods versus lake *Daphnia*, t = 9.13, d.f. = 115, $p<0.001$; lake *Daphnia* versus *D. magna*, t = 12.3, d.f. = 98, $p<0.001$).

Time budget. – These results can be used to develop a time budget of crappie feeding which should pre-

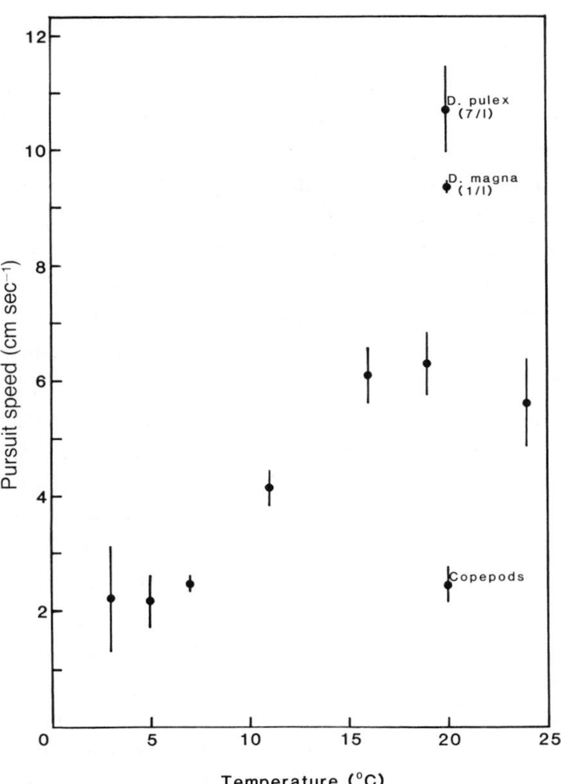

Fig. 6. Pursuit speeds as a function of temperature and prey type. The vertical bars indicate the 95% confidence interval.

Table 1. Time budget when feeding on small and large zooplankton at 20° C.

Small zooplankton		
Search	Successful	Unsuccessful
	1.3 sec	1.8 sec
Movement	Pursuit (5 cm)	Run (6 cm)
	0.85 sec	1.1 sec
Total time	2.15 sec	2.9 sec
Large zooplankton		
Search	Successful	Unsuccessful
	0.7 sec	1.2 sec
Movement	Pursuit (15 cm)	Run (12 cm)
	1.5 sec	0.81 sec
Total time	2.2 sec	2.0 sec

dict feeding rates. Table 1 shows a time budget for crappie feeding at 20° C. The two paths represent time required for successful search and time for unsuccessful search. To determine the choice of path, the probability of a prey being in the search volume must be determined. Given this probability is 25 percent, it would take four searches to locate a prey. The time required to eat one small prey would then be three times the unsuccessful search time of 2.9 sec, equalling 8.7 sec, plus one successful search time of 3.0 sec for a total of 11.7 sec.

The search volume of crappie at low visual densities of prey is not fully known, and thus the probability of a prey being in the search volume cannot be predicted. At high visual densities of prey every search results in a pursuit. When only the successful search path is followed, and assuming a pursuit distance of 5 cm, the proposed time budget predicts a feeding rate of 1674 prey h^{-1}. Observational data with a prey density of 70 l^{-1} at 100 lux yielded a total of 1440 attacks in 52.5 min, or 1646 h^{-1}.

Discussion

The time budget of white crappie feeding presented here provides a close approximation of actual feeding rate, thus supporting the accuracy of the estimated parameters. Further results consistent with the hypothesis of stationary search occurring only during the pause following a run or pursuit (Fig. 1) are given in detail in O'Brien et al. (1985). The search times shown in Figures 2 and 3 are fairly brief but increase by more than a factor of two as the temperature decreases from 24° to 7° C. This slowing of search at lower temperatures could decrease the feeding rate considerably because search is the major time component of the predation cycle at low prey density. It is intriguing that successful and unsuccessful search times are quite similar; our initial expectation was that unsuccessful search times would be considerably longer than successful ones. It seems possible that the crappie are adjusting the unsuccessful search time to be just a little longer than the average time for successful search. This would seem to be a good strategem, i.e., to wait only a little longer than average time to successful search, then give up and reposition.

Consistent with this hypothesis is the decrease in search time when crappie are feeding on large prey such as *Daphnia magna*. This decrease is not entirely due to more rapid location of prey; unsuccessful search times were also shorter when *D. magna* were the prey than when smaller prey were available. Because large prey such as *Daphnia magna* can be located more easily, the fish may be able to scan the search volume more rapidly. In lakes, white crappie commonly feed on large prey only at early dawn when *Chaoborus* are commonly present (O'Brien et al. 1984). *Chaoborus* remain in the upper waters of the reservoir for only a very brief time during early dawn before they migrate down to the bottom of the lake. Thus there is a real need for rapid and efficient feeding when crappie are exposed to large prey. The results presented here demonstrate both the speed and presumed efficiency of this feeding.

It is difficult to put the stationary search times presented here into a larger perspective because, to our knowledge, no one else has measured these times. We have made a few measurements with several other fish species and can report that, on average, bluegill sunfish, *Lepomis macrochirus*, have an even shorter stationary search period at a given temperature and prey size than do white crappie.

The search volume of white crappie, like the search time, appears to vary depending on prey type and the density. When feeding on *Daphnia galeata* at low visual density (Fig. 4), white crappie pursued prey within a forward directed semicircle but showed little tendency to pursue prey above or below the horizontal plane in which the fish was located. At higher visual densities, however, white crappie search volume was more constricted with pursuits being predominantly forward-directed. Luecke & O'Brien (1981) found that bluegill sunfish feeding on solitary 2.0 mm *Daphnia pulex* had a search volume shaped much like a forward directed hemisphere. Confer et al. (1978) and Dunbrack & Dill (1984) found the search volumes of lake charr *Salvelinus namaycush*, and coho salmon, *Oncorhynchus kisutch*, to be shaped somewhat like a forward directed hemisphere, but in the case of the lake charr it projected to the rear of the fish. All of the fish previously studied seemed to have more of a vertical component to the search volume than was true of the white crappie we have studied.

The mechanism which causes the white crappie search volume to be constricted under certain circumstances is not known. It may be that white crappie search from forward directed to lateral. At high visual densities, such as that represented by the upper panel of Figure 4, we would postulate that the fish always located a prey before completing the full 180° sweep, thus constricting the observed range of search. At low visual densities, the fish commonly finish the sweep and thus may locate prey within the entire forward-directed semicircle. Consistent with this hypothesis is the fact that almost all searches were successful at the high visual density (upper panel of Fig. 4), whereas many searches were not successful at the low visual density (lower panel of Fig. 4).

Swimming speed when a fish was repositioning was not different from pursuit speed, but both varied with temperature (Fig. 5, 6). Swimming speeds were low at low temperatures, about 2 cm^{-1} at 3° C, increased to about 6 cm^{-1} at 19–20° C, and then decreased at higher temperatures. The temperature where white crappie feeding is maximized is 20° C. For example, a stationary search-pursuit-stationary search-run sequence at 7° C takes 9.1 sec; at 20° C, the same sequence takes only 4.7 sec. Thus the feeding is almost twice as fast at the higher temperature. Because we do not know how respiration varies with temperature, we cannot assess net energy acquired. It would be of interest to know if 20° C is the temperature preferred by white crappie, but there is no published data on crappie temperature preference.

It seems clear from the results that given high prey densities white crappie can feed quite rapidly, around 1500 prey per hour. At such a rate the fish would not be food limited within the temperature range studied. However, field measured feeding rates of white crappie are a full order of magnitude less than this (O'Brien et al. 1984). From our analysis this suggests that unsuccessful search is the predominant activity of white crappie in lakes, and that visual density of prey limits white crappie feeding under most natural conditions.

Given that search is so important it is interesting to consider the possible advantages of the run and search mode of prey location compared with the sweep searching common to whitefish and grayling. One factor may be the fusiform shape of most salmonid fishes which allows them to be efficient swimmers. By comparison, the oblate shape of most sunfish allows them to be nimble turners but rather poor continuous cruisers (Alexander 1967). Furthermore, the placement of the pelvic and especially pectoral fins of sunfish make for efficient braking of forward movement, whereas the opposite is true of the salmonids (Webb 1984). This still leaves unclear why the salmonids search on the move while white crappie cannot or do not.

The run-search strategy has some advantages in situations of sparsely distributed small prey. By running into new water the fish increases the probability of seeing 'new' prey. Stopping to search may allow a greater search efficiency for small, hard-to-detect prey. A sweep searching fish could search a greater volume but presumably with lower search efficiency. Gendron & Staddon (1983) have theoretically shown that the speed of a sweep searcher should decline with more cryptic prey. In accord with this where prey are large, white crappie search times are attenuated, making their feeding behavior approach sweep searching. This suggests that time management may be a general tendency for optimizing feeding efficiency in planktivorous fish.

Acknowledgements

We thank Brad Loveless, K. Loving, and L. Johnson for assistance in data gathering and Greg Howick for assistance in data analysis. We thank Shirley Archinal for secretarial help and Marion O'Brien for timely editorial assistance. This work was supported by National Science Foundation grant DEB-80-21740 and University of Kansas General Research Fund grant 3288.

References cited

Alexander, R. McN. 1967. Functional design in fishes. Hutchinson University Library, London. 160 pp.

Curio, E. 1976. The ethology of predation. Springer-Verlag, New York. 250 pp.

Confer, J. L., G. L. Howick, M. H. Corzette, S. I. Kramer, S. Fitzgibbon & R. Landesberg. 1978. Visual predation by planktivores. Oikos 31: 27–37.

Dunbrack, R. L. & L. M. Dill. 1984. Three-dimensional prey reaction field of the juvenile coho salmon (*Oncorhynchus kisutch*). Can. J. Fish. Aquat. Sci. 41: 1176–1182.

Gendron, R. P. & J. E. R. Staddon. 1983. Searching for cryptic prey: the effect of search rate. Amer. Nat. 121: 172–186.

Holling, C. S. 1959. The components of predation as revealed by a study of small mammal predation of the European pine sawfly. Can. Ent. 91: 293–320.

Holling, C. S. 1966. The functional response of invertebrate predators to prey density. Mem. Entom. Soc. Can. 48: 5–86.

Luecke, C. & W. J. O'Brien 1981. Prey location volume of a planktivorous fish: a new measure of prey vulnerability. Can. J. Fish Aquat. Sci. 38: 1264–1270.

O'Brien, W. J., B. I. Evans & G. Howick. 1985. A new view of the predation cycle in a planktivorous fish. Can. J. Fish. Aquat. Sci. (in press).

O'Brien, W. J., B. Loveless & D. Wright. 1984. Feeding ecology of young white crappie in a Kansas reservoir. North Amer. J. Fish. Mgmt. 4: 341–349.

Pyke, G. H. 1984. Optimal foraging theory: a critical review. Ann. Rev. Ecol. Syst. 15: 523–575.

Webb, P. W. 1984. Body and fin form and strike tactics of four teleost predators attacking fathead minnow (*Pimephales promelas*) prey. Can. J. Fish. Aquat. Sci. 41: 157–165.

Wright, D. I. & W. J. O'Brien. 1984. The development and field test of a tactical model of the planktivorous feeding of white crappie (*Pomoxis annularis*). Ecol. Monogr. 54: 65–98.

Received 4.3.1985 *Accepted 10.9.1985*

Within-sample variabilities in stomach contents weight of fish – implications for field studies of consumption rate

Per-Arne Amundsen & Anders Klemetsen
Institute of biology/geology, University of Tromsø, P.O. Box 3085, Guleng, N-9001 Tromsø, Norway

Keywords: Field methods, Non-normality, Skewed distributions, Variance dependency, Arctic charr

Synopsis

Stomach contents weights in a study of Arctic charr (*Salvelinus alpinus* L.) were not normally distributed. This may have implications for the field estimation of consumption rates and also indicates that a more extensive handling of stomach content data is needed. Fish were sampled every 3 h throughout 24 h periods. The stomach contents showed very large variations within each 3 h sample. The distributions of stomach contents weights were skewed to the right, and the variance was dependent on the sample mean, with the variance to mean ratio equaling 1. Although mean values of stomach contents weight seemed to show diurnal variations, there were usually no significant differences between consecutive samples taken throughout each 24 h period, due to the large range in stomach contents weight within each sample. These findings question the accuracy of consumption estimates obtained using the method of Elliott & Persson (1978) and other related methods based on differences between consecutive sample mean values. Due to the large within-sample variations found in the current study, the method of Bajkov (1935) modified by Eggers (1979) is considered to give more reliable consumption estimates than the Elliott & Persson method. The generality of the findings is discussed, and it is pointed out that greater attention should be focused on within-sample variabilities in fish stomach contents studies.

Introduction

> *Everybody firmly believes in it because the mathematicians imagine it is a fact of observation, and observers that it is a theory of mathematics.*
> H. Poincare (1892) about normality, quoted from Gaddum (1945).

In recent years, considerable attention has been focused on the development of accurate methods for the estimation of food consumption by fish populations (reviewed by Mann 1978 and Windell 1978). The methods can be broadly divided into: (1) those in which food consumption is estimated indirectly from laboratory studies of energy requirements, nitrogen requirements or food–growth relationships of the fish; and (2) those in which food consumption is estimated directly from quantitative field measurements of stomach contents. Estimations of consumption based on field data embody fewer assumptions than indirect estimates, which involve the assumption that data collected in laboratory studies are applicable to natural populations (Mann 1978). There are good practical reasons for measuring the food consumption directly under field conditions (Elliott & Persson 1978), and during the 1970's several field methods were developed (e.g. Swenson & Smith 1973, Staples 1975 a, b, Eggers 1977, 1979, Thorpe 1977, Elliott & Persson 1978). These methods usually give estimates of the daily consumption, i.e. food consumption over a 24 h period. The Elliott & Persson (1978) model has been described as the best of these methods, being regarded as having a sound mathematical foundation (Cochran 1979, Elliott 1979, Persson 1982, 1983).

Diurnal variations in stomach content have been reported in several studies (e.g. Darnell & Meirotto 1962, Keast & Welsh 1968, Noble 1972, Swenson & Smith 1973, Clarke 1978, Godin 1981, Persson 1983, Allen & Wootton 1984), and the importance of taking these diurnal variations into account when estimating food consumption has been stressed (Darnell & Meirotto 1962, Swenson & Smith 1973). Many of the suggested field methods fulfil this requirement, in that differences in mean stomach contents weight between consecutive samples taken at intervals of a few hours through 24 h periods are incorporated into the estimate of food consumption. These methods have been used in a number of studies, and the results have been discussed in relation to diurnal periodicity of feeding (e.g. Swenson & Smith 1973, Thorpe 1977, Clarke 1978, Elliott & Persson 1978, Godin 1981, Cochran & Adelman 1982, Persson 1982, 1983, Allen & Wootton 1984). However, a detailed study of stomach contents weight of Arctic charr (*Salvelinus alpinus* L.) from a north Norwegian lake has revealed that there may be some inadequacies in the use of these methods, especially with regard to the statistical handling of the data. The distributions of stomach contents weight found in this study of charr indicate that the handling of stomach content data needs to be more extensive, and greater attention should be focused on within-sample variabilities. In the current paper, the generality and statistical limitations of these findings are discussed, and the consequences for the use of field methods for the examination of the feeding strategy of fishes are considered; the actual consumption estimates and other food relationships will be presented elsewhere (Amundsen & Klemetsen unpublished).

Materials and methods

The study was performed in Takvatn, an oligotrophic and dimictic lake situated in a birch wood landscape 214 meters above sea level in northern Norway (69° 07′ N, 19° 05′ E). The area of the lake is 14.2 km^2. There are two main basins, both of which have a maximum depth of about 70 m. The icefree season lasts from May/June to October/November, and there are nearly two months of midnight sun. The fish community of the lake is dominated by stunted, slowly growing charr, but brown trout (*Salmo trutta*) and threespined sticklebacks (*Gasterosteus aculeatus*) are also present.

Once a month from July to October 1980, and in June 1981, adult charr were sampled every 3 h throughout a 24 h period. Sampling was performed in the littoral zone using bottom gillnets. Sample sizes ranged from 6 to 58 (average 29), and on three occasions consecutive samples were pooled because individual sample sizes were too small. Over 95% of the fish sampled were 5 years or older with fork lengths ranging from 16 to 23 cm. The fish were carefully removed from the gillnets and immediately killed by a blow on the head. Each fish was weighed and measured, and the stomach removed and kept frozen until examined in the laboratory. The stomach contents were identified and sorted, surface moisture removed and dry weights (65° C for >48 h) and ash weights (540° C for >16 h) were determined. The weights of stomach contents were expressed as mg organic ash-free dry weight per g fresh weight of fish.

Results

The stomach contents weights showed very large variations within each 3 h sample, and both empty and totally filled stomachs were found in many of the samples. The distributions of stomach contents weight for the 3 h samples taken during a 24 h period in July and September are shown in Figure 1. The distributions were always strongly skewed. In each sample, many fish had little or no stomach content, while a small proportion had a large amount in the stomach. The same trends were seen

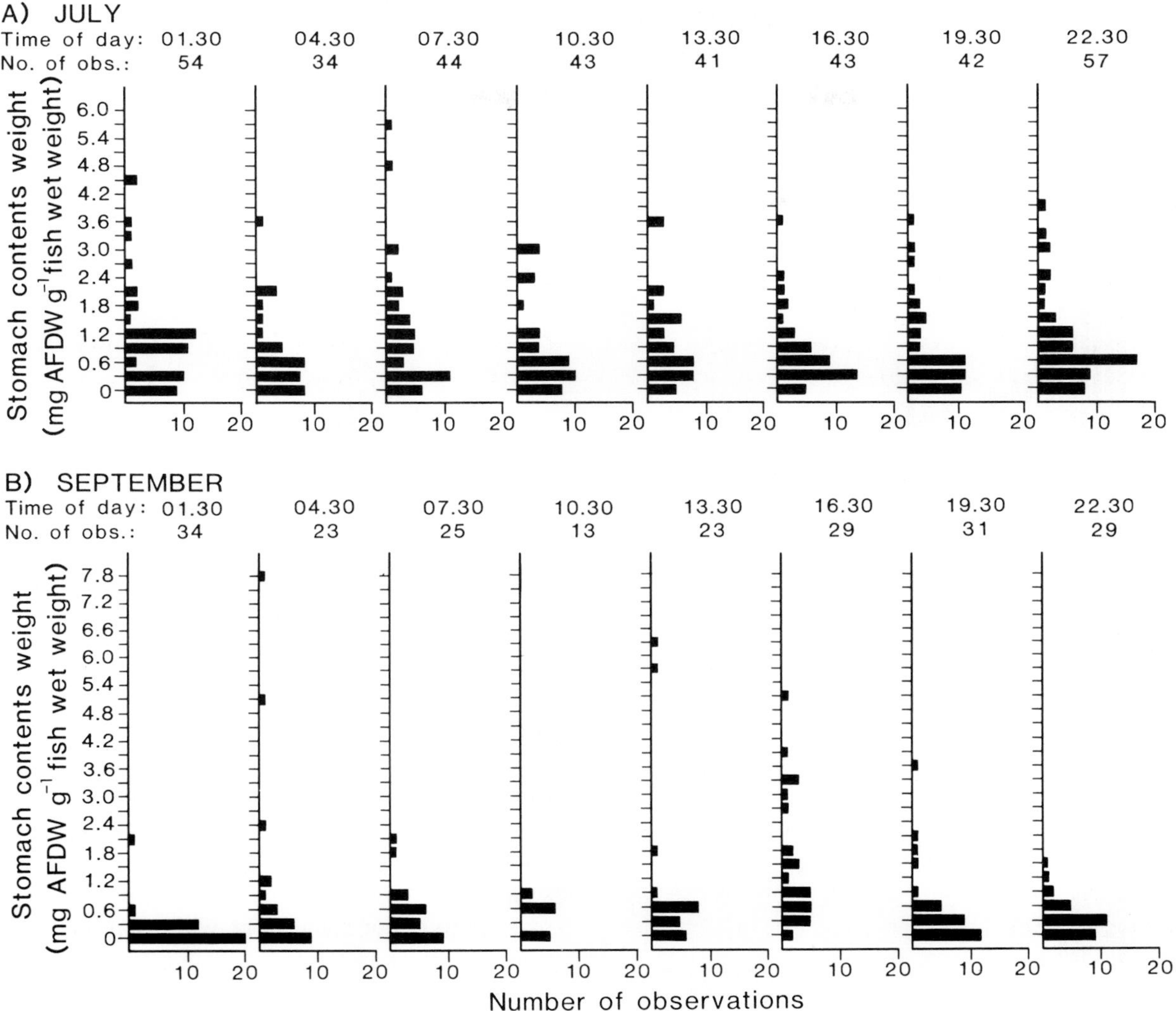

Fig. 1. Distributions of stomach contents weight (mg AFDW g^{-1} fish wet weight) of Arctic charr in Lake Takvatn in samples from (A) July and (B) September.

in the total distributions of weight of stomach contents for pooled data from each 24 h period (Fig. 2). The distributions were always skewed to the right, the values were obviously not normally distributed, and due to the skewness, the arithmetic means were always larger than the medians. The lack of normality in the data is confirmed by examination of the variance-to-mean ratio. The variance was clearly not independent of the sample mean, but increased in direct proportion, with the ratio equaling 1 (Fig. 3).

The confidence limits around mean stomach contents weights throughout the 24 h periods are wide (Fig. 4). Although the means appear to show diurnal variations, testing using non-parametric statistics (chi-square test) usually gave no significant differences between consecutive samples taken throughout each 24 h period (significant differences were found in less than 20% of the tests).

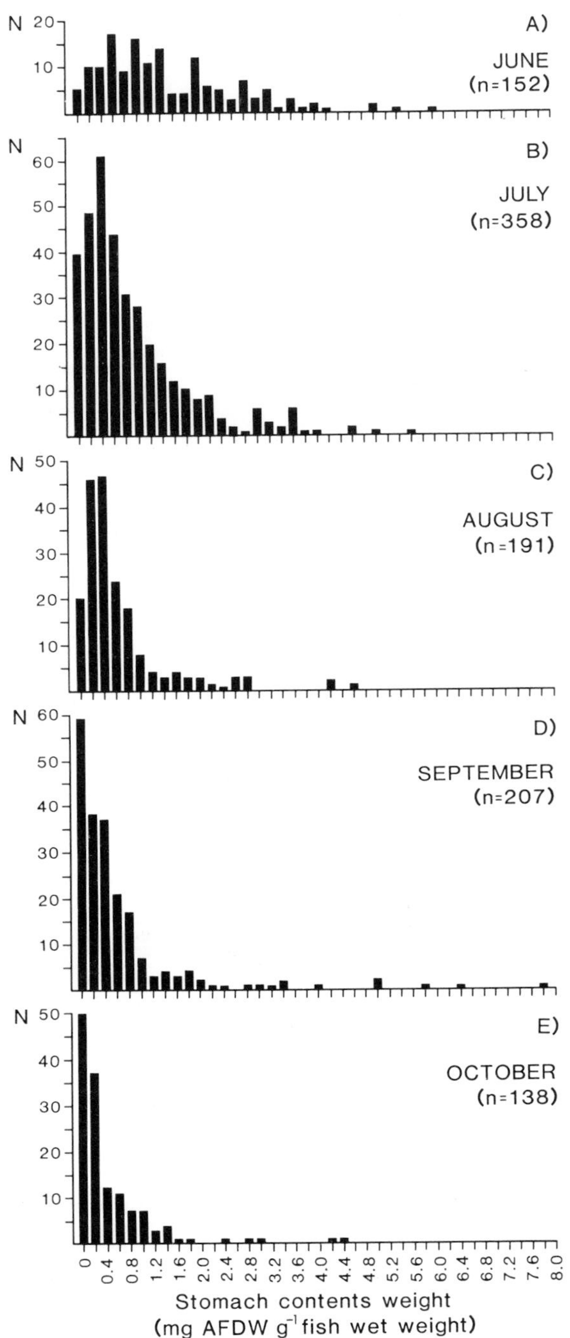

Fig. 2. Distributions of stomach contents weight (mg AFDW g^{-1} fish wet weight) of Arctic charr for pooled data from (A) June, (B) July, (C) August, (D) September and (E) October.

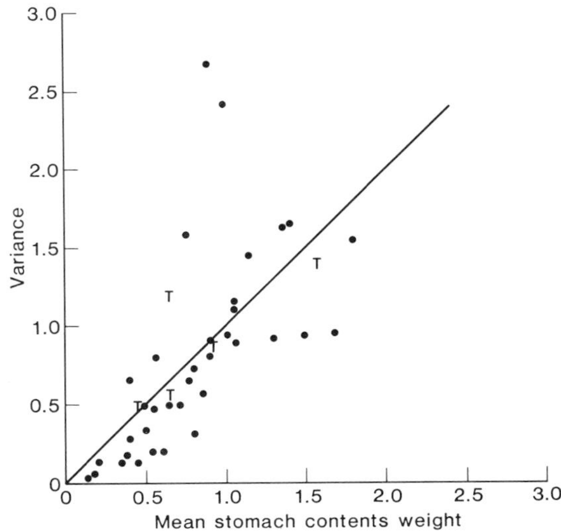

Fig. 3. Plot of variances against mean values for the distributions of stomach contents weight (mg AFDW g^{-1} fish wet weight) for each sample (●), and for pooled data from each month (T). Solid line indicates variance equal to mean.

Discussion

Information about within-sample variabilities in stomach contents weight is sparse. Careful examination of data presented in published studies of fish stomach contents suggests, however, that there are reasons to believe that large within-sample variation, skewness and variance-dependency are quite widespread phenomena. Jenkins & Green (1977) demonstrated the problems of large within-sample variations in stomach contents with regard to determination of feeding chronology of fishes, and used a study of *Myxocephalus octodecemspinosus* as an example. Furthermore, by refering to the studies of Keast & Welch (1968), Mathur & Robbins (1971) and Mathur (1973), Jenkins & Green (1977) stated that large intrasample variations are to be found quite frequently. Thorpe (1977) did not quantify within-sample variations, but pointed out that they were large. In other studies, the large variations of stomach contents can be inferred by the fact that the data display wide confidence limits (e.g. Clarke 1978, Allen & Wootton 1984).

Skewness to the right in distributions of stomach contents weight indicated by the median values

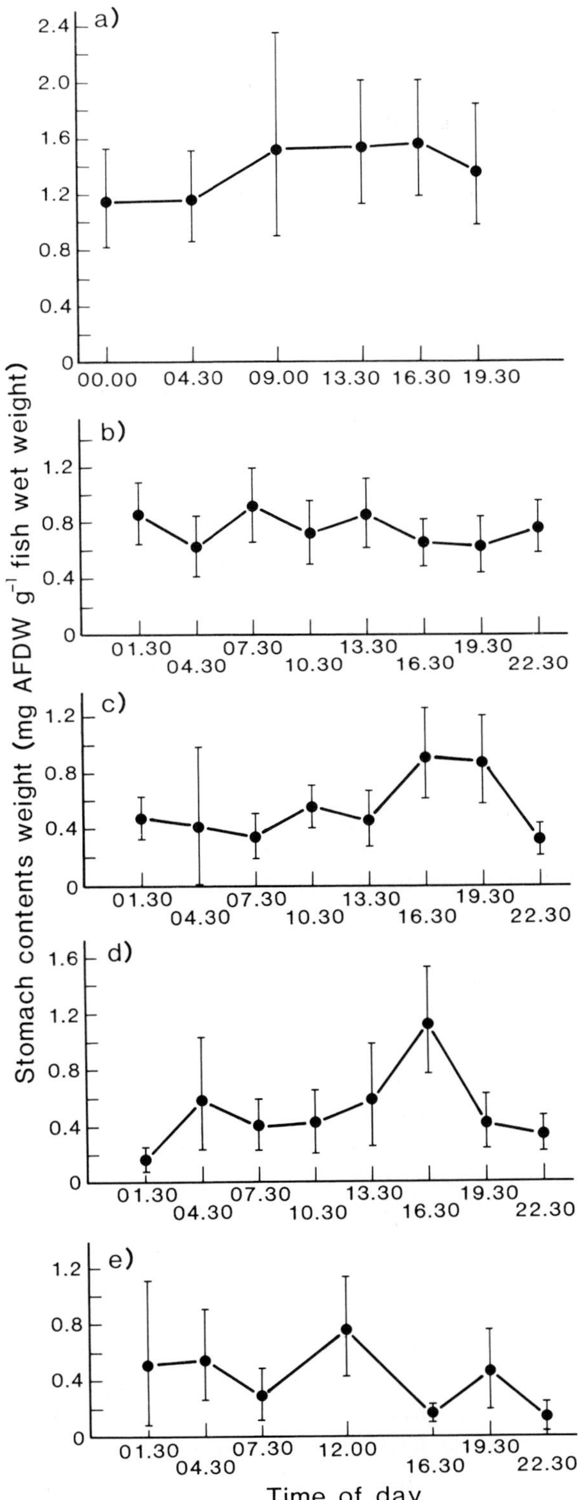

← *Fig. 4.* Geometric mean weights of stomach contents (mg AFDW g^{-1} fish wet weight) with 95% confidence limits for Arctic charr in Lake Takvatn. Samples taken throughout 24 h periods in (a) June, (b) July, (c) August, (d) September and (e) October.

being smaller than the means, have been reported by Clarke (1978) and Cochran & Adelman (1982). Another indication of skewness is a frequent occurrence of empty stomachs, and this is commonly reported in fish studies. The occurrence of variance-dependency is also related to skewness. The results presented by Lane et al. (1979) indicate that the variances in their data are proportional to the means. In addition, an examination of the raw data used by Allen & Wootton (1984), given in Allen (1980), clearly shows that variances are dependent upon the means.

With a few notable exceptions (e.g. Jenkins & Green 1977), the variabilities in the weight of stomach contents have been ignored. Very often stomach content results have been presented in terms of means without any indication of the confidence limits of the estimates (e.g. Darnell & Meirotto 1962, Noble 1972, Arntz 1973, Moriarity & Moriarity 1973, Griffiths 1976, Nakashima & Leggett 1978, Neveu 1980). Others have given some information about confidence intervals or variances, but have not taken them into account when performing calculations and making considerations based upon differences in mean values (e.g. Thorpe 1977, Lane et al. 1979, Allen 1980, Allen & Wootton 1984). Pooling of stomach contents of groups of fish has also been done, thereby giving a false impression of narrow ranges (e.g. Elliott & Persson 1978).

Despite the scarcity of published information about intrasample variabilities, the indications are that the large within-sample variation and skewed distributions of stomach contents weight found in our study of Arctic charr, may be a general phenomenon. There is a need for greater attention to be focused on within-sample variation and the distributions of stomach contents weight. It is of fundamental importance to take these factors into account, when choosing the most appropriate method for estimation of daily consumption. When using the Elliott & Persson (1978) model or other related methods, the estimation of daily consumption is founded on differences in average weights of stomach contents of consecutive samples taken throughout 24 h periods. The large within-sample variations of weight of stomach contents found in the present study cast doubts on the accuracy of consumption rate estimates based on these sample means, given the uncertainty as to how representative the sample means are as measures of the population averages. This uncertainty is clearly expressed in the wide ranges in confidence limits (Fig. 4), and it must be concluded that it is unsatisfactory to base estimates of feeding rate upon the differences between the means of the consecutive samples. This conclusion is supported by the fact that it was usually impossible to demonstrate any significant differences between the means of consecutive samples. This means that S_0 is not significantly different for S_t in the Elliott & Persson (1978) model:

$$C_t = \frac{(S_t - S_0 e^{-Rt})Rt}{1 - e^{-Rt}} , \tag{1}$$

where C_t is the amount of food consumed in t hours, S_0 the initial amount of food in the stomach, S_t the amount remaining after t hours and R the instantaneous gastric evacuation rate. Substituting $S_0 = S_t = S$ in this equation gives:

$$C_t = t * S * R . \tag{2}$$

This corresponds to Eggers (1979) modification of the Bajkov (1935) method, stating that:

$$C_{24} = 24 * S * R , \tag{3}$$

where C_{24} is the daily consumption and S is the mean stomach content over the 24 h period. The Bajkov/Eggers method is, therefore, to be regarded as the most appropriate method for estimation of daily consumption in this study. Being based on the average weight of stomach contents of the total distribution for each 24 h period, it is likely to give a far more robust estimate than the Elliott & Persson method. The use of the Elliott & Persson and other related methods in consumption studies of other fish populations can also be questionned since large intrasample variations in data appear to be present. Indications of uncertainty in using the Elliott & Persson method are clearly reflected in the result of some studies in that the estimates of food intake are negative over some intervals (Allen 1980, Cochran & Adelman 1982, Persson 1983).

The lack of normality and variance heterogeneity in data questions the use of parametric statistics, and the skewed distributions require a further statistical analysis. Due to the skewness, the arithmetic means are very sensitive to a few high values, and the medians may provide a more meaningful expression of average population values (Clarke 1978, Cochran & Adelman 1982). Also, transformation of the data may provide estimates of the average that are less sensitive to a few high values and, in the present study, the logarithmic transformation giving the geometric mean has been used. Frequency distributions skewed to the right are often rendered more symmetrical by this transformation, and the transformation is also quite likely to make the variance independent of the mean (Sokal & Rohlf 1981, p. 419). A variance-to-mean ratio equaling 1 indicates a Poisson distribution, and further examination of methods of expression based upon this form of distribution may prove to be fruitful.

The skewed distributions of stomach contents weight found in the current study could have a number of causes, one of them being related to feeding interactions in the fish population. The charr in Takvatn are stunted and this is probably due to food limitations (Amundsen & Klemetsen unpublished). Food limitations may possibly lead to a skewed distribution of stomach contents in the population. A quantitative interpretation of 'the principle of the smorgasbord' (Johnson 1980) can symbolize this: If the crowd is large and the table short, just a few will get hold of a lot of food, while most of the crowd will get little or none.

Additionally and/or alternately the distribution pattern of stomach contents weight may have a causal connection with the pattern of gastric evacuation. When the stomach evacuation pattern is exponential, the weight of food emptied in a given time will be large for fishes with high stomach contents, but for those fish with little food in their stomach, the weight emptied per unit time will be reduced. This will lead to a trend towards a fast turnover of fishes with high stomach contents, and an accumulation of fishes with low contents. Skewed distributions of stomach contents weight may, therefore, be a reflection of an exponential (or some other curvilinear) pattern of gastric evacuation.

In conclusion, it must be stressed that the large variations and skewed distributions of stomach contents found in the present study are not likely to be caused by small sample sizes. The sample sizes were larger than those used in previous studies designed to estimate food consumption of fish (e.g. Thorpe 1977, Elliott & Persson 1978, Godin 1981, Cochran & Adelman 1982, Allen & Wootton 1984). Furthermore, the variance-to-mean ratios were not correlated with the sample sizes. The large variations and skewed distributions seem, therefore, to be of a biological nature and may themselves constitute important information about the feeding regimes of the fishes.

Acknowledgements

Thanks to Per Grotnes for helpful discussions about the handling of the data, and to Roar Kristoffersen for assistance during the course of the study. Thanks also to Malcolm Jobling for critically reading the manuscript and correcting the English.

References cited

Allen, J.R.M. 1980. The estimation of natural feeding rate of the threespined stickleback (*Gasterosteus aculeatus* L.) (Pisces). Ph.D. Thesis, University of Wales, Aberystwyth. 341 pp.

Allen, J.R.M. & R.J. Wootton. 1984. Temporal patterns in diet and rate of food consumption of the three-spined stickleback (*Gasterosteus aculeatus* L.) in Llyn Frongoch, un upland Welsh lake. Freshw. Biol. 14: 335–346.

Arntz, W.E. 1973. Periodicity of diel food intake of cod *Gadus morhua* in the Kiel Bay. Oikos Suppl. 15: 138–145.

Bajkov, A.D. 1935. How to estimate the daily food consumption of fish under natural conditions. Trans. Amer. Fish. Soc. 65: 288–289.

Clarke, T.A. 1978. Diel feeding pattern of 16 species of mesopelagic fishes from Hawaiian waters. U.S. Fish. Bull. 76: 495–513.

Cochran, P.A. 1979. Comment on some recent methods for estimating food consumption by fish. J. Fish. Res. Board Can. 36: 1018.

Cochran, P.A. & I.R. Adelman. 1982. Seasonal aspects of daily ration and diet of largemouth bass, *Micropterus salmoides*, with an evaluation of gastric evacuation rates. Env. Biol. Fish. 7: 265–275.

Darnell, R.M. & R.R. Meirotto. 1962. Determination of feeding chronology in fishes. Trans. Amer. Fish. Soc. 9: 313–320.

Eggers, D.M. 1977. Factors in interpreting data obtained by diel sampling of fish stomachs. J. Fish. Res. Board Can. 34: 290–294.

Eggers, D.M. 1979. Comment on some recent methods for estimating food consumption by fish. J. Fish. Res. Board Can. 36: 1018–1019.

Elliott, J.M. 1979. Comments on some recent methods for estimating food consumption by fish. J. Fish. Res. Board Can. 36: 1020.

Elliott, J.M. & L. Persson. 1978. The estimation of daily rates of food consumption for fish. J. Anim. Ecol. 47: 977–993.

Gaddum, J.H. 1945. Lognormal distributions. Nature 156: 463–466.

Godin, J.-G.J. 1981. Daily patterns of feeding behavior, daily rations and diets of juvenile pink salmon (*Oncorhynchus gorbuscha*) in two marine bays of British Columbia. Can. J. Fish. Aquat. Sci. 38: 10–15.

Griffiths, W.E. 1976. Feeding and gastric evacuation rate in perch *Perca fluviatilis* L. Mauri Ora. 4: 19–34.

Jenkins, B.W. & J.M. Green. 1976. A critique of field methodology for determining fish feeding periodicity. Env. Biol. Fish. 1: 209–214.

Johnson, L. 1980. The arctic charr, *Salvelinus alpinus*. pp. 15–98. *In*: E.K. Balon (ed.) Charrs: Salmonid Fishes of the Genus *Salvelinus*, Dr W. Junk Publishers, the Hague.

Keast, A. & L. Welsh, 1968. Daily feeding periodicities, food uptake rates and dietary changes with hour of day in some lake fishes. J. Fish. Res. Board Can. 25: 1133–1144.

Lane, E.D., M.C.S. Kingsley & D.E. Thornton. 1979. Daily feeding and food conversion of the diamond turbot: an analysis based on field data. Trans. Amer. Fish. Soc. 108: 530–535.

Mann, K.H. 1978. Estimating the food consumption of fish in nature. pp. 250–273. *In*: S.D. Gerking (ed.) Ecology of Freshwater Fish Production, Blackwell, Oxford.

Mathur, D. 1973. Food habits and feeding chronology of the black banded darter, *Percina nirofasciata* (Agassiz) in Halawakee Creek, Alabama. Trans. Amer. Fish. Soc. 102: 48–55.

Mathur, D. & T.W. Robbins. 1971. Food habits and feeding chronology of young white crappie, *Pomoxis annularis* Rafinesque in Conowingo Reservoir. Trans. Amer. Fish. Soc. 100: 307–311.

Moriarity, C.M. & D.J.W. Moriarity. 1973. Quantitative estimation of the daily ingestion of phytoplankton by *Tilapia nilotica* and *Haplochromis nigripinnus* in Lake George, Uganda. J. Zool. 171: 15–23.

Nakashima, B.S. & W.C. Leggett. 1978. Daily ration of yellow perch (*Perca flavescens*) from Lake Memphremagog, Quebec-Vermont, with a comparison of methods for in situ determinations. J. Fish. Res. Board Can. 35: 1597–1603.

Neveu, A. 1980. Relations entre le benthos, la derive, le rhytme alimentaire, et le taux de consommation de truites communes (*S. trutta* L.) en canal experimental. Hydrobiologia 76: 217–228.

Noble, R.L. 1972. A method for direct estimation of total food consumption with application to young yellow perch. Prog. Fish-Cult. 34: 191–204.

Persson, L. 1982. Rate of food evacuation in roach (*Rutilus rutilus*) in relation to temperature, and the application of evacuation rate estimates for studies on the rate of food consumption. Freshw. Biol. 12: 203–210.

Persson, L. 1983. Food consumption and competition between age classes in a perch *Perca fluviatilis* population in a shallow eutropic lake. Oikos 40: 197–207.

Sokal, R.R. & F.J. Rohlf. 1981. Biometry. Second edition. W.H. Freeman & Co., San Francisco. 859 pp.

Staples, D.J. 1975a. Production biology of the upland bully *Philypnodon breviceps* Stokell in a small New Zealand lake. I. Life history, food, feeding and activity rhythms. J. Fish Biol. 7: 1–24.

Staples, D.J. 1975b. Production biology of the upland bully *Philypnodon breviceps* Stokell in a small New Zealand lake. III. Production, food consumption and efficiency of food utilization. J. Fish Biol. 7: 47–69.

Swenson, W.A. & L.L. Smith. 1973. Gastric digestion, food consumption, feeding periodicty, and food conversion efficiency in walleye (*Stizostedion vitreum vitreum*). J. Fish. Res. Board Can. 30: 1327–1336.

Thorpe, J.E. 1977. Daily ration of adult perch, *Perca fluviatilis* L. during summer in Loch Leven, Scotland. J. Fish Biol. 11: 55–68.

Windell, J.T. 1978. Estimating food consumption rates of fish populations. pp. 227–254. *In*: T. Bagenal (ed.) Methods for Assessment of Fish Production in Fresh Waters, Blackwell, Oxford.

Received 15.12.1984 *Accepted 3.7.1985*

An alternative to the fullness index

Bruce Herbold
Division of Wildlife and Fisheries Biology, University of California, Davis, CA 95616, U.S.A.

Keywords: Gut contents, Stomach, Fish, Methodology, *Pogononichthys macrolepidotus, Cottus asper, Morone saxatilis, Acanthogobius flavimanus*

Synopsis

The traditional fullness index exhibits many of the same undesirable features as the gonosomatic index. A recently proposed alternative is examined and refined. This alternative estimates the values for full stomachs at all lengths of a species from the maximum gut content masses observed when the data are sorted by the lengths of the fish from which they were taken. The actual gut content mass of each fish is then expressed as the proportion of the maximum amount expected for a fish of that size. This technique is flexible enough to be used for a variety of patterns of change in gut content mass with growth.

Introduction

Feeding intensity, foraging patterns, and environmental constraints on feeding are frequently inferred from data on fullness (recently, Durbin et al. 1983, Langton 1983, Daniels 1982). Since Blegvad (1917), fullness has been most frequently measured by a Fullness Index (FI) defined as:

$$FI = 10000 \times \text{stomach content mass/total body mass.}$$

Hureau (1969) proposed a similar 'L'indice de repletion' where the ratio of stomach content mass to body mass was multiplied by 100 instead of 10000. Many problems can be encountered with this sort of proportional index, as described by De Vlaming et al. (1982) for the gonosomatic index (GSI). These authors point out that the GSI is unreliable because: (1) the relationship between gonadal mass and body mass is not linear, (2) the regression of gonad mass on body mass has a non-zero intercept, (3) variance in gonadal mass is not independent of body size, and (4) gonad mass to body mass ratios change with stage of gonadal development. These problems may also apply to the traditional fullness index. Changes in diet may also obscure estimates of fullness among different sizes of fish. Another problem is that the relationship between gonad mass and body mass may be quite complex. For example, gonad mass increases with body size in small sea urchins but decreases with body size in larger sea urchins (Gonor 1972). Qualitative measures of fullness such as point methods (Hynes 1950) and similar methods (Pillay 1953, Ball 1961, Hunt & Jones 1972, Ryan 1984) provide a means of comparing changes in fullness for different species which are not biased by changes in body mass. However, these have been criticized as not comparable between studies since degree of fullness is a subjective estimate of each investigator (Berg 1979). In addition, this method is subject to severe biases if more than one observer collects the data or if a single observer is learning, over the course of the study, the degree to which a stomach can be packed.

Using the maximum observed gut content mass at various lengths of fish as a standard against which fishes of similar lengths are compared, Knight & Margraf (1982) briefly present a fullness index which avoids most of the problems of the traditional index and incorporates most of the advantages of the point methods. The purpose of this paper is to examine the FI as DeVlaming et al. (1982) examined the GSI to determine whether the same sorts of inappropriate assumptions undermine it. I also further refine and justify the index of Knight & Margraf (1982). I conclude by demonstrating the applicability of the modified index to a variety of species and situations.

Materials and methods

Two hundred and ten yellowfin goby, *Acanthogobius flavimanus*, 386 striped bass, *Morone saxatilis*, 181 prickly sculpin, *Cottus asper*, and 302 splittail, *Pogonichthys macrolepidotus*, were preserved in 10% formalin solutions during monthly trawl samples in Suisun marsh (Solano Co., California) between 1979 and 1983. Fish were taken from all months of the year, in various years. Fish longer than 10 cm were stunned to prevent regurgitation and the body cavities were opened to allow quicker preservation of gut contents. Standard lengths to the nearest 1 mm and total masses to the nearest 0.1 g were measured immediately before dissection. Stomach contents of gobies, bass, and sculpins were removed and weighed to the nearest 0.0001 g (wet mass). Since splittail lack stomachs, contents of the intestine to the first 180 bend were used instead.

Percent Fullness was calculated by: (1) sorting the gut content mass (gcm) data for all months by the standard length of the fish in which they occurred; (2) creating a data subset which consisted of the successive maxima in gcm and their associated standard lengths. Thus, the first entry was the smallest fish with a positive gcm, the second entry was the next largest fish in which the gcm exceeded that of the first entry and so on to the fish with the greatest gcm and its associated standard length; (3) this data subset, consisting of gcms and standard lengths was subjected to multiple regression analysis where the dependent variable was gcm and the independent variables were standard length to the first, second, and third powers; (4) the original data set was edited to remove those fish longer than the one with the maximum gcm or smaller than the one with the first positive gcm; (5) the regression equation was applied to the standard lengths of the edited data set in order to estimate the expected gut fullness at each standard length, and (6) finally, percent fullness was calculated as:

$$\text{Percent Fullness} = 100 * \frac{\text{(observed gut content mass)}}{\text{(expected maximum gut content mass)}}.$$

Although this was not a regression problem, the algorithms of regression were used since they are widely available. The goal was simply to describe a line through the successive maxima and similar results would be obtained using any other curve fitting algorithm. The cubic regression method used here allows the predicted line to be quite complex but larger size ranges of fish may require more variables or different approaches. Two consequences of the regression approach are that some of the gcm data will yield values for percent fullness greater than 100. These may represent distended stomachs but I did not record such observations. Following calculation of percent fullness the data may, if desired, be standardized to a maximum of unity. In addition, the lower size limit for the edited data may need to be adjusted to eliminate any prediction of negative gcm.

Results

The regression equation for gut content mass as a function of body mass (bm) was a poor predictor and had an intercept significantly different than zero for 3 of the 4 species (Table 1). Correlation of gut content mass and body mass was significant for all species (Table 1) but the variance in gcm was not independent of body size. The variance about the regression line increased drastically with body mass for all species (Fig. 1). For most species the variance within quartiles of body size doubled between each successive quartile (Table 1).

Table 1. Relationship of gut content mass with body mass. Values for intercepts are mean and one standard error.

Species	Sample size	Intercept	R^2	Variance within quartile of body mass			
				Q_1	Q_2	Q_3	Q_4
Yellowfin goby	210	-0.0174 ± 0.004**	0.22	33	93	250	620
Striped bass	386	-0.0017 ± 0.004	0.34	33	72	230	1110
Splittail	302	0.0443 ± 0.002**	0.13	36	96	240	430
Prickly sculpin	181	0.0081 ± 0.003**	0.28	97	91	230	660

** Probability of intercept equal to 0<0.01.

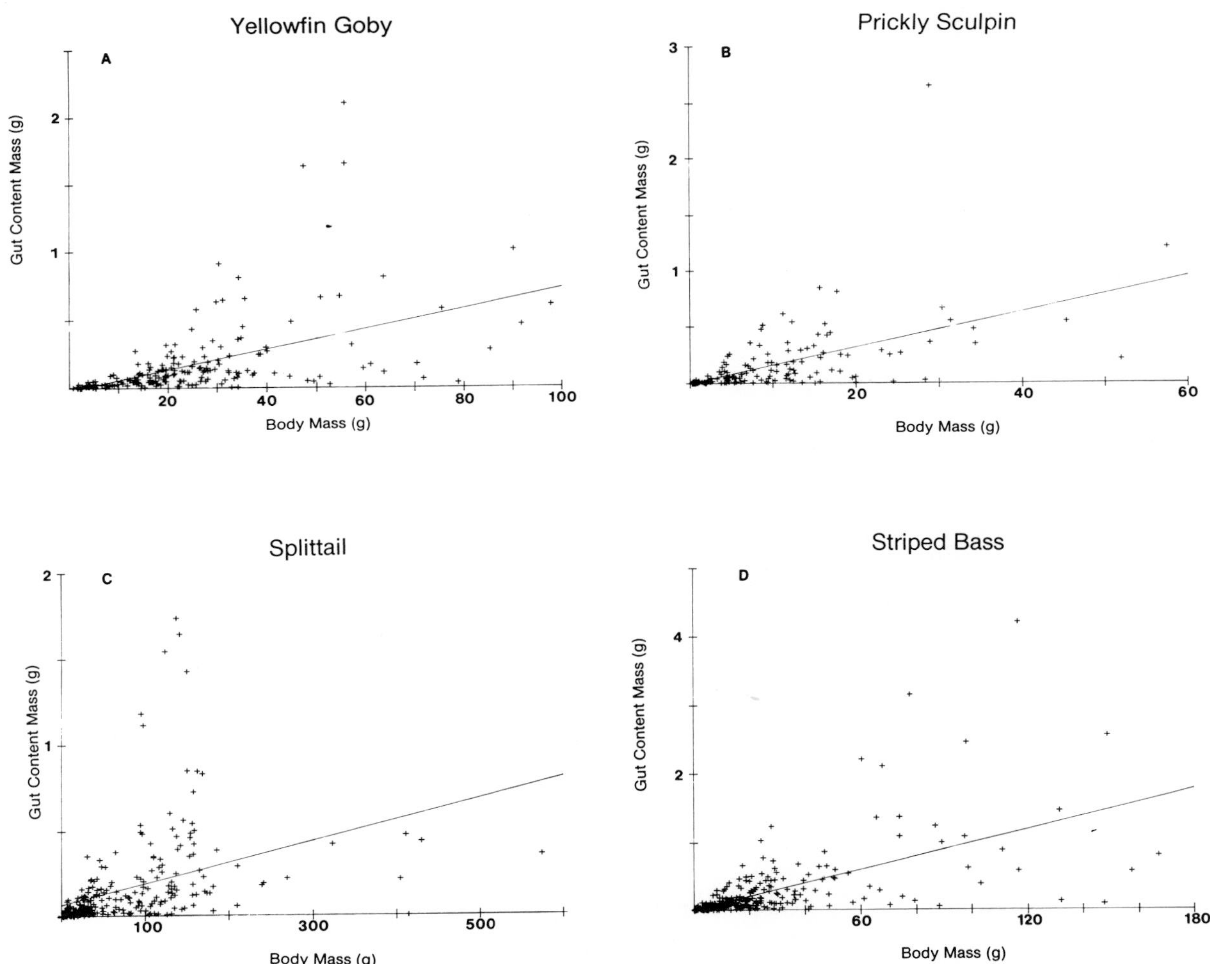

Fig. 1. Distribution of gut content mass and body mass in four species of fish.

Discussion

The results closely parallel those of De Vlaming et al. (1982) for the gonosomatic index. As with the gonosomatic index, the incorrect assumptions behind the fullness index give it very limited usefulness. Since the variance in gcm is not independent of body mass most parametric statistical tests can not be used unless an analysis of covariance is performed. A non-zero intercept for the regression between body size and stomach mass implies that the relationship between them cannot be accurately represented by a linear function. Finally, the low predictive value for the regression equation implies that the variables gcm and body mass are poor choices for quantifying stomach fullness.

Percent Fullness is an expression of the gcm of a fish as the proportion of the maximum expected gcm for a fish of that standard length. Standard length appeared to be a better estimate of fish size than bm for two reasons:

(1) The distribution of gcm with standard length (Fig. 2) was much more orderly than with body mass (Fig. 1), as indicated by the maximum gcm occurring in the longest rather than in the heaviest fish.

(2) If maximum gcm is a result of stomach geometry, then stomach volume would increase

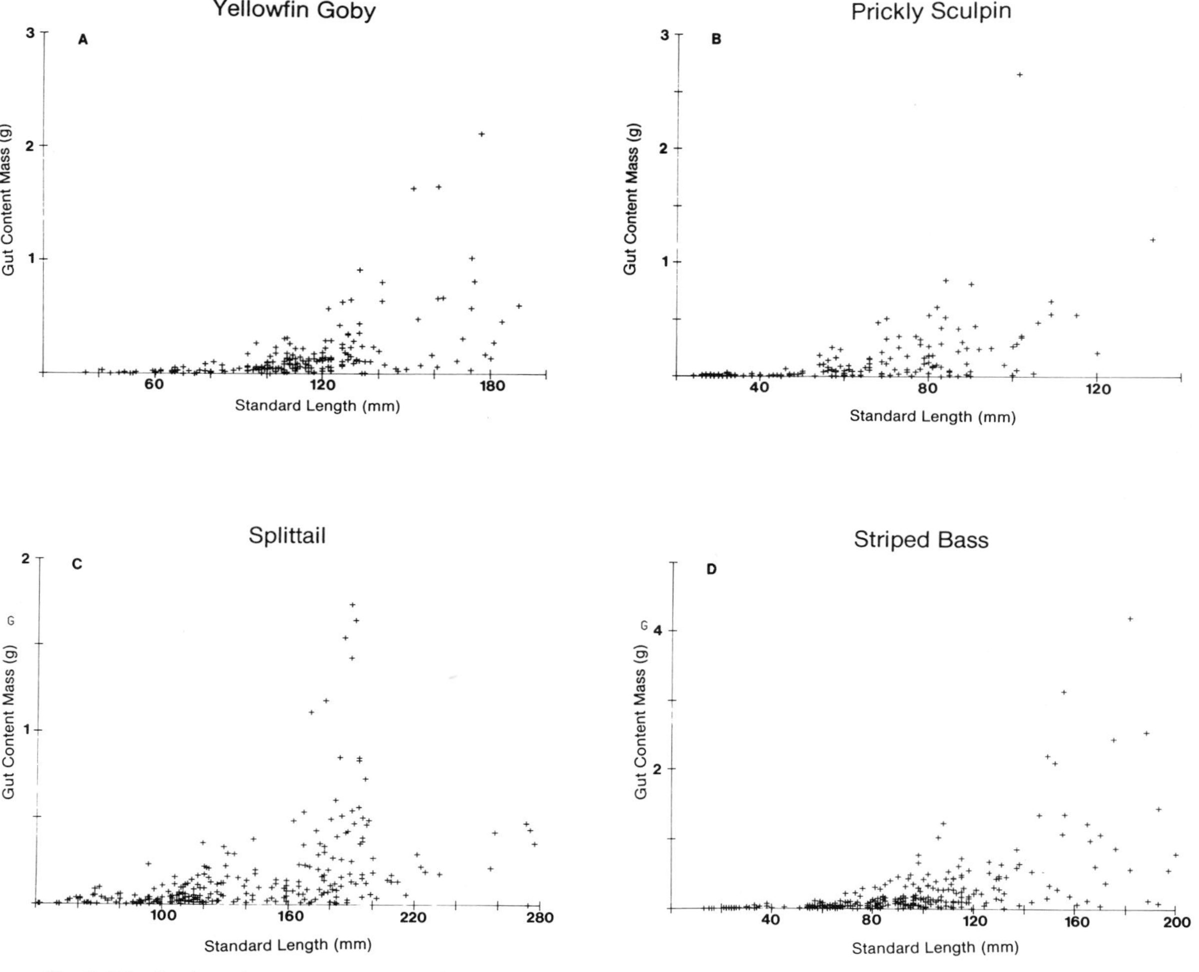

Fig. 2. Distribution of gut content mass and standard length in four species of fish. Same individuals as in Figure 1.

with fish length more directly than with body mass.

The most peculiar aspect of the relationship between gcm and bm was that the greatest gcm was found with fish of only intermediate body mass (Fig. 1). This unexpected result might be attributed to either of two factors. It may reflect an association between fullness and condition factor (condition factor = bm/standard length) where fish with less body stores tend to eat more. Conversely it may be a simple geometric effect in which two fish may have the same body mass but one is longer, with a correspondingly larger gut. In either case the traditional fullness index would be biased in counter-intuitive directions.

Another formulation of the fullness index, where gcm is divided by some power of the standard length, has been used. Unlike the original, this method removes the effects of varying body masses at the same length of fish and offers a better fit of the data to a regression line since the non-linearity of the curve is incorporated. This approach, however, assumes that the power function is the same for all species and the resulting number is difficult to compare between species.

Percent Fullness has several advantages for the quantification of fish stomach fullness: (1) it is independent of body size since each estimate can vary from zero to slightly more than one; (2) it takes into account the non-linear changes in gcm with body size; (3) the values for different species can be compared since they are all expressed as percentages of the maximum observed for each species. For example, one may wish to determine which species in a lake exhibits the greatest response to a manipulation. (4) The estimate of fullness is independent of the observer so that different workers or the same workers at different times will not bias the estimate; (5) different populations can be compared if the equations for expected maximum gcm are not significantly different, thus historical data may more safely be compared with more current data; and, (6) the estimation of percent fullness does not require any information not routinely gathered.

Restrictions on the use of Percent Fullness are that values should not be calculated for fish larger than that fish with the maximum stomach mass since the line may not be safely extrapolated. Comparison of values of Percent Fullness for fishes may be unsafe if the food items are different, particularly if they are of different densities. Calculation of the regression line describing maximum fullness will generally require a large sample size to assure that the maxima observed are reflecting size differences. This may be checked by examination of the coefficients of determination, sample sizes for this study yielded all R-squared >0.90. The samples must not only be numerous but they must be dispersed across the environmental feature of interest so that a test for differences can be performed once Percent Fullness has been calculated. Although the representation of fullness as a percent is intuitively appealing, statistical tests should be non-parametric or the data should be transformed to conform to normal distributions. Percent Fullness should not be transformed by an arcsine transformation unless they have been standardized to a maximum of unity. If there is evidence that maximum gut content mass is not an increasing function for all sizes the procedure might be altered by dividing the data into size increments and using the maximum for each size increment irrespective of the neighboring values.

Knight & Margraf (1982) used several techniques to construct the line describing maximum stomach fullness in walleye, *Stizostedion vitreum*, including the cubic regression method used here. They suggest that the exponential model they used is best because it is analogous to the weight-length model (Ricker 1975). As shown here, the relationship between standard length and gcm can be substantially different for different species. In some the cubic regression equation is very similar to a simple monotonically increasing function like the growth equation (Fig. 2a, d), while in others the use of a cubic coefficient allows recognition of a plateau in gcm across a range of lengths (Fig. 2, b, c). If the observed maxima in gcm are a function of stomach geometry, then a simple monotonic function is probably adequate in most cases. However, if foraging efficiency determines stomach fullness, then plateaus may reflect shifts in foraging behavior when fish become large enough to swim after more mobile prey or large enough to escape predation in richer environments.

Percent Fullness has several advantages over previously used methods to quantify degree of stomach fullness. The calculated values can accurately reflect the different relationships between body size and stomach fullness which occur in different species. However, since the final value is a percentage, trends through time or in different habitats can be compared for one or for several species. By publication of the coefficients of the regression equation for each species, results of various workers may be compared if the regression equations are not significantly different. It is free of several important biases which would tend to obscure the relationship between body size and gut content mass in the traditional fullness index.

Acknowledgements

I would like to thank Joseph J. Cech for reminding me that nothing biological behaves in a linear fashion, and Peter B. Moyle for apparently boundless support. Thanks are also due to several reviewers, particularly Robert J. Feller and Larry R. Brown whose comments have greatly reduced the incoherency of earlier drafts.

References cited

Ball, J.N. 1961. On the food of the brown trout of Llyn Tegid. Proc. Zool. Soc. London 137: 559–622.

Berg, J. 1979. Discussion of methods of investigating the food of fishes, with reference to a preliminary study of the prey of *Gobiusculus flavescens* (Gobiidae). Mar. Biol. 50: 263–273.

Blegvad, H. 1917. On the food of fish in Danish waters within the Skaw. Rep. Dan. biol. Stn. 24: 17–72.

Daniels, R.A. 1982. Feeding ecology of some fishes of the Antarctic Peninsula. U.S. Fish. Bull. 80: 575–588.

DeVlaming, V., G. Grossman & F. Chapman. 1982. On the use of the gonosomatic index. Comp. Biochem. Physiol. 73A: 31–39.

Durbin, E.G., A.G. Durbin, R.W. Langton & R.E. Bowman. 1983. Stomach contents of silver hake, *Merluccius bilinnearis*, and Atlantic cod *Gadus morhua*, and estimation of the daily ration. U.S. Fish. Bull. 81: 437–454.

Gonor, J.J. 1972. Gonad growth in the sea urchin, *Strongylocentrotus purpuratus* Stimpson) (Echinodermata: Echinoidea) and the assumptions of gonad index methods. J. Exp. Mar. Biol. Ecol. 10: 89–103.

Hunt, P.C. & J.W. Jones 1972. The food of brown trout in Llyn Alaw, Anglesey, North Wales. J. Fish Biol. 4: 333–352.

Hureau, J.C. 1979. Biologie comparee de quelques poissons antarctiques (Nototheneidae). Bull. Inst. oceanogr. Monaco 68: 1–44.

Hynes, H.B.N. 1950. The food of freshwater sticklebacks (*Gasterosteus aculeatus* and *Pygosteus pungitius*), with a review of methods used in studies of the food of fishes. J. Anim. Ecol. 19: 35–38.

Knight, R.L. & F.J. Margraf. 1982. Estimating stomach fullness in fish. N. Amer. J. Fish. Manag. 2: 413–414.

Langton, R.W. 1983. Food habits of yellowtail flounder, *Limanda ferruginae* (Storer), from off the northeastern United States. U.S. Fish. Bull. 81: 15–22.

Pillay, T.V.R. 1953. Studies on the food, feeding habits and alimentary tract of the grey mullet, *Mugil tade* Forskal. Proc. natn. Inst. Sci. India 19: 777–823.

Ricker, W.E. 1975. Computation and interpretation of biological statistics of fish populations. Fisheries Research Board of Canada, Bulletin 191. 382 pp.

Ryan, P.A. 1984. Diel and seasonal feeding activity of the short-finned eel, *Anguilla australis schmidtii*, in Lake Ellesmere, Canterbury, New Zealand. Env. Biol. Fish. 11: 229–234.

Received 14.12.1984 *Accepted 11.9.1985*

Contemporary studies on fish feeding: summary of GUTSHOP '84

Charles A. Simenstad[1] & Gregor M. Cailliet[2]
[1] *Fisheries Research Institute WH-10, University of Washington, Seattle, WA 98195, U.S.A.*
[2] *Moss Landing Marine Laboratories, P.O. Box 450, Moss Landing, CA 95039, U.S.A.*

As with any symposium, the essence of GUTSHOP '84 resides in the publication of selected papers, which have benefited from extemporaneous exchanges with peers during the workshop and subsequent comments and suggestions of reviewers and editors. But, despite their often insightful revelation of new ideas and alternative opinions, these in situ discussions are seldom published. As with previous proceedings, we have found much of this information deserving inclusion and have paraphrased the germane points in the following synopsis.

Methodology and statistics

A major point of discussion around the use of fish food habits data in marine fishery management models (Livingston) was the need to verify critical assumptions used in such functional relationships as consumption rates, diet selectivity, and growth coefficients and efficiency rates. For instance, it is difficult to incorporate prey size preference into biomass-based models despite the fact that empirical data increasingly support intense taxa and size selectivity. This constraint upon the applicability of these models argues for sensitivity analysis *before* large-scale sampling efforts are implemented. In particular, the relationship between the prey resource and the fish consumption parameters typically demands both encounter rate and selection functions, but prey resource availability data are seldom included in stock assessment surveys used to provide model input. If exploited fish stocks are to be managed using such ecosystem models, these sensitive data must be measured and verified.

Further suggestions to augment or refine the stomach contents mass: body length relationship as an index of fullness (Herbold) included: (1) conducting laboratory experiments feeding fish ad libitum or (2) injecting stomachs with latex to obtain the required maxima, and (3) ordering species or functional groups by stomach morphology, which should be representative across fish populations.

The perpetual dilemma of obtaining fish samples from deep water collections without regurgitation of the stomach contents (Bowman), and the biases introduced into the data, was discussed but not resolved. Suggestions for minimizing regurgitation (e.g. longer gear retrieval times, puncturing air bladders, killing or narcotizing the fish) were acknowledged to be generally impractical for intensive sampling at extreme (>SCUBA diving) depths. The fact that regurgitation is not always evident by external or internal examination further magnifies the need to construct a better sampling protocol to address this problem. It was agreed that the most obvious direction was toward laboratory experiments in pressure chambers to simulate deep-water collections.

Physiology of feeding and digestion

Gastric evacuation, and the numerical construct of this process, was a common topic of presentations

(Olson & Mullen, Jobling, Persson) as well as the discussions. The extent of our deliberations indicated that the processes affecting emptying or retention of fish stomach contents are more complex and subtle than those formulated by most of the common models. The sole consensus was that 'the ultimate answer is the question'. That is, if your intent is to describe a statistically predictable relationship for use as a component function in a larger model (e.g. consumption rate), the model which provides the highest confidence for your data is obviously the most appropriate. On the other hand, if the objective is to describe mathematically (simulate) the general process(es) of digestion and gastric evacuation, the effects of variables such as meal size and composition, multiple meals, individual and size variation, intraspecific competition (e.g. schooling), feeding chronology, post-prandial starvation, and the other subtle outcomes of fish feeding studies and experimentation must be considered.

There are a number of 'empirical' gastric evacuation models. Persson described three basic formulations: (1) exponential or (2) square root, describing weight- or volume-dependent relationships, and (3) rectilinear, describing surface area-dependent relationships; Jobling presented situations where a fourth, the linear model, was applicable. It was evident from both presentations, however, that real differences among them were small quantitatively, though not qualitatively. In addition, most of these models (except perhaps the linear) were developed and tested solely using data for pelagic planktivores and they are probably inappropriate for large carnivores with more intermittent feeding regimes. Some models (e.g. the square root) are also difficult to apply to any simple consumption rate estimation procedure. Everyone agreed that champions of any one model, especially one that deviates from the 'accepted' forms, should compare that model to the others for differences in the final estimate of consumption.

Ultimately, the predictions of both empirical and process models should converge, but theory is still a long way from reality. Many empirical gastric evacuation models are inappropriate for data having time lag effects at the initiation of digestion, and most investigators try to avoid these effects in their experiments (Persson). Yet, some of the more mechanistic models (Jobling) can explain lag phases which are evident under certain circumstances (e.g. carnivores eating large, high energy content particles) or are more intuitively acceptable (e.g. surface area-dependent model). Even more complex models may be envisoned, such as those incorporating the effects of microanatomical differences in enzymatic action. And, while the majority of the models describe the evacuation process through deterministic models, another unexplored approach suggested by Jim Anderson is to formulate evacuation as a stochastic process. He demonstrated that if evacuation follows a Poisson process, the frequency distribution of stomach contents follows the exponential distribution which is also predicted by the simple deterministic rate models.

Alternatively, fish may differ so much in the way that they process food that no one model will ever be acceptable as the 'uniform evacuation theory'. Obviously, in the case of pelagic planktivores feeding on uniform zooplankton assemblages, current empirical models may be refined enough to describe the observed change in stomach contents quantity over time as well as the actual process of its digestion. However, more mechanistic approaches may be required to address gastric evacuation and consumption rates in marine carnivores such as gadids feeding on benthic macroinvertebrates. This demands novel experimentation; for instance, both physical (fluid dynamics) and invertebrate (harpacticoid copepods, insect larvae) analogs were mentioned during the discussions as viable methods to measure gastric evacuation in fish.

Variability and non-normality in distributions of stomach contents data were also discussed as sources of problems in applying such models to consumption rate or daily ration estimates. An alternative approach recommended by Christopher Boggs and Bob Olson was to compare independent estimates of consumption rates, such as they have done for yellowfin tuna daily ration estimates using metabolic requirements and a cesium budget; their results (daily ration based upon stomach contents and gastric evacuation rate = 4% of body weight, and 5.2% and 6.7% based upon the other two methods, respectively) provided additional bounds of confidence around their consumption rates.

The utilization of detritus by *Cyprinodon* described by White et al. prompted the discussion of the actual nutrition pathway in these fishes, whether by direct (e.g. carbohydrates, free amino acids) or indirect (e.g. epibiota) sources. White et al.'s experimental treatment and interpretation of the ^{15}N enrichment results argued for epibionts while others (e.g. Bowen, *Science* 1980, 207: 1216–1218) have presented evidence for direct assimilation by some fishes. This suggests the need for more definitive information on the significance of digestive physiology and anatomy to diet composition in both detritivores and herbivores.

One point made during a brief discussion of Govoni's synthesis of the physiology of digestion in larval fishes reinforced the argument that, in terms of the embryological changes in digestive systems between larval and juvenile stages, 'metamorphosis' and 'transformation' are synonymous concepts.

Foraging behavior

Bence's experiments on feeding efficiency and selectivity in *Gambusia* included an important variant in tests of optimal foraging – diverse prey assemblages – which may be more representative of actual foraging situations. Since most tests of the popular optimal foraging models have been conducted using prey of similar behavior (e.g. same taxa, different sizes; same sizes, related taxa), there is an implicit assumption of no interaction between prey because the fish can utilize the same attack strategies. In these cases, a generalist strategy provides for the highest efficiency. When different attack strategies must be adopted for divergent prey taxa, however, the total feeding efficiency would be expected to be lower, and specialization a more viable strategy. This factor becomes especially pertinent when we wish to explain temporal shifts in foraging behavior with fish ontogeny and/or experience. Similarly, it was argued that the *potential* perception of prey quality by the fish, in terms of energy gain/expenditure, should also be incorporated into models and experiments by examining selection of prey of divergent evasive capabilities and energy content over time. We were not able to resolve, however, whether prey quality as *perceived* by the fish can actually be measured experimentally.

An interesting observation by O'Brien in their experiments with crappie and sunfish was that the fish apparently could assess or integrate simultaneously both distance and size of prey. In addition, when prey were relatively close (after approach), the fish tended to chose the absolute (actual) largest prey even if it was the apparent (relative to distance from fish, contrast, etc.) smallest prey. Variance in pursuit speed, i.e. higher for pursuit of absolute larger prey, also supported the fish being able to perceive absolute rather than apparent size.

David Noakes also made some particularly salient points in the final discussion session in which he sought to discriminate between considering behavior as a starting point toward a larger question (e.g. prey selection) versus examining behavior as a phenomenon in and of itself. As a consequence, the distinctions and assumptions associated with mathematical, statistical formulations describing (or simulating) the outcome of a behavior can vary significantly from those associated with the biological mechanisms which produce the effect. Fish draw on a broad range of available behaviors constrained mainly by their own capabilities (particularly morphological and physiological factors) and environmental circumstances. Much of this discussion, probably the most dialectic of the meeting, centered on our abilities and objectives to model, in any sense of the term, fish behavior. There were some of us who attacked this approach as basically inappropriate to understanding the varied, complex, and perhaps ultimately unpredictable behavioral plasticity observable in the real world. Others argued for the existence of fundamental underlying mechanisms which, although subject to a multitude of variables, are basically deterministic processes which can be described, perhaps mathematically, through hypothesis testing (which depends upon models). Perhaps the telling point made by one participant was that, although conceptually simple models may predict the consequence of complex behaviors, this outcome demands close interaction between the behavioralist observer and the modeler because the ultimate product, whether simple or complex, must be as 'realistic' as possible.

Optimal foraging theory, one popular model of these times, provided a focal point for this controversy. Some criticized the hypothesis in toto; others pointed out the typical problem of establishing the range of rejections of this model; while others argued that optimality is not usually treated as a hypothesis to be tested but rather an assumption of the model and the tests thereof are of the factors which a fish may be optimizing. We would add that optimality should always be placed in an ecological context; model constructs and the tests of their validity need some currency other than an esoteric rate of optimization. That is, successful fish growth and survival to reproduction are measurable currencies which, if optimal foraging theory is to have any significance to the real world, must be measured when testing the predictions of any model.

Perhaps a fruitful approach for future research will be examining intraspecific, individual behavioral variability, rather than continuing to look at all members of a species as a 'behavioral mean'. Thus, the breadth of an animal's behavioral repertoire, as well as its boundaries, should always be framed in the total context of functional morphology, physiology, environmental conditions, alternate-conflicting behaviors such as reproduction and predator avoidance, and the real choices of prey available to it. We must assess each of these constraints, some of which are fixed and some more pliant, before we can define the outcome of a behavior event. Reiterating David Noakes' summary analogy, fish must play a juggler's game with all these influences at one time or another and some of them all of the time. It is the objective of behavioral research to understand the rules by which the game is played.

Competition and resource partitioning

Two presentations summarized papers on resource partitioning which have been or will be published elsewhere, but which should nonetheless be consulted for their important contributions to this subject: (1) Gary Grossman (1982, Amer. Nat. 119: 611–637) and (2) Steven Ross (1986, Copeia in press).

One outcome of Steve Ross's review and the subsequent discussion was the need to recognize the philosophical difference between 'resource partitioning', wherein data are gathered to test the hypothesis that there is an active process maintaining observed differences in resource utilization, and 'differential resource utilization', which simply describes such differences in the absence of any mechanistic information. Obviously, the latter may or may not be explained subsequently by 'resource partitioning' theories.

A recurrent discussion after several papers involved our abilities to separate the influences of interference versus exploitative competition. Documentation of spatial or temporal separation in fish foraging does not preclude exploitative competition, even though the lack of interference may be substantiated. In testing exploitative competition, the standing stock, distribution, production, and relative availability of a mutually-exploited prey population must be measured in order to establish whether the common resource is limiting. That the often 'soft corroboration' of exploitative resource partitioning can be augmented by long-term data on the effects of niche shifts in non-coevolved carnivore assemblages was well illustrated by Larry Crowder's and Jim Kitchell's presentations on the Laurentian Great Lakes and Bruce Herbold's on the Suisan Marsh (unpublished).

Interference competition measured through microhabitat utilization may or may not be related to competition for food. A multidimensional approach, such as niche complimentarity between microhabitat use and diet (Ebeling) over broad ranges in fish densities and distributions, is usually necessary to test this form of competition. As indicated by earlier workshop papers (e.g. GUTSHOP '81: Ebeling & Laur, Hixon &

Brostoff) and by Ebeling's presentation on the Naples Reef embiotocid assemblage, evaluation of interference competition can be enhanced measurably by data on: (1) behavioral interactions (territoriality, dominance-subdominance) among fishes; (2) natural or experimental alteration of fish densities and distributions; (3) growth and fecundity of fish as function of foraging microhabitat; and (4) testing of niche overlaps against random models. Furthermore, resource separation does vary with taxonomic structure, particularly among confamilial taxa. Therefore, any interpretation of significant resource partitioning should be considered in the context of familial diversity within the subject assemblage.

Predation effects on prey behavior and ecology

An important point of discussion raised by Lennart Persson after Hixon's plenary presentation addressed the lack of studies which test any of the alternative hypotheses explaining a lack of prey response; in particular, that predator-prey and competitive interactions among the macroinvertebrate prey populations occur but mask each other. Similarly, later discussions illustrated the additional importance of corroborative data on standing stock and diet of other predators in the community, microhabitat structure relative to prey refugia, and overall community (primary) productivity. Although a complete suite of such information may exist for some lake systems, a general literature lacuna stresses the importance to interpretations of predator effects on community structure of material knowledge on: (1) competitive and trophic interactions among both prey and predator assemblages; and (2) microhabitat structure, diversity, and fish utilization.

Conclusions

Relative to the previous workshops, this meeting was most notable for exploring untrod ground on several topics: (1) the evolution of feeding and digestion organs and foraging strategies; (2) the physiological basis for digestion; (3) the causal 'psychology' behind feeding decisions; and (4) the diverse constraints upon and approaches to falsifying hypotheses of resource partitioning and competition.

Discussions of the evolutionary aspects of fish feeding involved: (1) development of different types of ontogenies in fishes as a consequence of different nutrient availability and feeding during early life history; and, (2) coevolution of fish and their prey assemblages. Relative to the first aspect, few investigations have extended beyond descriptive accounts of either morphology and physiology or ecology and behavior and they are infrequently linked in a holistic manner. It is also encouraging to see that the perspective of coevolution, perhaps the latest fashion of our profession, does lend credence to our intuitive models of predator-prey interactions. Well-documented, fortuitous situations such as exist for the Laurentian Great Lakes are rare, however, and more experiments need to be designed specifically toward this concept for other species, prey, habitats, and locations.

Despite significant advancement in technical approaches, studies of fish feeding are generally stuck in the chasm between description and process. Like the ancient prognosticators using chicken entrails to fortell the future, as fish witch doctors we are now forced to deal primarily with stomach contents as 'symptoms' describing the results of complex physiological or ecological processes such as gastric evacuation, feeding intensity and chronology, and prey selection and switching. The causal mechanisms, however, are seldom elucidated.

Fish feeding is a dynamic, flexible behavior affected by both external and internal influences within the context of evolutionary constraints (e.g. functional morphology). Numerous external factors are recognized. For example, it has been documented that stomach content biomass and composition typically vary dielly, are correlated with changes in light levels and prey availability, and often reflect differences in the fishes' distribution in the environment. Internal factors, however, are often hypothesized but rarely substantiated. We still do not know much about how these fish decide when to feed, upon which organisms to feed, or when they are losing their 'hunger' to do so. And we have few models of learning which could explain shifts in foraging efficiency or prey composition. Other, non-feeding influences (e.g. reproduction and potential predators) must also be evaluated as competing influences upon the fish's incentive and mode of feeding. Many of the presentations, notably those by Anderson (not included in these proceedings), Bence, Browman & Marcotte, Ebeling & Laur, Evans & O'Brien, Holbrook & Schmitt, Marcotte & Browman, Mittelbach, and Noakes attempted to address these perception/cognition processes. However, except for Marcotte & Browman, the potential underlying physiological and neurological structures were not discussed.

This may, in fact, be representative of the relatively untapped potential of coupling of physiology/neurophysiology to fish foraging behavior. For instance, models of optimal foraging strategy might be tested by direct measurement (e.g. respirometry) of fish energy budgets across spectra of prey sizes, energy contents, evasive behaviors, and densities relative to prey selection (switching) over time. Stein et al.'s (1984, Ecology 65: 702–715) experiment with sunfish feeding on different freshwater snails and that of Puckett & Dill (1985, Behavior 92: 97) related to territoriality are two prime examples.

This is not to suggest that resolution of the mechanisms behind fish foraging behavior precludes the value of describing and quantifying the diversity and complexity of behavior. These mechanisms, however, must be understood before we can define the internal constraints upon the models. The performance of models in generating realistic images of nature is, after all, dependent completely upon the validity of the conceptualizations, and the information upon which these are based. This constraint argues simply for the need of further investigation into learning and decision-making as the basal neurological processes of perception, cognition, and response. Still, the whole relevance of models will never be broadly acceptable unless they can reliably predict a behavioral outcome from varied permutations and combinations of external inputs and internal conditions. This is, to say the least, a demanding mandate.

One or our more important 'take-home' messages of the discussions on resource partitioning and competition might be that, while we are faced with size- and life history-structured communities, we have constrained our inferences on the control of their structure to information from adult organisms. In short, we are ignoring a lot, perhaps the majority, of the interactions which actually determine community structure. Similarly, we were challenged to place our relatively short-term data more into their evolutionary and historical context. Information such as the evolutionary origins, ontogenetic patterns, long-term environmental variability, and zoogeographic changes in demography should be incorporated for a better understanding of the present community state encompassed by our data.

If there was one theme in all the GUTSHOP '84 discussions, it was that valid generalizations about the feeding dynamics of fishes, whether they be behavioral, physiological, or ecological, cannot be made without knowledge of the fish's total potential niche and spectrum of foraging responses to its varied environments. In summary, we challenge everyone to consider and, if possible, incorporate these evolutionary, historical, neurophysiological, and developmental questions into their studies, such that we all may become better fish witch doctors.

Received 14.11.1985

Species and subject index

GUTSHOP '84 participants at Asilomar. Professional bias of this photographer caused the lamentable 'gut-contents resolution'.

→

Clockwise around Al Ebeling (center), starting with the left top corner: Charles (Si) Simenstad over Sandy Lipovsky; Peter Moyle, Gary Grossman and Larry Crowder; Greg Cailliet, 'innocent' in spite of the 'shark hat'; Eugene Balon; Lennart Persson facing Jim Kitchell and Gary Mittelbach; Chris Nuan, David Noakes, Jim Parrish and Steve Ross; in a US-Canada-Norway 'lean-in' Mark Saunders below Bob Dunbrack and Bob Emmett are separated by Luis Giguere from Waldo Wakefield, Malcolm Jobling, Larry Dill, Rich Brodeur and Bob Cowen; Greg Cailliet again, facing Kevin Hill, Debbie Molnar (of course) and Mark Saunders; Bob Feller watching Bruce Herbold at the blackboard; and finally, with tipping glass, George Boehlert, with whom the circle is complete.

Our logo is based on the recently quite overused print by Pieter Brueghel (1556) entitled 'Big fish eat little fish' reproduced above.